CMOS Digital Integrated Circuits

CMOS Digital Integrated Circuits

Analysis and Design

Third Edition

Sung-Mo (Steve) Kang
University of California at Santa Cruz

Yusuf Leblebici
Swiss Federal Institute of Technology - Lausanne

Boston   Burr Ridge, IL   Dubuque, IA   Madison, WI   New York   San Francisco   St. Louis
Bangkok   Bogotá   Caracas   Kuala Lumpur   Lisbon   London   Madrid   Mexico City
Milan   Montreal   New Delhi   Santiago   Seoul   Singapore   Sydney   Taipei   Toronto

McGraw-Hill Higher Education

A Division of The McGraw-Hill Companies

CMOS DIGITAL INTEGRATED CIRCUITS: ANALYSIS AND DESIGN
THIRD EDITION

Published by McGraw-Hill, a business unit of The McGraw-Hill Companies, Inc., 1221 Avenue of the Americas, New York, NY 10020. Copyright © 2003, 1999, 1996 by The McGraw-Hill Companies, Inc. All rights reserved. No part of this publication may be reproduced or distributed in any form or by any means, or stored in a database or retrieval system, without the prior written consent of The McGraw-Hill Companies, Inc., including, but not limited to, in any network or other electronic storage or transmission, or broadcast for distance learning.

Some ancillaries, including electronic and print components, may not be available to customers outside the United States.

This book is printed on acid-free paper.

International   1 2 3 4 5 6 7 8 9 0 QPF/QPF 0 9 8 7 6 5 4 3 2
Domestic   1 2 3 4 5 6 7 8 9 0 QPF/QPF 0 9 8 7 6 5 4 3 2

ISBN 0–07–246053–9
ISBN 0–07–119644–7 (ISE)

Publisher: *Elizabeth A. Jones*
Senior sponsoring editor: *Carlise Paulson*
Developmental editor: *Michelle L. Flomenhoft*
Executive marketing manager: *John Wannemacher*
Senior project manager: *Rose Koos*
Production supervisor: *Sherry L. Kane*
Media project manager: *Jodi K. Banowetz*
Senior media technology producer: *Phillip Meek*
Coordinator of freelance design: *Rick D. Noel*
Cover designer: *Sheilah Barrett*
Cover image: *DEC Alpha microprocessor chip photograph, courtesy Michael Davidson, Florida State University National High Magnetic Field Laboratory*
Compositor: *Interactive Composition Corporation*
Typeface: *10/12 Times Roman*
Printer: *Quebecor World Fairfield, PA*

Library of Congress Cataloging-in-Publication Data

Kang, Sung-Mo, 1945–
    CMOS digital integrated circuits : analysis and design / Sung-Mo (Steve) Kang, Yusuf Leblebici. —3rd ed.
        p.   cm.
    Includes bibliographical references and index.
    ISBN 0–07–246053–9 — ISBN 0–07–119644–7 (ISE)
    1. Metal oxide semiconductors, Complementary. 2. Digital integrated circuits. I. Leblebici, Yusuf. II. Title.

TK7871.99.M44 K36 2003
621.39'5—dc21   2002026558
                                                                 CIP

INTERNATIONAL EDITION ISBN 0–07–119644–7
Copyright © 2003. Exclusive rights by The McGraw-Hill Companies, Inc., for manufacture and export. This book cannot be re-exported from the country to which it is sold by McGraw-Hill. The International Edition is not available in North America.

www.mhhe.com

*To Myoung-A, Jennifer and Jeffrey,
and Anıl and Ebru*

BRIEF CONTENTS

About the Authors   x

Preface   xi

1 Introduction   1

2 Fabrication of MOSFETs   48

3 MOS Transistor   83

4 Modeling of MOS Transistors Using SPICE   146

5 MOS Inverters: Static Characteristics   169

6 MOS Inverters: Switching Characteristics and Interconnect Effects   218

7 Combinational MOS Logic Circuits   274

8 Sequential MOS Logic Circuits   323

9 Dynamic Logic Circuits   357

10 Semiconductor Memories   405

11 Low-Power CMOS Logic Circuits   481

12 BiCMOS Logic Circuits   521

13 Chip Input and Output (I/O) Circuits   558

14 Design for Manufacturability   585

15 Design for Testability   622

References   637

Index   641

CONTENTS

About the Authors   x

Preface   xi

Chapter 1
Introduction   1

1.1   Historical Perspective   1

1.2   Objective and Organization of the Book   5

1.3   A Circuit Design Example   8

1.4   Overview of VLSI Design Methodologies   18

1.5   VLSI Design Flow   21

1.6   Design Hierarchy   23

1.7   Concepts of Regularity, Modularity, and Locality   25

1.8   VLSI Design Styles   28

1.9   Design Quality   38

1.10   Packaging Technology   41

1.11   Computer-Aided Design Technology   43

Exercise Problems   45

Chapter 2
Fabrication of MOSFETs   48

2.1   Introduction   48

2.2   Fabrication Process Flow: Basic Steps   49

2.3   The CMOS n-Well Process   59

2.4   Layout Design Rules   66

2.5   Full-Custom Mask Layout Design   66

Exercise Problems   73

Chapter 3
MOS Transistor   83

3.1   The Metal Oxide Semiconductor (MOS) Structure   83

3.2   The MOS System under External Bias   87

3.3   Structure and Operation of MOS Transistor (MOSFET)   90

3.4   MOSFET Current-Voltage Characteristics   100

3.5   MOSFET Scaling and Small-Geometry Effects   115

3.6   MOSFET Capacitances   129

Exercise Problems   141

Chapter 4
Modeling of MOS Transistors Using SPICE   146

4.1   Introduction   146

4.2   Basic Concepts   147

4.3   The LEVEL 1 Model Equations   149

4.4   The LEVEL 2 Model Equations   153

4.5   The LEVEL 3 Model Equations   157

4.6   State-of-the-Art MOSFET Models   158

4.7   Capacitance Models   159

4.8   Comparison of the SPICE MOSFET Models   163

Appendix: Typical SPICE Model Parameters   165

Exercise Problems   168

Chapter 5
MOS Inverters: Static Characteristics   169

5.1   Introduction   169

5.2   Resistive-Load Inverter   177

5.3   Inverters with n-Type MOSFET Load   186

5.4   CMOS Inverter   197

Exercise Problems   214

Chapter **6**

**MOS Inverters: Switching
Characteristics and
Interconnect Effects** 218

6.1 Introduction 218
6.2 Delay-Time Definitions 220
6.3 Calculation of Delay Times 222
6.4 Inverter Design with Delay Constraints 230
6.5 Estimation of Interconnect Parasitics 241
6.6 Calculation of Interconnect Delay 252
6.7 Switching Power Dissipation of
        CMOS Inverters 260
        Appendix: Super Buffer Design 267
        Exercise Problems 270

Chapter **7**

**Combinational MOS Logic
Circuits** 274

7.1 Introduction 274
7.2 MOS Logic Circuits with Depletion
        nMOS Loads 275
7.3 CMOS Logic Circuits 287
7.4 Complex Logic Circuits 294
7.5 CMOS Transmission Gates (Pass Gates) 307
        Exercise Problems 317

Chapter **8**

Sequential MOS Logic Circuits 323

8.1 Introduction 323
8.2 Behavior of Bistable Elements 324
8.3 SR Latch Circuit 330
8.4 Clocked Latch and Flip-Flop Circuits 336
8.5 CMOS D-Latch and Edge-Triggered
        Flip-Flop 343
        Appendix: Schmitt Trigger Circuit 350
        Exercise Problems 353

Chapter **9**

Dynamic Logic Circuits 357

9.1 Introduction 357
9.2 Basic Principles of Pass Transistor
        Circuits 359
9.3 Voltage Bootstrapping 370
9.4 Synchronous Dynamic Circuit
        Techniques 374
9.5 Dynamic CMOS Circuit Techniques 379
9.6 High-Performance Dynamic CMOS
        Circuits 383
        Exercise Problems 400

Chapter **10**

Semiconductor Memories 405

10.1 Introduction 405
10.2 Dynamic Random Access Memory
         (DRAM) 410
10.3 Static Random Access Memory (SRAM) 438
10.4 Nonvolatile Memory 451
10.5 Flash Memory 464
10.6 Ferroelectric Random Access Memory
         (FRAM) 472
         Exercise Problems 475

Chapter **11**

**Low-Power CMOS Logic
Circuits** 481

11.1 Introduction 481
11.2 Overview of Power Consumption 482
11.3 Low-Power Design Through Voltage
         Scaling 493
11.4 Estimation and Optimization of Switching
         Activity 504
11.5 Reduction of Switched Capacitance 510
11.6 Adiabatic Logic Circuits 512
         Exercise Problems 520

Chapter **12**

BiCMOS Logic Circuits   521

12.1 Introduction   521
12.2 Bipolar Junction Transistor (BJT): Structure and Operation   524
12.3 Dynamic Behavior of BJTs   536
12.4 Basic BiCMOS Circuits: Static Behavior   543
12.5 Switching Delay in BiCMOS Logic Circuits   545
12.6 BiCMOS Applications   551
Exercise Problems   554

Chapter **13**

Chip Input and Output (I/O) Circuits   558

13.1 Introduction   558
13.2 ESD Protection   559
13.3 Input Circuits   562
13.4 Output Circuits and $L(di/dt)$ Noise   567
13.5 On-Chip Clock Generation and Distribution   571
13.6 Latch-Up and Its Prevention   576
Exercise Problems   583

Chapter **14**

Design for Manufacturability   585

14.1 Introduction   585
14.2 Process Variations   586

14.3 Basic Concepts and Definitions   588
14.4 Design of Experiments and Performance Modeling   594
14.5 Parametric Yield Estimation   602
14.6 Parametric Yield Maximization   607
14.7 Worst-Case Analysis   609
14.8 Performance Variability Minimization   615
Exercise Problems   618

Chapter **15**

Design for Testability   622

15.1 Introduction   622
15.2 Fault Types and Models   622
15.3 Controllability and Observability   626
15.4 Ad Hoc Testable Design Techniques   627
15.5 Scan-Based Techniques   630
15.6 Built-In Self Test (BIST) Techniques   632
15.7 Current Monitoring I_{DDQ} Test   635
Exercise Problems   636

References   637

Index   641

ABOUT THE AUTHORS

Sung-Mo "Steve" Kang received the Ph.D. degree in electrical engineering from the University of California at Berkeley. He has worked on CMOS VLSI design at AT&T Bell Laboratories at Murray Hill, NJ as supervisor and member of technical staff of high-end CMOS VLSI microprocessor design until 1985. Previously, he was department head and professor of electrical and computer engineering department at the University of Illinois at Urbana-Champaign. Currently, he is dean of the Baskin School of Engineering and professor of electrical engineering at the University of California at Santa Cruz. He was the founding editor-in-chief of the *IEEE Transactions on Very Large Scale Integration (VLSI) Systems* and has served on editorial boards of several IEEE and international journals. He is Fellow of IEEE, ACM, and AAAS and recipient of IEEE Third Millennium Medal, IEEE Graduate Teaching Technical Field Award, UC Berkeley Distinguished Alumnus Award, SRC Technical Excellence Award, IEEE Circuits and Systems Society Technical Achievement Award, Alexander von Humboldt U.S. Senior Scientist Award, IEEE CAS Darlington Prize Paper Award, KBS Award, and several other best paper awards.

Yusuf Leblebici received the Ph.D. degree in electrical and computer engineering from the University of Illinois at Urbana-Champaign. He was a visiting assistant professor of electrical and computer engineering at the University of Illinois at Urbana-Champaign, associate professor of electrical and electronics engineering at Istanbul Technical University, and associate professor of electrical and computer engineering at Worcester Polytechnic Institute. He also served as the microelectronics program coordinator at Sabanci University. Currently, he is full (chair) professor at the Swiss Federal Institute of Technology in Lausanne, Switzerland and director of the Microelectronic Systems Laboratory. His research interests include design of high-performance CMOS digital and mixed-signal integrated circuits, computer-aided design of VLSI systems, intelligent sensor interfaces, modeling and simulation of semiconductor devices, and VLSI reliability analysis. He is a Senior Member of IEEE, and recipient of the NATO Science Fellowship Award, the Young Scientist Award of the Turkish Scientific and Technological Research Council, and the Joseph Samuel Satin Distinguished Fellow Award of the Worcester Polytechnic Institute.

PREFACE

Complementary metal oxide semiconductor (CMOS) digital integrated circuits are the enabling technology for the modern information age. Because of their intrinsic features in low-power consumption, large noise margins, and ease of design, CMOS integrated circuits have been widely used to develop random access memory (RAM) chips, microprocessor chips, digital signal processor (DSP) chips, and application-specific integrated circuit (ASIC) chips. The popular use of CMOS circuits continues to grow with the increasing demands for low-power, low-noise integrated electronic systems in the development of portable computers, personal digital assistants (PDAs), portable phones, and multimedia agents.

Since the field of CMOS integrated circuits is very broad, it is conventionally divided into digital CMOS circuits and analog CMOS circuits. This book is focused on the CMOS digital integrated circuits. However, it should be noted that the boundary between classical digital and analog CMOS design is becoming increasingly blurred, especially with the challenges presented by very deep sub-micron (VDSM) fabrication technologies, very low operating voltages, and operating frequencies extending well into the GHz range. Therefore, we attempt to present the analysis and design of digital CMOS integrated circuits from an "analog" point-of-view, i.e., taking into account the analog, non-discrete nature of the devices and circuits that are used to implement digital functions.

The origins of this textbook date back to the early 1990s when both authors were intensively involved in undergraduate and graduate level teaching of digital IC fundamentals. At the University of Illinois at Urbana-Champaign, where both of us were teaching at the time, we tried some of the available textbooks on digital MOS integrated circuits for our senior-level technical elective course, ECE382—*Large Scale Integrated Circuit Design*. Students and instructors alike realized, however, that there was a need for a new book with more comprehensive treatment of CMOS digital circuits. Thus, our textbook project was initiated several years ago by assembling our own lecture notes. Since 1993, we have used evolving versions of this material at the University of Illinois at Urbana-Champaign, at Istanbul Technical University, at Worcester Polytechnic Institute, and at the Swiss Federal Institute of Technology in Lausanne. Both authors were very much encouraged by comments from their students, colleagues, and reviewers. The first edition of *CMOS Digital Integrated Circuits: Analysis and Design* was published in late 1995.

Soon after publishing the first edition, we saw the need for updating it to reflect many constructive comments we were receiving from instructors and students who used the textbook. We intended to include and update important topics such as low-power circuit design, interconnects in high-speed circuit design, as well as the deep

xi

sub-micron circuit design issues, and to provide more rigorous treatment of new developments in memory circuits. We also felt that in a very rapidly developing field such as CMOS digital circuits, the quality of a textbook can only be preserved by timely updates reflecting the current state-of-the-art. This realization has led us to embark on the extensive and continuous revision of our work, with the Second Edition appearing in 1998 and the Third Edition in 2002, to reflect recent advances in technology and in circuit design practices.

This book, *CMOS Digital Integrated Circuits: Analysis and Design,* is primarily intended as a comprehensive textbook at the senior level and first-year graduate level, as well as a reference for practicing engineers in the areas of integrated circuit design, digital design, and VLSI. Recognizing that the area of digital integrated circuit design is evolving at an increasingly faster pace, we have made our best effort to present up-to-date materials on all subjects covered. This book contains fifteen chapters; and we recognize that it would not be possible to cover rigorously all of this material in one semester. Thus, we would propose the following based on our teaching experience: At the undergraduate level, coverage of the first ten chapters would constitute sufficient material for a one-semester course on CMOS digital integrated circuits. Time permitting, some selected topics in Chapter 11, *Low-Power CMOS Logic Circuits,* Chapter 12*, BiCMOS Logic Circuits,* and Chapter 13, *Chip Input and Output (I/O) Circuits* can also be covered. Alternatively, this book can be used for a two-semester course, allowing a more detailed treatment of advanced issues, which are presented in the later chapters. At the graduate level, selected topics from the first ten chapters plus the last five chapters can be covered in one semester.

The first eight chapters of this book are devoted to a detailed treatment of the MOS transistor with all its relevant aspects; to the static and dynamic operation principles, analysis, and design of basic inverter circuits; and to the structure and operation of combinational and sequential logic gates. Note that the introduction chapter has been significantly expanded to include a detailed presentation of VLSI design methodologies. Since the digital IC design techniques discussed in the first half of this book are directly relevant for digital VLSI and ASIC design, we felt that the context should be presented at the very beginning of the book. The issues of on-chip interconnect modeling and interconnect delay calculation are covered extensively in Chapter 6, which provides a complete view of switching characteristics in digital integrated circuits. A separate chapter (Chapter 9) has been reserved for the treatment of *dynamic* logic circuits, which are used in state-of-the-art VLSI chips. Chapter 10 has been completely revised in both content and presentation; it offers an in-depth presentation of many state-of-the-art semiconductor memory circuits.

Recognizing the increasing importance of low-power circuit design, we dedicate one chapter (Chapter 11) to low-power CMOS logic circuits, which provides a comprehensive coverage of methodologies and design practices that are used to reduce the power dissipation of large-scale digital integrated circuits. BiCMOS digital circuit design is examined in Chapter 12, with a thorough coverage of bipolar transistor basics. In view of the continuing use of bipolar and BiCMOS circuits in very high-speed designs, we believe that at least one chapter should be allocated to cover the basics of bipolar transistors. Next, Chapter 13 provides a clear insight into the important subject of chip I/O design. Critical issues such as ESD protection, clock distribution, clock buffering, and latch-up phenomena are discussed in detail. Finally, the more advanced but very important topics of design for manufacturability and design for testability are covered in Chapters 14 and 15, respectively.

The authors have long debated the coverage of nMOS circuits in this book. We have concluded that some coverage should be provided for pedagogical reasons. Studying nMOS circuits will better prepare readers for analysis of other field effect transistor (FET) circuits such as GaAs circuits, the topology of which is quite similar to that of depletion-load nMOS circuits. Thus, to emphasize the *load* concept, which is still widely used in many areas in digital circuit design, we present basic depletion-load nMOS circuits along with their CMOS counterparts in several places throughout the book.

Although an immense amount of effort and attention to detail were expended to prepare the camera-ready manuscript, this book may still have some flaws and mistakes due to erring human nature. The authors would welcome and greatly appreciate suggestions and corrections from the readers, for the improvement of technical content as well as the presentation style.

ACKNOWLEDGMENTS FOR THE FIRST EDITION

Our colleagues have provided many constructive comments and encouragement for the completion of the first edition. Professor Timothy N. Trick, former head of the department of electrical and computer engineering at the University of Illinois at Urbana-Champaign, has strongly supported our efforts from the very beginning. The appointment of Sung-Mo Kang as an associate in the Center for Advanced Study at the University of Illinois at Urbana-Champaign helped to start the process.

Yusuf Leblebici acknowledges the full support and encouragement from the department of electrical and electronics engineering at Istanbul Technical University, where he introduced a new digital integrated circuits course based on the early version of this book and received very valuable feedback from his students. Yusuf Leblebici also thanks the ETA advanced Electronics Technologies Research and Development Foundation at Istanbul Technical University for their generous support.

Professor Elyse Rosenbaum and Professor Resve Saleh used the early versions of the manuscript as the textbook for ECE382 at Illinois and provided many helpful comments and corrections which have been fully incorporated with deep appreciation. Professor Elizabeth Brauer, currently at Northern Arizona University, has also done the same at the University of Kentucky.

The authors would like to express sincere gratitude to Professor Janak Patel of the University of Illinois at Urbana-Champaign for generously mentoring the authors in writing Chapter 16, *Design for Testability*. Professor Patel has provided many constructive comments and many of his expert views on the subject are reflected in this chapter. Professor Prith Banerjee of Northwestern University and Professor Farid Najm of the University of Illinois at Urbana-Champaign also provided many good comments. We would also like to thank Dr. Abhijit Dharchoudhury for his invaluable contribution to Chapter 15, *Design for Manufacturability*.

Professor Duran Leblebici of Istanbul Technical University, who is the father of the second author, reviewed the entire manuscript in its early development phase, and provided very extensive and constructive comments, many of which are reflected in the final version. Both authors gratefully acknowledge his support during all stages of this venture. We also thank Professor Cem Göknar of Istanbul Technical University, who offered very detailed and valuable comments on *Design for Testability*, and Professor Uğur Çilingiroğlu of the same university, who offered many excellent suggestions for improving the manuscript, especially the chapter on semiconductor memories.

Many of the authors' former and current students at the University of Illinois at Urbana-Champaign also helped in the preparation of figures and verification of circuits using SPICE simulations. In particular, Dr. James Morikuni, Dr. Weishi Sun, Dr. Pablo Mena, Dr. Jaewon Kim, Mr. Steve Ho, and Mr. Sueng-Yong Park deserve recognition. Ms. Lilian Beck and the staff members of the Publications Office in the department of electrical and computer engineering at the University of Illinois at Urbana-Champaign read the entire manuscript and provided excellent editorial comments.

The authors would also like to thank Dr. Masakazu Shoji of AT&T Bell Laboratories, Professor Gerold W. Neudeck of Purdue University, Professor Chin-Long Wey of Michigan State University, Professor Andrew T. Yang of the University of Washington, Professor Marwan M. Hassoun of Iowa State University, Professor Charles E. Stroud of the University of Kentucky, Professor Lawrence Pileggi of the University of Texas at Austin, and Professor Yu Hen Hu of the University of Wisconsin at Madison, who read all or parts of the manuscript and provided many valuable comments and encouragement.

The editorial staff of McGraw-Hill has been an excellent source of strong support from the beginning of this textbook project. The venture was originally initiated with the enthusiastic encouragement from the previous electrical engineering editor, Ms. Anne (Brown) Akay. Mr. George Hoffman, in spite of his relatively short association, was extremely effective and helped settle the details of the publication planning. During the last stage, the new electrical engineering editor, Ms. Lynn Cox, and Mr. John Morriss, Mr. David Damstra, and Mr. Norman Pedersen of the Editing Department were superbly effective and we enjoyed dashing with them to finish the last mile.

ACKNOWLEDGMENTS FOR THE SECOND EDITION

The authors are truly indebted to many individuals who, with their efforts and their help, made the second edition possible. We would like to thank Dr. Wolfgang Fichtner, President and CEO of ISE Integrated Systems Engineering, Inc., and the technical staff of ISE in Zurich, Switzerland for providing computer-generated cross-sectional color graphics of MOS transistors and CMOS inverters, which are featured in the color plates. The first author acknowledges the support provided by the U.S. Senior Scientist Research Award from the Alexander von Humbold Stiftung in Germany, which was very helpful for the second edition. The appointments of the second author as Associate Professor at Worcester Polytechnic Institute and as Visiting Professor at the Swiss Federal Institute of Technology in Lausanne, Switzerland have provided excellent environments for the completion of the revision project. The second author also thanks Professor Daniel Mlynek of the Swiss Federal Institute of Technology in Lausanne for his continuous encouragement and support. Many of the authors' former and current students at the University of Illinois at Urbana-Champaign, at the Swiss Federal Institute of Technology in Lausanne, and at Worcester Polytechnic Institute also helped in the preparation of figures and verification of circuits using SPICE simulations. In particular, Dr. James Stroming and Mr. Frank K. Gürkaynak deserve special recognition for their extensive and valuable efforts.

The authors would also like to thank Professor Charles Kime of the University of Wisconsin at Madison, Professor Gerold W. Neudeck of Purdue University, Professor D.E. Ioannou of George Mason University, Professor Subramanya Kalkur of the University of Colorado, Professor Jeffrey L. Gray of Purdue University, Professor Jacob Abraham of the University of Texas at Austin, Professor Hisham Z. Massoud of Duke University, Professor Norman C. Tien of Cornell University, Professor Rod Beresford of Brown University, Professor Elizabeth J. Brauer of Northern Arizona University, Professor Reginald J. Perry of Florida State University, and Professor Cem Göknar of Istanbul Technical University who read all or parts of the revised manuscript and provided their valuable comments and encouragement.

The editorial staff of McGraw-Hill has, as always, been wonderfully supportive from the beginning of the revision project. We thankfully recognize the contributions of our previous electrical engineering editor, Ms. Lynn Cox, and we appreciate the extensive efforts of Ms. Nina Kreiden, who helped the project get off the ground in its early stages. During the final stages of this project, Ms. Kelley Butcher, Ms. Karen Nelson, and Mr. Francis Owens have been extremely effective and helpful, and we enjoyed sharing this experience with them.

ACKNOWLEDGMENTS FOR THE THIRD EDITION

Several individuals have contributed with their time and their efforts to the third edition of our textbook. The authors would like to acknowledge the invaluable contribution of Dr. Seung-Moon Yoo who was instrumental in the extensive revision of the Memory chapter (Chapter 10). His technical insight, his meticulous attention to detail, and his very productive work are truly appreciated. The first author acknowledges the University of California at Santa Cruz for valuable support in his new position as Dean of the School of Engineering, and for enabling him to concentrate on the revision of the manuscript. The appointment of the second author as Full Professor at the Swiss Federal Institute of Technology in Lausanne, Switzerland has also provided an excellent environment for the completion of the project. The second author gratefully acknowledges Mme. Séverine Eggli for her valuable assistance in revisions, and for typing sections of the text. The authors thank Mr. Tom Vernier and the technical staff of the MOSIS organization for generously providing the SPICE BSIM parameters for TSMC 0.18 μm process that were extracted by MOSIS. The authors also acknowledge Dr. Michael W. Davidson of the Florida State University

National High Magnetic Field Laboratory, for providing the DEC Alpha chip microphotographs that appear on the cover.

The authors would like to thank the following individuals who read all or parts of the revised manuscript and provided their valuable comments and encouragement.

Professor Massoud Pedram, *University of Southern California*

Professor Eby G. Friedman, *University of Rochester*

Professor Chien-In Henry Chen, *Wright State University*

Professor Ivan Kourtev, *University of Pittsburgh*

Professor Dimitris E. Ioannou, *George Mason University*

Professor Thottam S. Kalkur, *University of Colorado at Colorado Springs*

Professor Yong-Bin Kim, *Northeastern University*

Professor Pratapa Reddy, *Rochester Institute of Technology*

Professor Hisham Z. Massoud, *Duke University*

Professor Resve A. Saleh, *University of British Columbia*

Professor Simon Foo, *Florida State University*

Professor David W. Parent, *San Jose State University*

Professor Jaime Ramirez-Angulo, *New Mexico State University*

Professor Nur Touba, *University of Texas at Austin*

Professor Nicholas C. Rumin, *McGill University*

The editorial staff of McGraw-Hill has again been very helpful and supportive throughout the entire revision project. This project started with the insightful initiative of Mr. Tom Casson, our publisher at McGraw-Hill. We would like to acknowledge his valuable support and encouragement. We thankfully recognize the contributions of Ms. Michelle Flomenhoft, Ms. Betsy Jones, and Ms. Rose Koos. We especially thank them for their helpful assistance during all stages of this complex project, and for their patience and persistence. We also acknowledge Mr. Rick Noel for creating the cover design of the third edition. We truly enjoyed sharing this experience with the entire McGraw-Hill team.

Finally, we would like to acknowledge the support from our families, Myoung-A (Mia), Jennifer, and Jeffrey Kang, and Anıl and Ebru Leblebici, for tolerating many of our physical and mental absences while we worked on the third edition of this book, and for providing us invaluable encouragement throughout the project.

Sung-Mo (Steve) Kang   **Yusuf Leblebici**
Santa Cruz, California   *Lausanne, Switzerland*
August 2002   *August 2002*

Introduction

1.1 Historical Perspective

The electronics industry has achieved a phenomenal growth over the last few decades, mainly due to the rapid advances in integration technologies and large-scale systems design. The use of integrated circuits in high-performance computing, telecommunications, and consumer electronics has been growing at a very fast pace. Typically, the required computational and information processing power of these applications is the driving force for the fast development of this field. Figure 1.1 gives an overview of the prominent trends in information technologies over the next decade. The current leading-edge technologies (such as low bit-rate video and cellular communications) already provide the end-users a certain amount of processing power and portability. This trend is expected to continue, with very important implications for VLSI and systems design. One of the most important characteristics of information services is their increasing need for very high processing power and bandwidth (in order to handle real-time video, for example). The other important characteristic is that the information services tend to become more personalized, which means that the information processing devices must be more intelligent and also be portable to allow more mobility. This trend towards portable, distributed system architectures is one of the main driving forces for system integration, even though it does not preclude a concurrent and equally important trend towards centralized, highly powerful information systems such as those required for network computing (NC) and video services.

As more and more complex functions are required in various data processing and telecommunications devices, the need to integrate these functions in a small package is also increasing. The level of integration as measured by the number of logic gates in a monolithic chip has been steadily rising for almost three decades, mainly due to the rapid progress in processing technology and interconnect technology. Table 1.1 shows the evolution of logic complexity in integrated circuits over the last three decades, and marks the *milestones* of each era. Here, the numbers for circuit complexity should be viewed only as representative measures to indicate the order-of-magnitude. A logic block can contain anywhere from 10 to 100 transistors,

Table 1.1 Evolution of logic complexity in integrated circuits

Era	Date	Complexity (# of logic blocks per chip)
Single transistor	1958	<1
Unit logic (one gate)	1960	1
Multi-function	1962	2–4
Complex function	1964	5–20
Medium Scale Integration (MSI)	1967	20–200
Large Scale Integration (LSI)	1972	200–2,000
Very Large Scale Integration (VLSI)	1978	2,000–20,000
Ultra Large Scale Integration (ULSI)	1989	20,000–?

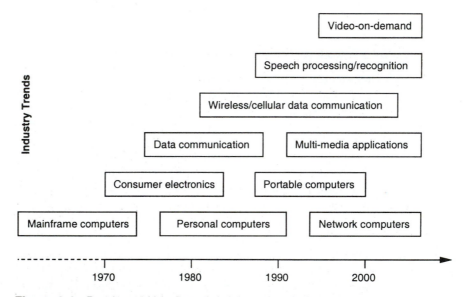

Figure 1.1 Prominent "driving" trends in information service technologies.

depending on the function. State-of-the-art ULSI chips, such as the *DEC Alpha* or the *INTEL Pentium,* contain 10 to 100 million transistors. Note that the term VLSI has been used continuously even for chips in the ULSI (Ultra Large Scale Integration) category, not necessarily abiding by the distinction in Table 1.1.

The monolithic integration of a large number of functions on a single chip usually provides:

■ Less area/volume and therefore, compactness
■ Less power consumption
■ Less testing requirements at system level
■ Higher reliability, mainly due to improved on-chip interconnects
■ Higher speed, due to significantly reduced interconnection length
■ Significant cost savings

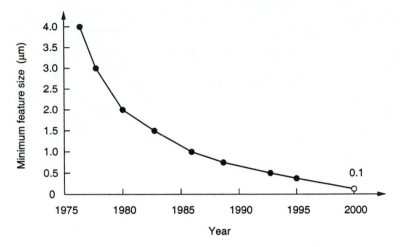

Figure 1.2 Evolution of minimum feature size in integrated circuits over time.

Therefore, the current trend of integration will continue in the foreseeable future. Advances in device manufacturing technology allow the steady reduction of minimum feature size (such as the minimum channel length of a transistor or an interconnect width realizable on chip). Figure 1.2 shows the evolution of the minimum feature size of transistors in integrated circuits, starting from the late 1970s. In 1980, at the beginning of the VLSI era, the typical minimum feature size was 2 μm, and a feature size of 0.3 μm was expected around the year 2000. The actual development of the technology, however, has far exceeded these expectations. A minimum feature size of 0.25 μm was achieved by 1995, and devices with a feature size of 0.18 μm were already the norm in 2001. The first 64-Mbit DRAM and the *INTEL Pentium* microprocessor chip containing more than 3 million transistors were already available by 1994, pushing the envelope of integration density. The first 4-Gbit DRAM based on 0.15 μm manufacturing technology was announced by NEC in early 1997. According to the International Technology Roadmap for Semiconductors (ITRS), MOS transistors with a feature size of 70 nm are expected to be available by 2008, allowing device densities of about 3 billion transistors per chip.

When comparing the integration density of integrated circuits, a clear distinction must be made between the memory chips and logic chips. Figure 1.3 shows the level of integration over time for memory and logic chips, starting in 1970. The number of transistors per chip has continued to increase at an exponential rate over the last three decades, effectively confirming Gordon Moore's prediction on the growth rate of chip complexity, which was made in the early 1960s (Moore's Law). It can be observed that in terms of transistor count, logic chips contain significantly fewer transistors in any given year mainly due to large consumption of chip area for complex interconnects. Memory circuits are highly regular, and thus more cells can be integrated with much less area for interconnects. This has also been one of the main reasons why the *rate of increase* of chip complexity (transistor count per chip) is consistently higher for memory circuits.

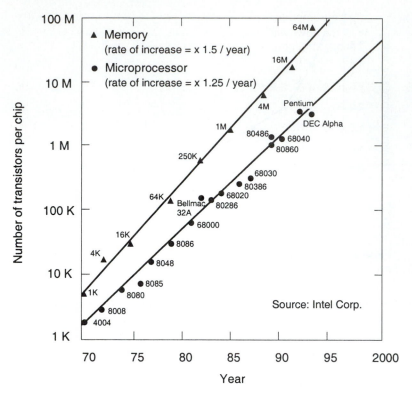

Figure 1.3 Level of integration versus time for memory chips and logic chips.

Digital CMOS (Complementary Metal Oxide Semiconductor) integrated circuits (ICs) have been the driving force behind Very Large Scale Integration (VLSI) for high-performance computing and other scientific and engineering applications. The demand for digital CMOS ICs will be continually strong due to salient features such as low power, reliable performance, circuit techniques for high speed such as using dynamic circuits, and ongoing improvements in processing technology.

It is now projected that the minimum feature size in CMOS ICs can decrease to 0.035 μm (35 nm) within a decade. With such a technology, the level of integration in a single chip can be on the order of several tens of billions of transistors for logic chips or even higher in the case of memory chips, which presents an immense challenge for chip developers in processing, design methodology, testing, and project management. Through the "divide-and-conquer" approach and more advanced design automation using computer-aided design (CAD) tools, ultra-large-scale problems should be solvable.

Bipolar and gallium arsenide (GaAs) circuits have been used for very high speed circuits, and this practice may continue. For instance, in Monolithic Microwave Integrated Circuits (MMICs), GaAs MESFET (MEtal Semiconductor Field Effect Transistor) technology has been highly successful. However, they are still not efficient for VLSI or Ultra Large Scale Integration (ULSI) due to processing difficulties and high power consumption, although for special applications their use may continue. As long as the downward scaling of CMOS technology remains strong, other technologies are likely to remain the technology of tomorrow.

1.2   Objective and Organization of the Book

The objective of this book is to help readers develop in-depth analytical and design capabilities in digital CMOS circuits and chips. The development of VLSI chips requires an interdisciplinary team of market experts, architects, logic designers, circuit and layout designers, packaging engineers, test engineers, and process and device engineers. Also essential are the computer aids for design automation and optimization. It is not possible to discuss the full spectrum of development issues in any single book. Therefore, this book concentrates on digital circuits and also presents related materials in processing and device principles essential to in-depth understanding of CMOS digital circuits.

Often readers can become lost in details and fail to see the global picture. For VLSI circuit design, however, it is important that the design be done in the context of global optimization with proper boundary conditions. In fact, the beauty of integrated circuits is that the final design goal is the concerted performance of all interconnected transistors, and not of individual transistors. Therefore, the *interconnect* issues are almost as important as the issues of individual transistors. No matter how well an individual transistor performs, if the technology fails to have equally good interconnects, the total performance can be very poor due to large parasitic capacitances and resistances; these translate into a large delay in the interconnection lines between transistors or logic gates.

This volume is intended as a comprehensive textbook for senior-level undergraduate students and first-year graduate students in an advanced course on digital circuit design. The material presented in this book should also be very useful to practicing VLSI design engineers. Most of the material presented has been taught over several years in undergraduate and graduate-level courses in many universities, including the department of electrical and computer engineering at the University of Illinois at Urbana-Champaign where both authors were affiliated. It is assumed that the readers of this book already have sufficient fundamental background on semiconductor devices, electronic circuit design and analysis, and logic theory. While the interactions among logic design, circuit design, and layout design are strongly emphasized throughout the text, the main focus is on transistor-level circuit design and analysis. This requires a fair amount of detailed current and voltage calculations, as well as a good understanding of how device characteristics affect overall circuit performances, such as propagation delay, noise margins, and power dissipation.

The relational ordering and extent of the topics covered in a typical digital integrated circuits course are depicted in Fig. 1.4. First, a fundamental knowledge of basic device physics is required to understand and use various MOSFET device models in circuit analysis. Following a review of device fundamentals, the emphasis will shift from single devices to simple two-transistor circuits, such as inverters, and then to more complex logic circuits. We will see that as we move to more advanced topics, the breadth of each subject also increases significantly. In fact, a large number of

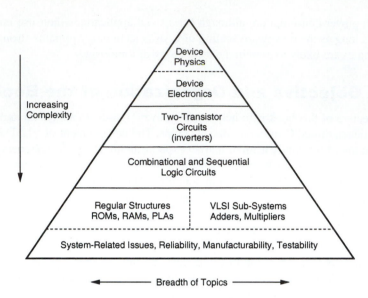

Figure 1.4   The ordering of topics covered in a typical digital integrated circuits course.

different variations may be considered for implementing complex circuits and systems. Consequently, we will examine representative examples for large-scale system implementations and compare their relative merits in terms of performance, reliability, and manufacturability.

The book begins with a review of fabrication-related issues. Representative integrated circuit fabrication techniques are summarized very briefly in the beginning, in order to establish a simple view of process flow and to provide the reader with the necessary terminologies related to processing. The level and the extent of MOS device physics covered in this book are specifically geared toward hands-on circuit design and analysis applications; hence, most of the device models used are relatively simple. The choice of simple device models imposes certain limitations on the accuracy; however, the emphasis is primarily on the clear understanding of basic design concepts and on the importance of generating meaningful estimates for circuit performance during the early design stages. The very important role of computer-aided circuit simulation tools in VLSI design is also well recognized. The book contains a large number of computer simulation examples and exercise problems based on SPICE (Simulation Program with Integrated Circuit Emphasis), which has become a de facto standard in transistor-level circuit simulation in a wide range of computing platforms. An entire chapter is devoted to the examination and comparison of MOSFET models implemented in SPICE, including the identification of various device model parameters. Computer simulation is, and will continue to be, an essential part of the design process, both for performance verification and for fine-tuning of circuits. However, the emphasis on simulation must be well-balanced with the emphasis on hands-on design and analytical estimates, so that the

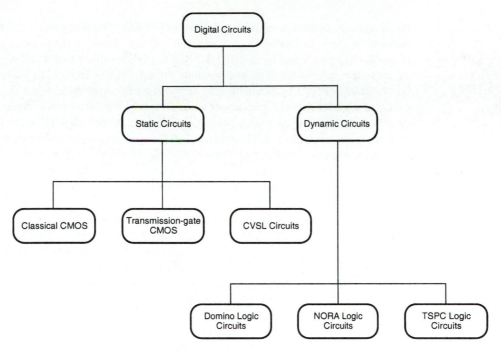

Figure 1.5 Classification of CMOS digital circuit types.

significance of the latter is not overwhelmed by the extensive use of computer-aided techniques.

The main focus of this book is on CMOS digital integrated circuits, but a significant amount of material on nMOS digital circuits is also presented. Although CMOS has become the technology of choice in many applications in recent years, the fundamental concepts of nMOS logic provide a strong basis both for the conceptual understanding and for the development of CMOS designs. Chapters 5 through 9 are devoted exclusively to the analysis and design of basic CMOS and also some nMOS digital circuits. Figure 1.5 shows a simple "family tree" for digital integrated circuits that clarifies the classification and relations among different types of circuits. Based on the fundamental operating principles, the circuits are classified into two main categories, i.e., static circuits and dynamic circuits. The static CMOS circuits are further divided into sub-categories such as classical (fully complementary) CMOS circuits, transmission-gate logic circuits, pass-transistor logic circuits and cascade voltage switch logic (CVSL) circuits. The dynamic CMOS circuits are divided into sub-categories such as domino logic, NORA, and true single-phase clock (TSPC) circuits.

In addition to transistor-level circuit design issues, the accurate prediction and reduction of interconnect parasitics has become a very significant topic in high-performance digital integrated circuits, especially for deep sub-micron technologies. A large portion of Chapter 6 is therefore devoted to interconnect effects. Semiconductor memories are covered in detail in Chapter 10. Specific emphasis is given to the design and operation of different static and dynamic memory types and to comparisons of their performance characteristics. One chapter of the book is devoted to bipolar transistors and to bipolar/BiCMOS digital circuits, which continue to play an important role in the high-performance digital circuits arena. The inclusion of bipolar-based circuits in this book may be puzzling to some readers, but the significance of BiCMOS design techniques cannot be neglected in a comprehensive text on digital design. One chapter is entirely dedicated to input/output (I/O) circuits and related issues, including ESD protection, level shifting, super-buffer design, and latch-up prevention. Finally, two chapters on design for manufacturability and design for testability cover many of the important topics, such as yield estimation, statistical design, and system testability, which deserve special attention in the context of large-scale integrated circuit design.

Since the second edition, a new chapter, *Low-Power CMOS Logic Circuits,* was added to the text. The increasing prominence of portable systems and the need to limit power consumption (and hence, heat dissipation) in very high density ULSI chips have led to rapid and very interesting developments in low-power design in recent years. In most cases, the requirements of low power consumption must be combined with the equally demanding tasks of higher integration density and higher circuit performance. In view of these developments, low-power design of digital integrated circuits deserves to be treated in a separate chapter. Here, various aspects of power consumption are discussed in detail and strategies are introduced to reduce the power dissipation.

The chapters are organized in order to allow several different variations of course plans and self-study programs. A number of chapters can be grouped together to accommodate a specific course syllabus, and others can be skipped without a significant loss of continuity. Each chapter contains a large number of solved problems and examples, integrated into the text to enhance the understanding of the material at hand. Also, a collection of problems, some of which are geared specifically for computer-based SPICE simulation, is provided at the end of each chapter.

1.3   A Circuit Design Example

To help form a global picture of the digital circuit design cycle, in this chapter we begin with a "once over lightly" design exercise wherein we, as circuit designers, start from a logic diagram along with design specifications. The logic circuit is first translated into a CMOS circuit and the initial layout is done. From the layout, all of the important parasitics are calculated by using a circuit extraction program. Once a full circuit description is obtained from the initial layout, we analyze the circuit for DC and transient performance by using the circuit-level simulation program, SPICE, and then compare the results with the given design specifications. If the initial design fails to meet any one of the specifications, which is the case in this exercise, we devise an improved circuit design to meet the design objective. Then the improved design will be implemented into a new layout and the design-analysis cycle will be repeated until all of the design specifications are met. The simplified flow of this circuit design procedure is illustrated in Fig. 1.6. Note that the topics covered in this textbook concern primarily the two important steps enclosed in the dotted box, namely, *VLSI design* and *design verification.*

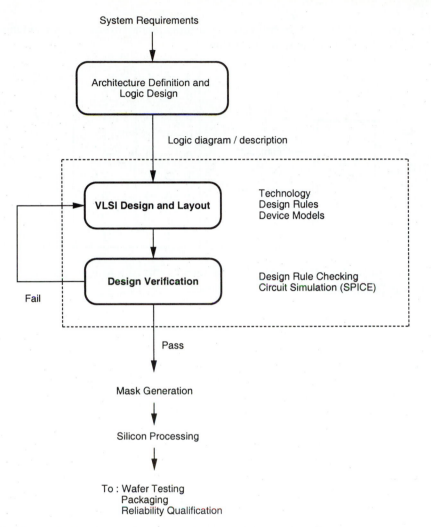

Figure 1.6   The flow of circuit design procedures.

EXAMPLE 1.1

In the following example, we will consider a one-bit binary full-adder circuit design using 0.8-μm, twin-well CMOS technology. The design specifications are

Propagation delay times of sum and carry_out signals < 1.2 ns   (worst case)

Transition delay times of sum and carry_out signals < 1.2 ns   (worst case)

Circuit area < 1500 μm^2

Dynamic power dissipation (@ $V_{DD} = 5$ V and $f_{max} = 20$ MHz) < 1 mW

We start our design by considering the Boolean description of the binary adder circuit. Let A and B represent the two input variables (addend bits), and let C represent the carry_in bit. The binary full adder is a three-input, two-output combinational circuit which satisfies the truth table below.

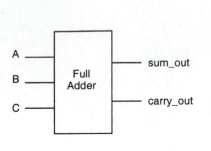

A	B	C	sum_out	carry_out
0	0	0	0	0
0	0	1	1	0
0	1	0	1	0
0	1	1	0	1
1	0	0	1	0
1	0	1	0	1
1	1	0	0	1
1	1	1	1	1

The sum_out and carry_out signals can be found as the following two combinational Boolean functions of the three input variables, A, B and C.

$$\text{sum\_out} = A \oplus B \oplus C$$
$$= ABC + A\bar{B}\bar{C} + \bar{A}\bar{B}C + \bar{A}C B$$
$$\text{carry\_out} = AB + AC + BC$$

A gate-level realization of these two functions is shown in Fig. 1.7. Note that instead of realizing the two functions independently, we use the carry_out signal to generate the sum output, since the output can also be expressed as

$$\text{sum\_out} = ABC + (A + B + C)\overline{\text{carry\_out}}$$

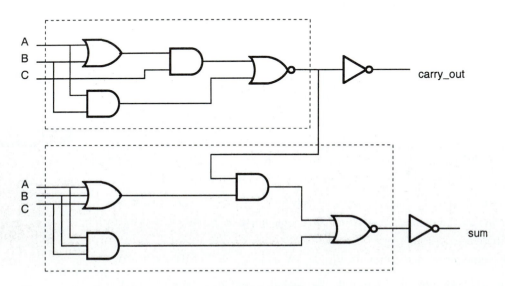

Figure 1.7   Gate-level schematic of the one-bit full-adder circuit.

This implementation will ultimately reduce the circuit complexity and, hence, save chip area. Also, we identify two separate sub-networks consisting of several gates (highlighted with dashed boxes) which will be utilized for the transistor-level realization of the full-adder circuit.

For translating the gate-level design into a transistor-level circuit description, we note that both the sum_out and the carry_out functions are represented by nested AND-OR-NOR structures in Fig. 1.7. Each such combined structure (complex logic gate) can be realized in CMOS as follows: the AND terms are implemented by *series-connected* nMOS transistors, and the OR terms are implemented by *parallel-connected* nMOS transistors. The input variables are applied to the gates of the nMOS (and the complementary pMOS) transistors. Thus, the nMOS net may consist of nested series-parallel connections of nMOS transistors between the output node and the ground. Once the nMOS part of a complex CMOS logic gate is realized, the corresponding pMOS net, which is connected between the output node and the power supply, is obtained as the *dual network* of the nMOS net. The resulting transistor-level design of the CMOS full-adder circuit is shown in Fig. 1.8. Note that the circuit contains a total of 14 nMOS and 14 pMOS transistors, together with the two CMOS inverters which are used to generate the outputs.

In this specific example, it can also be shown that the dual (pMOS) network is actually *equivalent* to the nMOS network for both the sum_out and the carry_out functions, which leads to a fully *symmetric* circuit topology. The alternate circuit diagram obtained by applying this principle of symmetry in shown in Fig. 1.9. Note that the Boolean functions realized by the circuits shown in Fig. 1.8 and Fig. 1.9 are identical; yet the symmetric circuit topology shown in Fig. 1.9 significantly simplifies the layout. These issues will be discussed in detail in Chapter 7.

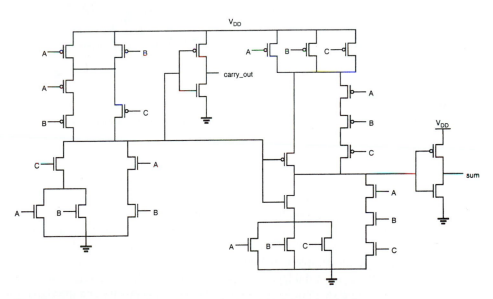

Figure 1.8 Transistor-level schematic of the one-bit full-adder circuit.

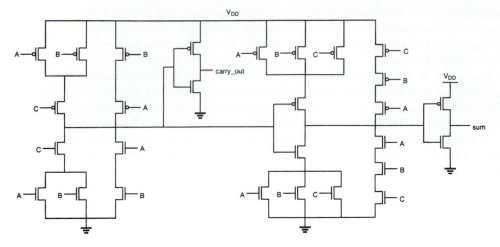

Figure 1.9 Alternate transistor-level schematic of the one-bit full-adder circuit (note that the nMOS and pMOS networks are completely symmetric).

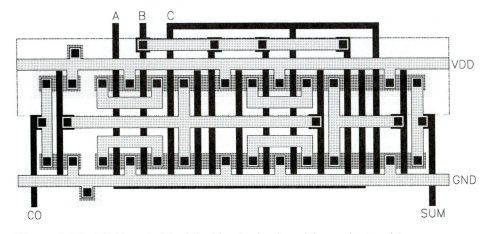

Figure 1.10 Initial layout of the full-adder circuit using minimum-size transistors.

Initially, we will design all nMOS and pMOS transistors with a (W/L) ratio of $(2\ \mu m/0.8\ \mu m)$, which is the minimum transistor size allowed in this particular technology. This initial sizing of transistors, which is obviously not an optimum solution, may be changed later depending on the performance characteristics of the adder circuit. Choosing minimum-size transistors in the initial design stage usually provides a simple, first-cut verification of the circuit functionality and helps the designer in developing a simple initial layout.

Next, the initial layout of the full-adder circuit is generated. Here we use a regular layout style in order to simplify the overall geometry and the signal routing. The initial layout using minimum-size transistors is shown in Fig. 1.10. Note that in this initial adder cell layout, all nMOS and pMOS transistors are placed in two parallel rows, between the horizontal power supply and the ground lines (metal). All polysilicon lines are laid out vertically. The area between the n-type and p-type diffusion regions is used for running local metal interconnections (routing). Also note that the diffusion regions of neighboring transistors have been merged as much as possible, in order to save chip area. The regular layout style used in this example also has the inherent advantage of being easily adaptable to computer-aided design (CAD). The silicon area occupied by this full-adder layout is $(21 \ \mu m \times 54 \ \mu m) = 1{,}134 \ \mu m^2$.

The designer must confirm, using an automatic *design rule checker* (DRC) tool, that none of the physical layout design rules are violated in this adder layout. This is usually done concurrently during the graphical entry of the layout. The next step is to extract the parasitic capacitances and resistances from the initial layout, and then to use a detailed circuit simulation tool (e.g. SPICE) to estimate the dynamic performance of the adder circuit. Thus, we are now in the *design verification* stage of the design-flow diagram shown in Fig. 1.6. The parasitic extraction tool reads in the physical layout file, analyzes the various mask layers to identify transistors, interconnects and contacts, calculates the parasitic capacitances and the parasitic resistances of these structures, and finally prepares a SPICE input file that accurately describes the circuit (see Chapter 4).

The extracted circuit file is now simulated using SPICE in order to determine its dynamic performance. The three input waveforms (A, B and C) are chosen so that all of the eight possible input combinations are applied consecutively to the full-adder circuit. Assuming that the outputs of this adder circuit may drive a similar circuit, both output nodes are loaded with capacitors which represent the typical input capacitance of a full adder. Figure 1.11 shows the simulated input and output waveforms. Unfortunately, the simulation results show that the circuit does not meet all of the design specifications. The propagation delay times of the sum_out and carry_out signals are found to violate the timing constraints, since the minimum-size transistors are not capable of properly driving the capacitive output loads. In particular, the worst-case delay is found to be about 2.0 ns, whereas the timing requirements dictate a maximum delay of 1.2 ns. Figure 1.12 shows the signal propagation delay of both inputs in detail during one of the worst-case input transitions. Design modifications will be necessary to correct this problem. Thus, we go back to the layout design stage.

One approach to increase switching speed, and thus, to reduce delay times, would be to increase the (W/L) ratios of all transistors in the circuit. However, increasing the transistor (W/L) ratios also increases the gate, source, and drain areas and, consequently, increases the parasitic capacitances loading the logic gates. Hence, the resizing of transistors is an iterative process which involves several cycles of consecutive layout modification, circuit extraction, and simulation. Since the carry_out signal is used to generate the sum output, reducing the delay in the carry_out stage should generally take a higher priority. Also, one should be careful to consider all possible input transitions: Optimizing the propagation delay for one particular input transition only may result in an unintended increase of propagations delays during other transitions.

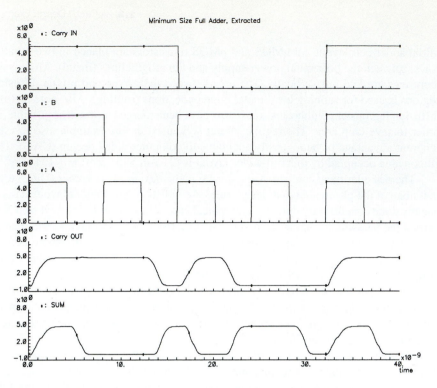

Figure 1.11 Simulated input and output waveforms of the full-adder circuit.

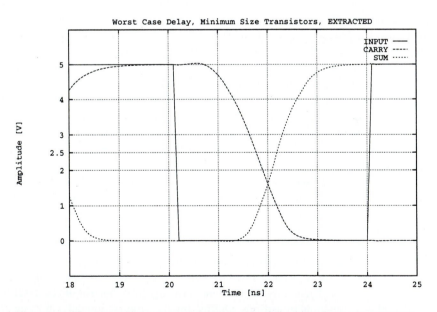

Figure 1.12 Simulated output waveforms of the full adder circuit with minimum transistor dimensions, showing the signal propagation delay during one of the worst-case transitions.

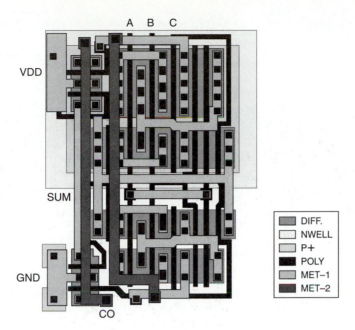

Figure 1.13 Modified layout of the full-adder circuit, with
optimized transistor dimensions.

While we resize the nMOS and pMOS transistors in the full-adder circuit to meet
the timing requirements, we can also reorganize the whole layout in order to achieve
a more compact placement, to increase silicon area utilization, and to reduce the interconnection parasitics within the cell. The resulting cell layout is shown in Fig. 1.13.
The new full-adder layout occupies an area of $(43 \, \mu\text{m} \times 90 \, \mu\text{m}) = 1{,}290 \, \mu\text{m}^2$, which
is about 14% larger than the initial layout (despite rather aggressive resizing of the
transistor dimensions) but still below the pre-set upper limit of $1{,}500 \, \mu\text{m}^2$.

For the optimized full-adder circuit, we find that all propagation and transition
(rise and fall) delay times are now within the specified limits, i.e., less than 1.2 ns.
Figure 1.14 shows the signal propagation delay of both inputs during the same worst-
case input transition depicted in Fig. 1.11. Note that the propagation delay is about
1.0 ns, a reduction of 50%. The dynamic power dissipation of this circuit is estimated
to be 460 μW. Thus, the circuit now satisfies the design specifications given in the
beginning.

The full-adder circuit designed in this example can now be used as the basic
building block of an 8-bit binary adder, which accepts two 8-bit binary numbers as
input and produces the binary sum at the output. The simplest such adder can be constructed by a cascade-connection of eight full adders, where each adder stage performs a two-bit addition, produces the corresponding sum bit, and passes the carry
output on to the next stage. Hence, this cascade-connected adder configuration is

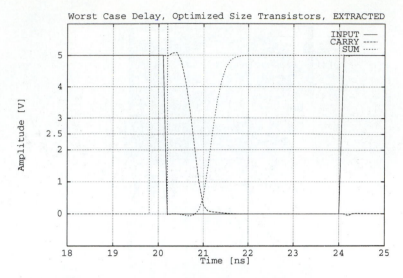

Figure 1.14 Simulated output waveforms of the full-adder circuit with optimized transistor dimensions, showing the signal propagation delay during the same worst-case transition.

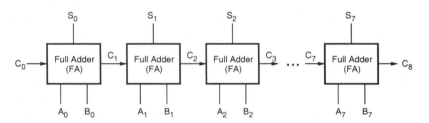

Figure 1.15 Block diagram of a carry ripple adder chain consisting of full adders.

called the carry ripple adder (Fig. 1.15). The overall speed of the carry ripple adder is obviously limited by the delay of the carry bits rippling through the carry chain; therefore, a fast carry_out response becomes essential for the overall performance of the adder chain.

Figure 1.16 shows the mask layout of a 4-bit section of the carry ripple adder circuit which is designed by simply cascading full-adder cells to form a regular array. Note that the input signals A_i and B_i are applied to a row of pins along the lower boundary of the array, while the output signals S_i (sum bits) are made available along the upper boundary of the array. This arrangement of the input and output pins simplifies signal routing by placing the *input bus* below, and the *output bus* above the adder array. Also note that no additional routing is necessary for the carry signal, since the carry_in and carry_out pin locations of consecutive full-adder cells are aligned against each other. Such structures are routinely used in circuits where a large number of arithmetic operations are required, such as arithmetic-logic units (ALUs) and digital signal processing (DSP) circuits. The overall performance of the multi-bit

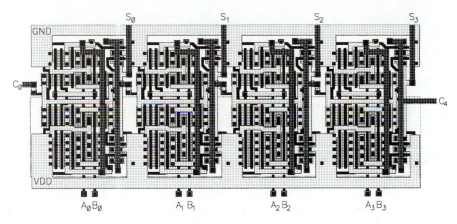

Figure 1.16 Mask layout of the 4-bit carry ripple adder array.

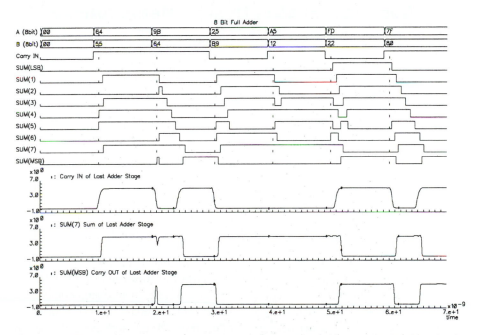

Figure 1.17 Simulated input and output waveforms of the 8-bit carry ripple adder circuit, showing a maximum signal propagation delay of about 7 ns.

adder can be further increased by various measures, and some of these issues will be discussed in later chapters.

The simulated input and output waveforms of the 8-bit binary adder circuit are shown in Fig. 1.17 for a series of sample input vectors. It can be seen that the sum bit of the last adder stage is typically generated last, and the overall delay can be as much as 7 ns.

This example has shown us that the design of CMOS digital integrated circuits involves a wide range of issues, from Boolean logic to gate-level design, to transistor-level design, to physical layout design, and to parasitics extraction followed by detailed circuit simulation for design tuning and performance verification. In essence, the final output of integrated circuit design is the mask data from which the actual circuit is fabricated. Thus, it is important to design the layout and, hence, the mask set such that the fabricated integrated circuits meet test specifications with a high yield.

To achieve such a goal, designers perform extensive simulations using computer models extracted from the layout data and iterate the design until simulated results meet the specifications with sufficient margins. In the following chapters, we will discuss the fabrication of MOS transistors using a set of masks, layout design rules, and electrical properties of MOS transistors and their computer models, before discussing the most basic CMOS inverter circuit.

1.4   Overview of VLSI Design Methodologies

As we have already pointed out earlier, the structural complexity of digital integrated circuits (usually expressed by the number of transistors per chip) has been increasing at an exponential rate over the last thirty years. This phenomenal growth rate has been sustained primarily by the constant advances in manufacturing technology, as well as by the increasing need for integrating more complex functions on chip. Answering the demands of the rapidly rising chip complexity has created significant challenges in many areas; practically hundreds of team members are involved in the development of a typical VLSI product, including the development of technology, computer-aided design (CAD) tools, chip design, fabrication, packaging, testing and reliability qualification. The efficient organization of these efforts under a well-structured system design methodology is essential for the development of economically viable VLSI products, in a timely manner. In this chapter, we will examine the overall flow of the design activities, important design concepts, various VLSI design styles, quality of design, and the CAD technology.

Generally speaking, logic chips such as microprocessor chips and digital signal processing (DSP) chips contain not only large arrays of memory (SRAM/DRAM) cells, but also many different functional units. As a result, their design complexity is considered much higher than that of memory chips, although advanced memory chips also contain some sophisticated logic functions. Notice that the design complexity of logic chips increases almost *exponentially* with the number of transistors to be integrated. This is translated into an increase in the design cycle time, which is the time period from the start of chip development until the mask-tape delivery time. The majority of this design cycle time is typically devoted to achieving the desired level of chip performance at an acceptable cost, which is essential for the economical success of any competitive commercial product. During the course of the design cycle the circuit performance can usually be increased by design improvements; more rapidly in the beginning, then more gradually until the performance finally saturates for the particular design style and technology being used.

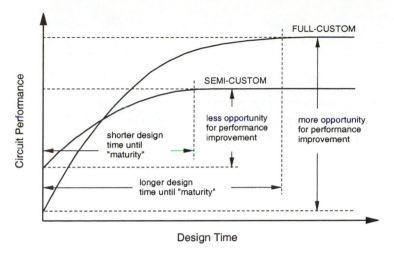

Figure 1.18 Impact of different VLSI design styles upon the design cycle time and the achievable circuit performance.

The level of circuit performance which can be reached within a certain design time strongly depends on the efficiency of the design methodologies, as well as on the design style.

This point is illustrated qualitatively in Fig. 1.18, where two different VLSI design styles are compared for their relative merits in the design of the same product. Using the full-custom design style (where the geometry and the placement of every transistor can be optimized individually) requires a longer time until design maturity can be reached, yet the inherent flexibility of adjusting almost every aspect of circuit design allows far more opportunity for circuit performance improvement during the design cycle. The final product typically has a high level of performance (e.g., high processing speed, low power dissipation) and the silicon area is relatively small because of better area utilization. But this comes at a larger cost in terms of design time. In contrast, using a semi-custom design style (such as standard-cell based design or FPGA) will allow a shorter design time until design maturity can be achieved. In the early design phase, the circuit performance can be even higher than that of a full-custom design, since some of the components used in semi-custom design are already optimized. But the semi-custom design style offers less opportunity for performance improvement, and the overall performance of the final product will inevitably be less than that of a full-custom design.

The choice of the particular design style for a VLSI product depends on the performance requirements, the technology being used, the expected lifetime of the product and the cost of the project. In the following sections, we will discuss the various aspects of different VLSI design styles and consider their impact upon circuit performance and overall cost.

In addition to the proper choice of a VLSI design style, there are other issues which must be addressed in view of the constantly evolving nature of the VLSI

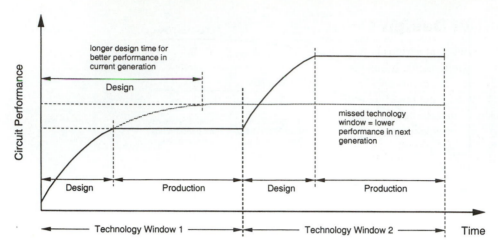

Figure 1.19 Progressive performance improvement of a VLSI product for each new generation of manufacturing technology. Shorter design cycle times are essential for economic viability.

manufacturing technologies. Approximately every two years, a new generation of technology is introduced, which typically allows for smaller device dimensions and consequently, higher integration density and higher performance. In order to make the best use of the current technology, the chip development time has to be short enough to allow the maturing of chip manufacturing and timely delivery of the product to customers. This may require that the level of logic integration and chip performance fall short of the level *achievable* with the current processing technology, as illustrated in Fig. 1.19.

It can be seen that the design cycle time of a successful VLSI product is kept shorter than what would be necessary for developing an optimum-performance chip, thus leaving enough time for the production and marketing of the chip during the current generation, or "technology window." When the next generation of manufacturing technology arrives, the design can be updated to take advantage of higher integration density and better performance. On the other hand, if the design time of a particular product is kept excessively long to achieve the highest possible performance for the current generation of technology, there is a danger of missing the next technology window. A longer design cycle time usually results in better overall performance, but this product must then remain viable in the market for a certain amount of time in order to recover the development costs. Thus, the advantages brought by the next generation of manufacturing technologies cannot be utilized, and the product becomes less competitive.

In reality, the design cycle of the next generation chips usually overlaps with the production cycle of the current generation chips, thereby assuring continuity. The use of sophisticated computer-aided design (CAD) tools and methodologies are also essential for reducing the design cycle time and for managing the increasing design complexity.

1.5 VLSI Design Flow

The design process, at various levels, is usually evolutionary in nature. It starts with a given set of requirements. Initial design is developed and tested against the requirements. When requirements are not met, the design has to be improved. If such improvement is either not possible or too costly, then a revision of requirements and an impact analysis must be considered. The Y-chart (first introduced by D. Gajski) shown in Fig. 1.20 illustrates a simplified design flow for most logic chips, using design activities on three different axes (domains) which resemble the letter "Y." In reality, there exist many feedback loops that are not shown for simplicity.

The Y-chart consists of three domains of representation, namely (i) behavioral domain, (ii) structural domain, and (iii) geometrical layout domain. The design flow starts from the *algorithm* that describes the behavior of the target chip. The corresponding *architecture of the processor* is first defined. It is mapped onto the chip surface by *floorplanning*. The next design evolution in the behavioral domain defines *finite state machines* (FSMs) which are structurally implemented with *functional modules* such as registers and arithmetic logic units (ALUs). These modules are then geometrically placed onto the chip surface using CAD tools for automatic *module placement* followed by routing, with a goal of minimizing the interconnects' area and signal delays. The third evolution starts with a behavioral *module description.*

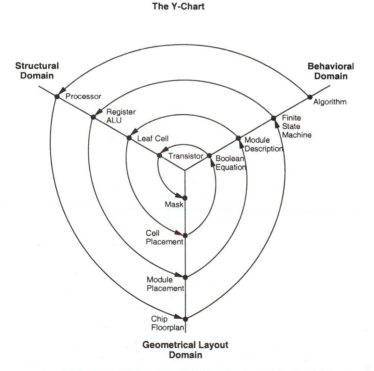

Figure 1.20 Simplified VLSI design flow in three domains (Y-chart representation).

Individual modules are then implemented with *leaf cells*. At this stage the chip is described in terms of logic gates (leaf cells), which can be placed and interconnected by using a *cell placement and routing* program. The last evolution involves a detailed *Boolean description* of leaf cells followed by a transistor level implementation of leaf cells and *mask generation*. In the standard-cell based design style, leaf cells are pre-designed (at the transistor level) and stored in a library for logic implementation, effectively eliminating the need for the transistor level design.

Figure 1.21 provides a more simplified view of the VLSI design flow, taking into account the various representations, or abstractions of design: behavioral, logic, circuit

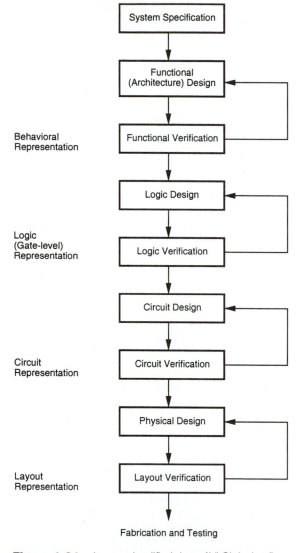

Figure 1.21   A more simplified view of VLSI design flow.

and mask layout. Note that the *verification* of design plays a very important role in every step during this process. The failure to properly verify a design in its early phases typically causes significant and expensive re-design at a later stage, which ultimately increases the time-to-market.

Although the design process has been described in linear fashion for simplicity, in reality there are many iterations, especially between any two neighboring steps, and occasionally even remotely separated pairs. Although top-down design flow appeals for design process control, in reality, there is no truly uni-directional top-down design flow. Both *top-down* and *bottom-up* approaches have to be combined for a successful design. For instance, if a chip designer defines an architecture without close estimation of the corresponding chip area, then it is very likely that the resulting chip layout exceeds the area limit of the available technology. In such a case, in order to fit the architecture into the allowable chip area, some functions may have to be removed and the design process must be repeated. Such changes may require significant modification of the original requirements. Thus, it is very important to feed forward low-level information to higher levels (bottom-up) as early as possible.

In the following, we will examine design methodologies and structured approaches which have been developed over the years to deal with both complex hardware and software projects. Regardless of the actual size of the project, the basic principles of structured design will improve the prospects of success. Some of the classical techniques for reducing the complexity of IC design are: Hierarchy, regularity, modularity and locality.

1.6 Design Hierarchy

The use of the hierarchy, or "divide and conquer" technique involves dividing a module into sub-modules and then repeating this operation on the sub-modules until the complexity of the smaller parts becomes manageable. This approach is very similar to software development wherein large programs are split into smaller and smaller sections until simple subroutines, with well-defined functions and interfaces, can be written. In the previous section, we have seen that the design of a VLSI chip can be represented in three domains. Correspondingly, a hierarchy structure can be described in each domain separately. However, it is important for the simplicity of design that the hierarchies in different domains be mapped into each other easily.

As an example of structural hierarchy, Fig. 1.22 shows the structural decomposition of a CMOS 4-bit adder into its components. The adder can be decomposed progressively into 1-bit adders, separate carry and sum circuits, and finally into individual logic gates. At this lower level of the hierarchy, the design of a simple circuit realizing a well-defined Boolean function is much easier to handle than at the higher levels of the hierarchy.

In the physical domain, partitioning a complex system into its various functional blocks will provide a valuable guide for the actual realization of these blocks on the chip. Obviously, the approximate shape and size (area) of each sub-module should be estimated in order to provide a useful floorplan. Figure 1.23 shows the hierarchical decomposition of the four-bit adder in physical description (geometrical layout)

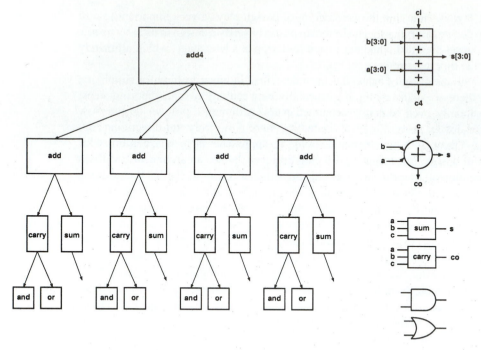

Figure 1.22 Structural decomposition of a 4-bit adder, showing the levels of hierarchy.

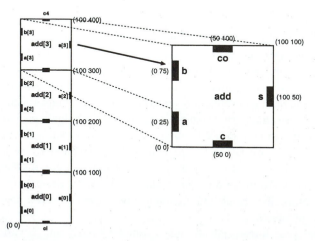

Figure 1.23 Decomposition of the 4-bit adder in the physical (geometrical) domain.

domain, resulting in a simple *floorplan*. This physical view describes the external geometry of the adder, the locations of input and output pins, and the pin locations that allow some signals (in this case the *carry* signals) to be transferred from one sub-block to the other without external routing. At lower levels of the physical hierarchy, the internal mask layout of each adder cell defines the locations and the connections of each transistor and wire. Figure 1.24 shows the full-custom layout of a 16-bit dynamic CMOS adder, and the sub-modules that describe the lower levels of its physical hierarchy. Here, the 16-bit adder consists of a cascade connection of four 4-bit adders, and each 4-bit adder can again be decomposed into its functional blocks such as the Manchester chain, carry/propagate circuits and the output buffers. Finally, Fig. 1.25 shows the structural hierarchy of the 16-bit adder. Note that there is a corresponding physical description for every module in the structural hierarchy, i.e., the components of the physical view closely match this structural view.

1.7 Concepts of Regularity, Modularity, and Locality

The hierarchical design approach reduces the design complexity by dividing the large system into several sub-modules. Usually, other design concepts and design approaches are also needed to simplify the process. Regularity means that the hierarchical decomposition of a large system should result in not only simple, but also similar blocks, as much as possible. A good example of regularity is the design of array structures consisting of identical cells—such as a parallel multiplication array. Regularity can exist at all levels of abstraction. For example, at the transistor level, uniformly sized transistors simplify the design and at the logic level, identical gate structures can be used. Figure 1.26 shows regular circuit-level designs of a 2-1 MUX (multiplexer) and a D-type edge-triggered flip flop. Note that both of these circuits were designed by using inverters and tri-state buffers only. If the designer has a small library of well-characterized basic building blocks, a number of different functions can be constructed by using this principle. Regularity usually reduces the number of different modules that need to be designed and verified, at all levels of abstraction.

Modularity in design means that the various functional blocks which make up the larger system must have *well-defined functions* and *interfaces*. Modularity allows that each block or module can be designed relatively independently from each other, since there is no ambiguity about the function and the signal interface of these blocks. All of the blocks can be combined with ease at the end of the design process, to form the large system. The concept of modularity enables the parallelization of the design process. The well-defined functionality and signal interface also allow the use of generic modules in various designs.

By defining well-characterized interfaces for each module in the system, we effectively ensure that the internals of each module become unimportant to the exterior modules. Internal details remain at the local level. The concept of locality also ensures that connections are mostly between neighboring modules, avoiding long-distance connections as much as possible. This last point is extremely important for avoiding long interconnect delays. Time-critical operations should be performed

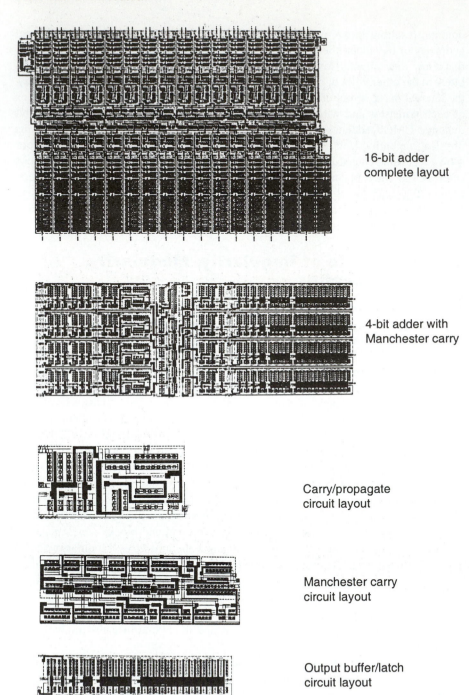

16-bit adder
complete layout

4-bit adder with
Manchester carry

Carry/propagate
circuit layout

Manchester carry
circuit layout

Output buffer/latch
circuit layout

Figure 1.24   Layout of a 16-bit adder, and the components of its physical hierarchy.

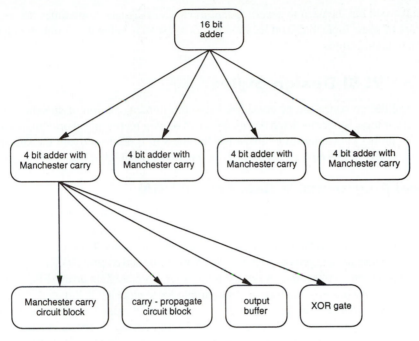

Figure 1.25 Structural hierarchy of the 16-bit adder circuit.

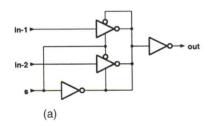

(a)

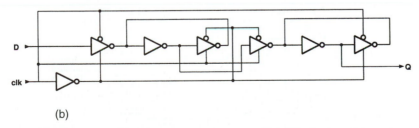

(b)

Figure 1.26 Regular design of (a) 2-1 MUX and (b) DFF, using inverters and tri-state buffers as basic building blocks.

locally, without the need to access distant modules or signals. Sometimes, the *replication* of some logic in distant location for local use may solve this problem in large system architectures.

1.8   VLSI Design Styles

Several design styles can be considered for chip implementation of specified algorithms or logic functions. Each design style has its own merits and shortcomings, and thus a proper choice has to be made by designers in order to provide the specified functionality at low cost and in a timely manner, as explained in Section 1.4.

Field Programmable Gate Array (FPGA)

Fully fabricated FPGA chips containing tens to hundreds of thousands, or even more, of logic gates with programmable interconnects, are available to users for their custom hardware programming to realize desired functionality. This design style provides a means for fast prototyping and also for cost-effective chip design, especially for low-volume applications. A typical field programmable gate array (FPGA) chip consists of I/O buffers, an array of configurable logic blocks (CLBs), and programmable interconnect structures. The programming of the interconnects is accomplished by programming of RAM cells whose output terminals are connected to the gates of MOS pass transistors. Thus, the signal routing between the CLBs and the I/O blocks is accomplished by setting the configurable switch matrices accordingly. The general architecture of an FPGA chip from Xilinx is shown in Fig. 1.27. A more detailed view showing the locations of switch matrices used for interconnect routing is given in Fig. 1.28.

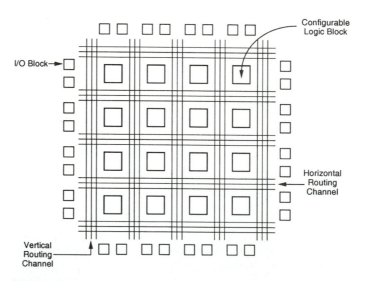

Figure 1.27   General architecture of Xilinx FPGAs.

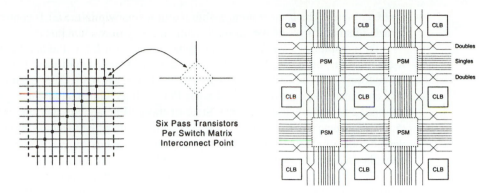

Figure 1.28 Detailed view of switch matrices and interconnection routing between CLBs.

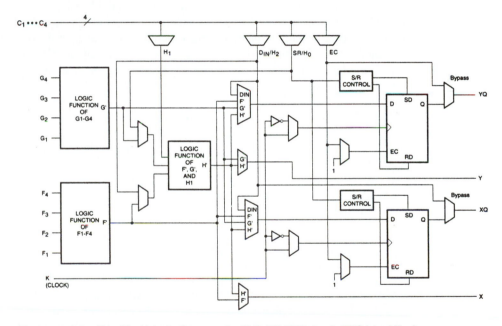

Figure 1.29 Simplified block diagram of a CLB (XC4000-family FPGA by Xilinx).

The simplified block diagram of a CLB (XC4000 family from Xilinx) is shown in Fig. 1.29. In this example, each CLB contains two independent 4-input combinational function generators, a clock signal terminal, user-programmable multiplexers and two flip-flops. The function generators, which are capable of realizing any arbitrarily defined Boolean function of their four inputs, are implemented as memory look-up tables. A third function generator can implement any Boolean function of its three inputs: F', G' and a third input from outside the CLB. Thus, the CLB offers significant flexibility of implementing a wide range of functions, with up to nine input variables. The user-programmable multiplexers within the CLB control the internal signal routing, and therefore, the functionality of the block.

The complexity of a FPGA chip is typically determined by the number of CLBs it contains. In the Xilinx XC4000 family of FPGAs, the size of the CLB array can range from 8×8 (64 CLBs) to 32×32 (1024 CLBs), where the latter example has an approximate gate count of 25,000. Larger FPGA chips with an equivalent gate count of about 200,000 are also available. State-of-the-art FPGA chips can support system clock frequencies up to hundreds of MHz. With the use of dedicated computer-aided design tools, the gate utilization rate (percentage of gates on the FPGA which are actually used in a particular design) can exceed 90%.

The typical design flow of an FPGA chip starts with the behavioral description of its functionality, using a hardware description language such as VHDL. The synthesized architecture is then *technology-mapped* (or partitioned) into circuits or logic cells. At this stage, the chip design is completely described in terms of available logic cells. Next, the *placement and routing* step assigns individual logic cells to FPGA sites (CLBs) and determines the routing patterns among the cells in accordance with the netlist. After routing is completed, the on-chip performance of the design can be simulated and verified before *downloading* the design for programming of the FPGA chip. The programming of the chip remains valid as long as the chip is powered-on, or until it is re-programmed.

The largest advantage of FPGA-based design is the very short *turn-around time,* i.e., the time required from the start of the design process until a functional chip is available. Since no physical manufacturing step is necessary for customizing the FPGA chip, a functional sample can be obtained almost as soon as the design is mapped into a specific technology. The typical price of FPGA chips is usually higher than other alternatives (such as gate array or standard cells) of the same design, but for small-volume production of ASIC chips and for fast prototyping, FPGA offers a very valuable option.

Gate Array Design

In terms of fast prototyping capability, the gate array (GA) ranks second after the FPGA with a typical turn-around time of a few days. While user programming is central to the design implementation of the FPGA chip, metal mask design and processing is used for GA. Gate array implementation requires a two-step manufacturing process: The first phase, which is based on generic (standard) masks, results in an array of *uncommitted* transistors on each GA chip. These uncommitted chips can be stored for later *customization,* which is completed by defining the metal interconnects between the transistors of the array (Fig. 1.30). Since the patterning of metallic interconnects is done at the end of the chip fabrication process, the turn-around time can still be short, a few days to a few weeks. Figure 1.31 shows a corner of a gate array chip which contains bonding pads on its left and bottom edges, diodes for I/O protection, nMOS transistors and pMOS transistors for chip output driver circuits adjacent to bonding pads, arrays of nMOS transistors and pMOS transistors, underpass wire segments, and power and ground buses along with contact windows.

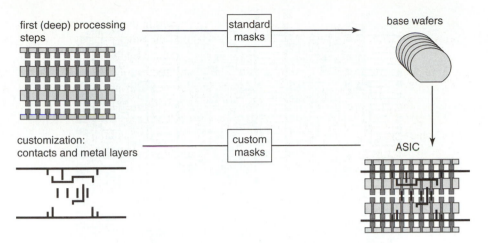

first (deep) processing steps

standard masks

base wafers

customization: contacts and metal layers

custom masks

ASIC

Figure 1.30 Basic processing steps required for gate array implementation.

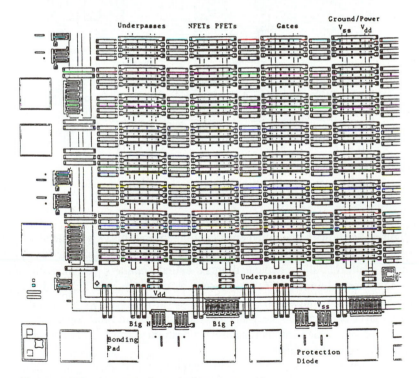

Underpasses   NFETs PFETs   Gates   Ground/Power
V_{ss} V_{dd}

Underpasses

V_{dd}

V_{ss}

Big N   Big P

Bonding Pad

Protection Diode

Figure 1.31 A corner of a typical gate array chip.
(Copyright © 1987 Prentice Hall, Inc.)

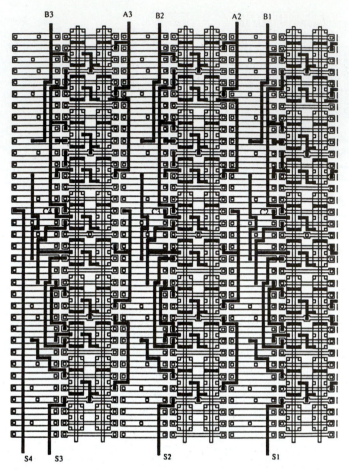

Figure 1.32 Metal mask design to realize a complex logic function on a channeled gate array (GA) platform.

Figure 1.32 shows a magnified portion of the internal array with metal mask design (metal lines highlighted in dark) to realize a complex logic function. Typical gate array platforms allow dedicated areas, called *channels,* for intercell routing between rows or columns of MOS transistors, as shown in Figs. 1.31 and 1.32. The availability of these routing channels simplifies the interconnections, even using one metal layer only. Interconnection patterns that perform basic logic gates can be stored in a *library,* which can then be used to customize rows of uncommitted transistors according to the netlist.

While most gate array platforms only contain rows of uncommitted transistors separated by routing channels, some other platforms also offer dedicated memory (RAM) arrays to allow a higher density where memory functions are required. Figure 1.33 shows the layout views of a conventional gate array and a gate array platform with two dedicated memory banks.

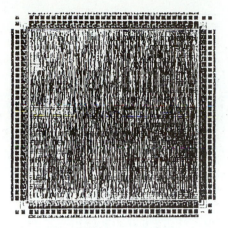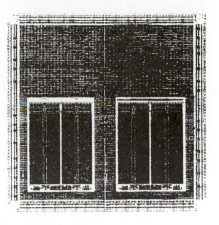

Figure 1.33 Layout views of a conventional GA chip (left) and a gate array with two embedded memory banks (right).

In most of the modern GAs, multiple metal layers are used for channel routing. With the use of multiple interconnect layers, the routing can also be achieved over the active cell areas; thus, the routing channels can be removed as in Sea-of-Gates (SOG) chips. Here, the entire chip surface is covered with *uncommitted* nMOS and pMOS transistors. As in the gate array case, neighboring transistors can be customized using a metal mask to form basic logic gates. For intercell routing, however, some of the uncommitted transistors must be sacrificed. This approach results in more flexibility for interconnections, and usually in a higher density. The basic platform of a SOG chip is shown in Fig. 1.34. Figure 1.35 offers a brief comparison between the channeled (GA) and the channelless (SOG) approaches.

In general, the GA chip utilization factor, as measured by the used chip area divided by the total chip area, is higher than that of the FPGA and so is the chip speed, since more customized design can be achieved with metal mask designs. The current gate array chips can implement as many as hundreds of thousands of logic gates.

Standard-Cells Based Design

The standard-cells based design is one of the most prevalent full custom design styles which require development of a full custom mask set. The standard cell is also called the polycell. In this design style, all of the commonly used logic cells are developed, characterized, and stored in a standard cell library. A typical library may contain a few hundred cells including inverters, NAND gates, NOR gates, complex AOI, OAI gates, D-latches, and flip-flops. Each gate type can be implemented in several *versions* to provide adequate driving capability for different fan-outs. For instance, the inverter gate can have standard size, double size, and quadruple size so that the chip designer can choose the proper size to achieve high circuit speed and layout

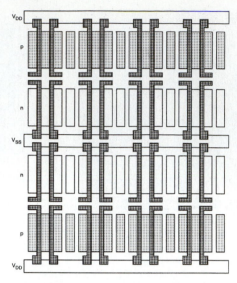

Figure 1.34 The typical platform of a
Sea-of-Gates (SOG) chip.

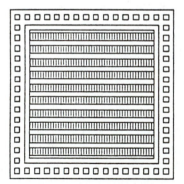

routing problem is simpler
OK with only one metal

flexibility in channel definition (position & width)
over-the-cell routing
higher packing density
RAM-compatible
supports variable-height cells & macrocells

Figure 1.35 Comparison between the channeled (GA) versus the channelless (SOG) designs.

density. Each cell is characterized according to several different characterization categories, including

■ Delay time versus load capacitance
■ Circuit simulation model
■ Timing simulation model

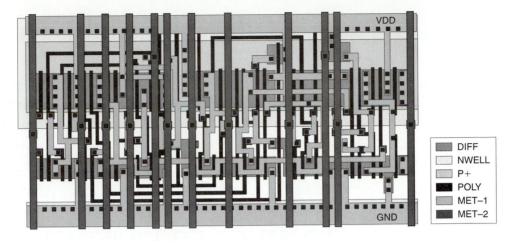

▓	DIFF
░	NWELL
▒	P+
███	POLY
▒	MET–1
▓	MET–2

Figure 1.36 A standard cell layout example.

- Fault simulation model
- Cell data for place-and-route
- Mask data

To enable automated placement of the cells and routing of inter-cell connections, each cell layout is designed with a fixed height, so that a number of cells can be abutted side-by-side to form *rows*. The power and ground rails typically run parallel to the upper and lower boundaries of the cell, thus, neighboring cells share a common power bus and a common ground bus. The input and output pins are located on the upper and lower boundaries of the cell. Figure 1.36 shows the layout of a typical standard cell. Notice that the nMOS transistors are located closer to the ground rail while the pMOS transistors are placed closer to the power rail.

Figure 1.37 shows a floorplan for standard-cell based design. Inside the I/O frame which is reserved for I/O cells, the chip area contains rows or columns of standard cells. Between cell rows are channels for dedicated intercell routing. As in the case of Sea-of-Gates with over-the-cell routing, the channel areas can be reduced or even removed provided that the cell rows offer sufficient routing space. The physical design and layout of logic cells ensure that when cells are placed into rows, their heights are matched and neighboring cells can be abutted side-by-side, which provides natural connections for power and ground lines in each row. The signal delay, noise margins, and power consumption of each cell should be also optimized with proper sizing of transistors using circuit simulation.

If a number of cells must share the same input and/or output signals, a common signal bus structure can also be incorporated into the standard-cells based chip layout. Figure 1.38 shows the simplified symbolic view of a case where a signal bus has been inserted between the rows of standard cells. Note that in this case the chip consists of two *blocks,* and power/ground routing must be provided from both sides of the layout area. Standard-cell based designs may consist of several such

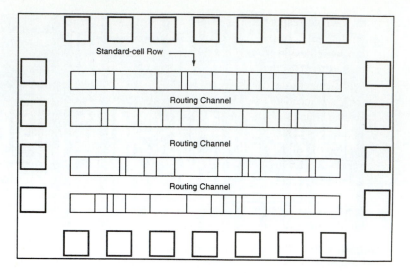

Figure 1.37 A simplified floorplan of standard-cells based design.

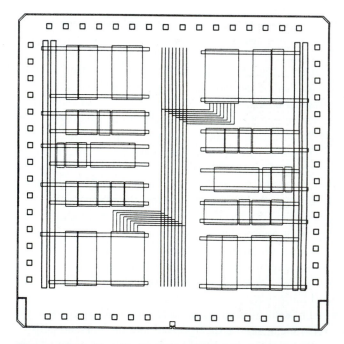

Figure 1.38 Simplified floorplan of a standard-cells based design, consisting of two separate blocks and a common signal bus.

macro-blocks, each corresponding to a specific unit of the system architecture such as clock generator, ALU, or control logic.

After chip logic design is done using standard cells from the library, the most challenging task is to place the individual cells into rows and interconnect them in a way that meets stringent design goals in circuit speed, chip area, and power consumption. Many advanced CAD tools for place-and-route have been developed and used to achieve such goals. Also from the chip layout, circuit models which include interconnect parasitics can be extracted and used for timing simulation and analysis to identify timing critical paths. For timing critical paths, proper gate sizing is often practiced to meet the timing requirements. In many VLSI chips, such as microprocessors and digital signal processing chips, standard-cell based design is used for complex control logic modules. Some full custom chips can be also implemented exclusively with standard cells.

Finally, Fig. 1.39 shows the detailed mask layout of a standard-cells based chip with an uninterrupted single block of cell rows, and three memory banks placed on one side of the chip. Notice that within the cell block, the separations between neighboring rows depend on the number of wires in the routing channel between the cell rows. If a high interconnect density can be achieved in the routing channel, the

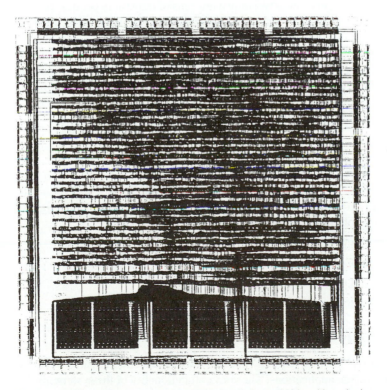

Figure 1.39 Mask layout of a standard-cells based chip with a single block of cells and three memory banks.

standard cell rows can be placed closer to each other, resulting in a smaller chip area. The availability of dedicated memory blocks also reduces the area, since the realization of memory elements using standard cells would occupy a much larger area.

Full Custom Design

Although the standard-cells based design style is sometimes called full custom design, in a strict sense, it is somewhat less than fully customized since the cells are pre-designed for general use and the same cells are utilized in many different chip designs. In a truly full-custom design, the entire mask design is done anew without use of any library. However, the development cost of such a design style is becoming prohibitively high. Thus, the concept of *design reuse* is becoming popular in order to reduce design cycle time and development cost. The most rigorous full custom design can be the design of a memory cell, be it static or dynamic. Since the same layout design is replicated, there would not be any alternative to high density memory chip design. For logic chip design, a good compromise can be achieved by using a combination of different design styles on the same chip, such as standard cells, data-path cells, and programmable logic arrays (PLAs). In real full-custom layout in which the geometry, orientation, and placement of every transistor is done individually by the designer, design productivity is usually very low—typically a few tens of transistors per day, per designer.

In digital CMOS VLSI, full-custom design is rarely used due to the high labor cost. Exceptions to this include the design of high-volume products such as memory chips, high-performance microprocessors, and FPGA masters. Figure 1.40 shows the full layout of the Intel Pentium microprocessor chip, which is a good example of a hybrid full-custom design. Here, one can identify four different design styles on one chip: memory banks (RAM cache), data-path units consisting of bit-slice cells, control circuitry mainly consisting of standard cells, and PLA blocks.

1.9   Design Quality

It is desirable to measure the quality of design in order to improve the chip design. Although no universally accepted metric exists to measure the design quality, the following criteria are considered to be important:

- Testability
- Yield and manufacturability
- Reliability
- Technology updateability

Testability

Developed chips are eventually inserted into printed circuit boards or multichip modules for system applications. The correct functionality of the system hinges upon the correct functionality of the chips used. Therefore, fabricated chips should be fully testable to ensure that all the chips passing the specified chip test can be inserted into the system, either in packaged or in bare die form, without causing failures. Such a

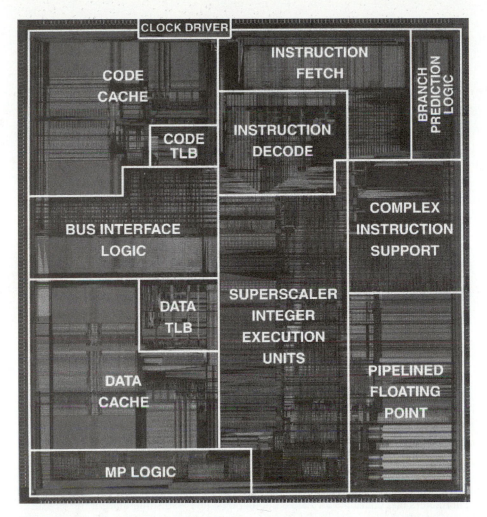

Figure 1.40 Mask layout of the Intel Pentium microprocessor chip, an example of a full-custom design where the rigorous design effort can be justified by the high performance and high production volume of the chip.
(Pentium® processor, provided courtesy of Intel Corporation.)

goal requires

- Generation of good test vectors
- Availability of reliable test fixture at speed
- Design of testable chip

In fact, some chip projects had to be abandoned after chip fabrication because of inadequate testability of the design. As the complexity of the chips increases with the increasing level of monolithic integration, additional circuitry for self testing has to be included to ensure that the fabricated chips can be fully tested. This translates into an increase in chip area and some speed penalty, but such a trade-off will become unavoidable in VLSI design. The design-for-testability issues will be discussed in Chapter 15.

Yield and Manufacturability

If we assume that the test procedure is flawless, the chip yield can be calculated by dividing the number of good tested chips by the total number of tested chips. However, this calculation may not correctly reflect the quality of the design or the processing. The most strict definition of the yield can be the number of good tested chips divided by the total number of chip sites available at the start of the wafer processing. However, since some wafers may be scrapped in the process line due to mishandling or for other reasons, such a metric may not reflect the design quality. Also, poor design of the wafer array for chips may cause some chips to fail routinely due to uncontrollable process variations and handling problems. On the other hand, poor chip design can cause processing problems and, therefore, drop-outs during the processing. In such a case, the first yield metric will overestimate the design quality. The chip yield can be further divided into the following subcategories:

■ Functional yield
■ Parametric yield

The functional yield is obtained by testing the functionality of the chip at a speed usually lower than the required chip speed. The functional test weeds out the problems of shorts, opens, and leakage current, and can detect logic and circuit design faults.

The parametric test is usually performed at the required speed on chips that passed the functional test. All the delay testing is performed at this stage. Poor design that failed to consider uncontrollable process variations which cause significant variations in chip performance may cause poor parametric yield, thus, significant manufacturing problems. In order to achieve high chip yield, chip designers should consider manufacturability of the chip by considering realistic fluctuations in device parameters that cause performance fluctuations, and also some margin for measurement uncertainty.

Reliability

The reliability of the chip depends on the design and process conditions. The major causes for chip reliability problems can be characterized into the following:

■ Electrostatic discharge (ESD) and electrical overstress (EOS)
■ Electromigration
■ Latch-up in CMOS I/O and internal circuits
■ Hot-carrier induced aging
■ Oxide breakdown
■ Single event upset
■ Power and ground bouncing
■ On-chip noise and crosstalk

Usually the wafer lots with poor yields also cause reliability problems. For example, when a particular wafer processing is poorly controlled, thus causing aluminum overetching, many chips on the wafer may suffer from open-circuited metallic interconnects. Some chips with severely overetched, but not fully open-circuited interconnects, may pass the test. But, under current stress, such interconnects can be open-circuited because of electromigration problems, causing chip and system failures in the field. Any good manufacturing practice should weed out such potential failures during the accelerated reliability test.

Nevertheless, for any specified process, chip design can be improved to overcome such process-dependent reliability problems. For example, knowing that aluminum overetching can occur, alert designers may choose to widen the metal width beyond the minimum width allowed. Similarly, to avoid the transistor aging problem due to hot-carrier aging, designers can improve the circuit reliability with proper sizing of transistors or by reducing the rise time of signals feeding into the nMOS transistor gate. The protection of I/O circuits against electrostatic damages (ESD) and latch-up is another example. Some of these specific points will be discussed in Chapter 13 for design of reliable I/O circuits.

Technology Updateability

Process technology development has progressed rapidly and as a result, the lifespan of a given technology generation has remained almost constant even for submicron technologies. Yet, the time pressure to develop increasingly more complex chips in a shorter time is constantly increasing. Under such circumstances, the chip products often have to be technology-updated to new design rules. Even without any change in the chip's functionality, the task of updating the mask to new design rules is very formidable. The so-called "dumb shrink" method whereby mask dimensions are scaled uniformly, is rarely practiced due to nonideal scaling of device feature sizes and technology parameters. Thus, the design style should be chosen such that the technology update of the chip or functional modules for design reuse can be achieved quickly with minimal cost. Designers can develop and use advanced CAD tools that can automatically generate the physical layout, the so-called *silicon compilation,* which meets the timing requirements with proper gate sizing or transistor sizing.

1.10 Packaging Technology

Novice designers often fail to give enough consideration to the packaging technology, especially in the early stages of chip development. However, many high-performance VLSI chips can fail stringent test specifications after packaging if chip designers have not included various effects of packaging constraints and parasitics in their design.

The numbers of ground planes and power planes and the bonding pads greatly affect the behaviors of the on-chip power and ground buses. Also the length of the bonding wire between the chip and the package and the lead length in the package determine the inductive voltage drop in the output circuit. An equally important consideration is thermal problems. Good packages should provide low thermal resistance and, hence, limited temperature increase beyond the ambient temperature due to power dissipation.

Since the choice of proper packaging technology is critical to the success of the chip development, chip designers should work closely with package designers from the start of the project, especially for full custom designs. Also, since the final cost of the packaged chip depends largely on the package cost itself, for low-cost chip development, designers must ensure enough design margins that the chips can function properly in low-cost packages with more parasitic effects and less thermal conductivity. Some of the important packaging concerns are

- Hermetic seals to prevent the penetration of moisture
- Thermal conductivity
- Thermal expansion coefficient
- Pin density
- Parasitic inductance and capacitance
- α-particle protection

Various types of packages are available for integrated circuit chips. Integrated circuit packages are generally classified by the method which is used to solder the package on the printed circuit board (PCB). The package pins can be introduced in holes drilled in the PCB; this method is called *pin-through-hole* (PTH). Alternatively, the package pins can be directly soldered on the PCB; this method is called *surface-mounted technology* (SMT).

PTH packages require that a precise hole be drilled in the PCB for each pin, which is not a cost-effective process. Moreover, holes usually require metal plating on their interior surface to ensure conductivity, and the lack of proper plating may cause yield and reliability problems. Nevertheless, PTH packages have the advantage that they can be soldered using a relatively inexpensive soldering process. In comparison, SMT packages are usually more cost- and space-effective, yet soldering of SMT packages on the PCB requires more expensive equipment.

Plastic has been the dominating material for IC packages for many years, although it has the disadvantage of being permeable to environmental moisture. Ceramic packages are used when power dissipation, performance, or environmental requirements justify the relatively higher cost. Some common IC package types are:

Dual In-line Packages (DIP)   This PTH package has been the most dominant IC package type for more than 20 years. DIP have the advantage of low cost but their dimensions can be prohibitive, especially for small, portable products. DIP are also characterized by their high interconnect inductances, which can lead to significant noise problems in high-frequency applications. The maximum pin count of DIP is typically limited to 64.

Pin Grid Array (PGA) Packages   This PTH package type offers a higher pin count (typically several hundreds) and higher thermal conductivity (hence, better power dissipation characteristics) compared to DIPs, especially when a passive or active heat sink is attached on the package. The PGA packages require a large PCB area, and the package cost is higher than DIP, especially for ceramic PGAs. More recently, Ball Grid Array (BGA) packages have been introduced to reduce parasitics for high performance chips.

Table 1.2 Some characteristics of 64–68 pin packages.

Parameter	Package type				
	DIP (ceramic)	DIP (plastic)	PGA	Leadless chip carrier	Leaded chip carrier
Max. lead R (Ω)	1.1	0.1	0.2	0.2	0.1
Max. lead C (pF)	7	4	2	2	2
Max. lead L (nH)	22	36	7	7	7
Thermal R (°C/W)	32	5	20	13	28
PCB area (cm^2)	18.7	18.7	6.45	6.45	6.45

Chip Carrier Packages (CCP)   This SMT package type is available in two variations, the *leadless chip carrier* and the *leaded chip carrier.* The leadless chip carrier is designed to be mounted directly on the PCB, and it can support a high pin count. The main drawback is the inherent difference in thermal coefficients between the chip carrier and the PCB, which can eventually cause mechanical stresses to occur on the surface of the PCB. The leaded chip carrier package solves this problem since the added leads can accommodate small dimension variations caused by the differences in thermal coefficients.

Quad Flat Packs (QFP)   This SMT package type is similar to leaded chip carrier packages, except that the leads extend outward rather than being bent under the package body. Ceramic and plastic QFPs with very high pin counts (up to 500) are becoming popular package types in recent years.

Multi-Chip Modules (MCM)   This IC package option can be used for special applications requiring very high performance, where multiple chips are assembled on a common substrate contained in a single package. Thus, a large number of critical interconnections between the chips can be made *within* the package. Advantages include significant savings of overall system size, reduced package lead counts, and faster operation since chips can be placed in very close proximity.

Table 1.2 provides some data on the characteristics of commonly used packages.

1.11   Computer-Aided Design Technology

Computer-aided design (CAD) tools are essential for timely development of integrated circuits. Although CAD tools cannot replace the creative and inventive parts of the design activities, the majority of time-consuming and computation intensive mechanistic parts of the design can be executed by using CAD tools. The CAD technology for VLSI chip design can be categorized into the following areas:

- High level synthesis
- Logic synthesis
- Circuit optimization
- Layout
- Simulation

■ Design rules checking
■ Formal verification

Synthesis Tools

The high-level synthesis tools using hardware description languages (HDLs), such as VHDL or Verilog, address the automation of the design phase in the top level of the design hierarchy. With an accurate estimation of lower level design features, such as chip area and signal delay, it can very effectively determine the types and quantities of modules to be included in the chip design. Many tools have also been developed for logic synthesis and optimization, and have been customized for particular design needs, especially for area minimization, low power, high speed, or their weighted combination.

Layout Tools

The tools for circuit optimization are concerned with transistor sizing for minimization of delays and with process variations, noise, and reliability hazards. The layout CAD tools include floorplanning, place-and-route, and module generation. Sophisticated layout tools are goal driven and include some degree of optimization functions. For example, timing-driven layout tools are intended to produce layouts which meet timing specifications.

Automatic cell placement and routing programs constitute a very important category of physical design automation tools. One of the most challenging tasks in cell-based design automation is achieving an optimum or near-optimum placement of all standard cells on chip, so that the signal routing between the cells can be accomplished with minimum interconnect area and minimum delay. Since the problem of finding the optimum placement for hundreds of thousands of standard cells is usually too costly to be solved as a formal geometrical placement problem, various heuristic methods (such as the min-cut algorithm, force-directed method, and the simulated annealing algorithm) are employed to find a near-optimum solution in most cases.

Once the physical locations of all the cells in a design are determined, an automatic routing tool is used to create the metal interconnections between the cell terminals, based on the gate netlist (formal gate-level circuit description). As in the case of automatic placement tools, a number of heuristic approaches are applied to minimize the computational burden of finding a near-optimum routing solution.

Simulation and Verification Tools

The simulation category, which is the most mature area of VLSI CAD, includes many tools ranging from circuit-level simulation (SPICE or its derivatives, such as HSPICE), timing level simulation, logic level simulation, and behavioral simulation. Many other simulation tools have also been developed for device-level simulation and process simulation for technology development. The aim of all simulation CAD tools is to determine if the designed circuit meets the required specifications, at all stages of the design process.

Logic simulation is performed mainly to verify the functionality of the circuit, i.e., to determine if the designed circuit actually has the desired logic behavior. The gate-level abstraction of the circuit structure is sufficient to verify logic functionality; detailed electrical operation of individual gates is not a concern of logic simulation. A number of test vectors (inputs) are applied to the circuit during logic simulation, and the outputs are compared with expected output patterns. While limited gate-delay information can also be incorporated in logic simulation tools, the detailed analysis of critical time-domain behavior is always accomplished by electrical simulation.

Circuit-level or electrical simulation tools are routinely used to determine nominal and worst-case gate delays, to identify delay-critical signal paths or elements, and to predict the influence of parasitic effects upon circuit behavior. To accomplish this, the current-voltage behavior of every transistor and every interconnect in the circuit is represented by a detailed physical model, and the coupled differential equations describing the temporal behavior of the circuit are solved in the time-domain. Hence, the computational cost of circuit-level simulation is several orders of magnitude higher than logic simulation. The identification of all parasitic capacitances and resistances from mask layout data (*layout extraction*) must be performed prior to circuit-level simulation, in order to obtain reliable information about the time-behavior of the circuit.

The design rules checking CAD category includes the tools for layout rules checking, electrical rules checking, and reliability rules checking. The layout rules checking program has been highly effective in weeding out potential yield problems and circuit malfunctions.

Exercise Problems

1.1 An ADD/SUBTRACT logic circuit is shown below. It performs the ADD operation for $P = 0$ and SUBTRACT for $P = 1$.

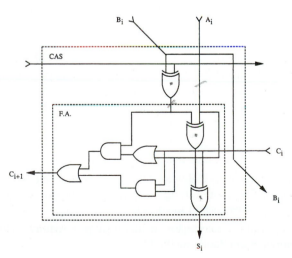

    a.   Draw an equivalent CMOS logic diagram by noting that most CMOS gates, except for the transmission gate and exclusive OR (XOR), are inverting. For example, the AND gate is implemented with NAND followed by an inverter.

    b.   By using the gate array platform given on page 47, implement the CMOS circuit as compactly as possible with the aspect ratio, which is the ratio of vertical dimension to horizontal dimension, as close to 1 as possible.

1.2   For the CMOS circuit in Problem 1.1,

    a.   First develop a small library of CMOS cells.

    b.   Place the cells into a single row and interconnect them with proper ordering such that the total interconnection wire length is minimized.

1.3   A measure of design productivity predicts the required engineer-months in terms of design implementation styles, such as repeated transistors (RPT), non-repeatable unique transistors (UNQ), PLA, RAM, and ROM transistors, the experience level of engineers (yr), the productivity improvement per year (D), and the design complexity (H). The formula proposed by Fey is

$$Engineer - Months(EM) = (1 + D)^{-yr}\left[A + Bk^{H}\right]$$

where the number k of equivalent transistors in the design is expressed by

$$k = UNQ + C \cdot RPT + E \cdot PLA + F\sqrt{RAM} + G\sqrt{ROM}$$

In this formula, the transistor count is in units of thousands and the coefficients A, B, C, D, E, F, G and H are model parameters which depend on the designers' experience and CAD tool support. The parameter (yr) represents the number of years since the extraction time of the model parameters. A set of sample values for these parameters is A = 0, B = 12, C = 0.13, D = 0.02, E = 0.37, F = 0.65, G = 0.08, and H = 1.13.

    a.   Discuss how one would extract the model parameters within a design organization.

    b.   A 24-bit floating-point processor has been designed using 20,500 repeated transistors, 10,500 unique transistors, 105,500 RAM transistors, and 150,200 ROM transistors. Calculate the expected engineer-months (*EM*) by assuming the experience year value of $yr = 3$. Note that the transistor counts in the formula are in units of thousands, for instance, UNQ = 10.5, not 10,500.

1.4   A large-scale fast prototyping system has been produced by using a very large array of field programmable logic arrays (FPGAs).

    a.   Discuss the pros (features) and cons (weaknesses) of such prototyping systems for proof of design concepts and verification in view of effort and speed performance of the design.

    b.   How would you compare the hardware prototyping method with the computer simulation method?

1.5 As the design complexity increases with increasing number of on-chip transistors, the on-chip noises have become more pronounced. Discuss the impact of packaging in suppressing on-chip noises in view of the numbers and strategic placement of ground and power pads, and the numbers of ground and power planes.

1.6 The testing of VLSI chips at speed has become increasingly more difficult due to undesirable parasitic effects in a testing environment. Also the cost of high-speed testing machines has become very high and, hence, in reality it has become difficult for smaller manufacturers to procure such equipment. Discuss what problem would the chip testing only at lower speed cause for systems houses which take such chips to develop systems at speed. What alternative ways can be used to ease the problem in the absence of at-speed testers?

1.7 Draft plans for developing a chip as a function of design turnaround time and development cost. In particular, what particular design style would be chosen when the customer requires that the chip be delivered in one month, six months, and one year, respectively?

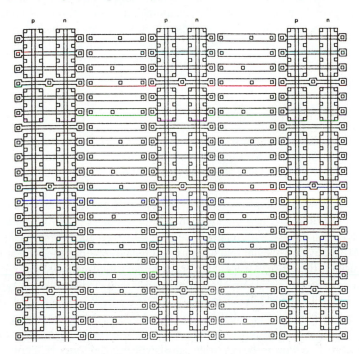

CMOS Gate Array Platform (for Problem 1.1b).

2 CHAPTER

Fabrication of MOSFETs

2.1 Introduction

In this chapter, the fundamentals of MOS chip fabrication will be discussed and the major steps of the process flow will be examined. It is not the aim of this chapter to present a detailed discussion of silicon fabrication technology, which deserves separate treatment in a dedicated course. Rather, the emphasis will be on the general outline of the process flow and on the interaction of various processing steps, which ultimately determine the device and the circuit performance characteristics. The following chapters show that there are very strong links between the fabrication process, the circuit design process, and the performance of the resulting chip. Hence, circuit designers must have a working knowledge of chip fabrication to create effective designs and to optimize the circuits with respect to various manufacturing parameters. Also, the circuit designer must have a clear understanding of the roles of various masks used in the fabrication process, and how the masks are used to define various features of the devices on-chip.

The following discussion will concentrate on the well-established CMOS fabrication technology, which requires that both n-channel (nMOS) and p-channel (pMOS) transistors be built on the same chip substrate. To accommodate both nMOS and pMOS devices, special regions must be created in which the semiconductor type is opposite to the substrate type. These regions are called *wells* or *tubs*. A p-well is created in an n-type substrate or, alternatively, an n-well is created in a p-type substrate. In the simple n-well CMOS fabrication technology presented here, the nMOS transistor is created in the p-type substrate, and the pMOS transistor is created in the n-well, which is built into the p-type substrate. In the twin-tub CMOS technology, additional tubs of the same type as the substrate can also be created for device optimization.

The simplified process sequence for the fabrication of CMOS integrated circuits on a p-type silicon substrate is shown in Fig. 2.1. The process starts with the creation of the n-well regions for pMOS transistors, by impurity implantation into the substrate. Then, a thick oxide is grown in the regions surrounding the nMOS and pMOS *active regions*. The thin gate oxide is subsequently grown on the surface through

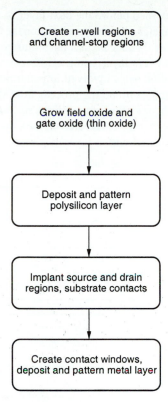

Figure 2.1 Simplified process sequence for the fabrication of the n-well CMOS integrated circuit with a single polysilicon layer, showing only major fabrication steps.

thermal oxidation. These steps are followed by the creation of n+ and p+ regions (source, drain, and channel-stop implants) and by final metallization (creation of metal interconnects).

The process flow sequence pictured in Fig. 2.1 may at first seem to be too abstract, since detailed fabrication steps are not shown. To obtain a better understanding of the issues involved in the semiconductor fabrication process, we first have to consider some of the basic steps in more detail.

2.2 Fabrication Process Flow: Basic Steps

Note that each processing step requires that certain areas are *defined* on chip by appropriate *masks*. Consequently, the integrated circuit may be viewed as a set of patterned layers of doped silicon, polysilicon, metal, and insulating silicon dioxide. In general, a layer must be patterned before the next layer of material is applied on the chip. The process used to transfer a pattern to a layer on the chip is called *lithography*. Since each layer has its own distinct patterning requirements, the lithographic sequence must be repeated for every layer, using a different mask.

To illustrate the fabrication steps involved in patterning silicon dioxide through optical lithography, let us first examine the process flow shown in Fig. 2.2. The sequence starts with the thermal oxidation of the silicon surface, by which an oxide layer of about 1 mm thickness, for example, is created on the substrate (Fig. 2.2(b)). The entire oxide surface is then covered with a layer of *photoresist,* which is essentially a

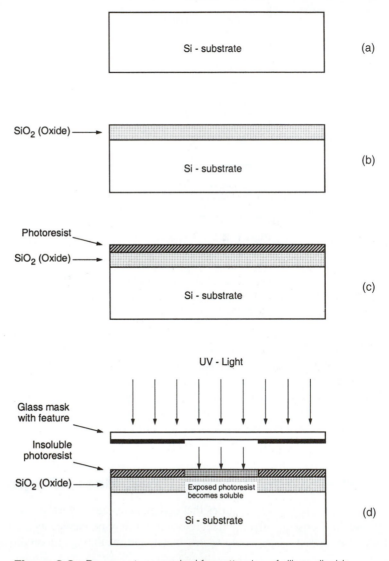

Figure 2.2   Process steps required for patterning of silicon dioxide.

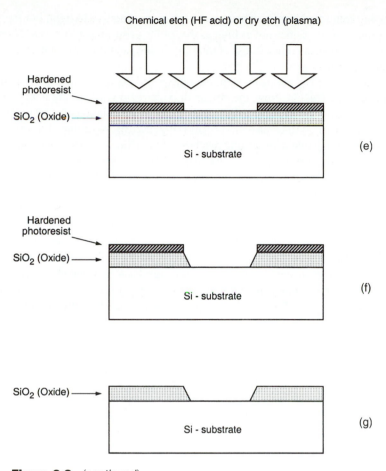

Figure 2.2 *(continued)*

light-sensitive, acid-resistant organic polymer, initially insoluble in the developing solution (Fig. 2.2(c)). If the photoresist material is exposed to ultraviolet (UV) light, the
exposed areas become soluble so that they are no longer resistant to etching solvents.
To selectively expose the photoresist, we have to cover some of the areas on the surface with a *mask* during exposure. Thus, when the structure with the mask on top is exposed to UV light, areas which are covered by the opaque features on the mask are
shielded. In the areas where the UV light can pass through, on the other hand, the photoresist is exposed and becomes soluble (Fig. 2.2(d)).

   The type of photoresist which is initially insoluble and becomes soluble after
exposure to UV light is called *positive photoresist*. The process sequence shown in
Fig. 2.2 uses positive photoresist. There is another type of photoresist which is initially soluble and becomes insoluble (hardened) after exposure to UV light, called
negative photoresist. If negative photoresist is used in the photolithography process,
the areas which are not shielded from the UV light by the opaque mask features become insoluble, whereas the shielded areas can subsequently be etched away by a

developing solution. Negative photoresists are more sensitive to light, but their photolithographic resolution is not as high as that of the positive photoresists. Therefore, negative photoresists are used less commonly in the manufacturing of high-density integrated circuits.

Following the UV exposure step, the unexposed portions of the photoresist can be removed by a solvent. Now, the silicon dioxide regions which are not covered by hardened photoresist can be etched away either by using a chemical solvent (HF acid) or by using a dry etch (plasma etch) process (Fig. 2.2(e)). Note that at the end of this step, we obtain an oxide window that reaches down to the silicon surface (Fig. 2.2(f)). The remaining photoresist can now be stripped from the silicon dioxide surface by using another solvent, leaving the patterned silicon dioxide feature on the surface as shown in Fig. 2.2(g).

The sequence of process steps illustrated in detail in Fig. 2.2 actually accomplishes a single pattern transfer onto the silicon dioxide surface, as shown in Fig. 2.3. The fabrication of semiconductor devices requires several such pattern transfers to be performed on silicon dioxide, polysilicon, and metal. The basic patterning process used in all fabrication steps, however, is quite similar to the one shown in Fig. 2.2. Also note that for accurate generation of high-density patterns required in sub-micron devices, *electron beam (E-beam) lithography* is used instead of optical lithography. In

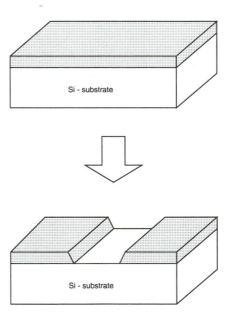

Figure 2.3 The result of a single lithographic patterning sequence on silicon dioxide, without showing the intermediate steps. Compare the unpatterned structure (top) and the patterned structure (bottom) with Fig. 2.2(b) and Fig. 2.2(g), respectively.

the following, the main processing steps involved in the fabrication of an n-channel MOS transistor on a p-type silicon substrate will be examined.

Fabrication of the nMOS Transistor

The process starts with the oxidation of the silicon substrate (Fig. 2.4(a)), in which a relatively thick silicon dioxide layer, also called field oxide, is created on the surface (Fig. 2.4(b)). Then, the field oxide is selectively etched to expose the silicon surface on which the MOS transistor will be created (Fig. 2.4(c)). Following this step, the surface is covered with a thin, high-quality oxide layer, which will eventually form the gate oxide of the MOS transistor (Fig. 2.4(d)). On top of the thin oxide layer, a layer of polysilicon (polycrystalline silicon) is deposited (Fig. 2.4(e)). Polysilicon is used both as gate electrode material for MOS transistors and also as an interconnect medium in

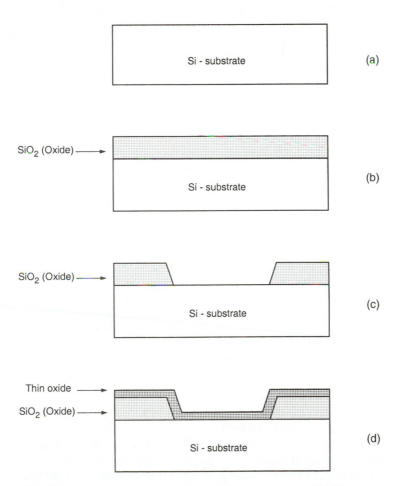

Figure 2.4 Process flow for the fabrication of an n-type MOSFET on p-type silicon.

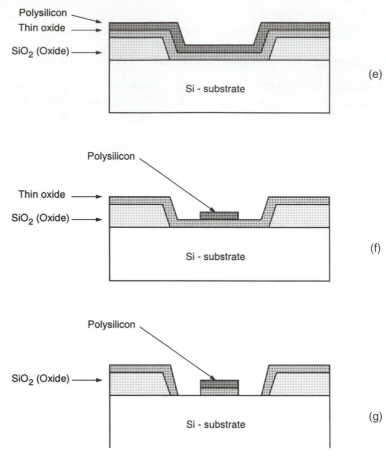

Figure 2.4 *(continued)*

silicon integrated circuits. Undoped polysilicon has relatively high resistivity. The resistivity of polysilicon can be reduced, however, by doping it with impurity atoms.

After deposition, the polysilicon layer is patterned and etched to form the interconnects and the MOS transistor gates (Fig. 2.4(f)). The thin gate oxide not covered by polysilicon is also etched away, which exposes the bare silicon surface on which the source and drain junctions are to be formed (Fig. 2.4(g)). The entire silicon surface is then doped with a high concentration of impurities, either through diffusion or ion implantation (in this case with donor atoms to produce n-type doping). Figure 2.4(h) shows that the doping penetrates the exposed areas on the silicon surface, ultimately creating two n-type regions (source and drain junctions) in the p-type substrate. The impurity doping also penetrates the polysilicon on the surface, reducing its resistivity. Note that the polysilicon gate, which is patterned *before* doping, actually defines the precise location of the channel region and, hence, the location of the source and the drain regions. Since this procedure allows very precise positioning of the two regions relative to the gate, it is also called the *self-aligned process*.

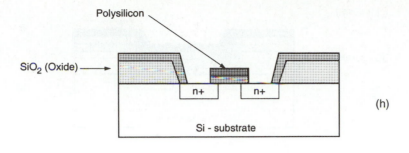

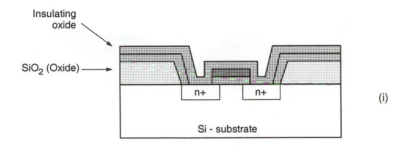

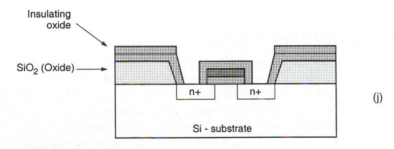

Figure 2.4 *(continued)*

Once the source and drain regions are completed, the entire surface is again covered with an insulating layer of silicon dioxide (Fig. 2.4(i)). The insulating oxide layer is then patterned in order to provide contact windows for the drain and source junctions (Fig. 2.4(j)). The surface is covered with evaporated aluminum which will form the interconnects (Fig. 2.4(k)). Finally, the metal layer is patterned and etched, completing the interconnection of the MOS transistors on the surface (Fig. 2.4(l)). Usually, a second (and third) layer of metallic interconnect can also be added on top of this structure by creating another insulating oxide layer, cutting contact (via) holes, depositing, and patterning the metal. The major process steps for the fabrication of an nMOS transistor on p-type silicon substrate are also illustrated in Plate 1 and Plate 2 on pages 75–76. The figures were generated using the DIOS™ multi-dimensional process simulator, a process simulation software by *ISE Integrated Systems Engineering AG,* Zurich, Switzerland.

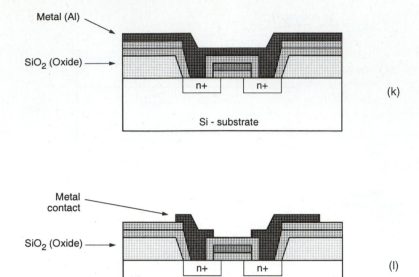

Metal (Al)

SiO$_2$ (Oxide)

n+   n+

Si - substrate

(k)

Metal contact

SiO$_2$ (Oxide)

n+   n+

Si - substrate

(l)

Figure 2.4   (*continued*)

Device Isolation Techniques

The MOS transistors that comprise an integrated circuit must be electrically *isolated* from each other during fabrication. Isolation is required to prevent unwanted conduction paths between the devices, to avoid creation of inversion layers outside the channel regions of transistors, and to reduce leakage currents. To achieve a sufficient level of electrical isolation between neighboring transistors on a chip surface, the devices are typically created in dedicated regions called *active areas,* where each active area is surrounded by a relatively thick oxide barrier called the *field oxide*.

One possible technique to create isolated active areas on silicon surface is first to grow a thick field oxide over the entire surface of the chip, and then to selectively etch the oxide in certain regions, to define the active areas. This fabrication technique, called *etched field-oxide isolation,* is already illustrated in Fig. 2.4(b) and Fig. 2.4(c). Here, the field oxide is selectively etched away to expose the silicon surface on which the MOS transistor will be created. Although the technique is relatively straightforward, it also has some drawbacks. The most significant disadvantage is that the thickness of the field oxide leads to rather large oxide steps at the boundaries between active areas and isolation (field) regions. When polysilicon and metal layers are deposited over such boundaries in subsequent process steps, the sheer height difference at the boundary can cause cracking of deposited layers, leading to chip failure. To prevent this, most manufacturers prefer isolation techniques that partially recess the field oxide into the silicon surface, resulting in a more planar surface topology.

Local Oxidation of Silicon (LOCOS)

The local oxidation of silicon (LOCOS) technique is based on the principle of selectively *growing* the field oxide in certain regions, instead of selectively etching away the active areas after oxide growth. Selective oxide growth is achieved by shielding the active areas with silicon nitride (Si_3N_4) during oxidation, which effectively inhibits oxide growth. The basic steps of the LOCOS process are illustrated in Fig. 2.5.

First, a thin pad oxide (also called stress-relief oxide) is grown on the silicon surface, followed by the deposition and patterning of a silicon nitride layer to mask (i.e., to define) the active areas (Fig. 2.5(a)). The thin pad oxide underneath the silicon

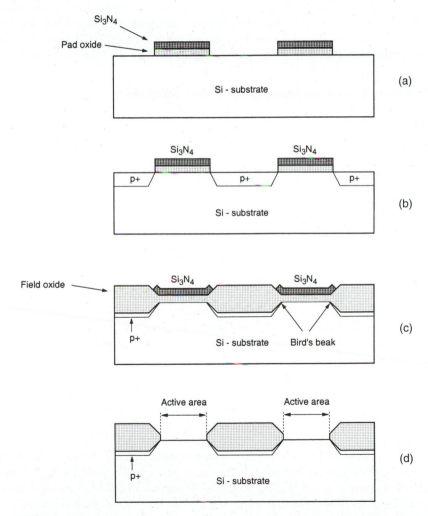

Figure 2.5 Basic steps of the LOCOS process to create oxide isolation around active areas.

nitride layer is used to protect the silicon surface from stress caused by nitride during the subsequent process steps. The exposed areas of the silicon surface, which will eventually form the isolation regions, are doped with a p-type impurity to create the channel-stop implants that surround the transistors (Fig. 2.5(b)). Next, a thick field oxide is grown in the areas not covered with silicon nitride, as shown in Fig. 2.5(c). Notice that the field oxide is partially recessed into the surface since the thermal oxidation process also consumes some of the silicon. Also, the field oxide forms a lateral extension under the nitride layer, called the *bird's beak* region. This lateral encroachment is mainly responsible for a reduction of the active area. The silicon nitride layer and the thin pad oxide layer are etched in the final step (Fig. 2.5(d)), resulting in active areas surrounded by the partially recessed field oxide.

The LOCOS process is a popular technique used for achieving field oxide isolation with a more planar surface topology. Several additional measures have also been developed over the years to control the lateral bird's beak encroachment, since this encroachment ultimately limits device scaling and device density in VLSI circuits.

Multilevel Interconnects and Metallization

In state-of-the-art CMOS processes, several (typically 4 to 8) metal layers are used for creating the interconnections between the transistors and for routing the power supply, signal and clock lines on the chip surface. The availability of multiple metal layers, where each layer is electrically isolated from the neighboring layers by a dielectric (SiO_2), allows higher integration densities for realization of complex structures and actually adds a third dimension that can be utilized very effectively in design. As mentioned earlier, the electrical connections between the layers are made by vias that are placed wherever such a contact is needed. Each via is formed by creating an opening (window) in the isolation oxide before each new metallization step, and filling the oxide opening with a special metal plug, usually tungsten. Following the creation of vias, a new layer of metal deposition and subsequent patterning forms the next metallization level.

It should be noted, however, that the chip surface is highly nonplanar as a result of previous process steps to create the transistors, including n-well and p-well formation, local oxidation, and gate formation. It is not desirable to deposit multiple metal interconnect lines directly on this uneven topography, because the deposited metal layer may exhibit localized thinning and discontinuities at uneven surface edges. Photolithography of nonplanar surfaces is also difficult and imprecise. The surface nonuniformity problems become even more pronounced with each additional metal layer, where successive accumulation of surface features, such as steps, may eventually result in pronounced hills and valleys on the chip surface.

To avoid the problems associated with nonplanar surface topology, the surface is usually planarized, or flattened, before each new metal deposition step. A fairly thick SiO_2 layer is first deposited on the wafer surface to cover all existing surface nonuniformities. The top surface of this thick oxide layer is then planarized using one of several techniques, such as glass reflow (heat treatment), etch-back, or Chemical-Mechanical Polishing (CMP). This last method (CMP), which involves actual

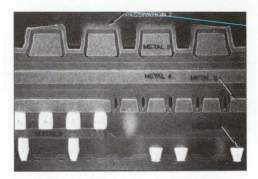

Figure 2.6 Actual cross-section of a modern IC (IBM's PowerPC chip). Note the multiple layers of metal for wiring above the silicon surface. The active parts of the transistors are barely visible at the bottom of the photograph. *(Reprinted with permission of Integrated Circuit Engineering Corporation.)*

polishing of the wafer surface using abrasive silica slurry, is being commonly adopted in the semiconductor industry in recent years. It allows the creation of successive layers of metal interconnections on chip, with each new layer being deposited on a planarized oxide surface. Figure 2.6 shows the scanning electron-microscope (SEM) cross-section of multilevel metal interconnects fabricated with CMP technique.

2.3 The CMOS n-Well Process

Having examined the basic process steps for pattern transfer through lithography and having gone through the fabrication procedure of a single n-type MOS transistor, we can now return to the generalized fabrication sequence of n-well CMOS integrated circuits, as shown in Fig. 2.1. In the following figures, some of the important process steps involved in the fabrication of a CMOS inverter will be shown by a top view of the lithographic masks and a cross-sectional view of the relevant areas.

The n-well CMOS process starts with a moderately doped (with impurity concentration typically less than 10^{15} cm^{-3}) p-type silicon substrate. Then, an initial oxide layer is grown on the entire surface. The first lithographic mask defines the n-well region. Donor atoms, usually phosphorus, are implanted through this window in the oxide.

Once the n-well is created, the active areas of the nMOS and pMOS transistors can be defined. Figures 2.7 through 2.12 illustrate the significant milestones that occur during the fabrication process of a CMOS inverter. The main process steps for the fabrication of a CMOS inverter are also illustrated in Plate 3, Plate 4 and Plate 5 on pages 77–79. The cross-sectional figures presented here were generated using the DIOS™ multi-dimensional process simulator and PROSIT™ 3-D structural modeling tool, by *ISE Integrated Systems Engineering AG*, Zurich, Switzerland.

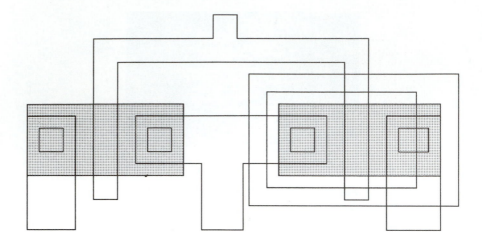

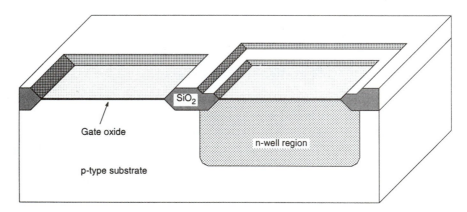

Figure 2.7 Following the creation of the n-well region, a thick field oxide is grown in the areas surrounding the transistor's active regions, and a thin gate oxide is grown on top of the active regions. The thickness and the quality of the gate oxide are two of the most critical fabrication parameters, since they strongly affect the operational characteristics of the MOS transistor, as well as its long-term reliability.
(*After* Atlas of IC Technologies, *by W. Maly.*)

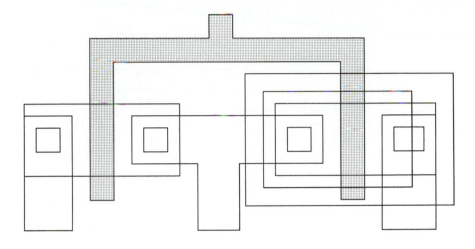

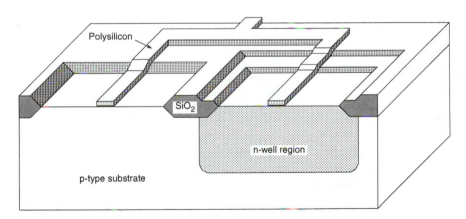

Figure 2.8 The polysilicon layer is deposited using chemical vapor deposition (CVD) and patterned by dry (plasma) etching. The created polysilicon lines will function as the gate electrodes of the nMOS and the pMOS transistors and their interconnects. Also, the polysilicon gates act as self-aligned masks for the source and drain implantations that follow this step.
(After Atlas of IC Technologies, *by W. Maly.)*

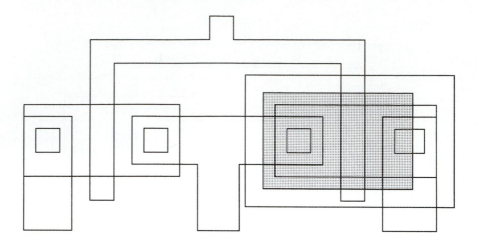

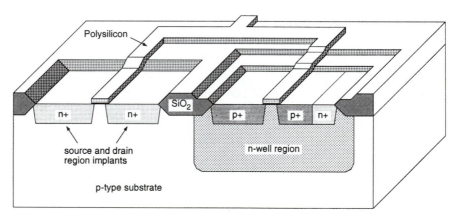

Figure 2.9 Using a set of two masks, the n+ and p+ regions are implanted into the substrate and into the n-well, respectively. Also, the ohmic contacts to the substrate and to the n-well are implanted in this process step.
(*After* Atlas of IC Technologies, *by W. Maly.*)

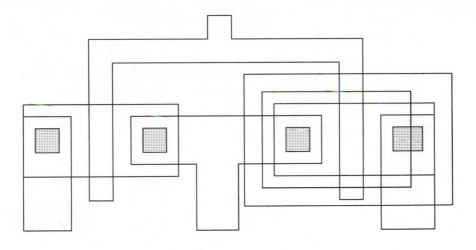

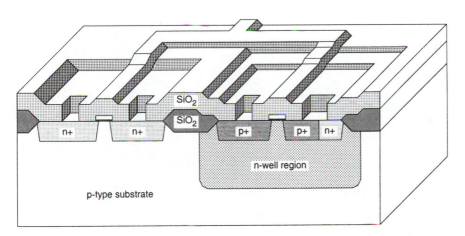

Figure 2.10 An insulating silicon dioxide layer is deposited over the entire wafer using CVD. Then, the contacts are defined and etched away to expose the silicon or polysilicon contact windows. These contact windows are necessary to complete the circuit interconnections using the metal layer, which is patterned in the next step. (*After* Atlas of IC Technologies, *by W. Maly.*)

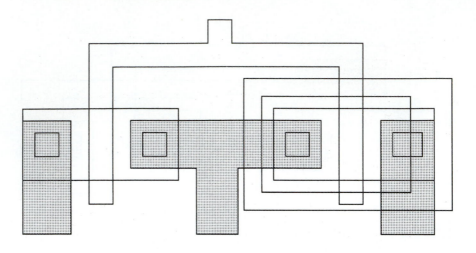

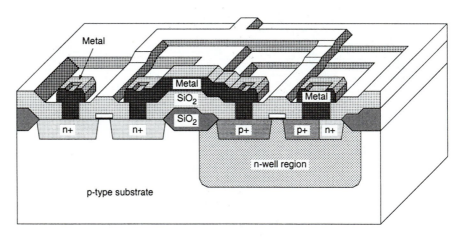

Figure 2.11 Metal (aluminum) is deposited over the entire chip surface using metal evaporation, and the metal lines are patterned through etching. Since the wafer surface is non-planar, the quality and the integrity of the metal lines created in this step are very critical and are ultimately essential for circuit reliability.
(After Atlas of IC Technologies, *by W. Maly.)*

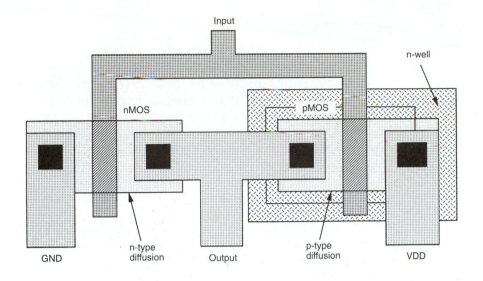

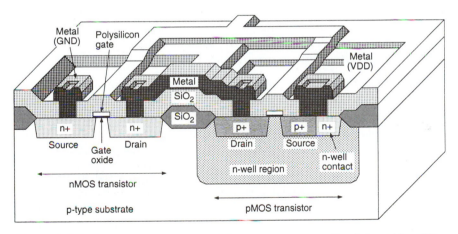

Figure 2.12 The composite layout and the resulting cross-sectional view of the chip, showing one nMOS and one pMOS transistor (in the n-well), and the polysilicon and metal interconnections. The final step is to deposit the passivation layer (for protection) over the chip, except over wire-bonding pad areas.
(*After* Atlas of IC Technologies, *by W. Maly.*)

2.4 Layout Design Rules

The physical mask layout of any circuit to be manufactured using a particular process must conform to a set of geometric constraints or rules, which are generally called layout design rules. These rules usually specify the minimum allowable line widths for physical objects on-chip such as metal and polysilicon interconnects or diffusion areas, minimum feature dimensions, and minimum allowable separations between two such features. If a metal line width is made too small, for example, it is possible for the line to break during the fabrication process or afterwards, resulting in an open circuit. If two lines are placed too close to each other in the layout, they may form an unwanted short circuit by merging during or after the fabrication process. The main objective of design rules is to achieve, for any circuit to be manufactured with a particular process, a high overall yield and reliability while using the smallest possible silicon area.

Note that there is usually a trade-off between higher yield, which is obtained through conservative geometries, and better area efficiency, which is obtained through aggressive, high-density placement of various features on the chip. The layout design rules which are specified for a particular fabrication process normally represent a reasonable optimum point in terms of yield and density. It must be emphasized, however, that the design rules do not represent strict boundaries which separate "correct" designs from "incorrect" ones. A layout which violates some of the specified design rules may still result in an operational circuit with reasonable yield, whereas another layout observing all specified design rules may result in a circuit which is not functional and/or has very low yield. To summarize, we can say, in general, that observing the layout design rules significantly increases the probability of fabricating a successful product with high yield.

The design rules are usually described in two ways:

1. Micron rules, in which the layout constraints such as minimum feature sizes and minimum allowable feature separations are stated in terms of absolute dimensions in micrometers, or,

2. Lambda rules, which specify the layout constraints in terms of a single parameter (λ) and thus allow linear, proportional scaling of all geometrical constraints.

Lambda-based layout design rules were originally devised to simplify the industry-standard micron-based design rules and to allow scaling capability for various processes. It must be emphasized, however, that most of the submicron CMOS process design rules do not lend themselves to straightforward linear scaling. The use of lambda-based design rules must therefore be handled with caution in submicron geometries. In the following, we present a sample set of the lambda-based layout design rules devised for the MOSIS (MOS Implementation System) CMOS process and illustrate the implications of these rules on a section of a simple layout which includes two transistors (Fig. 2.13). The complete set of MOSIS CMOS scalable design rules are also illustrated in Plate 6 and Plate 7 on pages 80–81.

2.5 Full-Custom Mask Layout Design

In this section, the basic mask layout principles for CMOS inverters and logic gates will be presented. The design of physical layout is very tightly linked to overall circuit performance (area, speed, and power dissipation) since the physical structure directly determines the transconductances of the transistors, the parasitic capacitances and resistances, and obviously, the silicon area which is used for a certain function. On the other hand, the detailed mask layout of logic gates requires a very intensive and time-consuming design effort, which is justifiable only in special circumstances where the area and/or the performance of the circuit must be optimized under very tight constraints. Therefore, automated layout generation (e.g., using a standard cell library, computer-aided placement-and-routing) is typically preferred for the design of most digital VLSI circuits. In order to judge the physical constraints and limitations, however, the VLSI designer must also have a good understanding of the physical mask layout process.

The physical (mask layout) design of CMOS logic gates is an iterative process which starts with the circuit topology (to realize the desired logic function) and the initial sizing of the transistors (to realize the desired performance specifications). At this point, the designer can only estimate the total parasitic load at the output node, based on the fan-out, the number of devices, and the expected length of the interconnection lines. If the logic gate contains more than 4 to 6 transistors, the topological graph representation and the Euler-path method allow the designer to determine the optimum ordering of the transistors (see Chapter 7). A simple stick diagram layout can now be drawn, showing the locations of the transistors, the local interconnections between the transistors, and the locations of the contacts.

After a topologically feasible layout is found, the mask layers are drawn (using a layout editor tool) according to the layout design rules. This procedure may require several small iterations in order to accommodate all the design rules, but the basic topology should not change very significantly. Following the final DRC (Design Rule Check), a circuit extraction procedure is performed on the finished layout to determine the actual transistor sizes, and more importantly, the parasitic capacitances at each node. The result of the extraction step is usually a detailed SPICE input file, which is automatically generated by the extraction tool. Now, the actual performance of the circuit can be determined by performing a SPICE simulation, using the extracted netlist. If the simulated circuit performance (e.g., transient response times or power dissipation) do not match the desired specifications, the layout must be modified and the whole process must be repeated. The layout modifications are usually concentrated on the width-to-length (W/L) ratios of the transistors (transistor resizing), since the (W/L) ratios of the transistors determine the device transconductance and the parasitic source/drain capacitances. The designer may also decide to change parts or all of the circuit topology in order to reduce the parasitics. The flow diagram of this iterative process is shown in Fig. 2.14.

CMOS Inverter Layout Design

In the following, the mask layout design of a CMOS inverter will be examined step-by-step, as an example for the application of layout design rules. First, we need to create the individual transistors according to the design rules. Assume that we attempt to design the inverter with minimum-size transistors. The width of the active

MOSIS Layout Design Rules (sample set)

Rule number	Description	λ-Rule
	Active area rules	
R1	Minimum active area width	3λ
R2	Minimum active area spacing	3λ
	Polysilicon rules	
R3	Minimum poly width	2λ
R4	Minimum poly spacing	2λ
R5	Minimum gate extension of poly over active	2λ
R6	Minimum poly-active edge spacing (poly outside active area)	1λ
R7	Minimum poly-active edge spacing (poly inside active area)	3λ
	Metal rules	
R8	Minimum metal width	3λ
R9	Minimum metal spacing	3λ
	Contact rules	
R10	Poly contact size	2λ
R11	Minimum poly contact spacing	2λ
R12	Minimum poly contact to poly edge spacing	1λ
R13	Minimum poly contact to metal edge spacing	1λ
R14	Minimum poly contact to active edge spacing	3λ
R15	Active contact size	2λ
R16	Minimum active contact spacing (on the same active region)	2λ
R17	Minimum active contact to active edge spacing	1λ
R18	Minimum active contact to metal edge spacing	1λ
R19	Minimum active contact to poly edge spacing	3λ
R20	Minimum active contact spacing (on different active regions)	6λ

Figure 2.13   Illustration of some of the typical MOSIS layout design rules.

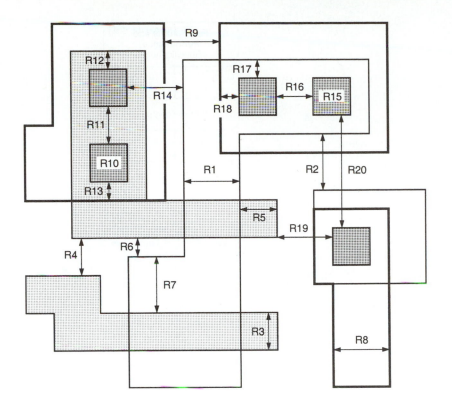

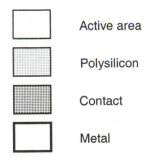

Active area

Polysilicon

Contact

Metal

Figure 2.13 *(continued)*

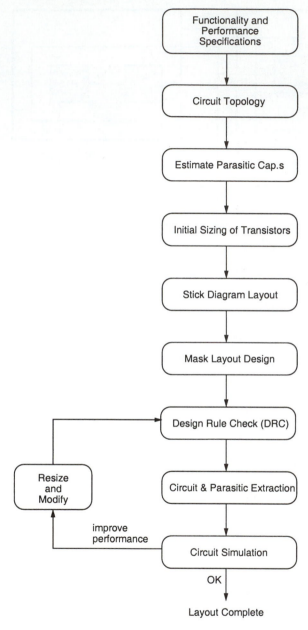

Figure 2.14 Typical design flow for the production of a mask layout.

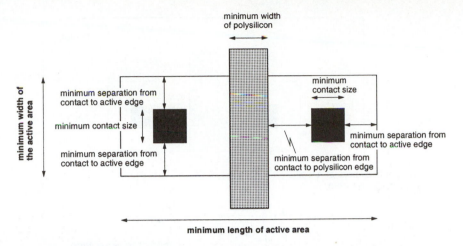

Figure 2.15 Design rules which determine the dimensions of a minimum-size transistor.

area is then determined by the minimum diffusion contact size (which is necessary for source and drain connections) and the minimum separation from diffusion contact to both active area edges. The width of the polysilicon line over the active area (which is the gate of the transistor) is typically taken as the minimum poly width (Fig. 2.15). Then, the minimum overall length of the active area is determined by the following sum: (minimum polysilicon width) + 2 × (minimum poly-to-contact spacing) + 2 × (minimum contact size) + 2 × (minimum spacing from contact to active area edge).

The pMOS transistor must be placed in an n-well region, and the minimum size of the n-well is dictated by the pMOS active area and the minimum n-well overlap over n+. The distance between the nMOS and the pMOS transistor is determined by the minimum separation between the n+ active area and the n-well (Fig. 2.16). The polysilicon gates of the nMOS and the pMOS transistors are usually aligned, so that the gate connections can be made with a single polysilicon line of least possible length. The reason for avoiding long polysilicon connections (as a general layout practice) is the fact that the large parasitic resistance and the parasitic capacitance of polysilicon lines may result in significant RC delays. Thus, even *local* signal connections are preferably made with metal lines as much as possible, and metal-polysilicon contacts are used to provide the electrical connection between the two layers, wherever necessary.

The final step in the mask layout is the local interconnections in metal for the output node, VDD and GND contacts (Fig. 2.17). The dimensions of metal lines in a mask layout are usually dictated by the minimum metal width and the minimum metal separation (between two neighboring lines, on the same level). Notice that in order to be biased properly, the n-well region must also have a VDD contact.

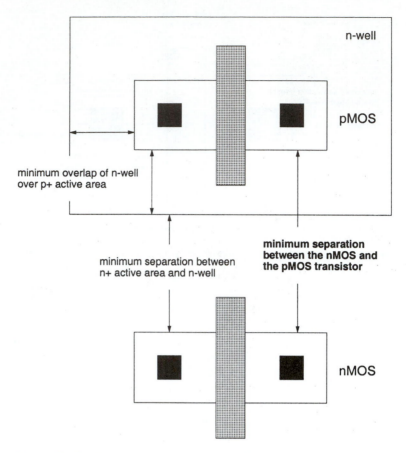

Figure 2.16 Design rules which determine the separation between the nMOS and the pMOS transistor of the CMOS inverter.

Having examined the main steps of a typical CMOS inverter mask layout design, we have to emphasize that this example obviously represents only one of many possibilities for the layout of this circuit. The layout design rules dictate a set of limitations for the mask geometry, yet the full-custom layout design process still allows a large number of variations in terms of device sizing, the placement of individual devices, and the routing of interconnections between the devices; even for a simple circuit consisting of only two transistors. Depending on the dominant design criteria and design constraints (minimization of overall silicon area, minimization of delay times, placement of input/output pins, etc.), one can choose a certain mask layout design over other alternatives. Some layout examples of CMOS inverters and simple logic gates are also presented in Plate 8 on page 82. Notice that the number of layout possibilities also tends to increase with circuit complexity, i.e., with the number of transistors involved in the design.

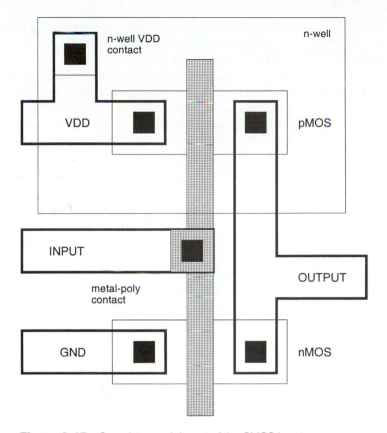

Figure 2.17 Complete mask layout of the CMOS inverter.

Exercise Problems

2.1 Design three masks for diffusion, window, and metal layers to realize a 1-kΩ resistor with n-type diffusion with sheet resistance of 100 Ω per square and lead metal lines. The minimum feature size allowed is 1 μm for line width and window opening. Also, 0.5-μm extension of metal feature and diffusion feature over the window opening is required.

2.2 Discuss whether the mask set in Problem 2.1 will realize the resistance exactly. What would be the effect of window design on the resistance? Discuss at least two other mechanisms in the processing that can make the processed resistance deviate from the target value even if the effect of the window is neglected.

2.3 Let us now assume that the variation in the resistance is affected only by the exact line feature sizes. Discuss the pros and cons of having the line width at a minimum as compared to making the width larger than the minimum width.

2.4   A method for reducing interconnection resistance in the polysilicon lines is to use silicide material deposited on top to form polycide. This process can reduce the nominal sheet resistance from 20 Ω to 2 Ω or even less. For the same purpose, silicide material is also deposited on top of source and drain diffusions in MOS transistors. Thus, a single-step deposition of silicide can be achieved on polysilicon gates and source and drain regions of MOS transistors. Discuss how such deposition can be achieved without causing electrical shorts between the gate and source or drain of an individual transistor.

2.5   In VLSI technologies with multiple layers of metallic interconnects, one of the most yield-limiting processes is the patterning of metal lines, especially when the surface features before metal deposition are not flat. To achieve flat surfaces, chemical mechanical polishing (CMP) has been introduced. Discuss the side effects of CMP on the circuit performance and on the process steps that follow CMP.

2.6   Discuss the difficulties in processing a window mask which contains a wide variation of window sizes, i.e., both very large rectangles and minimum-size square shapes. In particular, what problem would you expect when the process control is based on the monitoring of the smallest windows? How about the opposite case in which the process control is based on the largest window?

2.7   Photolithography has been the driving force behind massive processing of MOS chips at low cost. Despite significant improvements in both photolithography and photoresist materials, it has become increasingly more difficult to process very small, deep submicron feature sizes. Alternatives can be X-ray lithography or direct electron-beam writing. Discuss the difficulties inherent in such alternatives.

2.8   Consider a chip design using 10 mask levels. Suppose that each mask can be made with 98% yield. Determine the composite mask yield for the set of 10 masks. Would the processed chip yield be lower or higher than this composite yield? If your results are inconclusive, explain the reason.

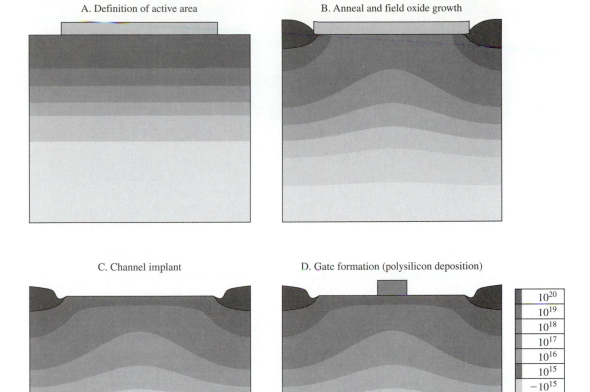

A. Definition of active area

B. Anneal and field oxide growth

C. Channel implant

D. Gate formation (polysilicon deposition)

| 10^{20} |
| 10^{19} |
| 10^{18} |
| 10^{17} |
| 10^{16} |
| 10^{15} |
| -10^{15} |
| -10^{16} |
| -10^{17} |
| Net[/cm3] |
| PO |
| OX |
| SI |

Plate 1 Fabrication steps for an nMOS transistor on p-type silicon substrate.
(Figures generated using DIOS™ multi-dimensional process simulator, courtesy of ISE Integrated Systems Engineering AG, CH-8005 Zurich, Switzerland.)

E. Spacer formation and S/D implant

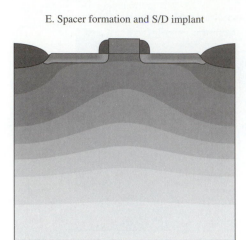

F. Oxide deposition

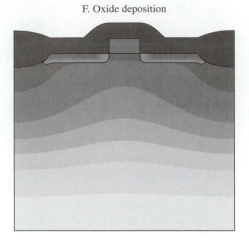

G. Contact hole etch

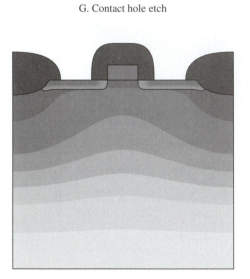

H. Metal deposition

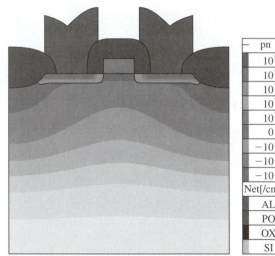

pn
10^{19}
10^{18}
10^{17}
10^{16}
10^{15}
0
-10^{15}
-10^{16}
-10^{17}
Net[/cm3]
AL
PO
OX
SI

Plate 2   Fabrication steps for an nMOS transistor on p-type silicon substrate (*continued*). *(Figures generated using DIOS™ multi-dimensional process simulator, courtesy of ISE Integrated Systems Engineering AG, CH-8005 Zurich, Switzerland.)*

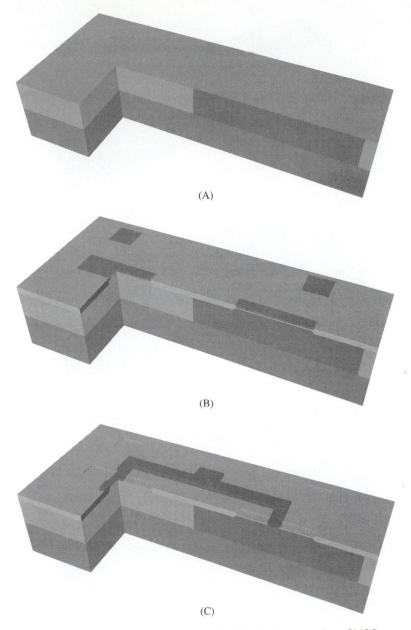

(A)

(B)

(C)

Plate 3   Cross-sectional view showing the fabrication steps for a CMOS inverter:
(A) Creation of the n-well.
(B) Definition of active areas and field oxide (FOX) growth.
(C) Polysilicon deposition.
(Figures generated using DIOS™ multi-dimensional process simulator and PROSIT™ 3-D structure modeling tool, courtesy of ISE Integrated Systems Engineering AG, CH-8005 Zurich, Switzerland.)

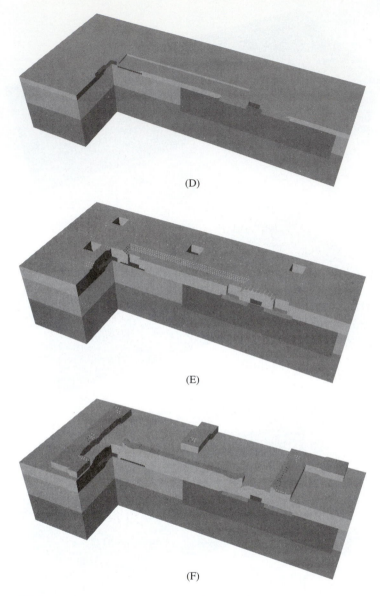

(D)

(E)

(F)

Plate 4 Cross-sectional view showing the fabrication steps for a CMOS inverter (*continued*).
(D) Source/drain region implantation.
(E) Oxide (SiO_2) deposition and contact hole etch.
(F) First-level metal (Metal-I) deposition.
(Figures generated using DIOS™ multi-dimensional process simulator and PROSIT™ 3-D structure modeling tool, courtesy of ISE Integrated Systems Engineering AG, CH-8005 Zurich, Switzerland.)

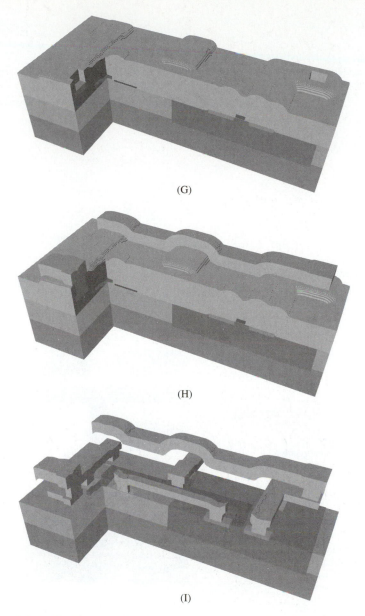

(G)

(H)

(I)

Plate 5   Cross-sectional view showing the fabrication steps for a
CMOS inverter (*continued*).
(G) Oxide (SiO$_2$) deposition and via hole etch.
(H) Second-level metal (Metal-2) deposition.
(I) Completed inverter structure, shown without the oxide layers for
better visibility.
*(Figures generated using DIOS™ multi-dimensional process simulator
and PROSIT™ 3-D structure modeling tool, courtesy of ISE Integrated
Systems Engineering AG, CH-8005 Zurich, Switzerland.)*

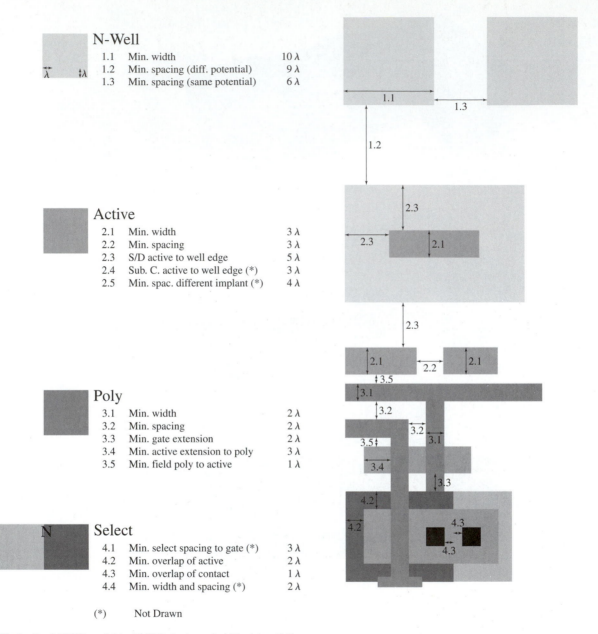

N-Well

1.1	Min. width	$10\,\lambda$
1.2	Min. spacing (diff. potential)	$9\,\lambda$
1.3	Min. spacing (same potential)	$6\,\lambda$

Active

2.1	Min. width	$3\,\lambda$
2.2	Min. spacing	$3\,\lambda$
2.3	S/D active to well edge	$5\,\lambda$
2.4	Sub. C. active to well edge (*)	$3\,\lambda$
2.5	Min. spac. different implant (*)	$4\,\lambda$

Poly

3.1	Min. width	$2\,\lambda$
3.2	Min. spacing	$2\,\lambda$
3.3	Min. gate extension	$2\,\lambda$
3.4	Min. active extension to poly	$3\,\lambda$
3.5	Min. field poly to active	$1\,\lambda$

Select

4.1	Min. select spacing to gate (*)	$3\,\lambda$
4.2	Min. overlap of active	$2\,\lambda$
4.3	Min. overlap of contact	$1\,\lambda$
4.4	Min. width and spacing (*)	$2\,\lambda$

(*)   Not Drawn

Plate 6   MOSIS scalable CMOS design rules (Revision 7.2).

Contact

5.1	Exact contact size	2λ
5.2	Min. poly overlap	1.5λ
5.3	Min. spacing	2λ
5.4	Min. spacing to gate	2λ

6.1	Exact contact size	2λ
6.2	Min. active overlap	1.5λ
6.3	Min. spacing	2λ
6.4	Min. spacing to gate	2λ

Metal1

7.1	Min. width	3λ
7.2.a	Min. spacing	3λ
7.3	Min. overlap of any contact	1λ

Via1

8.1	Exact size	2λ
8.2	Min. spacing	3λ
8.3	Min. overlap by metal1	1λ
8.4	Min. spacing to contact	2λ
8.5	Min. spac. to poly or act. edge	2λ

Metal2

9.1	Min. width	3λ
9.2.a	Min. spacing	4λ
9.3	Min. overlap to via1	1λ

(*)   Not Drawn

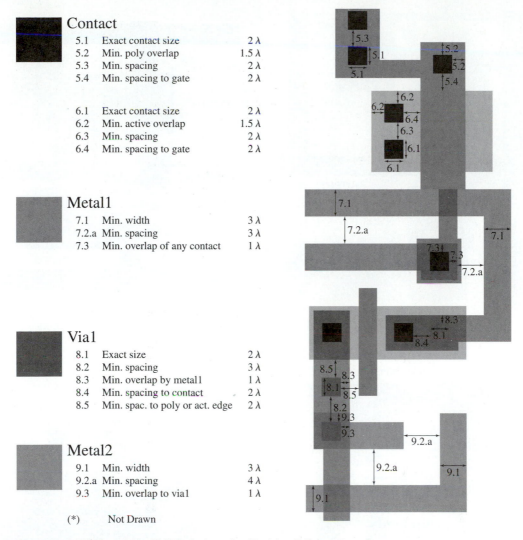

Plate 7 MOSIS scalable CMOS design rules (Revision 7.2) (*continued*).

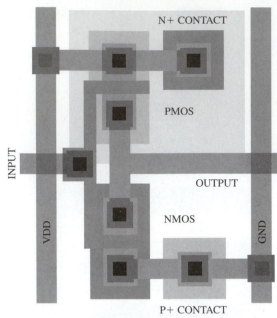

CMOS INVERTER (version 1)

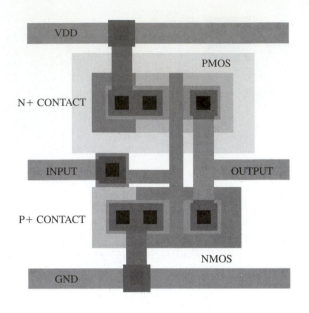

CMOS INVERTER (version 2)

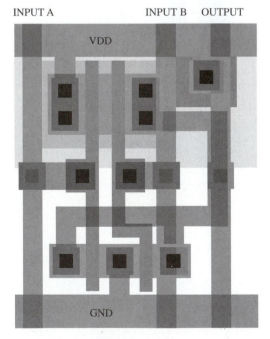

CMOS NOR2 GATE

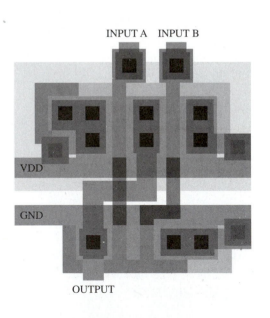

CMOS NAND2 GATE

Plate 8 Layout examples of CMOS inverters and simple logic gates.

CHAPTER 3

MOS Transistor

The *MOS Field Effect Transistor* (MOSFET) is the fundamental building block of MOS and CMOS digital integrated circuits. Compared to the bipolar junction transistor (BJT), the MOS transistor occupies a relatively smaller silicon area, and its fabrication used to involve fewer processing steps. The technological advantages, together with the relative simplicity of MOSFET operation, have helped make the MOS transistor the most widely used switching device in LSI and VLSI circuits. In this chapter, we will examine the basic structure and the electrical behavior of nMOS (n-channel MOS), as well as pMOS (p-channel MOS) devices. The nMOS transistor is used as the primary switching device in virtually all digital circuit applications, whereas the pMOS transistor is used mostly in conjunction with the nMOS device in CMOS circuits. However, the basic operation principles of both nMOS and pMOS transistors are very similar to each other.

This chapter starts with a detailed investigation of the basic electrical and physical properties of *Metal Oxide Semiconductor* (MOS) systems, upon which the MOSFET structure is based. We will consider the effects of external bias conditions on charge distribution in the MOS system and on the conductance of free carriers. It will be shown that, in field effect devices, the current flow is controlled by externally applied electric fields, and that the operation depends only on the majority carrier flow between two device terminals. Next, the current-voltage characteristics of MOS transistors will be examined in detail, including physical limitations imposed by small device geometries and various second-order effects observed in MOSFETs. Note that these considerations will be particularly important for the overall performance of large-scale digital circuits built by using small-geometry MOSFET devices.

3.1 The Metal Oxide Semiconductor (MOS) Structure

We will start our investigation by considering the electrical behavior of the simple two-terminal MOS structure shown in Fig. 3.1. Note that the structure consists of three layers: The metal *gate electrode*, the insulating oxide (SiO_2) layer, and the p-type bulk

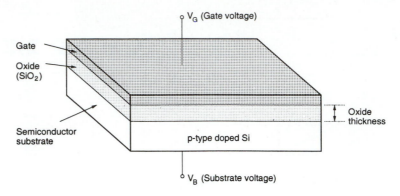

Figure 3.1   Two-terminal MOS structure.

semiconductor (Si), called the *substrate*. As such, the MOS structure forms a capacitor, with the gate and the substrate acting as the two terminals (plates) and the oxide layer as the dielectric. The thickness of the silicon dioxide layer is usually between 10 nm and 50 nm. The carrier concentration and its local distribution within the semiconductor substrate can now be manipulated by the external voltages applied to the gate and substrate terminals. A basic understanding of the bias conditions for establishing different carrier concentrations in the substrate will also provide valuable insight into the operating conditions of more complicated MOSFET structures.

Consider first the basic electrical properties of the semiconductor (Si) substrate, which acts as one of the electrodes of the MOS capacitor. The equilibrium concentrations of mobile carriers in a semiconductor always obey the *Mass Action Law* given by

$$n \cdot p = n_i^2 \tag{3.1}$$

Here, n and p denote the mobile carrier concentrations of electrons and holes, respectively, and n_i denotes the intrinsic carrier concentration of silicon, which is a function of the temperature T. At room temperature, i.e., $T = 300$ K, n_i is approximately equal to 1.45×10^{10} cm^{-3}. Assuming that the substrate is uniformly doped with an acceptor (e.g., boron) concentration N_A, the equilibrium electron and hole concentrations in the p-type substrate are approximated by

$$n_{po} \cong \frac{n_i^2}{N_A}$$
$$p_{po} \cong N_A \tag{3.2}$$

The doping concentration N_A is typically on the order of 10^{15} to 10^{16} cm^{-3}; thus, it is much greater than the intrinsic carrier concentration n_i. Note that the bulk electron and hole concentrations given in (3.2) are valid in the regions farther away from the *surface,* where the semiconductor substrate and the oxide layer meet. The conditions on the surface, however, are far more significant for the electrical behavior and the operation of the MOS system, and we will discuss these conditions in more detail.

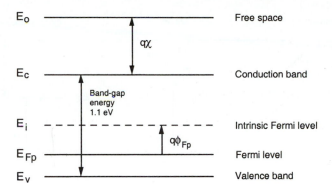

Figure 3.2 Energy band diagram of a p-type silicon substrate.

The energy band diagram of the p-type substrate is shown in Fig. 3.2. The band-gap between the *conduction band* and the *valence band* for silicon is approximately 1.1 eV. The location of the *equilibrium Fermi level E_F* within the band-gap is determined by the doping type and the doping concentration in the silicon substrate. The *Fermi potential ϕ_F*, which is a function of temperature and doping, denotes the difference between the intrinsic Fermi level E_i, and the Fermi level E_F.

$$\phi_F = \frac{E_F - E_i}{q} \tag{3.3}$$

For a p-type semiconductor, the Fermi potential can be approximated by

$$\phi_{Fp} = \frac{kT}{q} \ln \frac{n_i}{N_A} \tag{3.4}$$

whereas for an n-type semiconductor (doped with a donor concentration N_D), the Fermi potential is given by

$$\phi_{Fn} = \frac{kT}{q} \ln \frac{N_D}{n_i} \tag{3.5}$$

Here, k denotes the Boltzmann constant and q denotes the unit (electron) charge. Note that the definitions given in (3.4) and (3.5) result in a positive Fermi potential for n-type material, and a negative Fermi potential for p-type material. We will use this convention throughout the text. The *electron affinity* of silicon, which is the potential difference between the conduction band level and the vacuum (free-space) level, is denoted by $q\chi$ in Fig. 3.2. The energy required for an electron to move from the Fermi level into free space is called the *work function $q\Phi_S$*, and is given by

$$q\Phi_S = q\chi + (E_c - E_F) \tag{3.6}$$

The insulating silicon dioxide layer between the silicon substrate and the gate has a large band-gap of about 8 eV and an electron affinity of about 0.95 eV. On the other

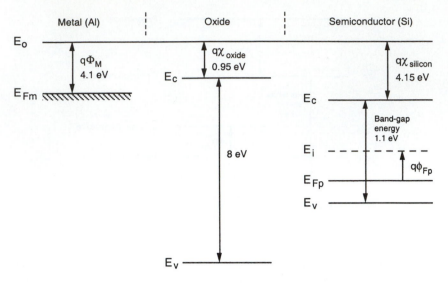

Figure 3.3   Energy band diagrams of the components that make up the MOS system.

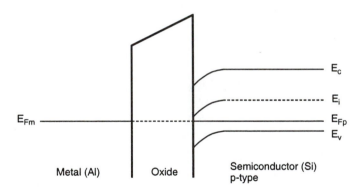

Figure 3.4   Energy band diagram of the combined MOS system.

hand, the work function $q\Phi_M$ of an aluminum gate is about 4.1 eV. Figure 3.3 shows the energy band diagrams of metal, oxide, and semiconductor layers in a MOS system as three separate components.

Now consider that the three components of the ideal MOS system are brought into physical contact. The Fermi levels of all three materials must line up, as they form the MOS capacitor shown in Fig. 3.1. Because of the work-function difference between the metal and the semiconductor, a voltage drop occurs across the MOS system. Part of this built-in voltage drop occurs across the insulating oxide layer. The rest of the voltage drop (potential difference) occurs at the silicon surface next to the silicon-oxide interface, forcing the energy bands of silicon to bend in this region. The resulting combined energy band diagram of the MOS system is shown in Fig. 3.4. Notice that the equilibrium Fermi levels of the semiconductor (Si) substrate and the metal gate are at the same potential. The bulk Fermi level is not significantly affected by the band bending, whereas the surface Fermi level moves closer to the intrinsic Fermi (mid-gap) level. The Fermi potential at the surface, also called *surface potential* ϕ_S, is smaller in magnitude than the bulk Fermi potential ϕ_F.

EXAMPLE 3.1

Consider the MOS structure that consists of a p-type doped silicon substrate, a silicon dioxide layer, and a metal (aluminum) gate. The equilibrium Fermi potential of the doped silicon substrate is given as $q\phi_{F_p} = 0.2$ eV. Using the electron affinity for silicon and the work function for aluminum given in Fig. 3.3, calculate the built-in potential difference across the MOS system. Assume that the MOS system contains no other charges in the oxide or on the silicon-oxide interface.

First, we have to calculate the work function for the doped silicon, which is given by (3.6). Since the electron affinity of silicon is 4.15 eV, the work function $q\Phi_S$ is found as

$$q\Phi_S = 4.15 \text{ eV} + 0.75 \text{ eV} = 4.9 \text{ eV}$$

Now calculate the work function difference between the silicon substrate and the aluminum gate. Note that the work function of aluminum is given as 4.1 eV in Fig. 3.3. Thus, the built-in potential difference across this MOS system is

$$q\Phi_M - q\Phi_S = 4.1 \text{ eV} - 4.9 \text{ eV} = -0.8 \text{ eV}$$

If a voltage corresponding to this potential difference is applied externally between the gate and the substrate, the bending of the energy bands near the surface can be compensated; i.e., the energy bands become "flat." Thus, the voltage defined by

$$V_{FB} = \Phi_M - \Phi_S$$

is called the *flat-band* voltage.

3.2 The MOS System under External Bias

We now turn our attention to the electrical behavior of the MOS structure under externally applied bias voltages. Assume that the substrate voltage is set at $V_B = 0$, and let the gate voltage be the controlling parameter. Depending on the polarity and the magnitude of V_G, three different operating regions can be observed for the MOS system: *accumulation, depletion,* and *inversion.*

If a negative voltage V_G is applied to the gate electrode, the holes in the p-type substrate are attracted to the semiconductor-oxide interface. The majority carrier concentration near the surface becomes larger than the equilibrium hole concentration in the substrate; hence, this condition is called carrier *accumulation* on the surface (Fig. 3.5). Note that in this case, the oxide electric field is directed towards the gate electrode. The negative surface potential also causes the energy bands to bend upward near the surface. While the hole density near the surface increases as a result of

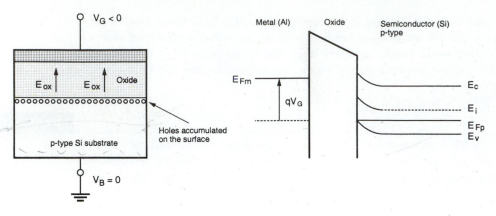

Figure 3.5 The cross-sectional view and the energy band diagram of the MOS structure operating in accumulation region.

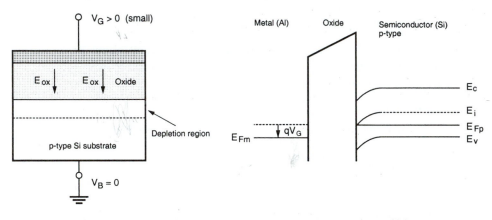

Figure 3.6 The cross-sectional view and the energy band diagram of the MOS structure operating in depletion mode, under small gate bias.

the applied negative gate bias, the electron (minority carrier) concentration decreases as the negatively charged electrons are pushed deeper into the substrate.

Now consider the next case in which a small positive gate bias V_G is applied to the gate electrode. Since the substrate bias is zero, the oxide electric field will be directed towards the substrate in this case. The positive surface potential causes the energy bands to bend downward near the surface, as shown in Fig. 3.6. The majority carriers, i.e., the holes in the substrate, will be repelled back into the substrate as a result of the positive gate bias, and these holes will leave negatively charged fixed acceptor ions behind. Thus, a *depletion* region is created near the surface. Note that under this bias condition, the region near the semiconductor-oxide interface is nearly devoid of all mobile carriers.

The thickness x_d of this depletion region on the surface can easily be found as a function of the surface potential ϕ_s. Assume that the mobile hole charge in a thin horizontal layer parallel to the surface is

$$dQ = -q \cdot N_A \cdot dx \tag{3.7}$$

The *change* in surface potential required to displace this charge sheet dQ by a distance x_d away from the surface can be found by using the Poisson equation.

$$d\phi_s = -x \cdot \frac{dQ}{\varepsilon_{Si}} = \frac{q \cdot N_A \cdot x}{\varepsilon_{Si}} dx \tag{3.8}$$

Integrating (3.7) along the vertical dimension (perpendicular to the surface) yields

$$\int_{\phi_F}^{\phi_s} d\phi_s = \int_0^{x_d} \frac{q \cdot N_A \cdot x}{\varepsilon_{Si}} dx \tag{3.9}$$

$$\phi_s - \phi_F = \frac{q \cdot N_A \cdot x_d^2}{2\varepsilon_{Si}} \tag{3.10}$$

Thus, the depth of the depletion region is

$$x_d = \sqrt{\frac{2\varepsilon_{Si} \cdot |\phi_s - \phi_F|}{q \cdot N_A}} \tag{3.11}$$

and the depletion region charge density, which consists solely of fixed acceptor ions in this region, is given by the following expression

$$Q = -q \cdot N_A \cdot x_d = -\sqrt{2q \cdot N_A \cdot \varepsilon_{Si} \cdot |\phi_s - \phi_F|} \tag{3.12}$$

The amount of this depletion region charge plays a very important role in the analysis of threshold voltage, as we will examine shortly.

To complete our qualitative overview of different bias conditions and their effects upon the MOS system, consider next a further increase in the positive gate bias. As a result of the increasing surface potential, the downward bending of the energy bands will increase as well. Eventually, the mid-gap energy level E_i becomes smaller than the Fermi level E_{Fp} on the surface, which means that the substrate semiconductor in this region becomes n-type. Within this thin layer, the electron density is larger than the majority hole density, since the positive gate potential attracts additional minority carriers (electrons) from the bulk substrate to the surface (Fig. 3.7). The n-type region created near the surface by the positive gate bias is called the *inversion layer*, and this condition is called *surface inversion*. It will be seen that the thin inversion layer on the surface with a large mobile electron concentration can be utilized for conducting current between two terminals of the MOS transistor.

As a practical definition, the surface is said to be *inverted* when the density of mobile electrons on the surface becomes equal to the density of holes in the bulk (p-type) substrate. This condition requires that the surface potential has the same magnitude, but the reverse polarity, as the bulk Fermi potential ϕ_F. Once the surface

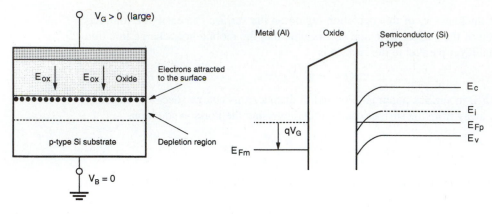

Figure 3.7   The cross-sectional view and the energy band diagram of the MOS structure in surface inversion, under larger gate bias voltage.

is inverted, any further increase in the gate voltage leads to an increase of mobile electron concentration on the surface, but not to an increase of the depletion depth. Thus, the depletion region depth achieved at the onset of surface inversion is also equal to the maximum depletion depth, x_{dm}, which remains constant for higher gate voltages. Using the inversion condition $\phi_s = -\phi_F$, the maximum depletion region depth at the onset of surface inversion can be found from (3.11) as follows:

$$x_{dm} = \sqrt{\frac{2 \cdot \varepsilon_{Si} \cdot |2\phi_F|}{q \cdot N_A}} \tag{3.13}$$

The creation of a conducting surface inversion layer through externally applied gate bias is an essential phenomenon for current conduction in MOS transistors. In the following section, we will examine the structure and the operation of the MOS Field Effect Transistor (MOSFET).

3.3   Structure and Operation of MOS Transistor (MOSFET)

The basic structure of an n-channel MOSFET is shown in Fig. 3.8. This four-terminal device consists of a p-type substrate, in which two n^+ diffusion regions, the drain and the source, are formed. The surface of the substrate region between the drain and the source is covered with a thin oxide layer, and the metal (or polysilicon) gate is deposited on top of this gate dielectric. The midsection of the device can easily be recognized as the basic MOS structure which was examined in the previous sections. The two n^+ regions will be the current-conducting terminals of this device. Note that the device structure is completely symmetrical with respect to the drain and source regions; the different roles of these two regions will be defined only in conjunction with the applied terminal voltages and the direction of the current flow.

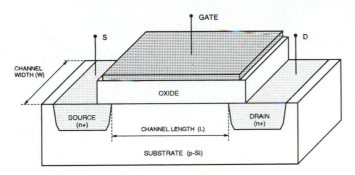

Figure 3.8 The physical structure of an n-channel enhancement-type MOSFET.

A conducting *channel* will eventually be formed through applied gate voltage in the section of the device between the drain and the source diffusion regions. The distance between the drain and source diffusion regions is the *channel length L,* and the lateral extent of the channel (perpendicular to the length dimension) is the *channel width W.* Both the channel length and the channel width are important parameters which can be used to control some of the electrical properties of the MOSFET. The thickness of the oxide layer covering the channel region, t_{ox}, is also an important parameter.

A MOS transistor which has no conducting channel region at zero gate bias is called an *enhancement-type* (or *enhancement-mode*) MOSFET. If a conducting channel already exists at zero gate bias, on the other hand, the device is called a *depletion-type* (or *depletion-mode*) MOSFET. In a MOSFET with p-type substrate and with n^+ source and drain regions, the channel region to be formed on the surface is n-type. Thus, such a device with p-type substrate is called an *n-channel MOSFET*. In a MOSFET with n-type substrate and with p^+ source and drain regions, on the other hand, the channel is p-type and the device is called a *p-channel MOSFET*.

The abbreviations used for the device terminals are: G for the gate, D for the drain, S for the source, and B for the substrate (or body). In an n-channel MOSFET, the source is defined as the n^+ region which has a *lower* potential than the other n^+ region, the drain. By convention, all terminal voltages of the device are defined with respect to the source potential. Thus, the gate-to-source voltage is denoted by V_{GS}, the drain-to-source voltage is denoted by V_{DS}, and the substrate-to-source voltage is denoted by V_{BS}. Circuit symbols for both n-channel and p-channel enhancement-type MOSFETs are shown in Fig. 3.9. While the four-terminal symbolic representation shows all external terminals of the device, the simple three-terminal representation will also be used extensively. Note that in the simple MOSFET circuit symbol, the small arrow always marks the source terminal.

Consider first the n-channel enhancement-type MOSFET shown in Fig. 3.8. The simple operation principle of this device is: *control the current conduction between the source and the drain, using the electric field generated by the gate voltage as a control variable.* Since the current flow in the channel is also controlled by the

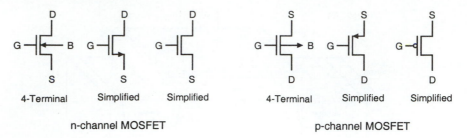

Figure 3.9   Circuit symbols for n-channel and p-channel enhancement-type MOSFETs.

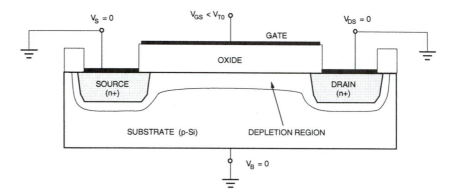

Figure 3.10   Formation of a depletion region in an n-channel enhancement-type MOSFET.

drain-to-source voltage and by the substrate voltage, the current can be considered a function of these external terminal voltages. We will examine in detail the functional relationships between the channel current (also called the *drain current*) and the terminal voltages. In order to start current flow between the source and the drain regions, however, we have to form a conducting channel first.

The simplest bias condition that can be applied to the n-channel enhancement-type MOSFET is shown in Fig. 3.10. The source, the drain, and the substrate terminals are all connected to ground. A positive gate-to-source voltage V_{GS} is then applied to the gate in order to create the conducting channel underneath the gate. With this bias arrangement, the channel region between the source and the drain diffusions behaves exactly the same as for the simple MOS structure we examined in Section 3.2. For small gate voltage levels, the majority carriers (holes) are repelled back into the substrate, and the surface of the p-type substrate is depleted. Since the surface is devoid of any mobile carriers, current conduction between the source and the drain is not possible.

Now assume that the gate-to-source voltage is further increased. As soon as the surface potential in the channel region reaches $-\phi_{Fp}$, surface inversion will be established, and a conducting n-type layer will form between the source and the drain diffusion regions (Fig. 3.11). This channel now provides an electrical connection

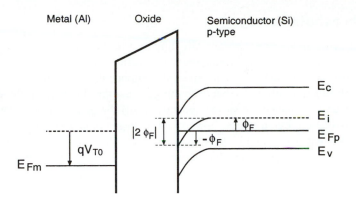

Figure 3.11 Band diagram of the MOS structure underneath the gate, at surface inversion. Notice the band bending by $|2\phi_F|$ at the surface.

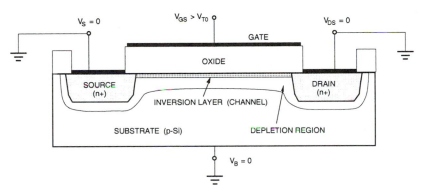

Figure 3.12 Formation of an inversion layer (channel) in an n-channel enhancement-type MOSFET.

between the two n^+ regions, and it allows current flow, as long as there is a potential difference between the source and the drain terminal voltages (Fig. 3.12). The bias conditions for the onset of surface inversion and for the creation of the conducting channel are therefore very significant for MOSFET operation.

The value of the gate-to-source voltage V_{GS} needed to cause surface inversion (to create the conducting channel) is called the *threshold voltage* V_{T0}. Any gate-to-source voltage smaller than V_{T0} is not sufficient to establish an inversion layer; thus, the MOSFET can conduct no current between its source and drain terminals unless $V_{GS} > V_{T0}$. For gate-to-source voltages larger than the threshold voltage, on the other hand, a larger number of minority carriers (electrons) are attracted to the surface, which ultimately contribute to channel current conduction. Also note that increasing the gate-to-source voltage above and beyond the threshold voltage will not affect the surface potential and the depletion region depth. Both quantities will remain approximately constant and equal to their values attained at the onset of surface inversion.

The Threshold Voltage

In the following, physical parameters affecting the threshold voltage of a MOS structure will be examined by considering the various components of V_{T0}. For all practical purposes, we can identify four physical components of the threshold voltage: (i) the work function difference between the gate and the channel, (ii) the gate voltage component to change the surface potential, (iii) the gate voltage component to offset the depletion region charge, and (iv) the voltage component to offset the fixed charges in the gate oxide and in the silicon-oxide interface. The analysis will be carried out for an n-channel device, but the results are applicable to p-channel devices as well, with minor modifications.

The work function difference Φ_{GC} between the gate and the channel reflects the built-in potential of the MOS system, which consists of the p-type substrate, the thin silicon dioxide layer, and the gate electrode. Depending on the gate material, the work function difference is

$$\Phi_{GC} = \phi_F(substrate) - \phi_M \qquad \text{for metal gate} \qquad (3.14)$$

$$\Phi_{GC} = \phi_F(substrate) - \phi_F(gate) \qquad \text{for polysilicon gate} \qquad (3.15)$$

This first component of the threshold voltage accounts for part of the voltage drop across the MOS system that is built-in. Now, the externally applied gate voltage must be changed to achieve surface inversion, i.e., to change the surface potential by $-2\phi_F$. This will be the second component of the threshold voltage.

Another component of the applied gate voltage is necessary to offset the depletion region charge, which is due to the fixed acceptor ions located in the depletion region near the surface. We can calculate the depletion region charge density at surface inversion ($\phi_s = -\phi_F$) using (3.12).

$$Q_{B0} = -\sqrt{2q \cdot N_A \cdot \varepsilon_{Si} \cdot |-2\phi_F|} \qquad (3.16)$$

Note that if the substrate (body) is biased at a different voltage level than the source, which is at ground potential (reference), then the depletion region charge density can be expressed as a function of the source-to-substrate voltage V_{SB}.

$$Q_B = -\sqrt{2q \cdot N_A \cdot \varepsilon_{S_i} \cdot |-2\phi_F + V_{SB}|} \qquad (3.17)$$

The component that offsets the depletion region charge is then equal to $-Q_B/C_{ox}$, where C_{ox} is the gate oxide capacitance per unit area.

$$C_{ox} = \frac{\varepsilon_{ox}}{t_{ox}} \qquad (3.18)$$

Finally, we must consider the influence of a nonideal physical phenomenon which we have neglected until now. There always exists a fixed positive charge density Q_{ox} at the interface between the gate oxide and the silicon substrate, due to impurities and/or lattice imperfections at the interface. The gate voltage component that is necessary to offset this positive charge at the interface is $-Q_{ox}/C_{ox}$. Now, we can combine all of these voltage components to find the threshold voltage. For zero substrate bias, the threshold voltage V_{T0} is expressed as follows:

$$V_{T0} = \Phi_{GC} - 2\phi_F - \frac{Q_{B0}}{C_{ox}} - \frac{Q_{ox}}{C_{ox}} \qquad (3.19)$$

For nonzero substrate bias, on the other hand, the depletion charge density term must be modified to reflect the influence of V_{SB} upon that charge, resulting in the following generalized threshold voltage expression.

$$V_T = \Phi_{GC} - 2\phi_F - \frac{Q_B}{C_{ox}} - \frac{Q_{ox}}{C_{ox}} \qquad (3.20)$$

The generalized form of the threshold voltage can also be written as

$$V_T = \Phi_{GC} - 2\phi_F - \frac{Q_{B0}}{C_{ox}} - \frac{Q_{ox}}{C_{ox}} - \frac{Q_B - Q_{B0}}{C_{ox}} = V_{T0} - \frac{Q_B - Q_{B0}}{C_{ox}} \qquad (3.21)$$

Note that in this case, the threshold voltage differs from V_{T0} only by an additive term. This substrate-bias term is a simple function of the material constants and of the source-to-substrate voltage V_{SB}.

$$\frac{Q_B - Q_{B0}}{C_{ox}} = -\frac{\sqrt{2q \cdot N_A \cdot \varepsilon_{Si}}}{C_{ox}} \cdot \left(\sqrt{|-2\phi_F + V_{SB}|} - \sqrt{|2\phi_F|}\right) \qquad (3.22)$$

Thus, the most general expression of the threshold voltage V_T can be found as follows:

$$V_T = V_{T0} + \gamma \cdot \left(\sqrt{|-2\phi_F + V_{SB}|} - \sqrt{|2\phi_F|}\right) \qquad (3.23)$$

where the parameter γ

$$\gamma = \frac{\sqrt{2q \cdot N_A \cdot \varepsilon_{Si}}}{C_{ox}} \qquad (3.24)$$

is the *substrate-bias* (or *body-effect*) coefficient.

The threshold voltage expression given in (3.23) can be used both for n-channel and p-channel MOS transistors. One must be careful, however, since some of the terms and coefficients in this equation have different polarities for the n-channel (nMOS) case and for the p-channel (pMOS) case. The reason for this polarity difference is that the substrate semiconductor is p-type in an n-channel MOSFET and n-type in a p-channel MOSFET. Specifically,

- ■ The substrate Fermi potential ϕ_F is *negative in nMOS, positive in pMOS*.
- ■ The depletion region charge densities Q_{B0} and Q_B are *negative in nMOS, positive in pMOS*.
- ■ The substrate bias coefficient γ is *positive in nMOS, negative in pMOS*.
- ■ The substrate bias voltage V_{SB} is *positive in nMOS, negative in pMOS*.

Typically, the threshold voltage of an enhancement-type n-channel MOSFET is a positive quantity, whereas the threshold voltage of a p-channel MOSFET is negative.

EXAMPLE 3.2

Calculate the threshold voltage V_{T0} at $V_{SB} = 0$, for a polysilicon gate n-channel MOS transistor, with the following parameters: substrate doping density $N_A = 10^{16}$ cm^{-3}, polysilicon gate doping density $N_D = 2 \times 10^{20}$ cm^{-3}, gate oxide thickness $t_{ox} = 500$ Å, and oxide-interface fixed charge density $N_{ox} = 4 \times 10^{10}$ cm^{-2}.

First, calculate the Fermi potentials for the p-type substrate and for the n-type polysilicon gate:

$$\phi_F(substrate) = \frac{kT}{q} \ln\left(\frac{n_i}{N_A}\right) = 0.026 \text{ V} \cdot \ln\left(\frac{1.45 \cdot 10^{10}}{10^{16}}\right) = -0.35 \text{ V}$$

Since the doping density of the polysilicon gate is very high, the heavily doped n-type gate material is expected to be degenerate. Thus, we may assume that the Fermi potential of the polysilicon gate is approximately equal to the conduction band potential, i.e., $\phi_F(gate) = 0.55$ V. Now, calculate the work function difference between the gate and the channel:

$$\Phi_{GC} = \phi_F(substrate) - \phi_F(gate) = -0.35 \text{ V} - 0.55 \text{ V} = -0.90 \text{ V}$$

The depletion region charge density at $V_{SB} = 0$ is found as follows:

$$Q_{B0} = -\sqrt{2 \cdot q \cdot N_A \cdot \varepsilon_{Si} \cdot |-2\phi_F(substrate)|}$$
$$= -\sqrt{2 \cdot 1.6 \cdot 10^{-19} \cdot 10^{16} \cdot 11.7 \cdot 8.85 \cdot 10^{-14} |-2 \cdot 0.35|}$$
$$= -4.82 \cdot 10^{-8} \text{ C/cm}^2$$

The oxide-interface charge is:

$$Q_{ox} = q \cdot N_{ox} = 1.6 \cdot 10^{-19} \text{ C} \times 4 \cdot 10^{10} \text{ cm}^{-2} = 6.4 \cdot 10^{-9} \text{ C/cm}^2$$

The gate oxide capacitance per unit area is calculated using the dielectric constant of silicon dioxide and the oxide thickness t_{ox}.

$$C_{ox} = \frac{\varepsilon_{ox}}{t_{ox}} = \frac{3.97 \cdot 8.85 \cdot 10^{-14} \text{ F/cm}}{500 \cdot 10^{-8} \text{ cm}} = 7.03 \cdot 10^{-8} \text{ F/cm}^2$$

Now, we can combine all components and calculate the threshold voltage.

$$V_{T0} = \Phi_{GC} - 2\phi_F(substrate) - \frac{Q_{B0}}{C_{ox}} - \frac{Q_{ox}}{C_{ox}}$$
$$= -0.90 - (-0.70) - (-0.69) - 0.09 = 0.40 \text{ V}$$

In this simplified analysis, the doping concentrations of the source and the drain diffusion regions and the geometry (physical dimensions) of the channel region have no influence upon the threshold voltage V_{T0}.

Note that the exact value of the threshold voltage of an actual MOS transistor cannot be determined using (3.23) in most practical cases, due primarily to uncertainties and variations of the doping concentrations, the oxide thickness, and the fixed oxide-interface charge. The nominal value and the statistical range of the threshold voltage for any MOS process are ultimately determined by direct measurements, which will be described later in Section 3.4. In most MOS fabrication processes, the threshold voltage can be adjusted by selective dopant ion implantation into the channel region of the MOSFET. For n-channel MOSFETs, the threshold voltage is *increased* (made more positive) by adding extra p-type impurities (acceptor ions). Alternatively, the threshold voltage of the n-channel MOSFET can be *decreased* (made more negative) by implanting n-type impurities (dopant ions) into the channel region.

The amount of change in the threshold voltage as a result of extra implants can be approximated as follows. Let the density of implanted impurities be represented by N_I [cm^{-2}]. Assume that all implanted ions are electrically active, i.e., each ion contributes to the depletion region charge. Then, the threshold voltage V_{T0} at zero substrate bias ($V_{SB} = 0$) will be shifted by an amount of qN_I/C_{ox}. This approximation obviously neglects the variation of the substrate Fermi level ϕ_F as the result of extra implants, but it nevertheless provides a fair estimate for the threshold voltage shift.

EXERCISE 3.1

Consider the following p-channel MOSFET process:

Substrate doping $N_D = 10^{15}$ cm^{-3}, polysilicon gate doping density $N_D = 10^{20}$ cm^{-3}, gate oxide thickness $t_{ox} = 650$ Å, and oxide-interface charge density $N_{ox} = 2 \times 10^{10}$ cm^{-2}. Use $\varepsilon_{Si} = 11.7\varepsilon_0$ and $\varepsilon_{ox} = 3.97\varepsilon_0$ for the dielectric coefficients of silicon and silicon-dioxide, respectively.

(a) Calculate the threshold voltage V_{T0}, for $V_{SB} = 0$.
(b) Determine the type and the amount of channel ion implantation which are necessary to achieve a threshold voltage of $V_{T0} = -2$ V.

Note that, using selective ion implantation into the channel, the threshold voltage of an n-channel MOSFET can also be made negative. This means that the resulting nMOS transistor will have a conducting channel at $V_{GS} = 0$, enabling current flow between its source and drain terminals as long as V_{GS} is larger than the negative threshold voltage. Such a device is called a *depletion-type* (or *normally on*) n-channel MOSFET. We will see several practical applications for depletion-type nMOS transistors in the design of MOS digital circuits. Except for its negative threshold voltage, the depletion-type n-channel MOSFET exhibits the same electrical behavior as the enhancement-type n-channel MOSFET. Figure 3.13 shows the conventional circuit symbols used for depletion-type n-channel MOSFETs.

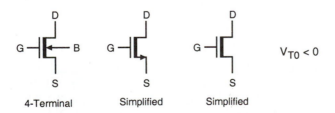

4-Terminal   Simplified   Simplified

Figure 3.13 Circuit symbols for n-channel depletion-type MOSFETs.

EXAMPLE 3.3

Consider the n-channel MOSFET process given in Example 3.2. In several digital circuit applications, the condition $V_{SB} = 0$ cannot be guaranteed for all transistors. We will examine in this example how a nonzero source-to-substrate voltage V_{SB} affects the threshold voltage of the MOS transistor.

First, we must calculate the substrate-bias coefficient γ using the process parameters given in Example 3.2.

$$\gamma = \frac{\sqrt{2 \cdot q \cdot N_A \cdot \varepsilon_{Si}}}{C_{ox}} = \frac{\sqrt{2 \cdot 1.6 \cdot 10^{-19} \cdot 10^{16} \cdot 11.7 \cdot 8.85 \cdot 10^{-14}}}{7.03 \cdot 10^{-8}}$$

$$= 0.82 \text{ V}^{\frac{1}{2}}$$

Now we compute and plot the threshold voltage V_T as a function of the source-to-substrate voltage V_{SB}. The voltage V_{SB} will be assumed to vary between zero and 5 V.

$$V_T = V_{T0} + \gamma \cdot \left(\sqrt{|-2\phi_F + V_{SB}|} - \sqrt{|2\phi_F|}\right)$$

$$= 0.40 + 0.82 \cdot \left(\sqrt{0.7 + V_{SB}} - \sqrt{0.7}\right)$$

It is seen that the threshold voltage variation is about 1.3 V over this range, which could present serious design problems if neglected. We will see in the following chapters that the substrate-bias effect is unavoidable in most digital circuits and that the circuit designer usually must take appropriate measures to account for and/or to compensate for the threshold voltage variations.

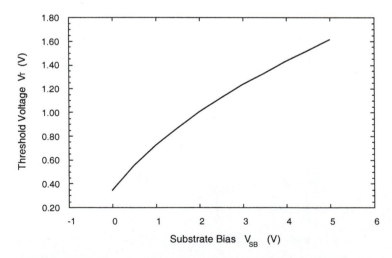

Variation of the threshold voltage as a function of the source-to-substrate voltage.

MOSFET Operation: A Qualitative View

The basic structure of the n-channel MOS (nMOS) transistor built on a p-type substrate was shown in Fig. 3.8. The MOSFET consists of a MOS capacitor with two p-n junctions placed immediately adjacent to the *channel* region that is controlled by the MOS gate. The carriers, i.e., electrons in an nMOS transistor, enter the structure through the source contact (S), leave through the drain (D), and are subject to the control of the gate (G) voltage. To ensure that both p-n junctions are reverse-biased initially, the substrate potential is kept lower than the other three terminal potentials.

We have seen that when $0 < V_{GS} < V_{T0}$, the gated region between the source and the drain is depleted; no carrier flow can be observed in the channel. As the gate voltage is increased beyond the threshold voltage ($V_{GS} > V_{T0}$), however, the midgap energy level at the surface is pulled below the Fermi level, causing the surface potential ϕ_s to turn positive and to *invert* the surface (Fig. 3.12). Once the inversion layer is established on the surface, an n-type conducting channel forms between the source and the drain, which is capable of carrying the drain current.

Next, the influence of drain-to-source bias V_{DS} and different modes of drain current flow will be examined for an nMOS transistor with $V_{GS} > V_{T0}$. At $V_{DS} = 0$, thermal equilibrium exists in the inverted channel region, and the drain current I_D is equal to zero (Fig. 3.14(a)). If a small drain voltage $V_{DS} > 0$ is applied, a drain current proportional to V_{DS} will flow from the source to the drain through the conducting channel. The inversion layer, i.e., the channel, forms a continuous current path from the source to the drain. This operation mode is called the *linear mode*, or the *linear region*. Thus, in linear region operation, the channel region acts as a voltage-controlled resistor. The electron velocity in the channel for this case is usually much lower than the drift velocity limit. Note that as the drain voltage is increased, the inversion layer charge and the channel depth at the drain end start to decrease. Eventually, for $V_{DS} = V_{DSAT}$, the inversion charge at the drain is reduced to zero, which is called the *pinch-off* point (Fig. 3.14(b)).

Beyond the pinch-off point, i.e., for $V_{DS} > V_{DSAT}$, a depleted surface region forms adjacent to the drain, and this depletion region grows toward the source with increasing drain voltages. This operation mode of the MOSFET is called the *saturation mode* or the *saturation region*. For a MOSFET operating in the saturation region, the effective channel length is reduced as the inversion layer near the drain vanishes, while the channel-end voltage remains essentially constant and equal to V_{DSAT} (Fig. 3.14(c)). Note that the pinched-off (depleted) section of the channel absorbs most of the excess voltage drop ($V_{DS} - V_{DSAT}$) and a high-field region forms between the channel-end and the drain boundary. Electrons arriving from the source to the channel-end are injected into the drain-depletion region and are accelerated toward the drain in this high electric field, usually reaching the drift velocity limit. The pinch-off event, or the disruption of the continuous channel under high drain bias, characterizes the saturation mode operation of the MOSFET.

The influence of these operating conditions upon the external (terminal) current-voltage characteristics of the MOS transistor will be examined in the following section. A good understanding of these relationships, and of the factors involved therein, will be essential for the design and analysis of MOS digital circuits.

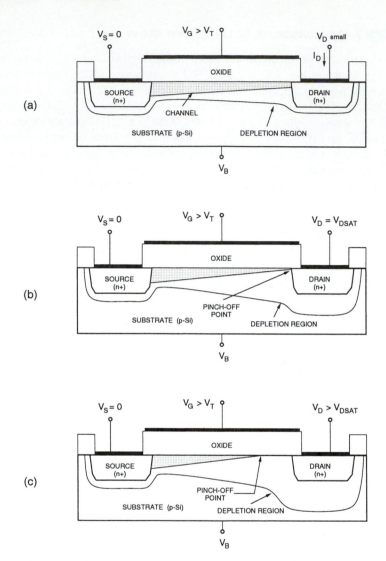

Figure 3.14 Cross-sectional view of an n-channel (nMOS) transistor, (a) operating in the linear region, (b) operating at the edge of saturation, and (c) operating beyond saturation.

3.4 MOSFET Current-Voltage Characteristics

The analytical derivation of the MOSFET current-voltage relationships for various bias conditions requires that several approximations be made to simplify the problem. Without these simplifying assumptions, analysis of the actual three-dimensional MOS system would become a very complex task and would prevent the derivation of closed-form current-voltage equations. In the following, we will use the *gradual channel approximation* (GCA) for establishing the MOSFET current-voltage relationships, which will effectively reduce the analysis to a one-dimensional current-flow problem. This will allow us to devise relatively simple current equations that agree well with experimental results. As in every approximate approach, however, the GCA also has its limitations, especially for small-geometry MOSFETs. We will investigate the most significant limitations and examine some of the possible remedies.

Gradual Channel Approximation

To begin with the current-flow analysis, consider the cross-sectional view of the n-channel MOSFET operating in the linear mode, as shown in Fig. 3.15. Here, the source and the substrate terminals are connected to ground, i.e., $V_S = V_B = 0$. The gate-to-source voltage (V_{GS}) and the drain-to-source voltage (V_{DS}) are the external parameters controlling the drain (channel) current I_D. The gate-to-source voltage is set to be larger than the threshold voltage V_{T0} to create a conducting inversion layer between the source and the drain. We define the coordinate system for this structure such that the x-direction is perpendicular to the surface, pointing down into the substrate, and the y-direction is parallel to the surface. The y-coordinate origin ($y = 0$) is at the source end of the channel. The *channel voltage* with respect to the source will be denoted by $V_c(y)$. Now assume that the threshold voltage V_{T0} is constant along the entire channel region, between $y = 0$ and $y = L$. In reality, the threshold voltage changes along the channel since the channel voltage is not constant. Next, assume that the electric field component E_y along the y-coordinate is *dominant* compared to the electric field component E_x along the x-coordinate. This assumption will allow us to reduce the current-flow problem in the channel to the y-dimension only. Note that the boundary conditions for the channel voltage

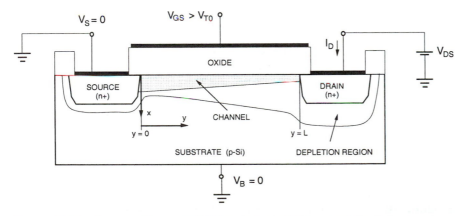

Figure 3.15 Cross-sectional view of an n-channel transistor, operating in linear region.

V_c are:

$$V_c(y = 0) = V_S = 0$$
$$V_c(y = L) = V_{DS} \qquad (3.25)$$

Also, it is assumed that the entire channel region between the source and the drain is inverted, i.e.,

$$V_{GS} \geq V_{T0}$$
$$V_{GD} = V_{GS} - V_{DS} \geq V_{T0} \qquad (3.26)$$

The channel current (drain current) I_D is due to the electrons in the channel region traveling from the source to the drain under the influence of the lateral electric field component E_y. Since the current flow in the channel is primarily governed by the lateral drift of the mobile electron charge in the surface inversion layer, we will consider the amount and the bias-voltage dependence of this inversion layer in more detail.

Let $Q_I(y)$ be the total mobile electron charge in the surface inversion layer. This charge can be expressed as a function of the gate-to-source voltage V_{GS} and of the channel voltage $V_c(y)$ as follows:

$$Q_I(y) = -C_{ox} \cdot [V_{GS} - V_c(y) - V_{T0}] \qquad (3.27)$$

Figure 3.16 shows the spatial geometry of the surface inversion layer and indicates its significant dimensions. Note that the thickness of the inversion layer tapers off as we move from the source to the drain, since the gate-to-channel voltage causing surface inversion is smaller at the drain end.

Now consider the incremental resistance dR of the differential channel segment shown in Fig. 3.16. Assuming that all mobile electrons in the inversion layer have a constant *surface mobility* μ_n, the incremental resistance can be expressed as

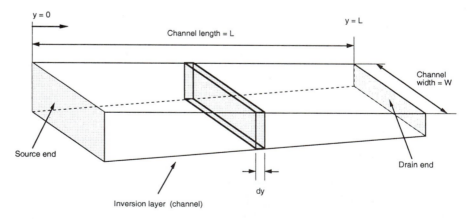

Figure 3.16   Simplified geometry of the surface inversion layer (channel region).

follows. Note that the minus sign is due to the negative polarity of the inversion layer charge Q_I.

$$dR = -\frac{dy}{W \cdot \mu_n \cdot Q_I(y)} \tag{3.28}$$

The *electron surface mobility* μ_n used in (3.28) depends on the doping concentration of the channel region, and its magnitude is typically about one-half of that of the bulk electron mobility. We will assume that the channel current density is uniform across this segment. According to our one-dimensional model, the channel (drain) current I_D flows between the source and the drain regions in the y-coordinate direction. Applying Ohm's law for this segment yields the voltage drop along the incremental segment dy, in the y-direction.

$$dV_c = I_D \cdot dR = -\frac{I_D}{W \cdot \mu_n \cdot Q_I(y)} \cdot dy \tag{3.29}$$

This equation can now be integrated along the channel, i.e., from $y = 0$ to $y = L$, using the boundary conditions given in (3.25).

$$\int_0^L I_D \cdot dy = -W \cdot \mu_n \int_0^{V_{DS}} Q_I(y) \cdot dV_c \tag{3.30}$$

The left-hand side of this equation is simply equal to LI_D. The integral on the right-hand side is evaluated by replacing $Q_I(y)$ with (3.27). Thus,

$$I_D \cdot L = W \cdot \mu_n \cdot C_{ox} \int_0^{V_{DS}} (V_{GS} - V_c - V_{T0}) \cdot dV_c \tag{3.31}$$

Assuming that the channel voltage V_c is the only variable in (3.31) that depends on the position y, the drain current is found as follows.

$$I_D = \frac{\mu_n \cdot C_{ox}}{2} \cdot \frac{W}{L} \cdot \left[2 \cdot (V_{GS} - V_{T0}) V_{DS} - V_{DS}^2\right] \tag{3.32}$$

Equation (3.32) represents the drain current I_D as a simple second-order function of the two external voltages, V_{GS} and V_{DS}. This current equation can also be rewritten as

$$I_D = \frac{k'}{2} \cdot \frac{W}{L} \cdot \left[2 \cdot (V_{GS} - V_{T0}) V_{DS} - V_{DS}^2\right] \tag{3.33}$$

or

$$I_D = \frac{k}{2} \cdot \left[2 \cdot (V_{GS} - V_{T0}) V_{DS} - V_{DS}^2\right] \tag{3.34}$$

where the parameters k and k' are defined as

$$k' = \mu_n \cdot C_{ox} \tag{3.35}$$

and

$$k = k' \cdot \frac{W}{L} \tag{3.36}$$

The drain current equation given in (3.33) is the simplest analytical approximation for the MOSFET current-voltage relationship. Note that, in addition to the process-dependent constants k' and V_{T0}, the current-voltage relationship is also affected by the device dimensions, W and L. In fact, we will see that the ratio of W/L is one of the most important design parameters in MOS digital circuit design. Now, we must determine the *region of validity* for this equation and what this means for the practical use of the equation.

EXAMPLE 3.4

For an n-channel MOS transistor with $\mu_n = 600 \text{ cm}^2/\text{V}\cdot\text{s}$, $C_{ox} = 7 \cdot 10^{-8}$ F/cm^2, $W = 20 \ \mu\text{m}$, $L = 2 \ \mu\text{m}$ and $V_{T0} = 1.0$ V, examine the relationship between the drain current and the terminal voltages.

First, calculate the parameter k:

$$k = \mu_n \cdot C_{ox} \cdot \frac{W}{L} = 600 \text{ cm}^2/\text{V}\cdot\text{s} \times 7 \cdot 10^{-8} \text{ F/cm}^2 \times \frac{20 \ \mu\text{m}}{20 \ \mu\text{m}} = 0.42 \text{ mA/V}^2$$

Now, the current-voltage equation (3.34) can be written as follows.

$$I_D = 0.21 \text{ mA/V}^2 \left[2 \cdot (V_{GS} - 1.0) \cdot V_{DS} - V_{DS}^2 \right]$$

To examine the effect of the gate-to-source voltage and the drain-to-source voltage upon the drain current, we will plot I_D as a function of V_{DS}, for different (constant) values of V_{GS}. It can easily be seen that the second-order current-voltage equation given above produces a set of inverted parabolas for each constant V_{GS} value.

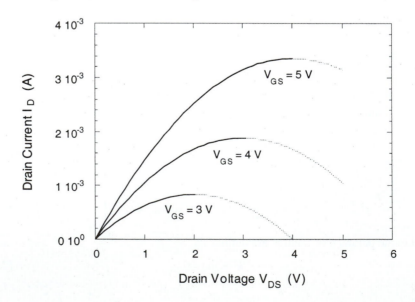

The drain current-drain voltage curves shown on page 104 reach their peak value for $V_{DS} = V_{GS} - V_{T0}$. Beyond this maximum, each curve exhibits a *negative* differential conductance, which is not observed in actual MOSFET current-voltage measurements (section shown by the dashed lines). We must remember now that the drain current equation (3.32) has been derived under the following voltage assumptions,

$$V_{GS} \geq V_{T0}$$

$$V_{GD} = V_{GS} - V_{DS} \geq V_{T0}$$

which guarantee that the entire channel region between the source and the drain is inverted. This condition corresponds to the *linear* operating mode for the MOSFET, which was examined qualitatively in Section 3.4. Hence, the current equation (3.32) is valid only for the linear mode operation. Beyond the linear region boundary, i.e., for V_{DS} values *larger* than $V_{GS} - V_{T0}$, the MOS transistor will be assumed to be in *saturation*. A different current-voltage expression will be necessary for the MOSFET operating in this region.

Example 3.4 shows that the current equation (3.32) is not valid beyond the linear region/saturation region boundary, i.e., for

$$V_{DS} \geq V_{DSAT} = V_{GS} - V_{T0} \tag{3.37}$$

Also, drain current measurements with constant V_{GS} show that the current I_D does not show much variation as a function of the drain voltage V_{DS} beyond the saturation boundary, but rather remains approximately constant around the peak value reached for $V_{DS} = V_{DSAT}$. This saturation drain current level can be found simply by substituting (3.37) for V_{DS} in (3.32).

$$I_D(sat) = \frac{\mu_n \cdot C_{ox}}{2} \cdot \frac{W}{L} \cdot \left[2 \cdot (V_{GS} - V_{T0}) \cdot (V_{GS} - V_{T0}) - (V_{GS} - V_{T0})^2 \right]$$

$$= \frac{\mu_n \cdot C_{ox}}{2} \cdot \frac{W}{L} \cdot (V_{GS} - V_{T0})^2 \tag{3.38}$$

Thus, the drain current I_D becomes a function only of the gate-to-source voltage V_{GS}, beyond the saturation boundary. Note that this constant saturation current approximation is not very accurate in reality, and that the saturation-region drain current continues to have a certain dependence on the drain voltage. For simple hand calculations, however, (3.38) provides a sufficiently accurate approximation of the MOSFET drain (channel) current in saturation.

Figure 3.17 shows the typical drain current versus drain voltage characteristics of an n-channel MOSFET, as described by the current equations (3.32) and (3.38). The parabolic boundary between the linear and the saturation regions is indicated here by the dashed line. The current-voltage characteristics of the MOS transistor can also be visualized by plotting the drain current as a function of the gate voltage, as shown in Fig. 3.18. This $I_D - V_{GS}$ transfer characteristic in saturation mode ($V_{DS} > V_{DSAT}$) provides a simple view of the drain current increasing as a second-order function of the gate-to-source voltage (cf. Equation (3.38)). The current is obviously equal to zero for any gate voltage smaller than the threshold voltage V_{T0}.

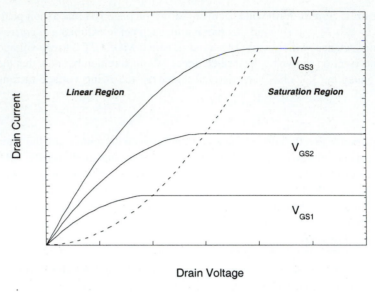

Figure 3.17   Basic current-voltage characteristics of an n-channel MOS transistor.

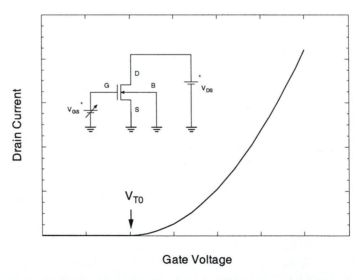

Figure 3.18   Drain current of the n-channel MOS transistor as a function of the gate-to-source voltage V_{GS}, with $V_{DS} > V_{DSAT}$ (transistor in saturation).

Channel Length Modulation

Next, we will examine the mechanisms of channel pinch-off and current flow in saturation mode in more detail. Consider the inversion layer charge Q_I that represents the total mobile electron charge on the surface, given by (3.27). The inversion layer charge at the source end of the channel is

$$Q_I(y = 0) = -C_{ox} \cdot (V_{GS} - V_{T0}) \tag{3.39}$$

and the inversion layer charge at the drain end of the channel is

$$Q_I(y = L) = -C_{ox} \cdot (V_{GS} - V_{T0} - V_{DS}) \tag{3.40}$$

Note that at the edge of saturation, i.e., when the drain-to-source voltage reaches V_{DSAT},

$$V_{DS} = V_{DSAT} = V_{GS} - V_{T0} \tag{3.41}$$

the inversion layer charge at the drain end becomes zero, according to (3.40). In reality, the channel charge does not become exactly equal to zero (remember that the GCA is just a simple approximation of the actual conditions in the channel), but it indeed becomes very small.

$$Q_I(y = L) \approx 0 \tag{3.42}$$

Thus, we can state that under the bias condition given in (3.41), the channel is *pinched-off* at the drain end, i.e., at $y = L$. The onset of the saturation mode operation in the MOSFET is signified by this pinch-off event. If the drain-to-source voltage V_{DS} is increased even further beyond the saturation edge so that $V_{DS} > V_{DSAT}$, an even larger portion of the channel becomes pinched-off.

Consequently, the *effective channel length* (the length of the inversion layer where GCA is still valid) is reduced to

$$L' = L - \Delta L \tag{3.43}$$

where ΔL is the length of the channel segment with $Q_I = 0$ (Fig. 3.19). Hence, the pinch-off point moves from the drain end of the channel toward the source with increasing drain-to-source voltages. The remaining portion of the channel between the pinch-off point and the drain will be in depletion mode. Since $Q_I(y) = 0$ for $L' < y < L$, the channel voltage at the pinch-off point remains equal to V_{DSAT}, i.e.,

$$V_c(y = L') = V_{DSAT} \tag{3.44}$$

The electrons traveling from the source toward the drain traverse the inverted channel segment of length L', and then they are injected into the depletion region of length ΔL that separates the pinch-off point from the drain edge. As seen in Fig. 3.19, we can represent the inverted portion of the surface by a shortened channel, with a channel-end voltage of V_{DSAT}. The gradual channel approximation is valid in this region; thus, the channel current can be found using (3.38).

$$I_D(sat) = \frac{\mu_n \cdot C_{ox}}{2} \cdot \frac{W}{L'} \cdot (V_{GS} - V_{T0})^2 \tag{3.45}$$

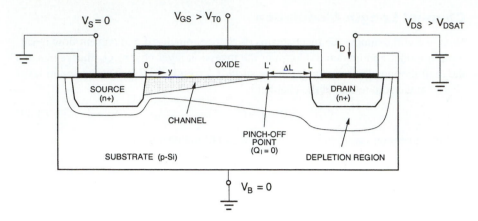

Figure 3.19   Channel length modulation in an n-channel MOSFET operation in saturation mode.

Note that this current equation corresponds to a MOSFET with effective channel length L', operating in saturation. Thus, (3.45) accounts for the actual shortening of the channel, also called *channel length modulation*. Since $L' < L$, the saturation current calculated by using (3.45) will be larger than that found by using (3.38) under the same bias conditions. As L' decreases with increasing V_{DS}, the saturation mode current I_D (*sat*) will also increase with V_{DS}. By approximating the effective channel length $L' = L - \Delta L$ as a function of the drain bias voltage, we can modify (3.45) to reflect this drain voltage dependence. First, rewrite the saturation current as follows:

$$I_D(sat) = \left(\cfrac{1}{1 - \cfrac{\Delta L}{L}} \right) \cdot \frac{\mu_n \cdot C_{ox}}{2} \cdot \frac{W}{L} \cdot (V_{GS} - V_{T0})^2 \qquad (3.46)$$

The first term of this saturation current expression accounts for the channel modulation effect, while the rest of this expression is identical to (3.38). It can be shown that the channel length shortening ΔL is actually proportional to the square root of $(V_{DS} - V_{DSAT})$.

$$\Delta L \propto \sqrt{V_{DS} - V_{DSAT}} \qquad (3.47)$$

To simplify the analysis even further, we will use the following empirical relation between ΔL and the drain-to-source voltage instead:

$$1 - \frac{\Delta L}{L} \approx 1 - \lambda \cdot V_{DS} \qquad (3.48)$$

Here, λ is an empirical model parameter, and is called the *channel length modulation coefficient*. Assuming that $\lambda V_{DS} \ll 1$, the saturation current given in (3.45) can now be written as:

$$I_D(sat) = \frac{\mu_n \cdot C_{ox}}{2} \cdot \frac{W}{L} \cdot (V_{GS} - V_{T0})^2 \cdot (1 + \lambda \cdot V_{DS}) \qquad (3.49)$$

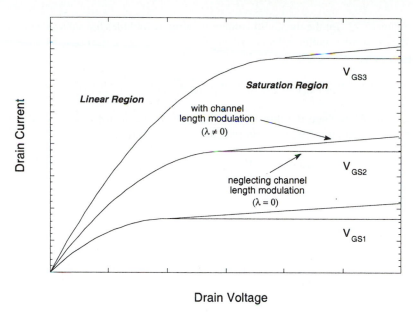

Figure 3.20 Current-voltage characteristics of an n-channel MOS transistor, including the channel length modulation effect.

This simple current equation prescribes a linear drain-bias dependence for the saturation current in MOS transistors, determined by the empirical parameter λ. Although this rough approximation does not accurately reflect the physical relationship between the channel length shortening ΔL and the drain bias, (3.49) can be used with sufficient confidence for most first-order hand calculations. The drain current versus drain-to-source voltage characteristics of an n-channel MOSFET, obtained by using (3.32) for the linear region and (3.49) for the saturation region, are shown in Fig. 3.20. The saturation mode current increases linearly with V_{DS} instead of remaining constant. The slope of the current-voltage curve in the saturation region is determined by the channel length modulation coefficient λ.

Substrate Bias Effect

Note that the derivation of linear-mode and saturation-mode current-voltage characteristics in the previous pages has been done under the assumption that the substrate potential is equal to the source potential, i.e., $V_{SB} = 0$. Consequently, the zero-substrate bias threshold voltage V_{T0} has been used in the current equations. In many digital circuit applications, on the other hand, the source potential of an nMOS transistor can be larger than the substrate potential, which results in a positive source-to-substrate voltage $V_{SB} > 0$. In this case, the influence of the nonzero V_{SB} upon the current characteristics must be accounted for. Recall that the general expression (3.23) for the threshold voltage V_T already includes the substrate bias term and, hence, it reflects the influence of the nonzero source-to-substrate voltage upon the device characteristics.

$$V_T(V_{SB}) = V_{T0} + \gamma \cdot \left(\sqrt{|2\phi_F| + V_{SB}} - \sqrt{|2\phi_F|}\right) \tag{3.50}$$

We can simply replace the threshold voltage terms in linear-mode and saturation-mode current equations with the more general $V_T(V_{SB})$ term.

$$I_D(lin) = \frac{\mu_n \cdot C_{ox}}{2} \cdot \frac{W}{L} \cdot \left[2 \cdot (V_{GS} - V_T(V_{SB}))V_{DS} - V_{DS}^2\right] \tag{3.51}$$

$$I_D(sat) = \frac{\mu_n \cdot C_{ox}}{2} \cdot \frac{W}{L} \cdot (V_{GS} - V_T(V_{SB}))^2 \cdot (1 + \lambda \cdot V_{DS}) \tag{3.52}$$

In general, we will use only the term V_T instead of $V_T(V_{SB})$ to express the general (substrate-bias dependent) threshold voltage. As already demonstrated in Example 3.3, the substrate-bias effect can significantly change the value of the threshold voltage and, hence, the current capability of the MOSFET. With this modification, we finally arrive at a complete first-order characterization of the drain (channel) current as a nonlinear function of the terminal voltages.

$$I_D = f(V_{GS}, V_{DS}, V_{BS}) \tag{3.53}$$

In the following, we will repeat the current-voltage equations derived under the first-order gradual channel approximation (GCA), both for n-channel and for p-channel MOS transistors. Figure 3.21 shows the polarities of applied terminal voltages and the drain current directions. Note that the threshold voltage V_T and the terminal voltages V_{GS}, V_{DS}, and V_{SB} are all *negative* for the pMOS transistor. The parameter μ_p denotes the surface hole mobility in the pMOSFET.

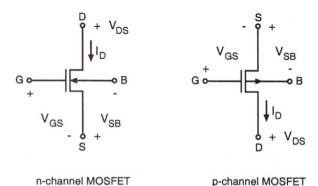

n-channel MOSFET   p-channel MOSFET

Figure 3.21   Terminal voltages and currents of the nMOS and the pMOS transistor.

Current-voltage equations of the n-channel MOSFET:

$$I_D = 0, \qquad \text{for} \quad V_{GS} < V_T \tag{3.54}$$

$$I_D(lin) = \frac{\mu_n \cdot C_{ox}}{2} \cdot \frac{W}{L} \cdot \left[2 \cdot (V_{GS} - V_T)V_{DS} - V_{DS}^2\right] \qquad \text{for} \quad V_{GS} \geq V_T$$

$$\text{and} \quad V_{DS} < V_{GS} - V_T \tag{3.55}$$

$$I_D(sat) = \frac{\mu_n \cdot C_{ox}}{2} \cdot \frac{W}{L} \cdot (V_{GS} - V_T)^2 \cdot (1 + \lambda \cdot V_{DS}) \qquad \text{for} \quad V_{GS} \geq V_T$$

$$\text{and} \quad V_{DS} \geq V_{GS} - V_T \tag{3.56}$$

Current-voltage equations of the p-channel MOSFET:

$$I_D = 0, \qquad \text{for} \quad V_{GS} > V_T \tag{3.57}$$

$$I_D(lin) = \frac{\mu_p \cdot C_{ox}}{2} \cdot \frac{W}{L} \cdot \left[2 \cdot (V_{GS} - V_T)V_{DS} - V_{DS}^2\right] \qquad \text{for} \quad V_{GS} \leq V_T$$

$$\text{and} \quad V_{DS} > V_{GS} - V_T \tag{3.58}$$

$$I_D(sat) = \frac{\mu_p \cdot C_{ox}}{2} \cdot \frac{W}{L} \cdot (V_{GS} - V_T)^2 \cdot (1 + \lambda \cdot V_{DS}) \qquad \text{for} \quad V_{GS} \leq V_T$$

$$\text{and} \quad V_{DS} \leq V_{GS} - V_T \tag{3.59}$$

Measurement of Parameters

The MOSFET current-voltage equations (3.54) through (3.59), together with the general threshold voltage expression (3.50), are very useful for simple, first-order calculations of the currents and voltages in the nMOS and pMOS transistors. Because of several simplifications and approximations involved in their derivation, however, the accuracy of these current-voltage equations is fairly limited. To exploit the simplicity of the equations and to achieve the maximum possible accuracy in calculations, the parameters appearing in the current equations must be determined carefully, through experimental measurements. The model parameters that are used in (3.50) and in (3.54) through (3.59) are the zero-bias threshold voltage V_{T0}, the substrate-bias coefficient γ, the channel length modulation coefficient λ, and the following transconductance parameters:

$$k_n = \mu_n \cdot C_{ox} \cdot \frac{W}{L} \tag{3.60}$$

$$k_p = \mu_p \cdot C_{ox} \cdot \frac{W}{L} \tag{3.61}$$

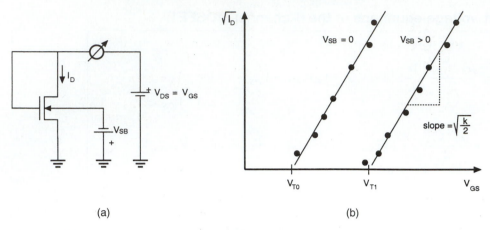

Figure 3.22 (a) Test circuit arrangement and (b) measured data for experimental determination of the parameters k_n, V_{T0}, and γ.

In the following section, some simple measurements for an enhancement-type n-channel MOSFET will be described for the determination of these parameters. First, consider the test circuit setup shown in Fig. 3.22(a). The source-to-substrate voltage V_{SB} is set at a constant value, and the drain current is measured for different values of the gate-to-source voltage V_{GS}. Since the drain and the gate of the transistor are at the same potential, $V_{DS} = V_{GS}$. Hence, the saturation condition $V_{DS} > V_{GS} - V_T$ is always satisfied, i.e., the nMOS transistor shown in Fig. 3.22(a) operates in saturation mode. Neglecting the channel length modulation effect for simplicity, the drain current is described by

$$I_D(sat) = \frac{k_n}{2} \cdot (V_{GS} - V_{T0})^2 \qquad (3.62)$$

Now, the square root of the drain current can be written as a *linear* function of the gate-to-source voltage.

$$\sqrt{I_D} = \sqrt{\frac{k_n}{2}} \cdot (V_{GS} - V_{T0}) \qquad (3.63)$$

If the square root of the measured drain current values is plotted against the gate-to-source voltage, the slope and the voltage-axis intercept of the resulting curve(s) can determine the parameters k_n, V_{T0}, and γ. Figure 3.22(b) shows the measured drain current vs. gate voltage curves, obtained for different values of substrate bias. By extrapolating the curves to zero-drain-current (voltage-axis intercept point), we can find the threshold voltage V_T that corresponds to each V_{SB} value. The voltage-axis intercept of the curve with $V_{SB} = 0$ gives the zero-bias threshold voltage, V_{T0}. Note that these *extrapolated threshold voltage* values do not exactly match the

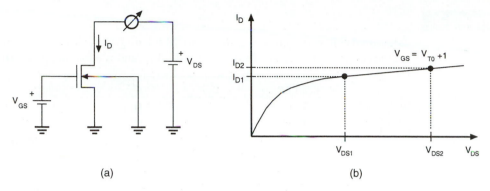

Figure 3.23 (a) Test circuit arrangement and (b) measured data for experimental determination of the channel length modulation coefficient λ.

threshold voltage values usually measured in the production environment, at a certain nonzero drain current. They can rather be viewed as fitting parameters for the current-voltage equations. The slope of each curve is equal to the square root of $(k_n/2)$. Thus, the transconductance parameter k_n can simply be calculated from this slope.

Next, consider the extrapolated threshold voltage values, obtained from voltage axis intercepts at nonzero substrate bias voltage. Using one of the available V_{SB} values, the substrate bias coefficient γ can be found from

$$\gamma = \frac{V_T(V_{SB}) - V_{T0}}{\sqrt{|2\phi_F| + V_{SB}} - \sqrt{|2\phi_F|}} \tag{3.64}$$

The experimental measurement of the channel length modulation coefficient λ requires a different test circuit setup, as shown in Fig. 3.23(a). The gate-to-source voltage V_{GS} is set to $V_{T0} + 1$. The drain-to-source voltage is chosen sufficiently large ($V_{DS} > V_{GS} - V_{T0}$) that the transistor operates in the saturation mode. The saturation drain current is then measured for two different drain voltage values, V_{DS1} and V_{DS2}. Note that the drain current in the saturation mode is given by

$$I_D(sat) = \frac{k_n}{2} \cdot (V_{GS} - V_{T0})^2 \cdot (1 + \lambda \cdot V_{DS}) \tag{3.65}$$

Since $V_{GS} = V_{T0} + 1$, the ratio of the measured drain current values I_{D1} and I_{D2} is

$$\frac{I_{D2}}{I_{D1}} = \frac{1 + \lambda \cdot V_{DS2}}{1 + \lambda \cdot V_{DS1}} \tag{3.66}$$

which can be used to calculate the channel length modulation coefficient λ. This is in fact equivalent to calculating the *slope* of the drain current versus drain voltage curve in the saturation region, as shown in Fig. 3.23(b). Specifically, the slope is $(\lambda k_n/2)$.

EXAMPLE 3.5

Measured voltage and current data for a MOSFET are given below. Determine the type of the device, and calculate the parameters k_n, V_{T0}, and γ. Assume $\phi_F = -0.3$ V.

V_{GS} (V)	V_{DS} (V)	V_{SB} (V)	I_D (μA)
3	3	0	97
4	4	0	235
5	5	0	433
3	3	3	59
4	4	3	173
5	5	3	347

First, the MOS transistor is on $(I_D > 0)$ for $V_{GS} > 0$ and $V_{DS} > 0$. Thus, the transistor must be an n-channel MOSFET. Assume that the transistor is enhancement-type and, therefore, operating in saturation mode for $V_{GS} = V_{DS}$. Neglecting the channel length modulation effect, the saturation mode current is written as

$$I_D = \frac{k_n}{2} \cdot (V_{GS} - V_T)^2 \quad \Leftrightarrow \quad \sqrt{I_D} = \sqrt{\frac{k_n}{2}} \cdot (V_{GS} - V_T)$$

Let (V_{GS1}, I_{D1}) and (V_{GS2}, I_{D2}) be any two current-voltage pairs obtained from the table. Then, the square-root of the transconductance parameter k_n can be calculated.

$$\sqrt{\frac{k_n}{2}} = \frac{\sqrt{I_{D1}} - \sqrt{I_{D2}}}{V_{GS1} - V_{GS2}} = \frac{\sqrt{433 \ \mu A} - \sqrt{97 \ \mu A}}{5 \ V - 3 \ V} = 5.48 \times 10^{-3} \ A^{1/2}/V$$

Thus, the transconductance parameter of this n-channel MOSFET is:

$$k_n = 2 \cdot (5.48 \times 10^{-3})^2 = 60 \times 10^{-6} \ A/V^2 = 60 \ \mu A/V^2$$

The *extrapolated threshold voltage* V_{T0} at zero substrate bias can be found by calculating the x-axis intercept of the *square-root of* (I_D) versus V_{GS} curve.

$$V_{T0} = V_{GS} - \sqrt{\frac{2 \cdot I_D}{k_n}} = 1.2 \ V$$

To find the substrate bias coefficient γ, we must first determine the threshold voltage V_T at the source-to-substrate voltage of 3 V. Using one of the current-voltage data pairs corresponding to $V_{SB} = 3$ V, V_T can be calculated as follows:

$$V_T(V_{SB} = 3 \ V) = V_{GS} - \sqrt{\frac{2 \cdot I_D}{k_n}} = 4 \ V - \sqrt{\frac{2 \cdot 173 \ \mu A}{60 \ \mu A/V^2}} = 1.6 \ V$$

Finally, the substrate bias coefficient is found as:

$$\gamma = \frac{V_T(V_{SB} = 3 \ V) - V_{T0}}{\sqrt{|2\phi_F| + V_{SB}} - \sqrt{|2\phi_F|}} = \frac{1.6 \ V - 1.2 \ V}{\sqrt{0.6 \ V + 3 \ V} - \sqrt{0.6 \ V}} = 0.36 \ V^{1/2}$$

3.5 MOSFET Scaling and Small-Geometry Effects

The design of high-density chips in MOS VLSI (Very Large Scale Integration) technology requires that the packing density of MOSFETs used in the circuits is as high as possible and, consequently, that the sizes of the transistors are as small as possible. The reduction of the size, i.e., the dimensions of MOSFETs, is commonly referred to as *scaling*. It is expected that the operational characteristics of the MOS transistor will change with the reduction of its dimensions. Also, some physical limitations eventually restrict the extent of scaling that is practically achievable. There are two basic types of size-reduction strategies: *full scaling* (also called constant-field scaling) and *constant-voltage scaling*. Both types of scaling approaches will be shown to have unique effects upon the operating characteristics of the MOS transistor. In the following, we will examine in detail the scaling strategies and their effects, and we will also consider some of the physical limitations and small-geometry effects that must be taken into account for scaled MOSFETs.

Scaling of MOS transistors is concerned with systematic reduction of overall dimensions of the devices as allowed by the available technology, while preserving the geometric ratios found in the larger devices. The proportional scaling of all devices in a circuit would certainly result in a reduction of the total silicon area occupied by the circuit, thereby increasing the overall functional density of the chip. To describe device scaling, we introduce a constant *scaling factor S > 1*. All horizontal and vertical dimensions of the *large-size* transistor are then divided by this scaling factor to obtain the scaled device. The extent of scaling that is achievable is obviously determined by the fabrication technology and more specifically, by the minimum feature size. Table 3.1 below shows the recent history of reducing feature sizes for the typical CMOS gate-array process. It is seen that a new generation of manufacturing technology replaces the previous one about every two or three years, and the down-scaling factor S of the minimum feature size from one generation to the next is about 1.2 to 1.5.

We consider the proportional scaling of all three dimensions by the same scaling factor S. Figure 3.24 shows the reduction of key dimensions on a typical MOSFET, together with the corresponding increase of the doping densities.

The primed quantities in Fig. 3.24 indicate the scaled dimensions and doping densities. It is easy to recognize that the scaling of all dimensions by a factor of $S > 1$ leads to the reduction of the area occupied by the transistor by a factor of S^2. To better understand the effects of scaling upon the current-voltage characteristics of the MOSFET, we will examine two different scaling options in the following sections.

Table 3.1 Reduction of the minimum feature size (minimum dimensions that can be defined and manufactured on chip) over the years, for a typical CMOS gate-array process

Year	1985	1987	1989	1991	1993	1995	1997	1999
Feature size (μm)	2.5	1.7	1.2	1.0	0.8	0.5	0.35	0.25

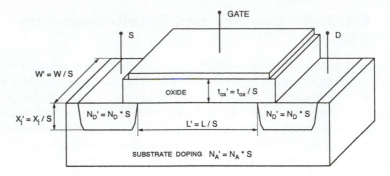

Figure 3.24 Scaling of a typical MOSFET by a scaling factor of *S*.

Full Scaling (Constant-Field Scaling)

This scaling option attempts to preserve the magnitude of internal electric fields in the MOSFET, while the dimensions are scaled down by a factor of *S*. To achieve this goal, all potentials must be scaled down proportionally, by the same scaling factor. Note that this potential scaling also affects the threshold voltage V_{T0}. Finally, the Poisson equation describing the relationship between charge densities and electric fields dictates that the charge densities must be *increased* by a factor of *S* in order to maintain the field conditions. Table 3.2 lists the scaling factors for all significant dimensions, potentials, and doping densities of the MOS transistor.

Now consider the influence of full scaling described here upon the current-voltage characteristics of the MOS transistor. It will be assumed that the surface mobility μ_n is not significantly affected by the scaled doping density. The gate oxide capacitance per unit area, on the other hand, is changed as follows.

$$C'_{ox} = \frac{\varepsilon_{ox}}{t'_{ox}} = S \cdot \frac{\varepsilon_{ox}}{t_{ox}} = S \cdot C_{ox} \qquad (3.67)$$

The aspect ratio W/L of the MOSFET will remain unchanged under scaling. Consequently, the transconductance parameter k_n will also be scaled by a factor of *S*. Since

Table 3.2 Full scaling of MOSFET dimensions, potentials, and doping densities

Quantity	Before scaling	After scaling
Channel length	L	$L' = L/S$
Channel width	W	$W' = W/S$
Gate oxide thickness	t_{ox}	$t'_{ox} = t_{ox}/S$
Junction depth	x_j	$x'_j = x_j/S$
Power supply voltage	V_{DD}	$V'_{DD} = V_{DD}/S$
Threshold voltage	V_{T0}	$V'_{T0} = V_{T0}/S$
Doping densities	N_A	$N'_A = S \cdot N_A$
	N_D	$N'_D = S \cdot N_D$

all terminal voltages are scaled down by the factor S as well, the linear-mode drain current of the scaled MOSFET can now be found as:

$$I'_D(lin) = \frac{k'_n}{2} \cdot \left[2 \cdot (V'_{GS} - V'_T) \cdot V'_{DS} - V'^2_{DS}\right]$$

$$= \frac{S \cdot k_n}{2} \cdot \frac{1}{S^2} \cdot \left[2 \cdot (V_{GS} - V_T) \cdot V_{DS} - V^2_{DS}\right] = \frac{I_D(lin)}{S} \quad (3.68)$$

Similarly, the saturation-mode drain current is also reduced by the same scaling factor.

$$I'_D(sat) = \frac{k'_n}{2} \cdot (V'_{GS} - V'_T)^2 = \frac{S \cdot k_n}{2} \cdot \frac{1}{S^2} \cdot (V_{GS} - V_T)^2 = \frac{I_D(sat)}{S} \quad (3.69)$$

Now consider the power dissipation of the MOSFET. Since the drain current flows between the source and the drain terminals, the instantaneous power dissipated by the device (before scaling) can be found as:

$$P = I_D \cdot V_{DS} \quad (3.70)$$

Notice that full scaling reduces both the drain current and the drain-to-source voltage by a factor of S; hence, the power dissipation of the transistor will be reduced by the factor S^2.

$$P' = I'_D \cdot V'_{DS} = \frac{1}{S^2} \cdot I_D \cdot V_{DS} = \frac{P}{S^2} \quad (3.71)$$

This significant reduction of the power dissipation is one of the most attractive features of full scaling. Note that with the device area reduction by S^2 discussed earlier, we find the *power density* per unit area remaining virtually unchanged for the scaled device.

Finally, consider the gate oxide capacitance defined as $C_g = W L C_{ox}$. It will be shown later in Section 3.6 that charging and discharging of this capacitance plays an important role in the transient operation of the MOSFET. Since the gate oxide capacitance C_g is scaled down by a factor of S, we can predict that the transient characteristics, i.e., the charge-up and charge-down times, of the scaled device will improve accordingly. In addition, the proportional reduction of all dimensions on-chip will lead to a reduction of various parasitic capacitances and resistances as well, contributing to the overall performance improvement. Table 3.3 summarizes the changes in key device characteristics as a result of full (constant-field) scaling.

Table 3.3 Effects of full scaling upon key device characteristics

Quantity	Before scaling	After scaling
Oxide capacitance	C_{ox}	$C'_{ox} = S \cdot C_{ox}$
Drain current	I_D	$I'_D = I_D/S$
Power dissipation	P	$P' = P/S^2$
Power density	$P/Area$	$P'/Area' = P/Area$

Constant-Voltage Scaling

While the full scaling strategy dictates that the power supply voltage and all terminal voltages be scaled down proportionally with the device dimensions, the scaling of voltages may not be very practical in many cases. In particular, the peripheral and interface circuitry may require certain voltage levels for all input and output voltages, which in turn would necessitate multiple power supply voltages and complicated level-shifter arrangements. For these reasons, constant-voltage scaling is usually preferred over full scaling.

In constant-voltage scaling, all dimensions of the MOSFET are reduced by a factor of S, as in full scaling. The power supply voltage and the terminal voltages, on the other hand, remain unchanged. The doping densities must be increased by a factor of S^2 in order to preserve the charge-field relations. Table 3.4 shows the constant-voltage scaling of key dimensions, voltages, and densities. Under constant-voltage scaling, the changes in device characteristics are significantly different compared to those in full scaling, as we will demonstrate. The gate oxide capacitance per unit area C_{ox} is increased by a factor of S, which means that the transconductance parameter is also increased by S. Since the terminal voltages remain unchanged, the linear mode drain current of the scaled MOSFET can be written as:

$$I_D'(lin) = \frac{k_n'}{2} \cdot \left[2 \cdot (V_{GS}' - V_T') \cdot V_{DS}' - V_{DS}'^2 \right]$$

$$= \frac{S \cdot k_n}{2} \cdot \left[2 \cdot (V_{GS} - V_T) \cdot V_{DS} - V_{DS}^2 \right] = S \cdot I_D(lin) \qquad (3.72)$$

Also, the saturation-mode drain current will be increased by a factor of S after constant-voltage scaling. This means that the drain current *density* (current per unit area) is increased by a factor of S^3, which may cause serious reliability problems for the MOS transistor.

$$I_D'(sat) = \frac{k_n'}{2} \cdot (V_{GS}' - V_T')^2 = \frac{S \cdot k_n}{2} \cdot (V_{GS} - V_T)^2 = S \cdot I_D(sat) \qquad (3.73)$$

Next, consider the power dissipation. Since the drain current is increased by a factor of S while the drain-to-source voltage remains unchanged, the power dissipation of

Table 3.4   Constant-voltage scaling of MOSFET dimensions, potentials, and doping densities

Quantity	Before scaling	After scaling
Dimensions	W, L, t_{ox}, x_j	reduced by S ($W' = W/S, \ldots$)
Voltages	V_{DD}, V_T	remain unchanged
Doping densities	N_A, N_D	increased by S^2 ($N_A' = S^2 \cdot N_A, \ldots$)

Table 3.5 Effects of constant-voltage scaling upon key device characteristics

Quantity	Before scaling	After scaling
Oxide capacitance	C_{ox}	$C'_{ox} = S \cdot C_{ox}$
Drain current	I_D	$I'_D = S \cdot I_D$
Power dissipation	P	$P' = S \cdot P$
Power density	$P/Area$	$P'/Area' = S^3 \cdot (P/Area)$

the MOSFET increases by a factor of S.

$$P' = I'_D \cdot V'_{DS} = (S \cdot I_D) \cdot V_{DS} = S \cdot P \tag{3.74}$$

Finally, the power density (power dissipation per unit area) is found to increase by a factor of S^3 after constant-voltage scaling, with possible adverse effects on device reliability (Table 3.5).

To summarize, constant-voltage scaling may be preferred over full (constant-field) scaling in many practical cases because of the external voltage-level constraints. It must be recognized, however, that constant-voltage scaling increases the drain current density and the power density by a factor of S^3. This large increase in current and power densities may eventually cause serious reliability problems for the scaled transistor, such as electromigration, hot-carrier degradation, oxide breakdown, and electrical over-stress.

As the device dimensions are systematically reduced through full scaling or constant-voltage scaling, various physical limitations become increasingly more prominent, and ultimately restrict the amount of feasible scaling for some device dimensions. Consequently, scaling may be carried out on a certain subset of MOSFET dimensions in many practical cases. Also, the simple gradual channel approximation (GCA) used for the derivation of current-voltage relationships does not accurately reflect the effects of scaling in smaller-size transistors. The current equations have to be modified accordingly. In the following, we will briefly investigate some of these small-geometry effects.

Short-Channel Effects

As a working definition, a MOS transistor is called a short-channel device if its channel length is on the same order of magnitude as the depletion region thicknesses of the source and drain junctions. Alternatively, a MOSFET can be defined as a short-channel device if the effective channel length L_{eff} is approximately equal to the source and drain junction depth x_j. The short-channel effects that arise in this case are attributed to two physical phenomena: (i) the limitations imposed on electron drift characteristics in the channel, and (ii) the modification of the threshold voltage due to the shortening channel length.

Note that the lateral electric field E_y along the channel increases, as the effective channel length is decreased. While the electron drift velocity v_d in the channel is proportional to the electric field for lower field values, this drift velocity tends to saturate at high channel electric fields. For channel electric fields of $Ey = 10^5$ V/cm and higher, the electron drift velocity in the channel reaches a saturation value of about $v_d(sat) = 10^7$ cm/s. This velocity saturation has very significant implications upon the current-voltage characteristics of the short-channel MOSFET. Consider the saturation-mode drain current, under the assumption that carrier velocity in the channel has already reached its limit value. The effective channel length L_{eff} will be reduced due to channel-length shortening.

$$I_D(sat) = W \cdot v_d(sat) \cdot \int_0^{L_{eff}} q \cdot n(x) \cdot dx = W \cdot v_d(sat) \cdot |Q_I| \qquad (3.75)$$

Since the channel-end voltage is equal to V_{DSAT}, the saturation current can be found as follows:

$$I_D(sat) = W \cdot v_d(sat) \cdot C_{ox} \cdot V_{DSAT} \qquad (3.76)$$

Carrier velocity saturation actually reduces the saturation-mode current below the current value predicted by the conventional long-channel current equations. The current is no longer a quadratic function of the gate-to-source voltage V_{GS}, and it is virtually independent of the channel length. Also note that under these conditions, the device is defined to be in saturation when the carrier velocity in the channel approaches about 90% of its limit value.

In short-channel MOS transistors, the carrier velocity in the channel is also a function of the normal (vertical) electric-field component E_x. Since the vertical field influences the scattering of carriers (collisions suffered by the carriers) in the surface region, the surface mobility is reduced with respect to the bulk mobility. The dependence of the surface electron mobility on the vertical electric field can be expressed by the following empirical formula:

$$\mu_n(eff) = \frac{\mu_{no}}{1 + \Theta \cdot E_x} = \frac{\mu_{no}}{1 + \dfrac{\Theta \varepsilon_{ox}}{t_{ox}\, \varepsilon_{Si}} \cdot (V_{GS} - V_c(y))} \qquad (3.77)$$

where μ_{no} is the low-field surface electron mobility and Θ is an empirical factor. For a simple estimation of field-related mobility reduction, (3.77) can be approximated by

$$\mu_n(eff) = \frac{\mu_{no}}{1 + \eta \cdot (V_{GS} - V_T)} \qquad (3.78)$$

where η is also an empirical coefficient.

Next, we consider the modification of the threshold voltage due to short-channel effects. The threshold voltage expression (3.23) was derived for a long-channel MOSFET. Specifically, the channel depletion region was assumed to be created only by the applied gate voltage, and the depletion regions associated with the drain and source pn-junctions were neglected. The shape of this gate-induced bulk (channel) depletion region was assumed to be rectangular, extending from the source to the

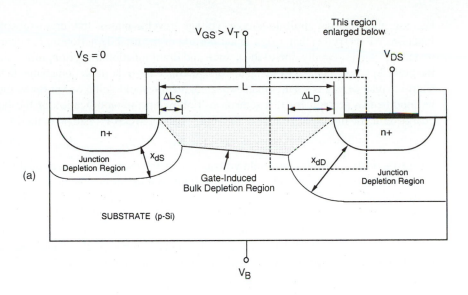

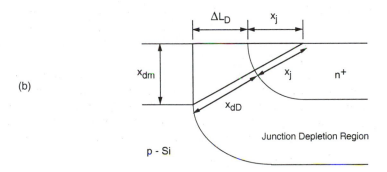

Figure 3.25 (a) Simplified geometry of the MOSFET channel region, with gate-induced bulk depletion region and the pn-junction depletion regions. (b) Close-up view of the drain diffusion edge.

drain. In short-channel MOS transistors, however, the n^+ drain and source diffusion regions in the p-type substrate induce a significant amount of depletion charge; consequently, the long-channel threshold voltage expression derived earlier overestimates the depletion charge supported by the gate voltage. The threshold voltage value found by using (3.23) is therefore larger than the actual threshold voltage of the short-channel MOSFET.

Figure 3.25(a) shows the simplified geometry of the gate-induced bulk depletion region and the pn-junction depletion regions in a short-channel MOS transistor. Note that the bulk depletion region is assumed to have an asymmetric trapezoidal shape, instead of a rectangular shape, to represent accurately the gate-induced charge. The drain depletion region is expected to be larger than the source depletion region because the positive drain-to-source voltage reverse-biases the drain-substrate junction. We recognize that a significant portion of the total depletion region charge under the gate is actually due to the source and drain junction depletion, rather than the bulk depletion induced by the gate voltage. Since the bulk depletion charge in the short-channel device is smaller than expected, the threshold voltage expression must be modified to account for this reduction. Following the modification of the bulk charge term, the threshold voltage of the short-channel MOSFET can be written as

$$V_{T0}(short\ channel) = V_{T0} - \Delta V_{T0} \tag{3.79}$$

where V_{T0} is the zero-bias threshold voltage calculated using the conventional long-channel formula (3.23) and ΔV_{T0} is the threshold voltage shift (reduction) due to the short-channel effect. The reduction term actually represents the amount of charge differential between a rectangular depletion region and a trapezoidal depletion region.

Let ΔL_S and ΔL_D represent the lateral extent of the depletion regions associated with the source junction and the drain junction, respectively. Then, the bulk depletion region charge contained within the trapezoidal region is

$$Q_{B0} = -\left(1 - \frac{\Delta L_S + \Delta L_D}{2L}\right) \cdot \sqrt{2 \cdot q \cdot \varepsilon_{Si} \cdot N_A \cdot |2\phi_F|} \tag{3.80}$$

To calculate ΔL_S and ΔL_D, we will use the simplified geometry shown in Fig. 3.25(b). Here, x_{dS} and x_{dD} represent the depth of the pn-junction depletion regions associated with the source and the drain, respectively. The edges of the source and drain diffusion regions are represented by quarter-circular arcs, each with a radius equal to the junction depth, x_j. The vertical extent of the bulk depletion region into the substrate is represented by x_{dm}. The junction depletion region depths can be approximated by

$$x_{dS} = \sqrt{\frac{2 \cdot \varepsilon_{Si}}{q \cdot N_A} \cdot \phi_0} \tag{3.81}$$

$$x_{dD} = \sqrt{\frac{2 \cdot \varepsilon_{Si}}{q \cdot N_A} \cdot (\phi_0 + V_{DS})} \tag{3.82}$$

with the junction built-in voltage

$$\phi_0 = \frac{kT}{q} \cdot ln\left(\frac{N_D \cdot N_A}{n_i^2}\right) \tag{3.83}$$

From Fig. 3.25(b), we find the following relationship between ΔL_D and the depletion region depths.

$$(x_j + x_{dD})^2 = x_{dm}^2 + (x_j + \Delta L_D)^2 \tag{3.84}$$

$$\Delta L_D^2 + 2 \cdot x_j \cdot \Delta L_D + x_{dm}^2 - x_{dD}^2 - 2 \cdot x_j \cdot x_{dD} = 0 \tag{3.85}$$

Solving for ΔL_D, we obtain:

$$\Delta L_D = -x_j + \sqrt{x_j^2 - \left(x_{dm}^2 - x_{dD}^2\right) + 2x_j x_{dD}} \cong x_j \cdot \left(\sqrt{1 + \frac{2x_{dD}}{x_j}} - 1\right) \tag{3.86}$$

Similarly, the length ΔL_S can also be found as follows:

$$\Delta L_S \cong x_j \cdot \left(\sqrt{1 + \frac{2x_{dS}}{x_j}} - 1 \right) \tag{3.87}$$

Now, the amount of threshold voltage reduction ΔV_{T0} due to short-channel effects can be found as:

$$\Delta V_{T0} = \frac{1}{C_{ox}} \cdot \sqrt{2q\varepsilon_{Si} N_A |2\phi_F|} \cdot \frac{x_j}{2L} \cdot \left[\left(\sqrt{1 + \frac{2x_{dS}}{x_j}} - 1 \right) + \left(\sqrt{1 + \frac{2x_{dD}}{x_j}} - 1 \right) \right] \tag{3.88}$$

The threshold voltage shift term is proportional to (x_j/L). As a result, this term becomes more prominent for MOS transistors with shorter channel lengths, and it approaches zero for long-channel MOSFETs where $L \gg x_j$. The following example illustrates the variation of the threshold voltage as a function of channel length in short-channel devices.

EXAMPLE 3.6

Consider an n-channel MOS process with the following parameters: substrate doping density $N_A = 10^{16}$ cm^{-3}, polysilicon gate doping density N_D (gate) $= 2 \times 10^{20}$ cm^{-3}, gate oxide thickness $t_{ox} = 50$ nm, oxide-interface fixed charge density $N_{ox} = 4 \times 10^{10}$ cm^{-2}, and source and drain diffusion doping density $N_D = 10^{17}$ cm^{-3}. In addition, the channel region is implanted with p-type impurities (impurity concentration $N_I = 2 \times 10^{11}$ cm^{-2}) to adjust the threshold voltage. The junction depth of the source and drain diffusion regions is $x_j = 1.0$ μm.

Plot the variation of the zero-bias threshold voltage V_{T0} as a function of the channel length (assume that $V_{DS} = V_{SB} = 0$). Also find V_{T0} for $L = 0.7$ μm, $V_{DS} = 5$ V, and $V_{SB} = 0$.

First, we have to find the zero-bias threshold voltage using the conventional formula (3.23). The threshold voltage *without* the channel implant was already calculated for the same process parameters in Example 3.2, and was found to be $V_{T0} = 0.40$ V. The additional p-type channel implant will increase the threshold voltage by an amount of qN_I/C_{ox}. Thus, we find the long-channel zero-bias threshold voltage for the process described above as

$$V_{T0} = 0.40 \text{ V} + \frac{q \cdot N_I}{C_{ox}} = 0.40 \text{ V} + \frac{1.6 \times 10^{-19} \cdot 2 \times 10^{11}}{7.03 \times 10^{-8}} = 0.855 \text{ V}$$

Next, the amount of threshold voltage reduction due to short-channel effects must be calculated using (3.88). The source and drain junction built-in voltage is

$$\phi_0 = \frac{kT}{q} \cdot \ln \left(\frac{N_D \cdot N_A}{n_i^2} \right) = 0.026 \text{ V} \cdot \ln \left(\frac{10^{17} \cdot 10^{16}}{2.1 \times 10^{20}} \right) = 0.76 \text{ V}$$

For zero drain bias, the depth of source and drain junction depletion regions is found as

$$x_{dS} = x_{dD} = \sqrt{\frac{2 \cdot \varepsilon_{Si}}{q \cdot N_A} \cdot \phi_0} = \sqrt{\frac{2 \cdot 11.7 \cdot 8.85 \times 10^{-14}}{1.6 \times 10^{-19} \cdot 10^{16}} \cdot 0.76}$$

$$= 31.4 \times 10^{-6} \text{ cm} = 0.314 \ \mu m$$

Now, the threshold voltage shift ΔV_{T0} due to short-channel effects can be calculated as a function of the gate (channel) length L.

$$\Delta V_{T0} = \frac{1}{C_{ox}} \cdot \sqrt{2q\varepsilon_{Si}N_A|2\phi_F|} \cdot \frac{x_j}{2L} \cdot \left[\left(\sqrt{1 + \frac{2x_{dS}}{x_j}} - 1 \right) + \left(\sqrt{1 + \frac{2x_{dD}}{x_j}} - 1 \right) \right]$$

$$= \frac{4.82 \times 10^{-8} \text{ C/cm}^2}{7.03 \times 10^{-8} \text{ F/cm}^2} \cdot \frac{1.0 \ \mu m}{L} \cdot \left(\sqrt{1 + \frac{2 \cdot 0.314 \ \mu m}{1.0 \ \mu m}} - 1 \right)$$

Finally, the zero-bias threshold voltage is found as

$$V_{T0}(short \ channel) = 0.855 \text{ V} - 0.19 \text{ V} \cdot \frac{1}{L[\mu m]}$$

The following plot shows the variation of the threshold voltage with the channel length. The threshold voltage decreases by as much as 50% for channel lengths in the submicron range, while it approaches the value of 0.8 V for larger channel lengths.

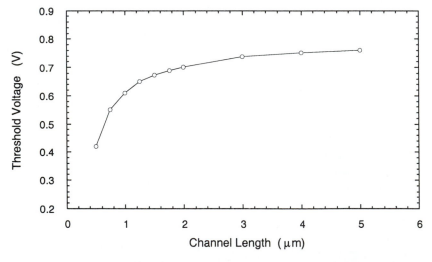

Since the conventional threshold voltage expression (3.23) is not capable of accounting for this drastic reduction of V_{T0} at smaller channel lengths, its application for short-channel MOSFETs must be carefully restricted.

Now, consider the variation of the threshold voltage with the applied drain-to-source voltage. Equation (3.82) shows that the depth of the drain junction depletion region increases with the voltage V_{DS}. For a drain-to-source voltage of $V_{DS} = 5$ V, the drain depletion depth is found as:

$$x_{dD} = \sqrt{\frac{2 \cdot \varepsilon_{Si}}{q \cdot N_A} \cdot (\phi_0 + V_{DS})}$$

$$= \sqrt{\frac{2 \cdot 11.7 \cdot 8.85 \times 10^{-14}}{1.6 \times 10^{-19} \cdot 10^{16}} \cdot (0.76 + 5.0)} = 0.863 \ \mu\text{m}$$

The resulting threshold voltage shift can be calculated by substituting x_{dD} found above in (3.88).

$$\Delta V_{T0} = \frac{1}{C_{ox}} \cdot \sqrt{2q\varepsilon_{Si}N_A|2\phi_F|} \cdot \frac{x_j}{2L} \cdot \left[\left(\sqrt{1 + \frac{2x_{dS}}{x_j}} - 1 \right) + \left(\sqrt{1 + \frac{2x_{dD}}{x_j}} - 1 \right) \right]$$

$$= \frac{4.82 \times 10^{-8}}{7.03 \times 10^{-8}} \cdot \frac{1.0}{2 \cdot 0.7} \cdot \left[\left(\sqrt{1 + \frac{2 \cdot 0.314}{1.0}} - 1 \right) + \left(\sqrt{1 + \frac{2 \cdot 0.863}{1.0}} - 1 \right) \right]$$

$$= 0.45 \ V$$

The threshold voltage of this short-channel MOS transistor is calculated as

$$V_{T0} = 0.855 \ \text{V} - 0.45 \ \text{V} = 0.405 \ \text{V}$$

which is significantly lower than the threshold voltage predicted by the conventional long-channel formula (3.23).

Narrow-Channel Effects

MOS transistors that have channel widths W on the same order of magnitude as the maximum depletion region thickness x_{dm} are defined as narrow-channel devices. Similar to the short-channel effects examined earlier, the narrow-channel MOSFETs also exhibit typical characteristics which are not accounted for by the conventional GCA analysis. The most significant narrow-channel effect is that the actual threshold voltage of such a device is *larger* than that predicted by the conventional threshold voltage formula (3.23). In the following, we will briefly review the physical reasons that cause this discrepancy. A typical cross-sectional view of a narrow-channel device is shown in Fig. 3.26. The oxide thickness in the channel region is t_{ox}, while the regions around the channel are covered by a thick *field oxide* (FOX). Since the gate electrode also overlaps with the field oxide as shown in Fig. 3.26, a relatively shallow depletion region forms underneath this FOX-overlap area as well. Consequently, the gate voltage must also support this additional depletion charge in order to establish the conducting channel. The charge contribution of this fringe depletion region to the overall channel depletion charge is negligible in wider devices. For MOSFETs with small channel widths, however, the actual threshold voltage increases as a result of this extra depletion charge.

$$V_{T0}(narrow \ channel) = V_{T0} + \Delta V_{T0} \tag{3.89}$$

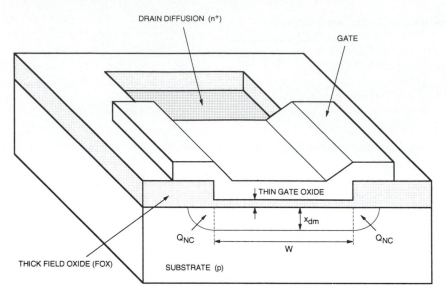

Figure 3.26 Cross-sectional view (across the channel) of a narrow-channel MOSFET. Note that Q_{NC} indicates the extra depletion charge due to narrow-channel effects.

The additional contribution to the threshold voltage due to narrow-channel effects can be modeled as follows:

$$\Delta V_{T0} = \frac{1}{C_{ox}} \cdot \sqrt{2q\varepsilon_{Si}N_A|2\phi_F|} \cdot \frac{\kappa \cdot x_{dm}}{W} \qquad (3.90)$$

where κ is an empirical parameter depending on the shape of the fringe depletion region. Assuming that the depletion region edges are modeled by quarter-circular arcs, for example, the parameter κ can be found as

$$\kappa = \frac{\pi}{2} \qquad (3.91)$$

The simple formula given in (3.90) can be modified for various device geometries and manufacturing processes, such as LOCOS, fully recessed LOCOS, and thick-field-oxide MOSFET process. In all cases, we recognize that the additional contribution to V_{T0} is proportional to (x_{dm}/W). The amount of threshold voltage increase becomes significant only for devices which have a channel width W on the same order of magnitude as x_{dm}. Finally, note that for minimum-geometry MOSFETs which have a small channel length *and* a small channel width, the threshold voltage variations due to short- and narrow-channel effects may tend to cancel each other out.

Other Limitations Imposed by Small-Device Geometries

In small-geometry MOSFETs, the characteristics of current flow in the channel between the source and the drain can be explained as being controlled by the two-dimensional electric field vector. The simple one-dimensional gradual channel approximation (GCA) assumes that the electric field components parallel to the surface and perpendicular to the surface are effectively decoupled and, therefore, cannot fully account for some of the observed device characteristics. These small-geometry device characteristics, however, may severely restrict the operating conditions of the transistor and impose limitations upon the practical utility of the device. Accurate identification and characterization of these small-geometry effects are crucial, especially for submicron MOSFETs.

One typical condition, which is due to the two-dimensional nature of channel current flow, is the *subthreshold conduction* in small-geometry MOS transistors. As already discussed in the previous sections, the current flow in the channel depends on creating and sustaining an inversion layer on the surface. If the gate bias voltage is not sufficient to invert the surface, i.e., $V_{GS} < V_{T0}$, the carriers (electrons) in the channel face a *potential barrier* that blocks the flow. Increasing the gate voltage reduces this potential barrier and, eventually, allows the flow of carriers under the influence of the channel electric field. This simple picture becomes more complicated in small-geometry MOSFETs, because the potential barrier is controlled by both the gate-to-source voltage V_{GS} and the drain-to-source voltage V_{DS}. If the drain voltage is increased, the potential barrier in the channel decreases, leading to *drain-induced barrier lowering* (DIBL). The reduction of the potential barrier eventually allows electron flow between the source and the drain, even if the gate-to-source voltage is lower than the threshold voltage. The channel current that flows under these conditions ($V_{GS} < V_{T0}$) is called the *subthreshold current*. Note that the GCA cannot account for any nonzero drain current I_D for $V_{GS} < V_{T0}$. Two-dimensional analysis of the small-geometry MOSFET yields the following approximate expression for the subthreshold current.

$$I_D(subthreshold) \cong \frac{q\,D_n W x_c n_0}{L_B} \cdot e^{\frac{q\phi_r}{kT}} \cdot e^{\frac{q}{kT}(A \cdot V_{GS} + B \cdot V_{DS})} \qquad (3.92)$$

Here, x_c is the subthreshold channel depth, D_n is the electron diffusion coefficient, L_B is the length of the barrier region in the channel, and ϕ_r is a reference potential. Note the exponential dependence of the subthreshold current on both the gate and the drain voltages. Identifying subthreshold conduction is very important for circuit applications where small amounts of current flow may significantly disturb the circuit operation.

We remember from the previous analysis that in small-geometry MOSFETs, the channel length is on the same order of magnitude as the source and drain depletion region thicknesses. For large drain-bias voltages, the depletion region surrounding the drain can extend farther toward the source, and the two depletion regions can eventually merge. This condition is termed *punch-through;* the gate voltage loses its control upon the drain current, and the current rises sharply once punch-through occurs. Being able to cause permanent damage to the transistor by localized melting of material, punch-through is obviously an undesirable condition, and should be prevented in normal circuit operation.

As some device dimensions, such as the channel length, are scaled down with each new generation, we find that some dimensions cannot be arbitrarily scaled because of physical limitations. One such dimension is the gate oxide thickness t_{ox}. The

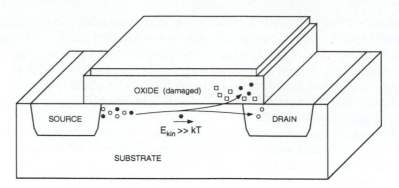

Figure 3.27 Hot-carrier injection into the gate oxide and resulting oxide damage.

reduction of t_{ox} by a scaling factor of S, i.e., building a MOSFET with $t'_{ox} = t_{ox}/S$, is restricted by processing difficulties involved in growing very thin, uniform silicon-dioxide layers. Localized sites of nonuniform oxide growth, also called *pinholes,* may cause electrical shorts between the gate electrode and the substrate. Another limitation on the scaling of t_{ox} is the possibility of *oxide breakdown.* If the oxide electric field perpendicular to the surface is larger than a certain *breakdown field,* the silicon-dioxide layer may sustain permanent damage during operation, leading to device failure.

Finally, we will consider another reliability problem caused by high electric fields within the device. We have seen that advances in VLSI fabrication technologies are primarily based on the reduction of device dimensions, such as the channel length, the junction depth, and the gate oxide thickness, without proportional scaling of the power supply voltage (constant-voltage scaling). This decrease in critical device dimensions to submicron ranges, accompanied by increasing substrate doping densities, results in a significant increase of the horizontal and vertical electric fields in the channel region. Electrons and holes gaining high kinetic energies in the electric field (*hot carriers*) may, however, be injected into the gate oxide, and cause permanent changes in the oxide-interface charge distribution, degrading the current-voltage characteristics of the MOSFET (Fig. 3.27). Since the likelihood of hot-carrier induced degradation increases with shrinking device dimensions, this problem was identified as one of the important factors that may impose strict limitations on maximum achievable device densities in VLSI circuits.

The channel hot-electron (CHE) effect is caused by electrons flowing in the channel region, from the source to the drain. This effect is more pronounced at large drain-to-source voltages, at which the lateral electric field in the drain end of the channel accelerates the electrons. The electrons arriving at the Si-SiO$_2$ interface with enough kinetic energy to surmount the surface potential barrier are injected into the oxide. Electrons and holes generated by impact ionization also contribute to the charge injection. Note that the channel hot-electron current and the subsequent damage in the gate oxide are localized near the drain junction (Fig. 3.27).

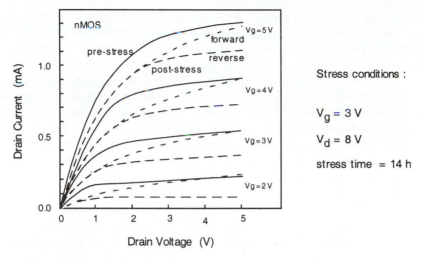

Figure 3.28 Typical drain current vs. drain voltage characteristics of an n-channel MOS transistor before and after hot-carrier induced oxide damage.

The hot-carrier induced damage in nMOS transistors has been found to result in either trapping of carriers on defect sites in the oxide or the creation of interface states at the silicon-oxide interface, or both. The damage caused by hot-carrier injection affects the transistor characteristics by causing a degradation in transconductance, a shift in the threshold voltage, and a general decrease in the drain current capability (Fig. 3.28). This performance degradation in the devices leads to the degradation of circuit performance over time. Hence, new MOSFET technologies based on smaller device dimensions must carefully account for the hot-carrier effects and also ensure reliable long-term operation of the devices.

Other reliability concerns for small-geometry devices include interconnect damage through electromigration, electrostatic discharge (ESD), and electrical overstress (EOS).

3.6 MOSFET Capacitances

The majority of the topics covered in this chapter has been related to the steady-state behavior of the MOS transistor. The current-voltage characteristics investigated here can be applied for investigating the DC response of MOS circuits under various operating conditions. In order to examine the transient (AC) response of MOSFETs and digital circuits consisting of MOSFETs, on the other hand, we have to determine the nature and the amount of parasitic capacitances associated with the MOS transistor.

The on-chip capacitances found in MOS circuits are in general complicated functions of the layout geometries and the manufacturing processes. Most of these capacitances are not lumped, but *distributed,* and their exact calculations would usually require complex, three-dimensional nonlinear charge-voltage models. In the following, we will develop simple approximations for the on-chip MOSFET capacitances that can be used in most hand calculations. These capacitance models are sufficiently accurate to represent the crucial characteristics of MOSFET charge-voltage behavior, and the equations are all based on fundamental semiconductor device theory, which should be familiar to most readers. We will also stress the distinction between the device-related capacitances and the interconnect capacitances. The capacitive contribution of metal interconnections between various devices is a very important component of the total parasitic capacitance observed in digital circuits. The estimation of this interconnect capacitance will be handled in Chapter 6.

Figure 3.29 shows the cross-sectional view and the top view (mask view) of a typical n-channel MOSFET. Until now, we concentrated on the cross-sectional view of the device, since we were primarily concerned with the flow of carriers within the MOSFET. As we study the parasitic device capacitances, we will have to become more familiar with the top view of the MOSFET. In this figure, the *mask length* (drawn length) of the gate is indicated by L_M, and the actual channel length is

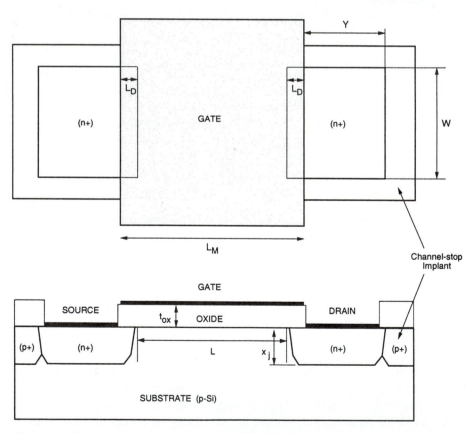

Figure 3.29 Cross-sectional view and the top view (mask view) of a typical n-channel MOSFET.

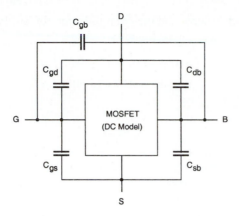

Figure 3.30 Lumped representation of the parasitic MOSFET capacitances.

indicated by L. The extent of both the gate-source and the gate-drain overlap are L_D; thus, the channel length is given by

$$L = L_M - 2 \cdot L_D \tag{3.93}$$

Note that the source and drain overlap region lengths are usually equal to each other because of the symmetry of the MOSFET structure. Typically, L_D is on the order of 0.1 μm. Both the source and the drain diffusion regions have a width of W. The typical diffusion region length is denoted by Y. Note that both the source diffusion region and the drain diffusion region are surrounded by a p^+ doped region, also called the channel-stop implant. As the name indicates, the purpose of this additional p^+ region is to prevent the formation of any unwanted (parasitic) channels between two neighboring n^+ diffusion regions, i.e., to ensure that the surface between two such regions cannot be inverted. Hence, the p^+ channel-stop implants act to electrically isolate neighboring devices built on the same substrate.

We will identify the parasitic capacitances associated with this typical MOSFET structure as lumped equivalent capacitances *observed* between the device terminals (Fig. 3.30), since such a lumped representation can be easily used to analyze the dynamic transient behavior of the device. The reader must always be reminded, however, that in reality most parasitic device capacitances are due to three-dimensional, distributed charge-voltage relations within the device structure. Based on their physical origins, the parasitic device capacitances can be classified into two major groups: *oxide-related capacitances* and *junction capacitances*. First, the oxide-related capacitances will be considered.

Oxide-related Capacitances

It was shown earlier that the gate electrode overlaps both the source region and the drain region at the edges. The two overlap capacitances that arise as a result of this structural arrangement are called $C_{GD}(overlap)$ and $C_{GS}(overlap)$, respectively.

Assuming that both the source and the drain diffusion regions have the same width W, the overlap capacitances can be found as

$$C_{GS}(overlap) = C_{ox} \cdot W \cdot L_D$$
$$C_{GD}(overlap) = C_{ox} \cdot W \cdot L_D \tag{3.94}$$

with

$$C_{ox} = \frac{\varepsilon_{ox}}{t_{ox}} \tag{3.95}$$

Note that both of these overlap capacitances do not depend on the bias conditions; i.e., they are voltage-independent.

Now consider the capacitances which result from the interaction between the gate voltage and the channel charge. Since the channel region is connected to the source, the drain, and the substrate, we can identify three capacitances between the gate and these regions, i.e., C_{gs}, C_{gd}, and C_{gb}, respectively. Notice that in reality, the gate-to-channel capacitance is distributed and voltage-dependent. Then, the gate-to-source capacitance C_{gs} is actually the gate-to-channel capacitance *seen* between the gate and the source terminals; the gate-to-drain capacitance C_{gd} is actually the gate-to-channel capacitance *seen* between the gate and the drain terminals. A simplified view of their bias-dependence can be obtained by observing the conditions in the channel region during cut-off, linear, and saturation modes.

In cut-off mode (Fig. 3.31(a)), the surface is not inverted. Consequently, there is no conducting channel that links the surface to the source and to the drain. Therefore, the gate-to-source and the gate-to-drain capacitances are both equal to zero: $C_{gs} = C_{gd} = 0$. The gate-to-substrate capacitance can be approximated by

$$C_{gb} = C_{ox} \cdot W \cdot L \tag{3.96}$$

In linear-mode operation, the inverted channel extends across the MOSFET, between the source and the drain (Fig. 3.31(b)). This conducting inversion layer on the surface effectively shields the substrate from the gate electric field; thus, $C_{gb} = 0$. In this case, the distributed gate-to-channel capacitance may be viewed as being shared equally between the source and the drain, yielding

$$C_{gs} \cong C_{gd} \cong \frac{1}{2} \cdot C_{ox} \cdot W \cdot L \tag{3.97}$$

When the MOSFET is operating in saturation mode, the inversion layer on the surface does not extend to the drain, but it is pinched off (Fig. 3.31(c)). The gate-to-drain capacitance component is therefore equal to zero ($C_{gd} = 0$). Since the source is still linked to the conducting channel, its shielding effect also forces the gate-to-substrate capacitance to be zero, $C_{gb} = 0$. Finally, the distributed gate-to-channel capacitance as seen between the gate and the source can be approximated by

$$C_{gs} \cong \frac{2}{3} \cdot C_{ox} \cdot W \cdot L \tag{3.98}$$

Table 3.6 lists a summary of the approximate oxide capacitance values in three different operating modes of the MOSFET. The variation of the distributed parasitic

Table 3.6 Approximate oxide capacitance values for three operating modes of the MOS transistor

Capacitance	Cut-off	Linear	Saturation
C_{gb} (total)	$C_{ox}WL$	0	0
C_{gd} (total)	$C_{ox}WL_D$	$\frac{1}{2}C_{ox}WL + C_{ox}WL_D$	$C_{ox}WL_D$
C_{gs} (total)	$C_{ox}WL_D$	$\frac{1}{2}C_{ox}WL + C_{ox}WL_D$	$\frac{2}{3}C_{ox}WL + C_{ox}WL_D$

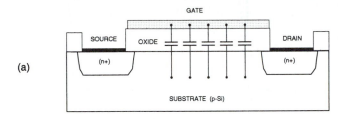

(a)

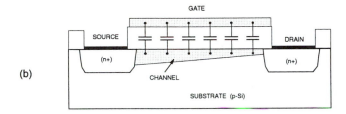

(b)

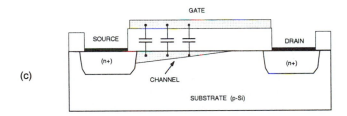

(c)

Figure 3.31 Schematic representation of MOSFET oxide capacitances during (a) cut-off, (b) linear, and (c) saturation modes.

oxide capacitances as functions of the gate-to-source voltage V_{GS} is also shown in Fig. 3.32.

Obviously, we have to combine the distributed C_{gs} and C_{gd} values found here with the relevant overlap capacitance values, in order to calculate the total capacitance between the external device terminals. It is also worth mentioning that the sum of all three voltage-dependent (distributed) gate oxide capacitances ($C_{gb} + C_{gs} + C_{gd}$) has a minimum value of 0.66 $C_{ox}WL$ (in saturation mode) and a maximum value of $C_{ox}WL$

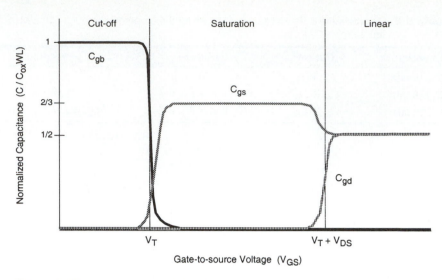

Figure 3.32   Variation of the distributed (gate-to-channel) oxide capacitances as functions of gate-to-source voltage V_{GS}.

(in cut-off and linear modes). For simple hand calculations where all three capacitances can be considered to be connected in parallel, a constant worst-case value of $C_{ox}W(L + 2L_D)$ can be used for the sum of MOSFET gate oxide capacitances.

Junction Capacitances

Now we consider the voltage-dependent source-substrate and drain-substrate junction capacitances, C_{sb} and C_{db}, respectively. Both of these capacitances are due to the depletion charge surrounding the respective source or drain diffusion regions embedded in the substrate. The calculation of the associated junction capacitances is complicated by the three-dimensional shape of the diffusion regions that form the source-substrate and the drain-substrate junctions. Note that both of these junctions are reverse-biased under normal operating conditions of the MOSFET and that the amount of junction capacitance is a function of the applied terminal voltages. Figure 3.33 shows the simplified, partial geometry of a typical n-channel enhancement MOSFET, focusing on the n-type diffusion region within the p-type substrate. The analysis to be carried out in the following will apply to both n-channel and p-channel MOS transistors.

As seen in Fig. 3.33, the n⁺ diffusion region forms a number of planar pn-junctions with the surrounding p-type substrate, indicated here with 1 through 5. The dimensions of the rectangular box representing the diffusion region are given as W, Y, and x_j. Abrupt (step) pn-junction profiles will be assumed for all junctions for simplicity. Also, comparing this three-dimensional view with Fig. 3.29, we recognize that three of the five planar junctions shown here (2, 3, and 4) are actually surrounded by the p⁺ channel-stop implant. The junction labeled (1) is facing the channel, and the bottom junction (5) is facing the p-type substrate, which has a doping density

Table 3.7 Types and areas of the pn-junctions shown in Fig. 3.33

Junction	Area	Type
1	$W \cdot x_j$	n^+/p
2	$Y \cdot x_j$	n^+/p^+
3	$W \cdot x_j$	n^+/p^+
4	$Y \cdot x_j$	n^+/p^+
5	$W \cdot Y$	n^+/p

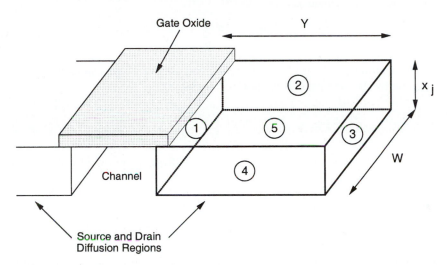

Figure 3.33 Three-dimensional view of the n^+ diffusion region within the p-type substrate.

of N_A. Since the p^+ channel-stop implant density is usually about $10 N_A$, the junction capacitances associated with these *sidewalls* will be different from the other junction capacitances (see Table 3.7). Note that in general, the actual shape of the diffusion regions as well as the doping profiles are much more complicated. However, this simplified analysis provides sufficient insight for the first-order estimation of junction-related capacitances.

To calculate the depletion capacitance of a reverse-biased abrupt pn-junction, consider first the depletion region thickness, x_d. Assuming that the n-type and p-type doping densities are given by N_D and N_A, respectively, and that the reverse bias voltage is given by V (negative), the depletion region thickness can be found as follows:

$$x_d = \sqrt{\frac{2 \cdot \varepsilon_{Si}}{q} \cdot \frac{N_A + N_D}{N_A \cdot N_D} \cdot (\phi_0 - V)} \qquad (3.99)$$

where the built-in junction potential is calculated as

$$\phi_0 = \frac{kT}{q} \cdot \ln \left(\frac{N_A \cdot N_D}{n_i^2} \right) \qquad (3.100)$$

Note that the junction is forward-biased for a *positive* bias voltage V, and reverse-biased for a *negative* bias voltage. The depletion-region charge stored in this area can be written in terms of the depletion region thickness, x_d.

$$Q_j = A \cdot q \cdot \left(\frac{N_A \cdot N_D}{N_A + N_D} \right) \cdot x_d = A \sqrt{ 2 \cdot \varepsilon_{Si} \cdot q \cdot \left(\frac{N_A \cdot N_D}{N_A + N_D} \right) \cdot (\phi_0 - V)}$$

(3.101)

Here, A indicates the junction area. The junction capacitance associated with the depletion region is defined as

$$C_j = \left| \frac{dQ_j}{dV} \right|$$

(3.102)

By differentiating (3.101) with respect to the bias voltage V, we can now obtain the expression for the junction capacitance as follows.

$$C_j(V) = A \cdot \sqrt{ \frac{\varepsilon_{Si} \cdot q}{2} \cdot \left(\frac{N_A \cdot N_D}{N_A + N_D} \right)} \cdot \frac{1}{\sqrt{\phi_0 - V}}$$

(3.103)

This expression can be rewritten in a more general form, to account for the junction grading.

$$C_j(V) = \frac{A \cdot C_{j0}}{\left(1 - \dfrac{V}{\phi_0} \right)^m}$$

(3.104)

The parameter m in (3.104) is called the *grading coefficient*. Its value is equal to $1/2$ for an abrupt junction profile, and $1/3$ for a linearly graded junction profile. Obviously, for an abrupt pn-junction profile, i.e., for $m = 1/2$, the equations (3.103) and (3.104) become identical. The zero-bias junction capacitance per unit area C_{j0} is defined as

$$C_{j0} = \sqrt{ \frac{\varepsilon_{Si} \cdot q}{2} \cdot \left(\frac{N_A \cdot N_D}{N_A + N_D} \right) \cdot \frac{1}{\phi_0}}$$

(3.105)

Note that the value of the junction capacitance C_j given by (3.104) ultimately depends on the external bias voltage that is applied across the pn-junction. Since the terminal voltages of a MOSFET will change during dynamic operation, accurate estimation of the junction capacitances under transient conditions is quite complicated; the instantaneous values of all junction capacitances will also change accordingly. The problem of estimating capacitance values under changing bias conditions can be simplified, if we calculate a large-signal average (linear) junction capacitance instead, which, by definition, is independent of the bias potential. This *equivalent large-signal capacitance* can be defined as follows:

$$C_{eq} = \frac{\Delta Q}{\Delta V} = \frac{Q_j(V_2) - Q_j(V_1)}{V_2 - V_1} = \frac{1}{V_2 - V_1} \cdot \int_{V_1}^{V_2} C_j(V) \, dV$$

(3.106)

Here, the reverse bias voltage across the pn-junction is assumed to change from V_1 to V_2. Hence, the equivalent capacitance C_{eq} is always calculated for a *transition between two known voltage levels.* By substituting (3.104) into (3.106), we obtain

$$C_{eq} = -\frac{A \cdot C_{j0} \cdot \phi_0}{(V_2 - V_1) \cdot (1 - m)} \cdot \left[\left(1 - \frac{V_2}{\phi_0}\right)^{1-m} - \left(1 - \frac{V_1}{\phi_0}\right)^{1-m} \right] \qquad (3.107)$$

For the special case of abrupt pn-junctions, equation (3.107) becomes

$$C_{eq} = -\frac{2 \cdot A \cdot C_{j0} \cdot \phi_0}{(V_2 - V_1)} \cdot \left[\sqrt{1 - \frac{V_2}{\phi_0}} - \sqrt{1 - \frac{V_1}{\phi_0}} \right] \qquad (3.108)$$

This equation can be rewritten in a simpler form by defining a dimensionless coefficient K_{eq}, as follows:

$$C_{eq} = A \cdot C_{j0} \cdot K_{eq} \qquad (3.109)$$

$$K_{eq} = -\frac{2\sqrt{\phi_0}}{V_2 - V_1} \cdot \left(\sqrt{\phi_0 - V_2} - \sqrt{\phi_0 - V_1} \right) \qquad (3.110)$$

where K_{eq} is the *voltage equivalence factor* (note that $0 < K_{eq} < 1$). Thus, the coefficient K_{eq} allows us to take into account the voltage-dependent variations of the junction capacitance. The accuracy of the large-signal equivalent junction capacitance C_{eq} found by using (3.109) and (3.110) is usually sufficient for most first-order hand calculations. Practical applications of the capacitance calculation methods discussed here will be illustrated in the following examples.

EXAMPLE 3.7

Consider a simple abrupt pn-junction, which is reverse-biased with a voltage V_{bias}. The doping density of the n-type region is $N_D = 10^{19}$ cm^{-3}, and the doping density of the p-type region is given as $N_A = 10^{16}$ cm^{-3}. The junction area is $A = 20 \, \mu\text{m} \times 20 \, \mu\text{m}$.

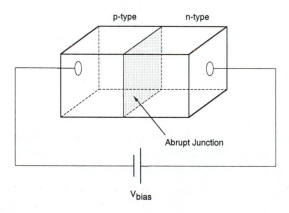

First, we will calculate the zero-bias junction capacitance per unit area, C_{j0}, for this structure. The built-in junction potential is found as

$$\phi_0 = \frac{kT}{q} \cdot \ln\left(\frac{N_A \cdot N_D}{n_i^2}\right) = 0.026 \text{ V} \cdot \ln\left(\frac{10^{16} \cdot 10^{19}}{2.1 \times 10^{20}}\right) = 0.88 \text{ V}$$

Using (3.105), we can calculate the zero-bias junction capacitance:

$$C_{j0} = \sqrt{\frac{\varepsilon_{Si} \cdot q}{2} \cdot \left(\frac{N_A \cdot N_D}{N_A + N_D}\right) \cdot \frac{1}{\phi_0}}$$

$$= \sqrt{\frac{11.7 \cdot 8.85 \times 10^{-14} \text{ F/cm} \cdot 1.6 \times 10^{-19} \text{ C}}{2} \cdot \left(\frac{10^{16} \cdot 10^{19}}{10^{16} + 10^{19}}\right) \cdot \frac{1}{0.88 \text{ V}}}$$

$$= 3.1 \times 10^{-8} \text{ F/cm}^2$$

Next, find the equivalent large-signal junction capacitance assuming that the reverse bias voltage changes from $V_1 = 0$ to $V_2 = -5$ V. The voltage equivalence factor for this transition can be found as follows:

$$K_{eq} = -\frac{2\sqrt{\phi_0}}{V_2 - V_1} \cdot \left(\sqrt{\phi_0 - V_2} - \sqrt{\phi_0 - V_1}\right)$$

$$= -\frac{2\sqrt{0.88}}{-5} \cdot \left(\sqrt{0.88 - (-5)} - \sqrt{0.88}\right) = 0.56$$

Then, the average junction capacitance can be found simply by using (3.109).

$$C_{eq} = A \cdot C_{j0} \cdot K_{eq} = 400 \times 10^{-8} \text{ cm}^2 \cdot 3.1 \times 10^{-8} \text{ F/cm}^2 \cdot 0.56 = 69 \text{ fF}$$

It was shown in Fig. 3.29 and Fig. 3.33 that the sidewalls of a typical MOSFET source or drain diffusion region are surrounded by a p^+ channel-stop implant, with a higher doping density than the substrate doping density N_A. Consequently, the sidewall zero-bias capacitance C_{j0sw}, as well as the sidewall voltage equivalence factor $K_{eq}(sw)$ will be different from those of the bottom junction. Assuming that the sidewall doping density is given by $N_A(sw)$, the zero-bias capacitance per unit area can be found as follows:

$$C_{j0sw} = \sqrt{\frac{\varepsilon_{Si} \cdot q}{2} \cdot \left(\frac{N_A(sw) \cdot N_D}{N_A(sw) + N_D}\right) \cdot \frac{1}{\phi_{0sw}}} \tag{3.111}$$

where ϕ_{0sw} is the built-in potential of the sidewall junctions. Since all sidewalls in a typical diffusion structure have approximately the same depth of x_j, we can define a zero-bias sidewall junction capacitance per unit length.

$$C_{jsw} = C_{j0sw} \cdot x_j \tag{3.112}$$

The sidewall voltage equivalence factor $K_{eq}(sw)$ for a voltage swing between V_1 and V_2 is defined as follows:

$$K_{eq}(sw) = -\frac{2\sqrt{\phi_{0sw}}}{V_2 - V_1} \cdot \left(\sqrt{\phi_{0sw} - V_2} - \sqrt{\phi_{0sw} - V_1}\right) \tag{3.113}$$

Combining the equations (3.111) through (3.113), the equivalent large-signal junction capacitance $C_{eq}(sw)$ for a sidewall of length (perimeter) P can be calculated as

$$C_{eq}(sw) = P \cdot C_{jsw} \cdot K_{eq}(sw) \tag{3.114}$$

EXAMPLE 3.8

Consider the n-channel enhancement-type MOSFET shown below. The process parameters are given as follows:

Substrate doping $\quad\quad N_A = 2 \times 10^{15}$ cm^{-3}
Source/drain doping $\quad N_D = 10^{19}$ cm^{-3}
Sidewall (p$^+$) doping $\quad N_A(sw) = 4 \times 10^{16}$ cm^{-3}
Gate oxide thickness $\quad t_{ox} = 45$ nm
Junction depth $\quad\quad\quad x_j = 1.0\ \mu$m

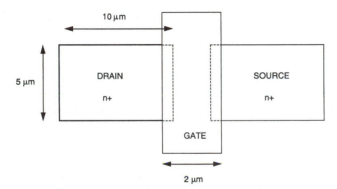

Note that both the source and the drain diffusion regions are surrounded by p$^+$ channel-stop diffusion. The substrate is biased at 0 V. Assuming that the drain voltage is changing from 0.5 V to 5 V, find the average drain-substrate junction capacitance C_{db}.

First, we recognize that three sidewalls of the rectangular drain diffusion structure form n$^+$/p$^+$ junctions with the p$^+$ channel-stop implant, while the bottom area and the sidewall facing the channel form n$^+$/p junctions. Start by calculating the built-in potentials for both types of junctions.

$$\phi_0 = \frac{kT}{q} \cdot \ln\left(\frac{N_A \cdot N_D}{n_i^2}\right) = 0.026\ \text{V} \cdot \ln\left(\frac{2 \times 10^{15} \cdot 10^{19}}{2.1 \times 10^{20}}\right) = 0.837\ \text{V}$$

$$\phi_{0sw} = \frac{kT}{q} \cdot \ln\left(\frac{N_A(sw) \cdot N_D}{n_i^2}\right) = 0.026\ \text{V} \cdot \ln\left(\frac{4 \times 10^{16} \cdot 10^{19}}{2.1 \times 10^{20}}\right) = 0.915\ \text{V}$$

Next, we calculate the zero-bias junction capacitances per unit area:

C_{j0}

$$= \sqrt{\frac{\varepsilon_{Si} \cdot q}{2} \cdot \left(\frac{N_A \cdot N_D}{N_A + N_D}\right) \cdot \frac{1}{\phi_0}}$$

$$= \sqrt{\frac{11.7 \cdot 8.85 \times 10^{-14} \text{ F/cm} \cdot 1.6 \times 10^{-19} \text{ C}}{2} \cdot \left(\frac{2 \times 10^{15} \cdot 10^{19}}{2 \times 10^{15} + 10^{19}}\right) \cdot \frac{1}{0.837 \text{ V}}}$$

$$= 1.41 \times 10^{-8} \text{ F/cm}^2$$

C_{j0sw}

$$= \sqrt{\frac{\varepsilon_{Si} \cdot q}{2} \cdot \left(\frac{N_A(sw) \cdot N_D}{N_A(sw) + N_D}\right) \cdot \frac{1}{\phi_{0sw}}}$$

$$= \sqrt{\frac{11.7 \cdot 8.85 \times 10^{-14} \text{ F/cm} \cdot 1.6 \times 10^{-19} \text{ C}}{2} \cdot \left(\frac{4 \times 10^{16} \cdot 10^{19}}{4 \times 10^{16} + 10^{19}}\right) \cdot \frac{1}{0.915 \text{ V}}}$$

$$= 6.01 \times 10^{-8} \text{ F/cm}^2$$

The zero-bias sidewall junction capacitance per unit length can also be found as follows.

$$C_{jsw} = C_{j0sw} \cdot x_j = 6.01 \times 10^{-8} \text{ F/cm}^2 \cdot 10^{-4} \text{ cm} = 6.01 \text{ pF/cm}$$

In order to take the given drain voltage variation into account, we must now calculate the voltage equivalence factors, K_{eq} and $K_{eq}(sw)$, for both types of junctions. This will allow us to find the average large-signal capacitance values.

$$K_{eq} = -\frac{2\sqrt{0.837}}{-5 - (-0.5)} \cdot \left(\sqrt{0.837 + 5} - \sqrt{0.837 + 0.5}\right) = 0.51$$

$$K_{eq}(sw) = -\frac{2\sqrt{0.915}}{-5 - (-0.5)} \cdot \left(\sqrt{0.915 + 5} - \sqrt{0.915 + 0.5}\right) = 0.53 \cong K_{eq}$$

The total area of the n^+/p junctions is calculated as the sum of the bottom area and the sidewall area facing the channel region.

$$A = (10 \times 5) \text{ } \mu\text{m}^2 + (5 \times 1) \text{ } \mu\text{m}^2 = 55 \text{ } \mu\text{m}^2$$

The total length of the n^+/p^+ junction perimeter, on the other hand, is equal to the sum of three sides of the drain diffusion area. Thus, the combined equivalent (average) drain-substrate junction capacitance can be found as follows:

$$\langle C_{db} \rangle = A \cdot C_{j0} \cdot K_{eq} + P \cdot C_{jsw} \cdot K_{eq}(sw)$$

$$= 55 \times 10^{-8} \text{ cm}^2 \cdot 1.41 \times 10^{-8} \text{ F/cm}^2 \cdot 0.51$$

$$+ 25 \times 10^{-4} \text{ cm} \cdot 6.01 \times 10^{-12} \text{ F/cm} \cdot 0.53 = 11.9 \times 10^{-15} \text{ F} = \underline{\underline{11.9 \text{ fF}}}$$

Exercise Problems

3.1   Consider a MOS system with the following parameters:

$t_{ox} = 200$ Å

$\phi_{GC} = -0.85$ V

$N_A = 2 \cdot 10^{15}$ cm^{-3}

$Q_{ox} = q2 \cdot 10^{11}$ C/cm^2

    a. Determine the threshold voltage V_{T0} under zero bias at room temperature $(T = 300$ K$)$. Note that $\varepsilon_{ox} = 3.97\varepsilon_0$ and $\varepsilon_{si} = 11.7\varepsilon_0$.

    b. Determine the type (p-type or n-type) and amount of channel implant (N_I/cm^2) required to change the threshold voltage to 0.8 V.

3.2   Consider a diffusion area which has the dimensions 10 μm $\times$ 5 μm, and the abrupt junction depth is 0.5 μm. Its n-type impurity doping level is $N_D = 1 \cdot 10^{20}$ cm^{-3} and the surrounding p-type substrate doping level is $N_A = 1 \cdot 10^{16}$ cm^{-3}. Determine the capacitance when the diffusion area is biased at 5 V and the substrate is biased at 0 V. In this problem, assume that there is no channel-stop implant.

3.3   Describe the relationship between the mask channel length, L_{mask}, and the electrical channel length, L. Are they identical? If not, how would you express L in terms of L_{mask} and other parameters?

3.4   How is the device junction temperature affected by the power dissipation of the chip and its package? Can you describe the relationship between the device junction temperature, ambient temperature, chip power dissipation, and packaging quality?

3.5   Describe the three main components of the load capacitance C_{load}, when a logic gate is driving other fan-out gates.

3.6   Consider the layout of an nMOS transistor shown in Fig. P3.6. The process parameters are:

$N_D = 2 \cdot 10^{20}$ cm^{-3}	$X_j = 0.5$ μm	$t_{ox} = 0.05$ μm
$N_A = 1 \cdot 10^{15}$ cm^{-3}	$L_D = 0.5$ μm	$V_{T0} = 0.8$ V

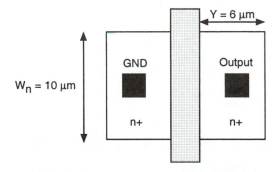

Figure P3.6

Channel stop doping $= 16.0 \times$ (p-type substrate doping)

Find the effective drain parasitic capacitance when the drain node voltage changes from 5 V to 2.5 V.

3.7   A set of *I-V* characteristics for an nMOS transistor at room temperature is shown below for different biasing conditions. Figure P3.7 shows the measurement setup.

Using the data, find: (a) the threshold voltage V_{T0}, (b) electron mobility μ_n, and (c) body effect coefficient gamma (γ).

Some of the parameters are given as: $W/L = 1.0$, $t_{ox} = 345$ Å, $|2\phi_F| = 0.64$ V.

V_{GS} (V)	V_{DS} (V)	V_{SB} (V)	I_D (μA)
4	4	0.0	256
5	5	0.0	441
4	4	2.6	144
5	5	2.6	256

3.8   Compare the two technology scaling methods, namely, (i) the constant electric-field scaling and (ii) the constant power-supply voltage scaling. In particular, show analytically by using equations how the delay time, power dissipation, and power density are affected in terms of the scaling factor, *S*.

To be more specific, what would happen if the design rules change from, say, 1 μm to $1/S$ μm ($S > 1$)?

3.9   A pMOS transistor was fabricated on an n-type substrate with a bulk doping density of $N_D = 10^{16}$ cm^{-3}, gate doping density (n-type poly) of $N_D = 10^{20}$ cm^{-3}, $Q_{ox}/q = 4 \cdot 10^{10}$ cm^{-2}, and gate oxide thickness of $t_{ox} = 0.1$ μm. Calculate the threshold voltage at room temperature for $V_{SB} = 0$. Use $\varepsilon_{si} = 11.7\varepsilon_0$.

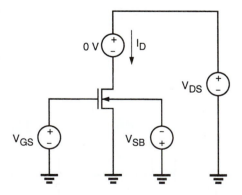

Figure P3.7

3.10 A depletion-type nMOS transistor has the following device parameters:

$$\mu_n = 500 \text{ cm}^2/\text{V} \cdot \text{s}$$
$$t_{ox} = 345 \text{ Å}$$
$$|2\phi_F| = 0.84 \text{ V}$$
$$W/L = 1.0$$

Some laboratory measurement results of the terminal behavior of this device are shown in the table below. Using the data in the table, find the missing value of the gate voltage in the last entry. Show all of the details of your calculation.

I_{DS} (μA)	V_D (V)	V_S (V)	V_B (V)	V_G (V)
50.0	3	0	0	0
40.0	5	3 V	0	?

3.11 Using the parameters given below, calculate the current through two nMOS transistors in series (see Figure P3.11), when the drain of the top transistor is tied to V_{DD}, the source of the bottom transistor is tied to $V_{SS} = 0$ and their gates are tied to V_{DD}. The substrate is also tied to $V_{SS} = 0$ V. Assume that $W/L = 10$ for both transistors.

$$k' = 25 \ \mu\text{A/V}^2$$
$$V_{T0} = 1.0 \text{ V}$$
$$\gamma = 0.39 \text{ V}^{1/2}$$
$$|2\phi_F| = 0.6 \text{ V}$$

Hint: The solution requires several iterations, and the body effect on threshold voltage has to be taken into account. Start with the KCL equation.

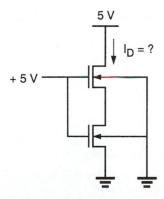

Figure P3.11

3.12 The following parameters are given for an nMOS process:

$t_{ox} = 500 \text{ Å}$

substrate doping $N_A = 1 \cdot 10^{16} \text{ cm}^{-3}$

polysilicon gate doping $N_D = 1 \cdot 10^{20} \text{ cm}^{-3}$

oxide-interface fixed-charge density $N_{ox} = 2 \cdot 10^{10} \text{ cm}^{-3}$

a. Calculate V_T for an unimplanted transistor.

b. What type and what concentration of impurities must be implanted to achieve $V_T = +2 \text{ V}$ and $V_T = -2 \text{ V}$?

3.13 Using the measured data given below, determine the device parameters V_{T0}, k, γ, and λ, assuming $2\phi_F = -0.6 \text{ V}$.

V_{GS} (V)	V_{DS} (V)	V_{BS} (V)	I_D (μA)
2	5	0	10
5	5	0	400
5	5	−3	280
5	8	0	480

3.14 Using the design rules specified in Chapter 2, sketch a simple layout of an nMOS transistor on grid paper. Use a minimum feature size of 3 μm. Neglect the substrate connection. After you complete the layout, calculate approximate values for C_g, C_{sb}, and C_{db}. The following parameters are given.

Substrate doping $N_A = 10^{16} \text{ cm}^{-3}$ $t_{ox} = 0.05 \ \mu$m

Drain/source doping $N_D = 10^{19} \text{ cm}^{-3}$ Junction depth $= 0.5 \ \mu$m

$W = 15 \ \mu$m   Sidewall doping $= 10^7 \text{ cm}^{-3}$

$L = 3 \ \mu$m   drain bias $= 0 \text{ V}$

3.15 Derive the current equation for a p-channel MOS transistor operating in the linear region, i.e., for $V_{SG} + V_{TP} > V_{SD}$.

3.16 An enhancement-type nMOS transistor has the following parameters:

$V_{T0} = 0.8 \text{ V}$

$\gamma = 0.2 \text{ V}^{1/2}$

$\lambda = 0.05 \text{ V}^{-1}$

$|2\phi_F| = 0.58 \text{ V}$

$k' = 20 \ \mu\text{A/V}^2$

a. When the transistor is biased with $V_G = 2.8 \text{ V}$, $V_D = 5 \text{ V}$, $V_S = 1 \text{ V}$, and $V_B = 0 \text{ V}$, the drain current is $I_D = 0.24 \text{ mA}$. Determine W/L.

b. Calculate I_D for $V_G = 5 \text{ V}$, $V_D = 4 \text{ V}$, $V_S = 2 \text{ V}$, and $V_B = 0 \text{ V}$.

c. If $\mu_n = 500 \text{ cm}^2/\text{V·s}$ and $C_g = C_{ox} \cdot W \cdot L = 1.0 \times 10^{-15} \text{ F}$, find W and L.

3.17 An nMOS transistor is fabricated with the following physical parameters:

$N_D = 10^{20} \text{ cm}^{-3}$

$N_A(\text{substrate}) = 10^{16} \text{ cm}^{-3}$

$N_A^+(\text{chan. stop}) = 10^{19} \text{ cm}^{-3}$

$W = 10 \ \mu\text{m}$

$Y = 5 \ \mu\text{m}$

$L = 1.5 \ \mu\text{m}$

$L_D = 0.25 \ \mu\text{m}$

$X_j = 0.4 \ \mu\text{m}$

a. Determine the drain diffusion capacitance for $V_{DB} = 5$ V and 2.5 V.

b. Calculate the overlap capacitance between gate and drain for an oxide thickness of $t_{ox} = 200$ Å.

4 CHAPTER

Modeling of MOS Transistors Using SPICE

SPICE (Simulation Program with Integrated Circuit Emphasis) is a general-purpose circuit simulator which is used very widely both in the microelectronics industry and in educational institutions as an essential computer-aided design (CAD) tool for circuit design. After over three decades of running on various platforms around the world, it can be regarded as the de facto standard in circuit simulation. Engineers and circuit designers using SPICE know how critical the input *models* for transistors are to obtain simulation outputs matching the experimental data. In the fast-advancing field of VLSI design, a sound knowledge of physical models used to describe the behavior of transistors and knowledge of various device parameters are essential for performing detailed circuit simulations and for optimizing design. This chapter will describe the physical aspects of various MOSFET models used in SPICE, and discuss the model equations as well as the model parameters. Also, practical comparisons among the different MOSFET models available in SPICE will be offered, which may help the users to select the most appropriate device model for a given simulation task. It will be assumed that the reader already has a working knowledge of SPICE, the structure of circuit input files, and the use of .MODEL description statements.

4.1   Introduction

The SPICE software that was distributed by UC Berkeley beginning in late 1970s had three built-in MOSFET models: LEVEL 1 (MOS1) is described by a square-law current-voltage characteristic, LEVEL 2 (MOS2) is a detailed analytical MOSFET model, and LEVEL 3 (MOS3) is a semi-empirical model. Both MOS2 and MOS3 include second-order effects such as the short-channel threshold voltage, subthreshold conduction, scattering-limited velocity saturation, and charge-controlled capacitances. More recently, the BSIM3 (Berkeley Short-Channel IGFET Model) version has also been added to the list of available models in the public release to allow more accurate characterization of sub-micron MOSFET characteristics. Commercial releases of SPICE (such as PSPICE and HSPICE) typically contain a much larger selection of refined device models. The level (type) of the MOSFET model to be used in a particular simulation task is declared on the .MODEL statement. In addition, the user can describe a large number of model parameters in this statement. Information about the geometry of a particular device, such as the channel length, channel width, and source and drain areas, is usually given on the element description line of that device. Some typical MOSFET element description lines and .MODEL statements are given below.

```
M1   3   1   0   0   NMOD   L=1U   W=10U   AD=120P   PD=42U

MDEV32   14   9   12   5   PMOD   L=1.2U   W=20U

.MODEL   NMOD   NMOS   (LEVEL=1   VTO=1.4   KP=4.5E-5   CBD=5PF   CBS=2PF)

.MODEL   PMOD   PMOS   (VTO=-2   KP=3.0E-5   LAMBDA=0.02   GAMMA=0.4
+                      CBD=4PF   CBS=2PF   RD=5   RS=3   CGDO=1PF
+                      CGSO=1PF   CGBO=1PF)
```

4.2 Basic Concepts

In the following, we will examine the various model equations and the model parameters associated with the built-in MOSFET models. We will also discuss the usual (normal) range for the parameter values and the default values of the model parameters which are already incorporated into SPICE. A list of all MOSFET model parameters is given in Table 4.1.

The equivalent circuit structure of the NMOS LEVEL 1 model, which is the default MOSFET model in SPICE, is shown in Fig. 4.1. This basic structure is also typical for the LEVEL 2 and LEVEL 3 models. Note that the voltage-controlled current source I_D determines the steady-state current-voltage behavior of the device, while the voltage-controlled (nonlinear) capacitors connected between the terminals represent the parasitic oxide-related and junction capacitances. The source-substrate and the drain-substrate junctions, which are reverse-biased under normal operating conditions, are represented by ideal diodes in this equivalent circuit. Finally, the parasitic source and drain resistances are represented by the resistors R_D and R_S, respectively, connected between the drain current source and the respective terminals.

The basic geometry of an MOS transistor can be described by specifying the nominal channel (gate) length L and the channel width W, both of which are indicated on the element description line. The channel width W is, by definition, the *width* of the area covered by the thin gate oxide. Note that the effective channel length L_{eff} is defined as the distance on the surface between the two (source and drain) diffusion regions. Thus, in order to find the effective channel length, the gate-source overlap distance and the gate-drain overlap distance must be subtracted from the nominal (mask) gate length specified on the device description line. The amount of gate overlap over the source and the drain can be specified by using the *lateral diffusion* coefficient L_D in SPICE. For modeling p-channel MOS transistors, the

Table 4.1 Model Parameters for the MOST

Symbol	SPICE keyword	LEVEL	Parameter description	Default value	Typical value	Units
Parameters of the MOST						
V_{T0}	VTO	1–3	Zero-bias threshold voltage	1.0	1.0	V
KP	KP	1–3	Transconductance parameter	2×10^{-5}	3×10^{-5}	A/V^2
γ	GAMMA	1–3	Body-effect parameter	0.0	0.35	V$^{1/2}$
$2\phi_F$	PHI	1–3	Surface inversion potential	0.6	0.65	V
λ	LAMBDA	1, 2	Channel-length modulation	0.0	0.02	V^{-1}
t_{ox}	TOX	1–3	Thin oxide thickness	1×10^{-7}	1×10^{-7}	m
N_b	NSUB	1–3	Substrate doping	0.0	1×10^{15}	cm^{-3}
N_{SS}	NSS	2, 3	Surface state density	0.0	1×10^{10}	cm^{-2}
N_{FS}	NFS	2, 3	Surface-fast state density	0.0	1×10^{10}	cm^{-2}
N_{eff}	NEFF	2	Total channel charge coefficient	1	5	
X_j	XJ	2, 3	Metallurgical junction depth	0.0	1×10^{-6}	m
X_{jl}	LD	1–3	Lateral diffusion	0.0	0.8×10^{-6}	m
T_{PG}	TPG	2, 3	Type of gate material	1	1	
μ_0	UO	1–3	Surface mobility	600	700	cm^2/(V·s)
U_c	UCRIT	2	Critical electric field for mobility	1×10^4	1×10^4	V/cm
U_e	UEXP	2	Exponential coefficient for mobility	0.0	0.1	
U_t	UTRA	2	Transverse field coefficient	0.0	0.5	
v_{max}	VMAX	2, 3	Maximum drift velocity of carriers	0.0	5×10^4	m/sec
	XQC	2, 3	Coefficient of channel charge share	0.0	0.4	
δ	DELTA	2, 3	Width effect on threshold voltage	0.0	1.0	
η	ETA	3	Static feedback on threshold voltage	0.0	1.0	
θ	THETA	3	Mobility modulation	0.0	0.05	V^{-1}
A_F	AF	1–3	Flicker-noise exponent	1.0	1.2	
K_F	KF	1–3	Flicker-noise coefficient	0.0	1×10^{-26}	
Parameters of parasitic effects						
I_s	IS	1–3	Bulk-junction saturation current	1×10^{-14}	1×10^{-15}	A
J_s	JS	1–3	Bulk-junction saturation current per square meter	0.0	1×10^{-8}	A
ϕ_j	PB	1–3	Bulk-junction potential	0.80	0.75	V
C_j	CJ	1–3	Zero-bias bulk capacitance per square meter	0.0	2×10^{-4}	F/m^2
M_j	MJ	1–3	Bulk-junction grading coefficient	0.5	0.5	
$C_j sw$	CJSW	1–3	Zero-bias perimeter capacitance per meter	0.0	1×10^{-9}	F/m
$M_j sw$	MJSW	1–3	Perimeter capacitance grading coefficient	0.33	0.33	
	FC	1–3	Bulk-junction forward-bias coefficient	0.5	0.5	
C_{GBO}	CGBO	1–3	Gate-bulk overlap capacitance per meter	0.0	2×10^{-10}	F/m
C_{GDO}	CGDO	1–3	Gate-drain overlap capacitance per meter	0.0	4×10^{-11}	F/m
C_{GSO}	CGSO	1–3	Gate-source overlap capacitance per meter	0.0	4×10^{-11}	F/m
R_D	RD	1–3	Drain ohmic resistance	0.0	10.	Ω
R_S	RS	1–3	Source ohmic resistance	0.0	10.	Ω
R_{sh}	RSH	1–3	Source and drain sheet resistance	0.0	30.	Ω

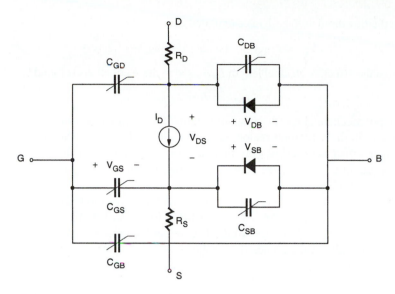

Figure 4.1 Equivalent circuit structure of the LEVEL 1 MOSFET model in SPICE.

direction of the dependent current source, the polarities of the terminal voltages, and the directions of the two diodes representing the source-substrate and the drain-substrate junctions must be reversed. Otherwise, the equations to be presented in the following sections apply to p-channel MOSFETs as well.

4.3 The LEVEL 1 Model Equations

The LEVEL 1 model is the simplest current-voltage description of the MOSFET, which is basically the GCA-based quadratic model originally conceived by Sah in the early 1960s and later developed by Shichman and Hodges. The equations used for the LEVEL 1 n-channel MOSFET model in SPICE are as follows.

Linear Region

$$I_D = \frac{k'}{2} \cdot \frac{W}{L_{eff}} \cdot \left[2 \cdot (V_{GS} - V_T) V_{DS} - V_{DS}^2 \right] \cdot (1 + \lambda V_{DS}) \quad \text{for} \quad V_{GS} \geq V_T$$

$$\text{and} \quad V_{DS} < V_{GS} - V_T$$

$$(4.1)$$

Saturation Region

$$I_D = \frac{k'}{2} \cdot \frac{W}{L_{eff}} \cdot (V_{GS} - V_T)^2 \cdot (1 + \lambda \cdot V_{DS}) \quad \text{for} \quad V_{GS} \geq V_T$$

$$\text{and} \quad V_{DS} \geq V_{GS} - V_T \quad (4.2)$$

where the threshold voltage V_T is calculated as

$$V_T = V_{T0} + \gamma \cdot \left(\sqrt{|2\phi_F| + V_{SB}} - \sqrt{|2\phi_F|}\right) \tag{4.3}$$

Note that the effective channel length L_{eff} used in these equations is found as follows:

$$L_{eff} = L - 2 \cdot L_D \tag{4.4}$$

The empirical channel length modulation term $(1 + \lambda V_{DS})$ appears both in the linear region and in the saturation region equations, although the physical channel length shortening effect is observed only in saturation. This term is included in the linear region equations to ensure the continuity of the first-order derivatives at the linear-saturation region boundary.

Five electrical parameters completely characterize this model: k', V_{T0}, γ, $|2\phi_F|$, and λ. These parameters (KP, VTO, GAMMA, PHI, and LAMBDA, respectively) can be specified directly in the .MODEL statement, or some of them can be calculated from the physical parameters, as follows.

$$k' = \mu \cdot C_{ox}, \qquad \text{where} \quad C_{ox} = \frac{\varepsilon_{ox}}{t_{ox}} \tag{4.5}$$

$$\gamma = \frac{\sqrt{2 \cdot \varepsilon_{Si} \cdot q \cdot N_A}}{C_{ox}} \tag{4.6}$$

$$2\phi_F = 2\frac{kT}{q} \cdot \ln\left(\frac{n_i}{N_A}\right) \tag{4.7}$$

Thus, it is also possible to specify the physical parameters μ, t_{ox}, and N_A in the .MODEL statement instead of the electrical parameters, or a combination of both types of parameters. If a conflict occurs (for example, if both the electron mobility μ and the electrical parameter k' are specified in the .MODEL statement), the given value of the electrical parameter (in this case k') overrides the physical parameter. Typical variations of the drain current with the significant electrical model parameters KP, VTO, GAMMA, and LAMBDA are shown in Figs. 4.2 through 4.6. Nominal parameter values used in the simulations are:

$k' = 27.6\ \mu\text{A/V}^2$	KP = 27.6U
$V_{T0} = 1.0\ \text{V}$	VTO = 1
$\gamma = 0.53\ \text{V}^{1/2}$	GAMMA = 0.53
$2\phi_F = -0.58$	PHI = 0.58
$\lambda = 0$	LAMBDA = 0

The corresponding physical parameter values (which may be overridden by the electrical parameters listed above) are as follows:

$\mu_n = 800\ \text{cm}^2/\text{V·s}$	UO = 800
$t_{ox} = 100\ \text{nm}$	TOX = 100E-9
$N_A = 10^{15}\ \text{cm}^{-3}$	NSUB = 1E15
$L_D = 0.8\ \mu\text{m}$	LD = 0.8E-6

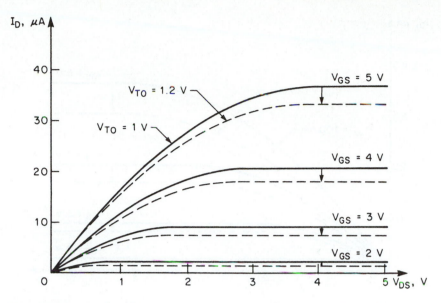

Figure 4.2 Variation of the drain current with model parameter VTO, for the LEVEL 1 model.
(Copyright © 1988 by McGraw-Hill, Inc.)

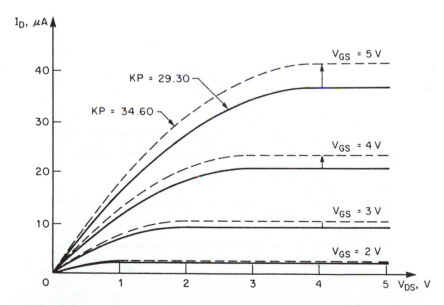

Figure 4.3 Variation of the drain current with model parameter KP, for the LEVEL 1 model.
(Copyright © 1988 by McGraw-Hill, Inc.)

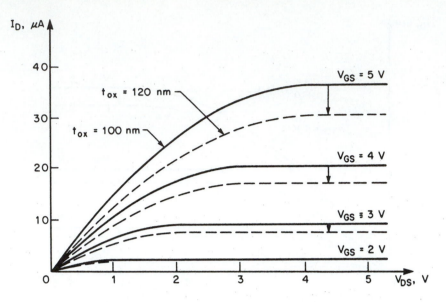

Figure 4.4   Variation of the drain current with model parameter TOX, for the
LEVEL 1 model.
(Copyright © 1988 by McGraw-Hill, Inc.)

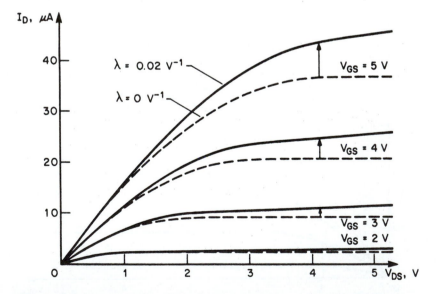

Figure 4.5   Variation of the drain current with parameter LAMBDA, for the
LEVEL 1 model.
(Copyright © 1988 by McGraw-Hill, Inc.)

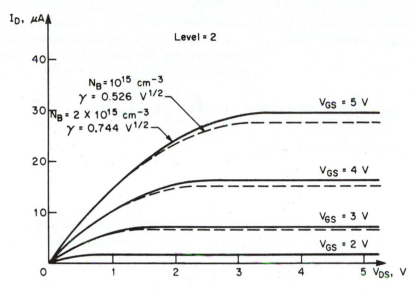

Figure 4.6 Variation of the drain current with parameter GAMMA, for the LEVEL 2 model.
(Copyright © 1988 by McGraw-Hill, Inc.)

In summary, for simple simulation problems the LEVEL 1 model offers a useful estimate of the circuit performance without using a large number of device model parameters. In the following two sections, we present basic equations for LEVEL 2 and LEVEL 3 models, although they are not identical to the SPICE-implemented versions.

4.4 The LEVEL 2 Model Equations

To obtain a more accurate model for the drain current, it is necessary to eliminate some of the simplifying assumptions made in the original GCA analysis. Specifically, the bulk depletion charge must be calculated by taking into account its dependence on the channel voltage. Solving the drain current equation using the voltage-dependent bulk charge term, the following current-voltage characteristics can be obtained:

$$
I_D = \frac{k'}{(1 - \lambda \cdot V_{DS})} \cdot \frac{W}{L_{eff}} \cdot \left\{ \left(V_{GS} - V_{FB} - |2\phi_F| - \frac{V_{DS}}{2} \right) \cdot V_{DS} \right.
$$
$$
\left. - \frac{2}{3} \cdot \gamma \cdot \left[(V_{DS} - V_{BS} + |2\phi_F|)^{3/2} - (-V_{BS} + |2\phi_F|)^{3/2} \right] \right\} \quad (4.8)
$$

Here, V_{FB} denotes the *flat-band* voltage of the MOSFET. Note that the corrective term for channel length modulation is present in the denominator in this current equation. The model equation (4.8) also includes a variation of the drain current with the parameter γ, even if the substrate-to-source voltage V_{BS} is equal to zero. The saturation condition is reached when the channel (inversion) charge at the drain end becomes equal to zero. From this definition, the saturation voltage V_{DSAT} can be calculated as

$$V_{DSAT} = V_{GS} - V_{FB} - |2\phi_F| + \gamma^2 \cdot \left(1 - \sqrt{1 + \frac{2}{\gamma^2} \cdot (V_{GS} - V_{FB})} \right) \qquad (4.9)$$

The saturation mode current is

$$I_D = I_{Dsat} \cdot \frac{1}{(1 - \lambda \cdot V_{DS})} \qquad (4.10)$$

where I_{Dsat} is calculated from (4.9) using $V_{DS} = V_{DSAT}$. The zero-bias threshold voltage V_{T0} corresponding to the LEVEL 2 model can be calculated from (4.8) as follows:

$$V_{T0} = \Phi_{GC} - \frac{q \cdot N_{SS}}{C_{ox}} + |2\phi_F| + \gamma \cdot \sqrt{|2\phi_F|} \qquad (4.11)$$

where Φ_{GC} denotes the gate-to-channel work function difference and N_{SS} denotes the fixed surface (interface) charge density. The LEVEL 2 model yields more accurate results than the simple LEVEL 1 model, but its accuracy is still not sufficient to achieve good agreement with experimental data, especially for short- and narrow-channel MOSFETs. Consequently, a number of semi-empirical corrections have been added to the basic equations, in order to improve their precision. Some of these enhancements will be discussed in the following.

Variation of Mobility with Electric Field

In the current equations presented above, the surface carrier mobility has been assumed constant, and its variation with applied terminal voltages has been neglected. This approximation simplifies the calculation of the drain current integral, but in reality, the surface mobility *decreases* with the increasing gate voltage. In order to simulate this mobility variation, which is primarily due to the scattering of carriers in the channel, the parameter k' is modified as follows:

$$k'(new) = k' \cdot \left(\frac{\varepsilon_{Si}}{\varepsilon_{ox}} \cdot \frac{t_{oc} \cdot U_c}{(V_{GS} - V_T - U_t \cdot V_{DS})} \right)^{U_e} \qquad (4.12)$$

Here, the parameter U_c represents the gate-to-channel critical field, the parameter U_t represents the contribution of the drain voltage to the gate-to-channel field, and U_e is an exponential fitting parameter. U_t is usually chosen between 0 and 0.5. This formula allows a good agreement between SPICE simulation results and experimental data, for long-channel MOSFETs.

Variation of Channel Length in Saturation Mode

Both the LEVEL 1 and the LEVEL 2 model equations use the empirical parameter λ to account for the channel length modulation effect in the saturation region. The LEVEL 2 model also offers the possibility of using a physical expression for calculating the channel length in saturation mode:

$$L'_{eff} = L_{eff} - \Delta L \tag{4.13}$$

where

$$\Delta L = \sqrt{\frac{2 \cdot \varepsilon_{Si}}{q \cdot N_A}} \cdot \left[\frac{V_{DS} - V_{DSAT}}{4} + \sqrt{1 + \left(\frac{V_{DS} - V_{DSAT}}{4} \right)^2} \right] \tag{4.14}$$

Thus, if the empirical channel length shortening (modulation) coefficient λ is not specified in the .MODEL statement, its value can be found as

$$\lambda = \frac{\Delta L}{L_{eff} \cdot V_{DS}} \tag{4.15}$$

The slope of the I_D-V_{DS} curve in saturation can be adjusted and fitted to experimental data by changing the substrate doping parameter N_A. In this case, however, other N_A-dependent electrical parameters such as $2\phi_F$ and γ must be specified separately in the .MODEL statement, since N_A is being used as a fitting parameter for the saturation mode slope.

Saturation of Carrier Velocity

The calculation of the saturation voltage V_{DSAT} in (4.9) is based on the assumption that the channel charge near the drain becomes equal to zero when the device enters saturation. This hypothesis is actually incorrect, since a minimum charge concentration greater than zero must exist in the channel, due to the carriers that sustain the saturation current. This minimum concentration depends on the speed of the carriers. Moreover, carriers in the channel usually reach the maximum speed limit; i.e., their velocity saturates, before the channel charge approaches zero. This maximum carrier speed in the channel is denoted as v_{max}. The inversion layer charge at the channel-end (for $V_{DS} = V_{DSAT}$) is found as

$$Q_{inv} = \frac{I_{Dsat}}{W \cdot v_{max}} \tag{4.16}$$

The value of the saturation voltage V_{DSAT} can also be found from this expression. If the parameter is specified in the .MODEL statement, the amount of channel length shortening (ΔL) in saturation is calculated by the following formula, instead of (4.14).

$$\Delta L = X_D \cdot \sqrt{\left(\frac{X_D \cdot v_{max}}{2 \cdot \mu} \right)^2 + V_{DS} - V_{DSAT}} - \frac{X_D^2 \cdot v_{max}}{2 \cdot \mu} \tag{4.17}$$

where

$$X_D = \sqrt{\frac{2 \cdot \varepsilon_{Si}}{q \cdot N_A \cdot N_{eff}}} \tag{4.18}$$

Here, the parameter N_{eff} is used as a fitting parameter. This model provides good agreement with experimental data for MOSFETs with long channel lengths. On the other hand, its implementation is complicated because of a discontinuity of the first derivative at the saturation region boundary. This difficulty may sometimes cause convergence problems in the Newton-Raphson algorithm for nonlinear equation solver.

Subthreshold Conduction

The basic model implemented in SPICE calculates the drift current in the channel when the surface potential is equal to or larger than $2\phi_F$, i.e., in strong surface inversion. In reality, as already explained in Chapter 3, a significant concentration of electrons exists near the surface for $V_{GS} < V_T$; therefore, there is a channel current even when the surface is not in strong inversion. This *subthreshold current* is due mainly to diffusion between the source and the channel is becoming an increasing concern for deep-sub-micron designs. The model implemented in SPICE introduces an exponential, semi-empirical dependence of the drain current on V_{GS} in the *weak inversion region*. A voltage V_{on} is defined as the boundary between the regions of weak and strong inversion (Fig. 4.7).

$$I_D(weak\ inversion) = I_{on} \cdot e^{(V_{GS}-V_{on}) \cdot \left(\frac{q}{nkT}\right)} \tag{4.19}$$

Here, I_{on} is the current in strong inversion for $V_{GS} = V_{on}$, and the voltage V_{on} is found as

$$V_{on} = V_T + \frac{nkT}{q} \tag{4.20}$$

where

$$n = 1 + \frac{q \cdot N_{FS}}{C_{ox}} + \frac{C_d}{C_{ox}} \tag{4.21}$$

The parameter N_{FS} is defined as the number of fast superficial states, and is used as a fitting parameter that determines the slope of the subthreshold current-voltage characteristics. C_d is the capacitance associated with the depletion region. It is clear that the model introduces a discontinuity for $V_{GS} = V_{on}$; therefore, the simulation of the transition region between weak and strong inversion is not very precise.

Other Small-Geometry Corrections

The current-voltage equations (4.8) through (4.10) have been obtained from a theory that does not take into account two-dimensional and small-geometry effects.

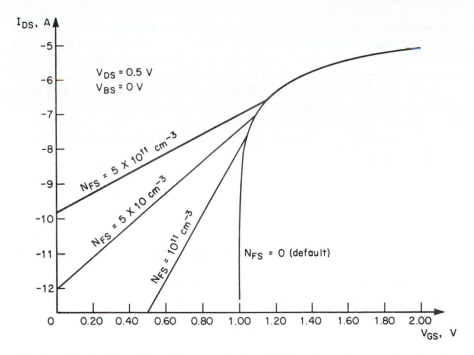

Figure 4.7 Variation of the drain current in the weak inversion region, as a function of the gate voltage and for different values of the parameter N_{FS}, in the LEVEL 2 model. *(Copyright © 1988 by McGraw-Hill, Inc.)*

Therefore, these equations do not include a link between the threshold voltage and the channel dimensions W and L. As already discussed in Chapter 3, however, the threshold voltage is a function of both device dimensions in small-geometry MOSFETs.

The model used in SPICE (LEVEL 2) incorporates essentially the same equations presented in Chapter 3 for short- and narrow-channel effects. The model parameters x_j and N_A can be used as fitting parameters for the short-channel effect, but it is difficult to obtain satisfactory results over a large range of channel lengths. Moreover, this model does not adequately explain the dependence of V_T on the drain voltage. For narrow-channel effects, the empirical parameter δ is used for fitting the experimental data. The calculation of narrow-channel threshold voltage variations can be disabled by specifying $\delta = 0$ in the .MODEL statement.

4.5 The LEVEL 3 Model Equations

The LEVEL 3 model has been developed for simulation of short-channel MOS transistors; it can represent the characteristics of MOSFETs quite precisely for channel lengths down to 2 μm. The current-voltage equations are formulated in the same way as for the LEVEL 2 model. However, the current equation in the linear region has been simplified with a Taylor series expansion of (4.8). This approximation allows the development of more manageable basic current equations compared to those for the LEVEL 2 model. The short-channel and other small-geometry effects are introduced in the threshold voltage and mobility calculations.

The majority of the LEVEL 3 model equations are empirical. The aims of using empirical equations instead of analytical models are both to improve the accuracy of the model and to limit the complexity of the calculations and, hence, the amount of required simulation time. The drain current in the linear region is given by the following equation:

$$I_D = \mu_s \cdot C_{ox} \cdot \frac{W}{L_{eff}} \cdot \left(V_{GS} - V_T - \frac{1 + F_B}{2} \cdot V_{DS} \right) \cdot V_{DS} \qquad (4.22)$$

where

$$F_B = \frac{\gamma \cdot F_s}{4 \cdot \sqrt{|2\phi_F| + V_{SB}}} + F_n \qquad (4.23)$$

The empirical parameter F_B expresses the dependence of the bulk depletion charge on the three-dimensional geometry of the MOSFET. Here, the parameters V_T, F_s, and μ_s are influenced by the short-channel effects, while the parameter F_n is influenced by the narrow-channel effects. The dependence of the surface mobility on the gate electric field is simulated as follows:

$$\mu_s = \frac{\mu}{1 + \theta \cdot (V_{GS} - V_T)} \qquad (4.24)$$

The LEVEL 3 model also includes a simple equation that accounts for the decrease in the effective mobility with the average lateral electric field:

$$\mu_{eff} = \frac{\mu_s}{1 + \mu_s \cdot \dfrac{V_{DS}}{v_{max} \cdot L_{eff}}} \qquad (4.25)$$

where μ_s is the surface mobility calculated from (4.24). The model for the weak inversion region is the same as that for the LEVEL 2 model.

4.6   State-of-the-Art MOSFET Models

BSIM—Berkeley Short-Channel IGFET Model

The recently developed LEVEL 4 (Berkeley Short-Channel IGFET Model, or BSIM) model is analytically simple and is based on a small number of parameters, which are normally extracted from experimental data. Its accuracy and efficiency make it one of the most popular SPICE MOSFET models at present, especially in the microelectronics industry. Currently, the BSIM3 version is widely used by many companies and silicon foundries to accurately model the electrical behavior of sub-micron MOSFETs that are manufactured with various sub-micron CMOS fabrication processes.

EKV (Enz-Krummenacher-Vittoz) Transistor Model

All of the previously examined MOSFET models consider the strong-inversion region of operation separately from the weak-inversion region, which is primarily responsible for the sub-threshold behavior of the transistor. The fact that these two regions are modeled with disjoint sets of equations, however, causes serious problems in the modeling of transistors operating at very low voltages as in many cases involving deep sub-micron CMOS technologies. A novel MOSFET model proposed by Enz, Krummenacher, and Vittoz (EKV Model) attempts to resolve this problem by using a unified view of the transistor operating regions, and by avoiding the use of disjoint equations in strong and weak inversion.

4.7 Capacitance Models

The SPICE MOSFET models account for the parasitic device capacitances by using separate sets of equations in cut-off, linear mode, and saturation mode. Oxide capacitances and depletion capacitances are calculated as nonlinear functions of the bias voltages, using the basic parasitic capacitance information (such as the zero-bias capacitance values) and the device geometry (such as the junction areas and perimeters) supplied by the user.

Gate Oxide Capacitances

SPICE uses a simple gate oxide capacitance model that represents the charge storage effect by three nonlinear two-terminal capacitors: C_{GB}, C_{GS}, and C_{GD}. The voltage dependence of these capacitances (cf. Meyer's capacitance model) is quite similar to that shown in Fig. 3.32 in Chapter 3. The geometry information required for the calculation of gate oxide capacitances are: gate oxide thickness TOX, channel width W, channel length L, and the lateral diffusion LD. This information is to be provided by the user in the respective device description line. The capacitances CGBO, CGSO, and CGDO, which are specified in the .MODEL statement, are the overlap capacitances between the gate and the other terminals *outside* the channel region.

If the parameter XQC is specified in the .MODEL statement, then SPICE uses a simplified version of the charge-controlled capacitance model proposed by Ward, instead of Meyer's model. Ward's model analytically calculates the charge in the gate and in the substrate. While this model avoids errors caused in the simulation in which some nodes in the network cannot change their charge, it may occasionally lead to convergence problems during simulation. The variation of oxide capacitances as functions of the gate voltage according to Ward's model are shown in Fig. 4.8.

Junction Capacitances

The parasitic capacitances of the source and drain diffusion regions are simulated in SPICE with the simple pn-junction model. Since both the source and the drain diffusion regions are surrounded by a p^+ doped sidewall (channel-stop implant), two

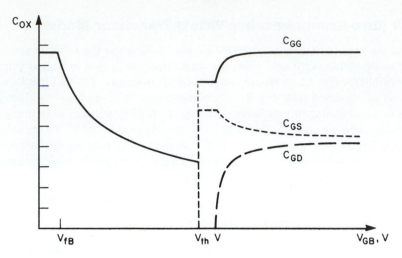

Figure 4.8   Oxide capacitances as functions of the gate-to-substrate voltage, according to Ward's capacitance model.
(Copyright © 1988 by McGraw-Hill, Inc.)

separate models are used for the bottom area depletion capacitance and for the side-wall area depletion capacitance.

$$C_{SB} = \frac{C_j \cdot AS}{\left(1 - \dfrac{V_{BS}}{\phi_0}\right)^{M_j}} + \frac{C_{jsw} \cdot PS}{\left(1 - \dfrac{V_{BS}}{\phi_0}\right)^{M_{jsw}}} \tag{4.26}$$

$$C_{DB} = \frac{C_j \cdot AD}{\left(1 - \dfrac{V_{BD}}{\phi_0}\right)^{M_j}} + \frac{C_{jsw} \cdot PD}{\left(1 - \dfrac{V_{BD}}{\phi_0}\right)^{M_{jsw}}} \tag{4.27}$$

Here, C_j is the zero-bias depletion capacitance per unit area at the bottom junction of the drain or the source diffusion region, and C_{jsw} is the zero-bias depletion capacitance per unit length at the sidewall junctions. For typical sidewall doping, which is about 10 times larger than the substrate doping concentration, C_{jsw} can be approximated by

$$C_{jsw} \cong \sqrt{10} \cdot C_j \cdot x_j \tag{4.28}$$

AS and *AD* are the source and the drain areas; *PS* and *PD* are the source and the drain perimeters, respectively (This geometry information must be specified in the corresponding device description line). Note that although in reality only three sides of a rectangular diffusion region are surrounded by the p$^+$ channel-stop implant, entire source and drain perimeters are usually specified as *PS* and *PD*. This tends to overestimate the total diffusion capacitance, but the difference is not very significant. Also, the built-in junction potential ϕ_0, which is actually a function of the doping densities, is assumed to be equal for both the bottom junction and the sidewall junction.

Finally, the parameters M_j and M_{jsw} denote the junction grading coefficients for the bottom and the sidewall junctions, respectively. Default values are $M_j = 0.5$ (assuming abrupt junction profile for the bottom area) and $M_{jsw} = 0.33$ (assuming linearly graded junction profile for the sidewalls).

EXAMPLE 4.1

The top view of an n-channel MOSFET is shown in the figure below. The process parameters for this device are:

$N_A = 10^{15}$ cm^{-3}

$N_A(sidewall) = 2.1 \times 10^{16}$ cm^{-3}

$N_D = 10^{20}$ cm^{-3}

$x_j = 0.8\ \mu$m

$t_{ox} = 600$ Å

$L_D = 0.5\ \mu$m

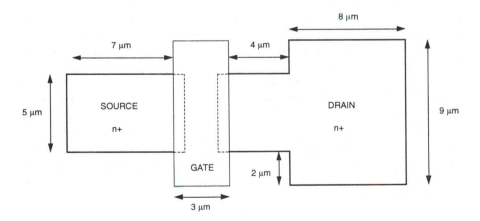

The zero-bias threshold voltage is measured as 0.85 V, and k' is determined to be 45 μA/V^2. The channel length modulation coefficient is $\lambda = 0.05$. The source, drain, gate, and substrate nodes of the device are labeled by node numbers 4, 6, 12, and 7, respectively. Prepare the device description line and the .MODEL line for SPICE simulation. Use the LEVEL 1 model, and avoid conflicting parameter definitions.

The gate oxide capacitance per unit area is

$$C_{ox} = \frac{\varepsilon_{ox}}{t_{ox}} = \frac{3.9 \cdot 8.85 \times 10^{-14}}{600 \times 10^{-8}} = 5.75 \times 10^{-8} \text{ F/cm}^2$$

The substrate bias coefficient (GAMMA) and the surface inversion potential (PHI) are found as follows:

$$\gamma = \frac{\sqrt{2 \cdot q \cdot N_A \cdot \varepsilon_{Si}}}{C_{ox}} = \frac{\sqrt{2 \cdot 1.6 \cdot 10^{-19} \cdot 10^{15} \cdot 11.7 \cdot 8.85 \cdot 10^{-14}}}{5.75 \times 10^{-8}} = 0.32 \text{ V}^{\frac{1}{2}}$$

$$|2 \cdot \phi_F(substrate)| = \left| 2 \cdot \frac{kT}{q} \ln \left(\frac{n_i}{N_A} \right) \right|$$

$$= \left| 2 \cdot 0.026 \text{ V} \cdot \ln \left(\frac{1.45 \cdot 10^{10}}{10^{15}} \right) \right| = 0.58 \text{ V}$$

Now we start to calculate the parameters needed for the parasitic capacitance descriptions. The built-in junction potential (PB) for the bottom diffusion area is

$$\phi_0 = \frac{kT}{q} \cdot \ln \left(\frac{N_A \cdot N_D}{n_i^2} \right) = 0.026 \text{ V} \cdot \ln \left(\frac{10^{20} \cdot 10^{15}}{2.1 \times 10^{20}} \right) = 0.878 \text{ V}$$

Note that this junction potential is used for calculating *all* junction capacitances, which results in an *overestimation* of the sidewall junction capacitance. The zero-bias depletion capacitances associated with the bottom junction (CJ) and the sidewall junctions (CJSW) are found as

$$C_{j0} = \sqrt{\frac{\varepsilon_{Si} \cdot q}{2} \cdot \left(\frac{N_A \cdot N_D}{N_A + N_D} \right) \cdot \frac{1}{\phi_0}}$$

$$= \sqrt{\frac{11.7 \cdot 8.85 \cdot 10^{-14} \text{ F/cm} \cdot 1.6 \cdot 10^{-19} \text{ C}}{2} \cdot \left(\frac{10^{20} \cdot 10^{15}}{10^{20} + 10^{15}} \right) \cdot \frac{1}{0.878}}$$

$$= 9.7 \times 10^{-9} \text{ F/cm}^2$$

C_{j0sw}

$$= x_j \cdot \sqrt{\frac{\varepsilon_{Si} \cdot q}{2} \cdot \left(\frac{N_A(sw) \cdot N_D}{N_A(sw) + N_D} \right) \cdot \frac{1}{\phi_0}}$$

$$= 0.8 \times 10^{-4} \cdot \sqrt{\frac{11.7 \cdot 8.85 \cdot 10^{-14} \cdot 1.6 \cdot 10^{-19}}{2} \cdot \left(\frac{10^{20} \cdot 2.1 \cdot 10^{16}}{10^{20} + 2.1 \cdot 10^{16}} \right) \cdot \frac{1}{0.878}}$$

$$= 3.56 \times 10^{-12} \text{ F/cm}$$

We will assume abrupt junction profiles for both the bottom junctions and the side-wall junctions; thus, MJ = 0.5 and MJSW = 0.5. The gate overlap capacitances per unit length (CGSO and CGDO) are calculated as:

$$C_{GSO} = C_{GDO} = C_{ox} \cdot L_D = 5.75 \times 10^{-8} \cdot 0.5 \times 10^{-4} = 2.87 \text{ pF/cm}$$

Finally, we calculate the area and the perimeter (in m^2 and m, respectively) of the source and the drain diffusion regions. Now, we can write the device description line and the .MODEL line that correspond to this device, as follows:

```
M1   6   12   4   7 NM1 W=5U L=3U LD=0.5U AS=37.5P PS=25U
+                              AD=94.5P PD=43U

.MODEL   NM1   NMOS  (VTO=0.85   KP=45U   LAMBDA=0.05   GAMMA=0.32
+                     PHI=0.58   PB=0.878   CJ=9.7E-5   CJSW=3.56E-10
+                     CGSO=2.87E-10   CGDO=2.87E-10     MJ=0.5
+                     MJSW=0.5)
```

Notice that PS and PD are calculated here by taking into account the *entire* perimeter of the respective diffusion region, including the boundaries facing the gate (channel). Similarly, AS and AD are simply taken as the bottom area of the respective diffusion region. This approach slightly diverges from the exact capacitance calculation formulas given in Chapter 3, and results in a minor *over-estimation* of the actual junction capacitances. Yet due to the relative simplicity of calculating the area and the perimeter of polygons (i.e., source and drain regions) defined in the mask layout, this method can easily be applied to automatic layout parasitic extraction.

The junction capacitances are very important not only for the correct description of the circuit, but also because the convergence of the variable time-step integration algorithms used in SPICE actually improves with the presence of node capacitances. Therefore, it is useful to define the areas of the source and the drain diffusions early in the design process, even if the actual areas and the perimeters in the SPICE circuit description file must be updated following the completion of the layout.

4.8   Comparison of the SPICE MOSFET Models

Now we can briefly reexamine the main differences between the MOSFET models presented in this chapter. This comparison may be useful for choosing the best model for a given circuit simulation task.

The LEVEL 1 model is usually not very precise because the GCA used in the derivation of model equations is too approximate and the number of fitting parameters is too small. The LEVEL 1 model is useful for a quick and rough estimate of the circuit performance without much accuracy.

The LEVEL 2 model can be used with differing complexities by adding the parameters relating to the various effects that this model supports. However, if all model

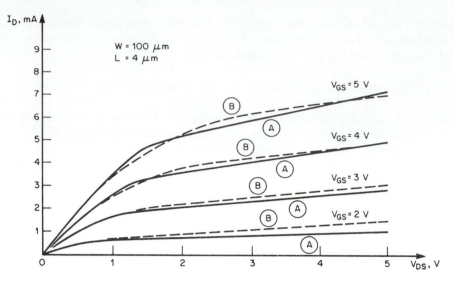

Figure 4.9　Drain current versus drain voltage characteristics of an n-channel MOSFET calculated with the LEVEL 2 model (A) and the LEVEL 3 model (B). The parameters common for both models are:
`VTO = 1, XJ = 1.0E-6, LD = 0.8E-6.`
The parameters of the LEVEL 2 model are:
`UO = 800, UCRIT = 5.0E4, UEXP = 0.15.`
The parameters of the LEVEL 3 model are: `UO = 850, THETA = 0.04.`
(Copyright © 1988 by McGraw-Hill, Inc.)

parameters are specified by the user, i.e., if the greatest level of complexity is obtained, this model requires a large amount of CPU time for the calculations. Moreover, the LEVEL 2 model may occasionally cause convergence problems in the Newton-Raphson algorithm used in SPICE.

A comparison between the LEVEL 2 and the LEVEL 3 models is interesting (Fig. 4.9). The LEVEL 3 model usually achieves about the same level of accuracy as the LEVEL 2 model, but the CPU time needed for model evaluation is less and the number of iterations are significantly fewer for the LEVEL 3 model. Consequently, the only disadvantage of the LEVEL 3 model is the complexity of calculating some of its parameters.

Using LEVEL 1 as a Fitting Model

The precision of the simplistic LEVEL 1 model is inadequate for representing the exact electrical behavior of short-channel MOS transistors, yet the simple model equations still make it very useful for hand calculations. To achieve a reasonable agreement between hand calculations and actual device/circuit performance, one possible approach is to use empirical (not physical) LEVEL 1 parameters. For example, applying the physical description (4.5) to calculate the value of KP will usually result in an overestimation of the drain current. Instead, all parameter values should be extracted from data, i.e., the model equations should be used purely as a fitting model. Due to the simplicity of model equations, it may not always be possible to achieve a good agreement with measured data. Yet, it can be shown that LEVEL 1 model parameters optimized to fit the data in the saturation region (e.g.: $V_{GS} = V_{DD}$, $V_{DS} > V_{DD} - V_T$) will work reasonably well for estimating the transient (switching) behavior of circuits. This is one of the reasons for using the simple LEVEL 1 equations in the analysis of various digital circuit examples in the following chapters.

Appendix

Typical SPICE Model Parameters

SPICE Level 2 model parameters for a typical 0.8 μm CMOS process:

```
.MODEL MODN NMOS LEVEL=2
+NLEV=0
+CGSO=0.350e-09    CGDO=0.350e-09    CGBO=0.150e-09
+CJ=0.300e-03      MJ=0.450e+00      CJSW=0.250e-09
+MJSW=0.330e+00    IS=0.000e+00      N=1.000e+00
+NDS=1.000e+12     VNDS=0.000e+00
+JS=0.010e-03      PB=0.850e+00      RSH=25.00e+00
+TOX=15.50e-09     XJ=0.080e-06
+VTO=0.850e+00     NFS=0.835e+12     NSUB=64.00e+15
+NEFF=10.00e+00    UTRA=0.000e+00
+UO=460.0e+00      UCRIT=38.00e+04   UEXP=0.325e+00
+VMAX=62.00e+03    DELTA=0.250e+00   KF=0.275e-25
+LD=0.000e-06      WD=0.600e-06      AF=1.500e+00
+BEX=-1.80e+00     TLEV=1.000e+00    TCV=1.400e-03


.MODEL MODP PMOS LEVEL=2
+NLEV=0
+CGSO=0.350e-09    CGDO=0.350e-09    CGBO=0.150e-09
+CJ=0.500e-03      MJ=0.470e+00      CJSW=0.210e-09
+MJSW=0.290e+00    IS=0.000e+00      N=1.000e+00
+NDS=1.000e+12     VNDS=0.000e+00
+JS=0.040e-03      PB=0.800e+00      RSH=47.00e+00
+TOX=15.00e-09     XJ=0.090e-06
+VTO=-.725e+00     NFS=0.500e+12     NSUB=32.80e+15
+NEFF=2.600e+00    UTRA=0.000e+00
+UO=160.0e+00      UCRIT=30.80e+04   UEXP=0.350e+00
+VMAX=61.00e+03    DELTA=0.950e+00   KF=0.470e-26
+LD=-.075e-06      WD=0.350e-06      AF=1.600e+00
+BEX=-1.50e+00     TLEV=1.000e+00    TCV=-1.80e-03
```

BSIM3 model parameters for the TSMC 0.18 μm CMOS process:

```
.MODEL CMOSN NMOS (                                      LEVEL    = 49
+VERSION = 3.1          TNOM    = 27           TOX      = 4.2E-9
+XJ      = 1E-7         NCH     = 2.3549E17    VTH0     = 0.3680296
+K1      = 0.5911252    K2      = 2.288938E-3  K3       = 1E-3
+K3B     = 1.9516573    W0      = 1E-7         NLX      = 1.686788E-7
+DVT0W   = 0            DVT1W   = 0            DVT2W    = 0
+DVT0    = 1.7464037    DVT1    = 0.4568438    DVT2     = -0.0181191
+U0      = 263.836679   UA      = -1.178099E-9 UB       = 1.749553E-18
+UC      = -9.76209E-12 VSAT    = 8.994393E4   A0       = 1.8288772
+AGS     = 0.3397201    B0      = -2.675178E-8 B1       = -1E-7
+KETA    = -3.47067E-3  A1      = 7.995817E-4  A2       = 1
+RDSW    = 108.6492521  PRWG    = 0.5          PRWB     = -0.2
+WR      = 1            WINT    = 2.64543E-10  LINT     = 1.338295E-9
+XL      = -2E-8        XW      = -1E-8        DWG      = 6.346367E-9
+DWB     = 2.756527E-9  VOFF    = -0.0790381   NFACTOR  = 2.3051491
+CIT     = 0            CDSC    = 2.4E-4       CDSCD    = 0
+CDSCB   = 0            ETA0    = 1.034575E-4  ETAB     = -4.486535E-3
+DSUB    = 0.0116456    PCLM    = 1.1328276    PDIBLC1  = 0.2376928
+PDIBLC2 = 6.786697E-3  PDIBLCB = 0.1          DROUT    = 0.677486
+PSCBE1  = 6.738022E10  PSCBE2  = 6.832776E-8  PVAG     = 0.1870951
+DELTA   = 0.01         RSH     = 6.7          MOBMOD   = 1
+PRT     = 0            UTE     = -1.5         KT1      = -0.11
+KT1L    = 0            KT2     = 0.022        UA1      = 4.31E-9
+UB1     = -7.61E-18    UC1     = -5.6E-11     AT       = 3.3E4
+WL      = 0            WLN     = 1            WW       = 0
+WWN     = 1            WWL     = 0            LL       = 0
+LLN     = 1            LW      = 0            LWN      = 1
+LWL     = 0            CAPMOD  = 2            XPART    = 0.5
+CGDO    = 7.58E-10     CGSO    = 7.58E-10     CGBO     = 1E-12
+CJ      = 9.906354E-4  PB      = 0.730091     MJ       = 0.3599246
+CJSW    = 2.273142E-10 PBSW    = 0.6198535    MJSW     = 0.1268548
+CJSWG   = 3.3E-10      PBSWG   = 0.6198535    MJSWG    = 0.1268548
+CF      = 0            PVTH0   = -3.683048E-3 PRDSW    = -1.4166565
+PK2     = 2.066895E-3  WKETA   = 2.06959E-3   LKETA    = 0.0251872
+PU0     = -1.4215545   PUA     = -3.53899E-11 PUB      = 6.764061E-25
+PVSAT   = 1.864733E3   PETA0   = 1E-4         PKETA    = -1.41807E-3
)
*
```

```
.MODEL CMOSP PMOS (                                    LEVEL     = 49
+VERSION = 3.1            TNOM      = 27               TOX       = 4.2E-9
+XJ       = 1E-7          NCH       = 4.1589E17        VTH0      = -0.4349298
+K1       = 0.6076257     K2        = 0.0240727        K3        = 0
+K3B      = 10.1162091    W0        = 1E-6             NLX       = 8.008684E-8
+DVT0W    = 0             DVT1W     = 0                DVT2W     = 0
+DVT0     = 0.4263447     DVT1      = 0.2825945        DVT2      = 0.1
+U0       = 118.2923681   UA        = 1.595982E-9      UB        = 1.109698E-21
+UC       = -1E-10        VSAT      = 1.682653E5       A0        = 1.6728458
+AGS      = 0.4152711     B0        = 1.855408E-6      B1        = 5E-6
+KETA     = 0.0180992     A1        = 0.5086627        A2        = 0.3747271
+RDSW     = 296.9038493   PRWG      = 0.5              PRWB      = -0.3874288
+WR       = 1             WINT      = 0                LINT      = 1.914706E-8
+XL       = -2E-8         XW        = -1E-8            DWG       = -2.400357E-8
+DWB      = 1.079858E-8   VOFF      = -0.097118        NFACTOR   = 1.8520072
+CIT      = 0             CDSC      = 2.4E-4           CDSCD     = 0
+CDSCB    = 0             ETA0      = 0.0163774        ETAB      = -0.1095661
+DSUB     = 0.7737497     PCLM      = 2.3031926        PDIBLC1   = 1.921807E-4
+PDIBLC2  = 0.0174673     PDIBLCB   = -9.975699E-4     DROUT     = 0
+PSCBE1   = 2.054597E9    PSCBE2    = 5.934159E-10     PVAG      = 15
+DELTA    = 0.01          RSH       = 7.5              MOBMOD    = 1
+PRT      = 0             UTE       = -1.5             KT1       = -0.11
+KT1L     = 0             KT2       = 0.022            UA1       = 4.31E-9
+UB1      = -7.61E-18     UC1       = -5.6E-11         AT        = 3.3E4
+WL       = 0             WLN       = 1                WW        = 0
+WWN      = 1             WWL       = 0                LL        = 0
+LLN      = 1             LW        = 0                LWN       = 1
+LWL      = 0             CAPMOD    = 2                XPART     = 0.5
+CGDO     = 6.74E-10      CGSO      = 6.74E-10         CGBO      = 1E-12
+CJ       = 1.124859E-3   PB        = 0.8637387        MJ        = 0.4237235
+CJSW     = 1.889062E-10  PBSW      = 0.6187797        MJSW      = 0.2845939
+CJSWG    = 4.22E-10      PBSWG     = 0.6187797        MJSWG     = 0.2845939
+CF       = 0             PVTH0     = 1.8347E-3        PRDSW     = 15.2709708
+PK2      = 2.005769E-3   WKETA     = 2.478814E-3      LKETA     = 1.457236E-3
+PU0      = -2.0661953    PUA       = -8.44317E-11     PUB       = 1E-21
+PVSAT    = -5.8202946    PETA0     = 1E-4             PKETA     = 2.75599E-3
)
```

BSIM3 parameters extracted by MOSIS (www.mosis.org).

Exercise Problems

4.1 Rewrite the SPICE code for the nMOS model in Example 4.1 for a graded junction using the following parameters:

- $N_A = 2 \cdot 10^{15}$ cm^{-3}
- $N_A (sidewall) = 2.1 \cdot 10^{16}$ cm^{-3}
- $N_D = 10^{19}$ cm^{-3}
- $X_j = 0.5$ μm
- $t_{ox} = 200$ Å
- $L_D = 0.2$ μm

4.2 A layout of nMOS transistors in the NAND2 gate is shown below. Write a SPICE description corresponding to the layout. The diffusion region between two polysilicon gates can be split equally between the two transistors.

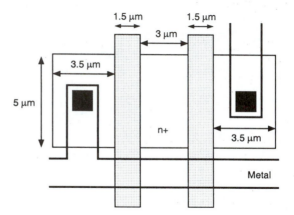

Figure P4.2

4.3 Using the SPICE LEVEL 1 MOSFET model equations, derive an expression for the sensitivity of the drain current I_D with respect to temperature. Calculate the sensitivity at room temperature $T = 300$ K by using the values in Example 4.1. For simplicity, assume that n_i is independent of temperature. Also verify your solution numerically by calculating I_D at 310 K and then $\Delta I_D / \Delta T$.

4.4 Explain why several versions of nMOS transistor models and pMOS transistor models coexist despite the fact that some models are more accurate than others.

CHAPTER 5

MOS Inverters: Static Characteristics

The inverter is the most fundamental logic gate that performs a Boolean operation on a single input variable. In this chapter, we will examine the DC (static) characteristics of various MOS inverter circuits. It will be seen later that many of the basic principles employed in the design and analysis of MOS inverters can be directly applied to more complex logic circuits, such as NAND and NOR gates, as well. The inverter design therefore forms a significant basis for digital circuit design. Consequently, the DC analysis of MOS inverters will be carried out rigorously, and in detail. While analytic techniques constitute most of the material presented in this chapter, numerical comparisons with circuit simulation using SPICE will also be emphasized, since computer-aided techniques are integral components of the analysis and design processes for digital circuits.

5.1 Introduction

The logic symbol and the truth table of the ideal inverter are shown in Fig. 5.1. In MOS inverter circuits, both the input variable A and the output variable B are represented by node voltages, referenced to the ground potential. Using *positive logic convention,* the Boolean (or logic) value of "1" can be represented by a high voltage of V_{DD}, and the Boolean (or logic) value of "0" can be represented by a low voltage of 0. The DC voltage transfer characteristic (VTC) of the ideal inverter circuit is shown in Fig. 5.2. The voltage V_{th} is called the *inverter threshold voltage*. Note that for any input voltage between 0 and $V_{th} = V_{DD}/2$, the output voltage is equal to V_{DD} (logic "1"). The output switches from V_{DD} to 0 when the input is equal to V_{th}. For any input voltage between V_{th} and V_{DD}, the output voltage assumes a value of 0 (logic "0"). Thus, an input voltage $0 \le V_{in} < V_{th}$ is interpreted by this ideal inverter as a logic "0," while an input voltage $V_{th} < V_{in} \le V_{DD}$ is interpreted as a logic "1." The DC characteristics of actual inverter circuits will obviously differ in various degrees from the ideal characteristic shown in Fig. 5.2. The accurate estimation and

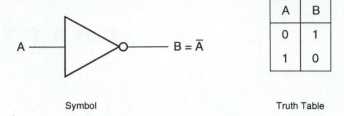

Symbol   Truth Table

A	B
0	1
1	0

Figure 5.1 Logic symbol and truth table of the inverter.

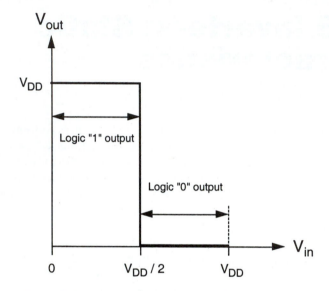

Figure 5.2 Voltage transfer characteristic (VTC) of the ideal inverter.

the manipulation of the shape of VTC for various inverter types are actually important parts of the design process.

Figure 5.3 shows the generalized circuit structure of an nMOS inverter. The input voltage of the inverter circuit is also the gate-to-source voltage of the nMOS transistor ($V_{in} = V_{GS}$), while the output voltage of the circuit is equal to the drain-to-source voltage ($V_{out} = V_{DS}$). The source and the substrate terminals of the nMOS transistor, also called the *driver* transistor, are connected to ground potential; hence, the source-to-substrate voltage is $V_{SB} = 0$. In this generalized representation, the *load* device is represented as a two-terminal circuit element with terminal current I_L and terminal voltage $V_L(I_L)$. One terminal of the load device is connected to the drain of the n-channel MOSFET, while the other terminal is connected to V_{DD}, the power supply voltage. We will see shortly that the characteristics of the inverter circuit actually depend very strongly upon the type and the characteristics of the load device. The output terminal of the inverter shown in Fig. 5.3 is connected to the input of another

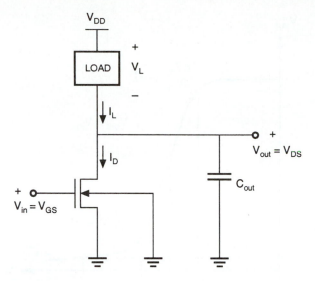

Figure 5.3 General circuit structure of an
nMOS inverter.

MOS inverter. Consequently, the next circuit *seen* by the output node can be represented as a lumped capacitance, C_{out}. Since the DC gate current of an MOS transistor
is negligible for all practical purposes, there will be no current flow into or out of the
input and output terminals of the inverter in DC steady state.

Voltage Transfer Characteristic (VTC)

Applying Kirchhoff's Current Law (KCL) to this simple circuit, we see that the load
current is always equal to the nMOS drain current.

$$I_D(V_{in}, V_{out}) = I_L(V_L) \tag{5.1}$$

The voltage transfer characteristic describing V_{out} as a function of V_{in} under DC
conditions can then be found by analytically solving (5.1) for various input voltage
values. The typical VTC of a realistic nMOS inverter is shown in Fig. 5.4. Upon
examination, we can identify a number of important properties of this DC transfer
characteristic.

The general shape of the VTC in Fig. 5.4 is qualitatively similar to that of the
ideal inverter transfer characteristic shown in Fig. 5.2. There are, however, several
significant differences that deserve special attention. For very low input voltage
levels, the output voltage V_{out} is equal to the high value of V_{OH} (*output high voltage*).
In this case, the driver nMOS transistor is in cut-off, and hence, does not conduct any
current. Consequently, the voltage drop across the load device is very small in magnitude, and the output voltage level is high. As the input voltage V_{in} increases, the
driver transistor starts conducting a certain drain current, and the output voltage
eventually starts to decrease. Notice that this drop in the output voltage level does not

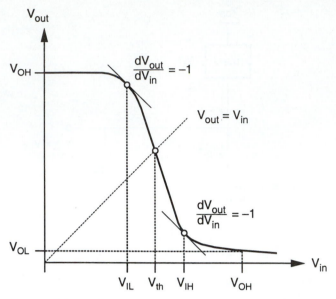

Figure 5.4   Typical voltage transfer characteristic (VTC) of a realistic nMOS inverter.

occur abruptly, such as the vertical drop assumed for the ideal inverter VTC, but rather gradually and with a finite slope. We identify two *critical voltage points* on this curve, where the slope of the $V_{out}(V_{in})$ characteristic becomes equal to -1, i.e.,

$$\frac{dV_{out}}{dV_{in}} = -1 \tag{5.2}$$

The smaller input voltage value satisfying this condition is called the *input low voltage* V_{IL}, and the larger input voltage satisfying this condition is called the *input high voltage* V_{IH}. Both of these voltages play significant roles in determining the noise margins of the inverter circuit, as we will discuss in the following sections. The physical justification for selecting these voltage points will also be examined in the context of noise immunity.

As the input voltage is further increased, the output voltage continues to drop and reaches a value of V_{OL} (*output low voltage*) when the input voltage is equal to V_{OH}. The *inverter threshold voltage* V_{th}, which is considered as the transition voltage, is defined as the point where $V_{in} = V_{out}$ on the VTC. Thus, a total of five critical voltages, V_{OL}, V_{OH}, V_{IL}, V_{IH}, and V_{th}, characterize the DC input-output voltage behavior of the inverter circuit. The functional definitions for the first four of these critical voltages are given below.

> V_{OH}: Maximum output voltage when the output level is logic "1"
>
> V_{OL}: Minimum output voltage when the output level is logic "0"
>
> V_{IL}: Maximum input voltage which can be *interpreted* as logic "0"
>
> V_{IH}: Minimum input voltage which can be *interpreted* as logic "1"

These definitions obviously imply that logic levels in digital circuits are not represented by quantized voltage values, but rather by voltage ranges corresponding to these logic levels. According to the definitions before, any input voltage level between the lowest available voltage in the system (usually the ground potential) and V_{IL} is interpreted as a logic "0" input, while any input voltage level between the highest available voltage in the system (usually the power supply voltage) and V_{IH} is interpreted as a logic "1" input. Any output voltage level between the lowest available voltage in the system and V_{OL} is interpreted as a logic "0" output, while any output voltage level between the highest available voltage in the system and V_{OH} is interpreted as a logic "1" output. These voltage ranges are also shown on the inverter voltage transfer characteristic shown in Fig. 5.4.

The ability of an inverter to interpret an input signal within a voltage range as either a logic "0" or as a logic "1" allows digital circuits to operate with a certain *tolerance* to external signal perturbations. This tolerance to variations in the signal level is especially valuable in environments in which circuit noise can significantly corrupt the signals. Here the *circuit noise* corresponds to unwanted signals that are coupled to some part of the circuit from neighboring lines (usually interconnection lines) by capacitive or inductive coupling, or from outside of the system. The result of this interference is that the signal level at one end of an interconnection line may be significantly different from the signal level at the other end.

Noise Immunity and Noise Margins

To illustrate the effect of noise on the circuit reliability, we will consider the circuit consisting of three cascaded inverters, as shown in Fig. 5.5. Assume that all inverters are identical, and that the input voltage of the first inverter is equal to V_{OH}, i.e., a logic "1." By definition, the output voltage of the first inverter will be equal to V_{OL}, corresponding to a logic "0" level. Now, this output signal is being transmitted to the next inverter input via an interconnect, which could be a metal or polysilicon line connecting the two gates. Since on-chip interconnects are generally prone to signal noise, the output signal of the first inverter will be perturbed during transmission. Consequently, the voltage level at the input of the second inverter will be either

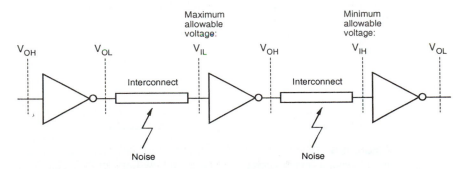

Figure 5.5 Propagation of digital signals under the influence of noise.

larger or smaller than V_{OL}. If the input voltage of the second inverter is smaller than V_{OL}, this signal will be interpreted correctly as a logic "0" input by the second inverter. On the other hand, if the input voltage becomes larger than V_{IL} as a result of noise, then it may not be interpreted correctly by the inverter. Thus, we conclude that V_{IL} is the maximum allowable voltage at the input of the second inverter, which is low enough to ensure a logic "1" output.

Now consider the signal transmission from the output of the second inverter to the input of the third inverter, assuming that the second inverter produces an output voltage level of V_{OH}. As in the previous case, this output signal will be perturbed because of noise interference, and the voltage level at the input of the third inverter will be different from V_{OH}. If the input voltage of the third inverter is larger than V_{OH}, this signal will be interpreted correctly as a logic "1" input by the third inverter. If the voltage level drops below V_{IH} due to noise, however, the input may not be interpreted as a logic "1." Consequently, V_{IH} is the minimum allowable voltage at the input of the third inverter which is high enough to ensure a logic "0" output.

These observations lead us to the definition of noise tolerances for digital circuits, called *noise margin* and denoted by NM. The noise immunity of the circuit increases with NM. Two noise margins will be defined: the noise margin for low signal levels (NM_L) and the noise margin for high signal levels (NM_H).

$$NM_L = V_{IL} - V_{OL} \tag{5.3}$$

$$NM_H = V_{OH} - V_{IH} \tag{5.4}$$

Figure 5.6 shows a graphical illustration of the noise margins. Here, the shaded areas indicate the valid regions of the input and output voltages, and the noise margins are shown as the amount of variation in the signal levels that can be allowed while the signal is transmitted from the output of one gate to the input of the next gate.

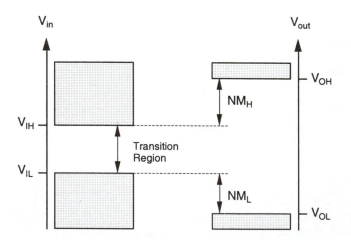

Figure 5.6 Definition of noise margins NM_L and NM_H. Note that the shaded regions indicate valid high and low levels for the input and output signals.

The selection of the two critical voltage points, V_{IL} and V_{IH}, using the slope condition (5.2) can now be justified based on noise considerations. We know that the output voltage V_{out} of an inverter under noise-free, steady-state conditions is a non-linear function of the input voltage, V_{in}. The shape of this function is described by the voltage transfer characteristic (VTC).

$$V_{out} = f(V_{in}) \tag{5.5}$$

If the input signal is perturbed from its nominal value because of external influences, such as noise, the output voltage will also deviate from its nominal level. Here, ΔV_{noise} represents the voltage perturbation that is assumed to affect the input voltage of the inverter.

$$V'_{out} = f(V_{in} + \Delta V_{noise}) \tag{5.6}$$

By using a simple first-order Taylor series expansion and by neglecting higher-order terms, we can express the perturbed output voltage V'_{out} as follows:

$$V'_{out} = f(V_{in}) + \frac{dV_{out}}{dV_{in}} \cdot \Delta V_{noise} + higher\ order\ terms\ (neglected) \tag{5.7}$$

Notice that here, $f(V_{in})$ represents the nominal (non-perturbed) output signal, and the term represents the *voltage gain* of the inverter at the nominal input voltage value V_{in}. Thus, the equation (5.7) can also be expressed as:

$$Perturbed\ Output = Nominal\ Output + Gain \times External\ Perturbation \tag{5.8}$$

If the magnitude of the voltage gain at the nominal input voltage V_{in} is smaller than unity, then the input perturbation is not amplified and, consequently, the output perturbation remains relatively small. Otherwise, with the voltage gain larger than unity, a small perturbation in the input voltage level will cause a rather large (amplified) perturbation in the output voltage. Hence, we define the boundaries of the valid input signal regions as the voltage points where the magnitude of the inverter voltage gain is equal to unity.

Finally, note that there is a voltage range between V_{IL} and V_{IH} (Fig. 5.6), corresponding to input voltage values that may not be processed correctly either as a logic "0" input or as a logic "1" input by the inverter. This region is called the *uncertain region* or, alternatively, the *transition region*. Ideally, the slope of the voltage transfer characteristic should be very large between V_{IL} and V_{IH}, because a narrow uncertain (transition) region obviously allows for larger noise margins. We will see that reducing the width of the uncertain region is one of the most important design objectives.

The preceding discussion of inverter static (DC) characteristics shows that the shape of the VTC in general and the noise immunity properties, in particular, are very significant criteria that ultimately dictate the design priorities. For any inverter circuit, the five critical voltage points V_{OL}, V_{OH}, V_{IL}, V_{IH}, and V_{th} fully determine the DC input-output voltage behavior, the noise margins, and the width and location of the transition region. Accurate estimation of these critical voltage points for different inverter designs will be an important task throughout this chapter.

Power and Area Considerations

In addition to these concerns, we can identify two other issues that play significant roles in inverter design: *power* consumption and the *chip area* occupied by the inverter circuit. About one million logic gates can be accommodated on a very large scale integrated (VLSI) chip using 0.5 μm MOS technology, and the circuit density is expected to increase even further in future-generation chips. Since each gate on the chip dissipates power and thus generates heat, the removal of this thermal energy, i.e., cooling of the chip, becomes an essential and usually very expensive task. Note that the junction temperature is given as $T_j = T_a + \theta P$, where T_a is the ambient temperature, θ is the thermal resistance, and P is the amount of power dissipated. Also, most portable systems, such as cellular communication devices and laptop and palmtop computers, operate from a limited power supply, and the extension of battery-based operation time is a significant design goal. Therefore, it is very important to reduce the amount of power dissipated by the circuit in both DC (usually standby) and dynamic operation.

The DC power dissipation of an inverter circuit can be calculated as the product of its power supply voltage and the amount of current drawn from the power supply during steady state or in standby mode. In deep sub-micron technologies, the increase in subthreshold currents cause more power consumption in standby mode, thus will shorten the battery life unless the trend can be reversed with novel circuit design or system design techniques.

$$P_{DC} = V_{DD} \cdot I_{DC} \tag{5.9}$$

Notice that the DC current drawn by the inverter circuit may vary depending on the input and output voltage levels. Assuming that the input voltage level corresponds to logic "0" during 50% of the operation time and to logic "1" during the other 50%, the overall DC power consumption of the circuit can be estimated as follows:

$$P_{DC} = \frac{V_{DD}}{2} \cdot [I_{DC}(V_{in} = low) + I_{DC}(V_{in} = high)] \tag{5.10}$$

It will be seen in the following sections that the DC power dissipation of different inverter designs varies significantly from each other, and that these differences may become a very important factor in the selection of a particular circuit type, e.g., CMOS vs. depletion-load nMOS, for a given design task.

To reduce the chip area occupied by the inverter circuit, it is necessary to reduce the area of the MOS transistors used in the circuit. As a practical measure, we use the gate area of the MOS transistor, i.e., the product of W and L. Thus, an MOS transistor has minimum area when both of the gate (channel) dimensions are made as small as possible within the constraints of the particular technology. It follows that the ratio of the gate width to gate length (W/L) should also be as close to unity as possible, in order to achieve minimum transistor area. This requirement, however, usually contradicts other design criteria, such as the noise margins, the output current driving capability, and the dynamic switching speed. We will observe in the following sections that the design of inverter circuits involves a detailed consideration and a trade-off of these criteria.

Our examination of different MOS inverter structures will begin with the resistive-load MOS inverter. The analysis of this simple circuit will help illustrate some of the basic aspects encountered in inverter design. We will briefly examine the nMOS depletion-load inverter to explore the basic characteristics of active-load digital circuits, and then devote most of our attention throughout the remainder of this chapter to the CMOS inverter.

5.2   Resistive-Load Inverter

The basic structure of the resistive-load inverter circuit is shown in Fig. 5.7. As in the general inverter circuit already examined in Fig. 5.3, an enhancement-type nMOS transistor acts as the driver device. The load consists of a simple linear resistor, R_L. The power supply voltage of this circuit is V_{DD}. Since the following analysis concentrates on the static behavior of the circuit, the output load capacitance is not shown in this figure.

As already noted in Section 5.1, the drain current I_D of the driver MOSFET is equal to the load current I_R, in DC steady-state operation. To simplify the calculations, the channel-length modulation effect will be neglected in the following; i.e., $I = 0$. Also, note that the source and the substrate terminals of the driver transistor are both connected to the ground; hence, $V_{SB} = 0$. Consequently, the threshold voltage of the driver transistor is always equal to V_{T0}. We start our analysis by identifying the various operating regions of the driver transistor under steady-state conditions.

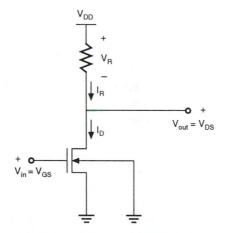

Figure 5.7   Resistive-load inverter circuit.

Table 5.1   Operating regions of the driver transistor in the resistive-load inverter

Input voltage range	Operating mode
$V_{in} < V_{T0}$	cut-off
$V_{T0} \leq V_{in} < V_{out} + V_{T0}$	saturation
$V_{in} \geq V_{out} + V_{T0}$	linear

For input voltages smaller than the threshold voltage V_{T0}, the transistor is in cut-off, and does not conduct any drain current. Since the voltage drop across the load resistor is equal to zero, the output voltage must be equal to the power supply voltage, V_{DD}. As the input voltage is increased beyond V_{T0}, the driver transistor starts conducting a nonzero drain current. Note that the driver MOSFET is initially in saturation, since its drain-to-source voltage ($V_{DS} = V_{out}$) is larger than ($V_{in} - V_{T0}$). Thus,

$$I_R = \frac{k_n}{2} \cdot (V_{in} - V_{T0})^2 \tag{5.11}$$

With increasing input voltage, the drain current of the driver also increases, and the output voltage V_{out} starts to drop. Eventually, for input voltages larger than $V_{out} + V_{T0}$, the driver transistor enters the linear operation region. At larger input voltages, the transistor remains in linear mode, as the output voltage continues to decrease.

$$I_R = \frac{k_n}{2} \cdot \left[2 \cdot (V_{in} - V_{T0}) \cdot V_{out} - V_{out}^2 \right] \tag{5.12}$$

The various operating regions of the driver transistor and the corresponding input-output conditions are listed in Table 5.1.

Figure 5.8 shows the voltage transfer characteristic of a typical resistive-load inverter circuit, indicating the operating modes of the driver transistor and the critical voltage points on the VTC. Now, we start with the calculation of the five critical voltage points, which determine the steady-state input-output behavior of the inverter.

Calculation of V_{OH}

First, we note that the output voltage V_{out} is given by

$$V_{out} = V_{DD} - R_L \cdot I_R \tag{5.13}$$

When the input voltage V_{in} is low, i.e., smaller than the threshold voltage of the driver MOSFET, the driver transistor is cut-off. Since the drain current of the driver transistor is equal to the load current, $I_R = I_D = 0$. It follows that the output voltage of the inverter under these conditions is:

$$V_{OH} = V_{DD} \tag{5.14}$$

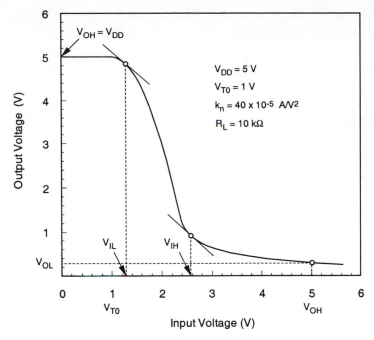

Figure 5.8 Typical VTC of a resistive-load inverter circuit. Important design parameters of the circuit are shown in the inset.

Calculation of V_{OL}

To calculate the output low voltage V_{OL}, we assume that the input voltage is equal to V_{OH}, i.e., $V_{in} = V_{OH} = V_{DD}$. Since $V_{in} - V_{T0} > V_{out}$ in this case, the driver transistor operates in the linear region. Also note that the load current I_R is

$$I_R = \frac{V_{DD} - V_{out}}{R_L} \tag{5.15}$$

Using Kirchhoff Current Law (KCL) for the output node, i.e., $I_R = I_D$, we can write the following equation:

$$\frac{V_{DD} - V_{OL}}{R_L} = \frac{k_n}{2} \cdot \left[2 \cdot (V_{DD} - V_{T0}) \cdot V_{OL} - V_{OL}^2 \right] \tag{5.16}$$

This equation yields a simple quadratic in V_{OL}, which is solved to find the value of the output low voltage.

$$V_{OL}^2 - 2 \cdot \left(V_{DD} - V_{T0} + \frac{1}{k_n R_L} \right) \cdot V_{OL} + \frac{2}{k_n R_L} \cdot V_{DD} = 0 \tag{5.17}$$

Note that of the two possible solutions of (5.17), we must choose the one that is physically correct, i.e., the value of the output low voltage must be between 0 and V_{DD}.

The solution of (5.17) is given below. It can be seen that the product $(k_n R_L)$ is one of the important design parameters that determine the value of V_{OL}.

$$V_{OL} = V_{DD} - V_{T0} + \frac{1}{k_n R_L} - \sqrt{\left(V_{DD} - V_{T0} + \frac{1}{k_n R_L}\right)^2 - \frac{2V_{DD}}{k_n R_L}} \qquad (5.18)$$

Calculation of V_{IL}

By definition, V_{IL} is the smaller of the two input voltage values at which the slope of the VTC becomes equal to (-1); i.e., $dV_{out}/dV_{in} = -1$. Simple inspection of Fig. 5.8 shows that when the input is equal to V_{IL}, the output voltage (V_{out}) is only slightly smaller than V_{OH}. Consequently, $V_{out} > V_{in} - V_{T0}$, and the driver transistor operates in saturation. We start our analysis by writing the KCL for the output node.

$$\frac{V_{DD} - V_{out}}{R_L} = \frac{k_n}{2} \cdot (V_{in} - V_{T0})^2 \qquad (5.19)$$

To satisfy the derivative condition, we differentiate both sides of (5.19) with respect to V_{in}, which results in the following equation:

$$-\frac{1}{R_L} \cdot \frac{dV_{out}}{dV_{in}} = k_n \cdot (V_{in} - V_{T0}) \qquad (5.20)$$

Since the derivative of the output voltage with respect to the input voltage is equal to (-1) at V_{IL}, we can substitute $dV_{out}/dV_{in} = -1$ in (5.20).

$$-\frac{1}{R_L} \cdot (-1) = k_n \cdot (V_{IL} - V_{T0}) \qquad (5.21)$$

Solving (5.21) for V_{IL}, we obtain

$$V_{IL} = V_{T0} + \frac{1}{k_n R_L} \qquad (5.22)$$

The value of the output voltage when the input is equal to V_{IL} can also be found by substituting (5.22) into (5.19), as follows:

$$V_{out}(V_{in} = V_{IL}) = V_{DD} - \frac{k_n R_L}{2} \cdot \left(V_{T0} + \frac{1}{k_n R_L} - V_{T0}\right)^2$$

$$= V_{DD} - \frac{1}{2k_n R_L} \qquad (5.23)$$

Calculation of V_{IH}

V_{IH} is the larger of the two voltage points on VTC at which the slope is equal to (-1). It can be seen from Fig. 5.8 that when the input voltage is equal to V_{IH}, the output voltage V_{out} is only slightly larger than the output low voltage V_{OL}. Hence,

$V_{out} < V_{in} - V_{T0}$, and the driver transistor operates in the linear region. The KCL equation for the output node is given below.

$$\frac{V_{DD} - V_{out}}{R_L} = \frac{k_n}{2} \cdot \left[2 \cdot (V_{in} - V_{T0}) \cdot V_{out} - V_{out}^2\right] \tag{5.24}$$

Differentiating both sides of (5.24) with respect to V_{in}, we obtain

$$-\frac{1}{R_L} \cdot \frac{dV_{out}}{dV_{in}} = \frac{k_n}{2} \cdot \left[2 \cdot (V_{in} - V_{T0}) \cdot \frac{dV_{out}}{dV_{in}} + 2V_{out} - 2V_{out} \cdot \frac{dV_{out}}{dV_{in}}\right] \tag{5.25}$$

Next, we can substitute $dV_{out}/dV_{in} = -1$ into (5.25), since the slope of the VTC is equal to (-1) also at $V_{in} = V_{IH}$.

$$-\frac{1}{R_L} \cdot (-1) = k_n \cdot \left[(V_{IH} - V_{T0}) \cdot (-1) + 2V_{out}\right] \tag{5.26}$$

Solving (5.26) for V_{IH} yields the following expression.

$$V_{IH} = V_{T0} + 2V_{out} - \frac{1}{k_n R_L} \tag{5.27}$$

Thus, we obtain two algebraic equations, (5.24) and (5.27), for two unknowns, V_{IH} and V_{out}. To determine the unknown variables, we substitute (5.27) into the current equation given by (5.24) above.

$$\frac{V_{DD} - V_{out}}{R_L} = \frac{k_n}{2} \cdot \left[2 \cdot \left(V_{T0} + 2V_{out} - \frac{1}{k_n R_L} - V_{T0}\right) \cdot V_{out} - V_{out}^2\right] \tag{5.28}$$

The positive solution of this second-order equation gives the output voltage V_{out} when the input is equal to V_{IH}.

$$V_{out}(V_{in} = V_{IH}) = \sqrt{\frac{2}{3} \cdot \frac{V_{DD}}{k_n R_L}} \tag{5.29}$$

Finally, V_{IH} can be found by substituting (5.29) into (5.27), as follows.

$$V_{IH} = V_{T0} + \sqrt{\frac{8}{3} \cdot \frac{V_{DD}}{k_n R_L}} - \frac{1}{k_n R_L} \tag{5.30}$$

The four critical voltage points V_{OL}, V_{OH}, V_{IL}, V_{IH} can now be used to determine the noise margins, NM_L and NM_H, of the resistive-load inverter circuit. In addition to these voltage points, which characterize the static input-output behavior, the inverter threshold voltage V_{th} may also be calculated in a straightforward manner. Note that the driver transistor operates in saturation mode at this point. Thus, the inverter threshold voltage can be found simply by substituting $V_{in} = V_{out} = V_{th}$ into (5.19), and by solving the resulting quadratic for V_{th}.

It can be seen from the preceding discussion that the term $(k_n R_L)$ plays an important role in determining the shape of the voltage transfer characteristic, and that

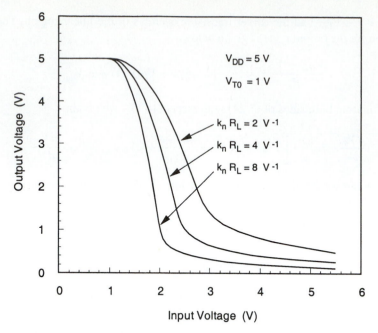

Figure 5.9   Voltage transfer characteristics of the resistive-load inverter, for different values of the parameter $(k_n R_L)$.

it appears as a critical parameter in expressions for V_{OL} (5.18), V_{IL} (5.22), and V_{IH} (5.30). Assuming that parameters such as the power supply voltage V_{DD} and the driver MOSFET threshold voltage V_{T0} are dictated by system- and processing-related constraints, the term $(k_n R_L)$ remains as the only design parameter which can be adjusted by the circuit designer to achieve certain design goals.

The output high voltage V_{OH} is determined primarily by the power supply voltage, V_{DD}. Among the other three critical voltage points, the adjustment of V_{OL} receives primary attention, while V_{IL} and V_{IH} are usually treated as secondary design variables. Figure 5.9 shows the VTC of a resistive-load inverter for different values of $(k_n R_L)$. Note that for larger $(k_n R_L)$ values, the output low voltage V_{OL} becomes smaller and that the shape of the VTC approaches that of the ideal inverter, with very large transition slope. Achieving larger $(k_n R_L)$ values in a design, however, may involve other trade-offs with the area and the power consumption of the circuit.

Power Consumption and Chip Area

The average DC power consumption of the resistive-load inverter circuit is found by considering two cases, $V_{in} = V_{OL}$ (*low*) and $V_{in} = V_{OH}$ (*high*). When the input voltage is equal to V_{OL}, the driver transistor is in cut-off. Consequently, if the leakage current is neglected, there is no steady-state current flow in the circuit

$(I_D = I_R = 0)$, and the DC power dissipation is equal to zero. When the input voltage is equal to V_{OH}, on the other hand, both the driver MOSFET and the load resistor conduct a nonzero current. Since the output voltage in this case is equal to V_{OL}, the current drawn from the power supply can be found as

$$I_D = I_R = \frac{V_{DD} - V_{OL}}{R_L} \tag{5.31}$$

Assuming that the input voltage is "low" during 50% of the operation time, and "high" during the remaining 50%, the average DC power consumption of the inverter can be estimated as follows:

$$P_{DC}(average) = \frac{V_{DD}}{2} \cdot \frac{V_{DD} - V_{OL}}{R_L} \tag{5.32}$$

EXAMPLE 5.1

Consider the following inverter design problem: Given $V_{DD} = 5\,\text{V}$, $k'_n = 30\,\mu\text{A/V}^2$, and $V_{T0} = 1\,\text{V}$, design a resistive-load inverter circuit with $V_{OL} = 0.2\,\text{V}$. Specifically, determine the (W/L) ratio of the driver transistor and the value of the load resistor R_L that achieve the required V_{OL}.

In order to satisfy the design specification on the output low voltage V_{OL}, we start our design by writing the relevant current equation. Note that the driver transistor is operating in the linear region when the output voltage is equal to V_{OL} and the input voltage is equal to $V_{OH} = V_{DD}$.

$$\frac{V_{DD} - V_{OL}}{R_L} = \frac{k'_n}{2} \cdot \frac{W}{L} \cdot \left[2 \cdot (V_{OH} - V_{T0}) \cdot V_{OL} - V_{OL}^2 \right]$$

Assuming $V_{OL} = 0.2\,\text{V}$ and using the given values for the power supply voltage, the driver threshold voltage and the driver transconductance k'_n, we obtain the following equation:

$$\frac{5 - 0.20}{R_L} = \frac{30 \times 10^{-6}}{2} \cdot \frac{W}{L} \cdot (2 \cdot 4 \cdot 0.20 - 0.20^2)$$

This equation can be rewritten as:

$$\frac{W}{L} \cdot R_L = 2.05 \times 10^5\,\Omega$$

At this point, we recognize that the designer has a choice of different (W/L) and R_L values, all of which satisfy the given design specification, $V_{OL} = 0.2\,\text{V}$. The selection of the pair of values to use for (W/L) and R_L in the final design ultimately depends on other considerations, such as the power consumption of the circuit and the silicon area. The next table lists some of the design possibilities, along with the average DC power consumption estimated for each design.

$\left(\dfrac{W}{L}\right)$ – Ratio	Load resistor $R_L\,[\mathrm{k}\Omega]$	DC power consumption $P_{DC,average}\,[\mu\mathrm{W}]$
1	205.0	58.5
2	102.5	117.1
3	68.4	175.4
4	51.3	233.9
5	41.0	292.7
6	34.2	350.8

It is seen that the power consumption increases significantly as the value of the load resistor R_L is decreased, and the (W/L) ratio is increased. If lowering the DC power consumption is the overriding concern, we may choose a small (W/L) ratio and a large load resistor. On the other hand, if the fabrication of the large load resistor requires a large silicon area, a clear trade-off exists between the DC power dissipation and the area occupied by the inverter circuit.

The chip area occupied by the resistive-load inverter circuit depends on two parameters, the (W/L) ratio of the driver transistor and the value of the resistor R_L. The area of the driver transistor can be approximated by the gate area, $(W \times L)$. Assuming that the gate length L is kept at its smallest possible value for the given technology, the gate area will be proportional to the (W/L) ratio of the transistor. The resistor area, on the other hand, depends very strongly on the technology which is used to fabricate the resistor on the chip.

We will briefly consider two possibilities for fabricating resistors using the standard MOS process: diffused resistor and polysilicon (undoped) resistor. The diffused resistor is fabricated, as the name implies, as an isolated n-type (or p-type) diffusion region with one contact on each end. The resistance is determined by the doping density of the diffusion region and the dimensions, i.e., the length-to-width ratio, of the resistor. Practical values of the diffusion-region sheet resistance range between 20 to 100 Ω/square. Consequently, very large length-to-width ratios would be required to achieve resistor values on the order of tens to hundreds of kΩ.

The placement of these resistor structures on chip, commonly in a serpentine shape for compactness, requires significantly more area than the driver MOSFET, as illustrated in Fig. 5.10(a). A resistive inverter with a large diffused load resistor is, therefore, not a practical component for VLSI applications.

An alternative approach to save silicon area is to fabricate the load resistor using undoped polysilicon. In conventional poly-gate MOS technology, the polysilicon structures forming the gates of transistors and the interconnect lines are heavily doped in order to reduce resistivity. The sheet resistivity of doped polysilicon interconnects and gates is about 20 to 40 Ω/square. If sections of the polysilicon are masked off from this doping step, on the other hand, the resulting *undoped* polysilicon layer has a very high sheet resistivity, in the order of 10 MΩ/square. Thus, very compact and very high-valued resistors can be fabricated using undoped poly layers (Fig. 5.10(b)). One drawback of this approach is that the resistance value can not be controlled very accurately, which results in large variations of the VTC. Consequently, resistive-load inverters with undoped poly resistors are not commonly used in logic gate circuits

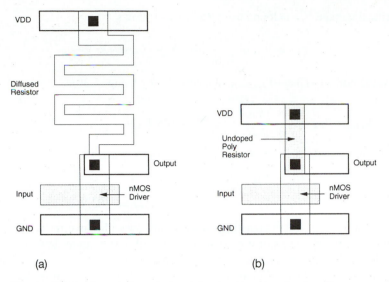

Figure 5.10 Sample layout of resistive-load inverter circuits with (a) diffused resistor and (b) undoped polysilicon resistor.

where certain design criteria, such as noise margins, are expected to be met. Simple inverter structures with large, undoped poly load resistors are used primarily in *low-power* Static Random Access Memory (SRAM) cells, where the primary emphasis is on the reduction of steady-state (DC) power consumption, and the operation of the memory circuit is not significantly affected by the variations of the inverter VTCs. This issue will be discussed in further detail in Chapter 10.

EXAMPLE 5.2

Consider a resistive-load inverter circuit with $V_{DD} = 5\,\text{V}$, $k'_n = 20\,\mu\text{A/V}^2$, $V_{T0} = 0.8\,\text{V}$, $R_L = 200\,\text{k}\Omega$, and $W/L = 2$. Calculate the critical voltages (V_{OL}, V_{OH}, V_{IL}, V_{IH}) on the VTC and find the noise margins of the circuit.

When the input voltage is low, i.e., when the driver nMOS transistor is cut-off, the output high voltage can be found as

$$V_{OH} = V_{DD} = 5\,\text{V}$$

Note that in this resistive-load inverter example, the transconductance of the driver transistor is $k_n = k'_n(W/L) = 40\,\mu\text{A/V}^2$ and, hence, $(k_n R_L) = 8\,\text{V}^{-1}$.

The output low voltage V_{OL} is calculated by using (5.18):

$$V_{OL} = V_{DD} - V_{T0} + \frac{1}{k_n R_L} - \sqrt{\left(V_{DD} - V_{T0} + \frac{1}{k_n R_L}\right)^2 - \frac{2V_{DD}}{k_n R_L}}$$

$$= 5 - 0.8 + \frac{1}{8} - \sqrt{\left(5 - 0.8 + \frac{1}{8}\right)^2 - \frac{2 \cdot 5}{8}}$$

$$= 0.147\,\text{V}$$

The critical voltage V_{IL} is found using (5.22), as follows.

$$V_{IL} = V_{T0} + \frac{1}{k_n R_L} = 0.8 + \frac{1}{8} = 0.925 \, \text{V}$$

Finally, the critical voltage V_{IH} can be calculated by using (5.30).

$$V_{IH} = V_{T0} + \sqrt{\frac{8}{3} \cdot \frac{V_{DD}}{k_n R_L}} - \frac{1}{k_n R_L} = 0.8 + \sqrt{\frac{8}{3} \cdot \frac{5}{8}} - \frac{1}{8} = 1.97 \, \text{V}$$

Now, the noise margins can be found, according to (5.3) and (5.4).

$$NM_L = V_{IL} - V_{OL} = 0.93 - 0.15 = 0.78 \, \text{V}$$

$$NM_H = V_{OH} - V_{IH} = 5.0 - 1.97 = 3.03 \, \text{V}$$

At this point, we can comment on the quality of this particular inverter design for DC operation. Notice that the noise margin NM_L found here is quite low, and it may eventually lead to misinterpretation of input signal levels. For better noise immunity, the noise margin for "low" signals should be at least about 25% of the power supply voltage V_{DD}, i.e., about 1.25 V for 5.0 V power supply.

5.3 Inverters with n-Type MOSFET Load

The simple resistive-load inverter circuit examined in the previous section is not a suitable candidate for most digital VLSI system applications, primarily because of the large area occupied by the load resistor. In this section, we will introduce inverter circuits which use an nMOS transistor as the *active load* device, instead of the linear load resistor. The main advantage of using a MOSFET as the load device is that the silicon area occupied by the transistor is usually smaller than that occupied by a comparable resistive load. Moreover, inverter circuits with active loads can be designed to have better overall performance compared to that of passive-load inverters. In a chronological view, the development of inverters with an enhancement-type MOSFET load precedes other active-load inverter types, since its fabrication process was perfected earlier.

Enhancement-Load nMOS Inverter

The circuit configurations of two inverters with enhancement-type load devices are shown in Fig. 5.11. Depending on the bias voltage applied to its gate terminal, the load transistor can be operated either in the saturation region or in the linear region. Both types of inverters have some distinct advantages and disadvantages from the circuit design point of view. The saturated enhancement-load inverter shown in Fig. 5.11(a) requires a single voltage supply and a relatively simple fabrication process, yet the V_{OH} level is limited to $V_{DD} - V_{T,load}$. The load device of the inverter circuit shown in Fig. 5.11(b), on the other hand, is always biased in the linear region. Thus, the V_{OH} level is equal to V_{DD}, resulting in higher noise margins compared to saturated enhancement-load inverter. The most significant drawback of

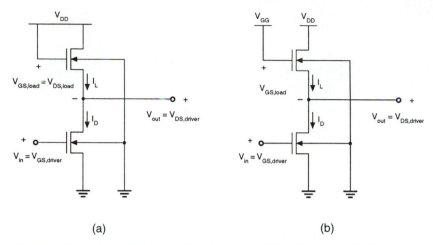

(a)   (b)

Figure 5.11 (a) Inverter circuit with saturated enhancement-type nMOS load. (b) Inverter with linear enhancement-type load.

this configuration is the use of two separate power supply voltages. In addition, both types of inverter circuits shown in Fig. 5.11 suffer from relatively high stand-by (DC) power dissipation; hence, enhancement-load nMOS inverters are not used in any large-scale digital applications.

Depletion-Load nMOS Inverter

Several of the disadvantages of the enhancement-type load inverter can be avoided by using a depletion-type nMOS transistor as the load device. The fabrication process for producing an inverter with an enhancement-type nMOS driver and a depletion-type nMOS load is slightly more complicated and requires additional processing steps, especially for the channel implant to adjust the threshold voltage of the load device. The resulting improvement of circuit performance and integration possibilities, however, easily justify the additional processing effort required for the fabrication of depletion-load inverters. The immediate advantages of implementing this circuit configuration are: (i) sharp VTC transition and better noise margins, (ii) single power supply, and (iii) smaller overall layout area. Also, judicious use of depletion transistors in CMOS circuits can help reduce standby (leakage) currents, but such discussion is beyond the scope of this chapter.

The circuit diagram of the depletion-load inverter circuit is shown in Fig. 5.12(a), and a simplified view of the circuit consisting of a nonlinear load resistor and a non-ideal switch (driver) in shown in Fig. 5.12(b). The driver device is an enhancement-type nMOS transistor, with $V_{T0,driver} > 0$, whereas the load is a depletion-type nMOS transistor, with $V_{T0,load} < 0$. The current-voltage equations to be used for the depletion-type load transistor are identical to those of the enhancement-type device, with the exception of the negative threshold voltage (cf. Section 3.3). The gate and the source nodes of the load transistor are connected; hence, $V_{GS,load} = 0$ always.

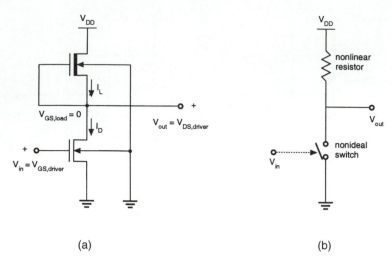

Figure 5.12   (a) Inverter circuit with depletion-type nMOS load. (b) Simplified equivalent circuit consisting of a nonlinear load resistor and a nonideal switch controlled by the input.

Since the threshold voltage of the depletion-type load is negative, the condition $V_{GS,load} > V_{T,load}$ is satisfied, and the load device always has a conducting channel regardless of the input and output voltage levels. Also note that both the driver transistor and the load transistor are built on the same p-type substrate, which is connected to the ground. Consequently, the load device is subject to the substrate-bias effect, so that its threshold voltage is a function of its source-to-substrate voltage, $V_{SB,load} = V_{out}$.

$$V_{T,load} = V_{T0,load} + \gamma \left(\sqrt{|2\phi_F| + V_{out}} - \sqrt{|2\phi_F|} \right) \tag{5.33}$$

The operating mode of the load transistor is determined by the output voltage level. When the output voltage is small; i.e., when $V_{out} < V_{DD} + V_{T,load}$, the load transistor is in saturation. Note that this condition corresponds to $V_{DS,load} > V_{GS,load} - V_{T,load}$. Then, the load current is given by the following equation.

$$I_{D,load} = \frac{k_{n,load}}{2} \cdot [-V_{T,load}(V_{out})]^2 = \frac{k_{n,load}}{2} \cdot |V_{T,load}(V_{out})|^2 \tag{5.34}$$

For larger output voltage levels, i.e., for $V_{out} > V_{DD} + V_{T,load}$, the depletion-type load transistor operates in the linear region. The load current in this case is

$$I_{D,load} = \frac{k_{n,load}}{2} \cdot \left[2|V_{T,load}(V_{out})| \cdot (V_{DD} - V_{out}) - (V_{DD} - V_{out})^2 \right] \tag{5.35}$$

The voltage transfer characteristic (VTC) of this inverter can be constructed by setting $I_{D,driver} = I_{D,load}$, $V_{GS,driver} = V_{in}$, and $V_{DS,driver} = V_{out}$, and by solving the corresponding current equations for $V_{out} = f(V_{in})$. Figure 5.13 shows the VTC

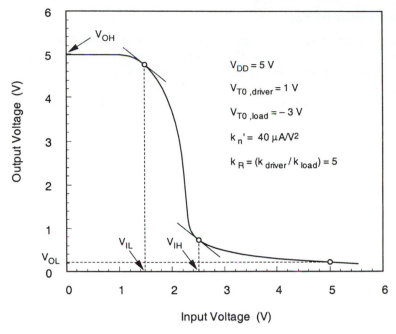

Figure 5.13 Typical VTC of a depletion-load inverter circuit.

of a typical depletion-load inverter, with $k'_{n,driver} = k'_{n,load}$. Next, we will consider the critical voltage points V_{OH}, V_{OL}, V_{IL}, and V_{IH} for this inverter circuit. The operating regions and the voltage levels of the driver and the load transistors at these critical points are listed below.

V_{in}	V_{out}	Driver operating region	Load operating region
V_{OL}	V_{OH}	cut-off	linear
V_{IL}	$\approx V_{OH}$	saturation	linear
V_{IH}	small	linear	*saturation*
V_{OH}	V_{OL}	linear	saturation

Calculation of V_{OH}

When the input voltage V_{in} is smaller than the driver threshold voltage V_{T0}, the driver transistor is turned off and does not conduct any drain current. Consequently, the load device, which operates in the linear region, also has zero drain current. Substituting V_{OH} for V_{out} in (5.35), and letting the load current $I_{D,load} = 0$, we obtain

$$I_{D,load} = \frac{k_{n,load}}{2} \cdot \left[2|V_{T,load}(V_{OH})| \cdot (V_{DD} - V_{OH}) - (V_{DD} - V_{OH})^2 \right] = 0$$
$$(5.36)$$

The only valid solution in the linear region is $V_{OH} = V_{DD}$.

Calculation of V_{OL}

To calculate the output low voltage V_{OL}, we assume that the input voltage V_{in} of the inverter is equal to $V_{OH} = V_{DD}$. Note that in this case, the driver transistor operates in the linear region while the depletion-type load is in saturation.

$$\frac{k_{driver}}{2} \cdot \left[2 \cdot (V_{OH} - V_{T0}) \cdot V_{OL} - V_{OL}^2 \right] = \frac{k_{load}}{2} \cdot \left[- V_{T,load}(V_{OL}) \right]^2 \quad (5.37)$$

This second-order equation in V_{OL} can be solved by temporarily neglecting the dependence of $V_{T,load}$ on V_{OL}, as follows.

$$V_{OL} = V_{OH} - V_{T0} - \sqrt{(V_{OH} - V_{T0})^2 - \left(\frac{k_{load}}{k_{driver}} \right) \cdot |V_{T,load}(V_{OL})|^2} \quad (5.38)$$

The actual value of the output low voltage can be found by solving the two equations (5.38) and (5.33) using numerical iterations. The iterative method converges rapidly because the actual value of V_{OL} is relatively small.

Calculation of V_{IL}

By definition, the slope of the VTC is equal to (-1), i.e., $dV_{out}/dV_{in} = -1$ when the input voltage is $V_{in} = V_{IL}$. Note that in this case, the driver transistor operates in saturation while the load transistor operates in the linear region. Applying KCL for the output node, we obtain the following current equation:

$$\frac{k_{driver}}{2} \cdot (V_{in} - V_{T0})^2$$
$$= \frac{k_{load}}{2} \cdot \left[2|V_{T,load}(V_{out})| \cdot (V_{DD} - V_{out}) - (V_{DD} - V_{out})^2 \right] \quad (5.39)$$

To satisfy the derivative condition at V_{IL}, we differentiate both sides of (5.39) with respect to V_{in}.

$$k_{driver} \cdot (V_{in} - V_{T0}) = \frac{k_{load}}{2} \cdot \left[2|V_{T,load}(V_{out})| \left(-\frac{dV_{out}}{dV_{in}} \right) + \right.$$
$$\left. 2(V_{DD} - V_{out}) \left(-\frac{dV_{T,load}}{dV_{in}} \right) - 2(V_{DD} - V_{out}) \left(-\frac{dV_{out}}{dV_{in}} \right) \right] \quad (5.40)$$

In general, we can assume that the term $(dV_{T,load}/dV_{in})$ is negligible with respect to the others. Substituting V_{IL} for V_{in}, and letting $dV_{out}/dV_{in} = -1$, we obtain V_{IL} as a function of the output voltage V_{out}.

$$V_{IL} = V_{T0} + \left(\frac{k_{load}}{k_{driver}} \right) \cdot \left[V_{out} - V_{DD} + |V_{T,load}(V_{out})| \right] \quad (5.41)$$

To account for the substrate-bias effect on $V_{T,load}$, this equation must be solved together with (5.33) and (5.39) using successive iterations.

Calculation of V_{IH}

V_{IH} is the larger of the two voltage points on the VTC at which the slope is equal to (-1). Since the output voltage corresponding to this operating point is relatively small, the driver transistor is in the linear region and the load transistor is in saturation.

$$\frac{k_{driver}}{2} \cdot [2 \cdot (V_{in} - V_{T0}) \cdot V_{out} - V_{out}^2] = \frac{k_{load}}{2} \cdot [-V_{T,load}(V_{out})]^2 \quad (5.42)$$

Differentiating both sides of (5.42) with respect to V_{in}, we obtain:

$$k_{driver} \cdot \left[V_{out} + (V_{in} - V_{T0})\left(\frac{dV_{out}}{dV_{in}}\right) - V_{out}\left(\frac{dV_{out}}{dV_{in}}\right)\right]$$

$$= k_{load} \cdot [-V_{T,load}(V_{out})] \cdot \left(\frac{dV_{T,load}}{dV_{out}}\right) \cdot \left(\frac{dV_{out}}{dV_{in}}\right) \quad (5.43)$$

Now, substitute $dV_{out}/dV_{in} = -1$ into (5.43), and solve for $V_{in} = V_{IH}$.

$$V_{IH} = V_{T0} + 2V_{out} + \left(\frac{k_{load}}{k_{driver}}\right) \cdot [-V_{T,load}(V_{out})] \cdot \left(\frac{dV_{T,load}}{dV_{out}}\right) \quad (5.44)$$

Note that the derivative of the load threshold voltage with respect to the output voltage cannot be neglected in this case.

$$\frac{dV_{T,load}}{dV_{out}} = \frac{\gamma}{2\sqrt{|2\phi_F| + V_{out}}} \quad (5.45)$$

The actual values of V_{IH} and the corresponding output voltage V_{out} are determined by solving (5.45) together with the current equation (5.42) and with the threshold voltage expression (5.33), using numerical iterations. A relatively accurate first-order estimate for V_{IH} can be obtained by assuming that the output voltage in this case is approximately equal to V_{OL}.

It is seen from this discussion that the critical voltage points, the general shape of the inverter VTC, and ultimately, the noise margins, are determined essentially by the threshold voltages of the driver and the load devices, and by the driver-to-load ratio (k_{driver}/k_{load}). Since the threshold voltages are usually set by the fabrication process, the driver-to-load ratio emerges as a primary design parameter which can be adjusted to achieve the desired VTC shape. Notice that with $k'_{n,driver} = k'_{n,load}$, the driver-to-load ratio is solely determined by the (W/L) ratios of the driver and the load transistors, i.e., by the device geometries. Figure 5.14 shows the VTCs of depletion-load inverter circuits with different driver-to-load ratios $k_R = (k_{driver}/k_{load})$.

One important observation is that, unlike in the enhancement-load inverter case, a sharp VTC transition and larger noise margins can be obtained with relatively small driver-to-load ratios. Thus, the total area occupied by a depletion-load inverter circuit with an acceptable circuit performance is expected to be much smaller than the area occupied by a comparable resistive-load or enhancement-load inverter.

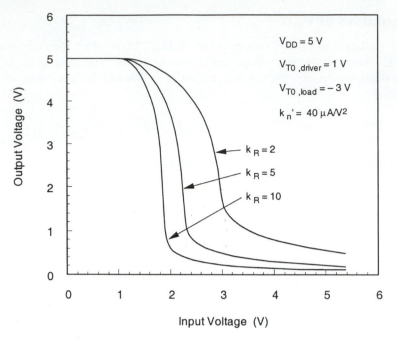

Figure 5.14   Voltage transfer characteristics of depletion-load inverters, with different driver-to-load ratios.

Design of Depletion-Load Inverters

Based on the VTC analysis given in the previous section, we can now consider the *design* of depletion-load inverters to satisfy certain DC performance criteria. In the broadest sense, the designable parameters in an inverter circuit are: (i) the power supply voltage V_{DD}, (ii) the threshold voltages of the driver and the load transistors, and (iii) the (W/L) ratios of the driver and the load transistors. In most practical cases, however, the power supply voltage and the device threshold voltages are dictated by other external constraints and by the fabrication process; thus, they cannot be adjusted for every individual inverter circuit to satisfy performance requirements. This leaves the (W/L) ratio of the transistors and more specifically, the driver-to-load ratio k_R, as the primary design parameter.

Note that the power supply voltage V_{DD} of the inverter circuit also determines the level of the output high voltage V_{OH}, since $V_{OH} = V_{DD}$. Of the remaining three critical voltages on the VTC, the output low voltage V_{OL} is usually the most significant design constraint. Designing the inverter to achieve a certain V_{OL} value will automatically set the other two critical voltages, V_{IL} and V_{IH}, as well. Equation (5.37) can be rearranged to calculate the driver-to-load ratio that achieves a target V_{OL} value, assuming that the power supply voltage and the threshold voltage values are set previously by independent design and processing constraints.

$$k_R = \frac{k_{driver}}{k_{load}} = \frac{|V_{T,load}(V_{OL})|^2}{2(V_{OH} - V_{T0})V_{OL} - V_{OL}^2} \tag{5.46}$$

Here, the driver-to-load ratio is given by

$$k_R = \frac{k'_{n,driver} \cdot \left(\dfrac{W}{L}\right)_{driver}}{k'_{n,load} \cdot \left(\dfrac{W}{L}\right)_{load}} \tag{5.47}$$

Since the channel doping densities and, consequently, the channel electron mobilities of the enhancement-type driver transistor and the depletion-type load transistor are not equal, we shall expect that $k'_{n,driver} \neq k'_{n,load}$, in general. Only if $k'_{n,driver} \approx k'_{n,load}$ can the driver-to-load ratio be reduced to

$$k_R = \frac{\left(\dfrac{W}{L}\right)_{driver}}{\left(\dfrac{W}{L}\right)_{load}} \tag{5.48}$$

Finally, note that the design procedure thus far determines the *ratio* of the driver and the load transconductances, but not the specific (W/L) ratio of each transistor. As a result, one can propose a number of designs with different (W/L) ratios for the driver (and for the load) device, each of which satisfies the driver-to-load ratio condition. The actual sizes of the driver and the load transistors are usually determined by other design constraints, such as the current-drive capability, the steady-state power dissipation, and the transient switching speed.

Power and Area Considerations

The steady-state DC power consumption of the depletion-load inverter circuit can be easily found by calculating the amount of current being drawn from the power supply, during the input-low state and the input-high state. When the input voltage is low, i.e., when the driver transistor is cut-off and $V_{out} = V_{OH} = V_{DD}$, there should be no significant current flow through the driver and the load transistors. Consequently, the inverter does not dissipate DC power under this condition when there is no subthreshold leakage current. When the input is at the high state with $V_{in} \approx V_{DD}$ and $V_{out} = V_{OL}$, on the other hand, both the driver and the load transistors conduct a significant current, given by

$$
\begin{aligned}
I_{DC}(V_{in} = V_{DD}) &= \frac{k_{load}}{2} \cdot [-V_{T,load}(V_{OL})]^2 \\
&= \frac{k_{driver}}{2} \cdot [2 \cdot (V_{OH} - V_{T0}) \cdot V_{OL} - V_{OL}^2]
\end{aligned} \tag{5.49}
$$

Assuming that the input voltage level is low during 50% of the operation time and high during the other 50%, the overall average DC power consumption of this circuit can be estimated as follows:

$$P_{DC} = \frac{V_{DD}}{2} \cdot \frac{k_{load}}{2} \cdot [-V_{T,load}(V_{OL})]^2 \tag{5.50}$$

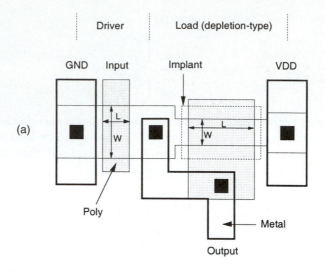

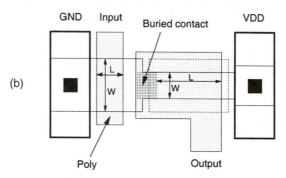

Figure 5.15 Sample layout of depletion-load inverter circuits (a) with output contact on diffusion and (b) with buried contact.

Figure 5.15(a) shows a simplified layout of the depletion-load inverter circuit. Note that the drain of the enhancement-type driver and the source of the depletion-type load transistor share a common n^+ diffusion region, which saves silicon area compared to two separate diffusion regions. The threshold voltage of the depletion-type load device is adjusted by a donor implant into the channel region. The width-to-length ratio of the driver transistor is seen to be larger than the width-to-length ratio of the load, which yields a driver-to-load ratio of about 4. Overall, this inverter circuit configuration is relatively compact, and it occupies a significantly smaller area compared to those for resistive-load or enhancement-load inverters with similar performance.

The area requirements of the depletion-load inverter circuit can be reduced even further by using a *buried contact* for connecting the gate and the source of the load transistor, as shown in Fig. 5.15(b). In this case, the polysilicon gate of the depletion-mode transistor makes a direct ohmic contact with the n^+ source diffusion. The contact window on the intermediate diffusion area can be omitted, which results in a further reduction of the total area occupied by the inverter.

EXAMPLE 5.3

Calculate the critical voltages (V_{OL}, V_{OH}, V_{IL}, V_{IH}) and find the noise margins of the following depletion-load inverter circuit:

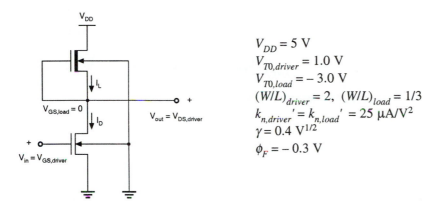

$$V_{DD} = 5 \text{ V}$$
$$V_{T0,driver} = 1.0 \text{ V}$$
$$V_{T0,load} = -3.0 \text{ V}$$
$$(W/L)_{driver} = 2, \ (W/L)_{load} = 1/3$$
$$k_{n,driver}' = k_{n,load}' = 25 \ \mu\text{A/V}^2$$
$$\gamma = 0.4 \text{ V}^{1/2}$$
$$\phi_F = -0.3 \text{ V}$$

First, the output high voltage is simply found according to (5.36) as $V_{OH} = V_{DD} = 5 \text{ V}$.

To calculate the output low voltage V_{OL}, we must solve (5.33) and (5.38) simultaneously, using numerical iterations. We start the iterations by assuming that the output voltage is equal to zero, thus letting $V_{T,load} = V_{T0,load} = -3 \text{ V}$. Solving (5.38) with this assumption yields a first-order estimate for V_{OL}.

$$V_{OL} = V_{OH} - V_{T0} - \sqrt{(V_{OH} - V_{T0})^2 - \left(\frac{k_{load}}{k_{driver}}\right) \cdot |V_{T,load}(V_{OL})|^2}$$

$$= 5 - 1 - \sqrt{(5-1)^2 - \left(\frac{1}{6}\right)|3|^2} = 0.192 \text{ V}$$

Now, the threshold voltage of the depletion-load device can be updated by substituting this output voltage into (5.33).

$$V_{T,load} = V_{T0,load} + \gamma\left(\sqrt{|2\phi_F| + V_{OL}} - \sqrt{|2\phi_F|}\right)$$

$$= -3 + 0.4\left(\sqrt{0.6 + 0.2} - \sqrt{0.6}\right) = -2.95 \text{ V}$$

Using this new value for $V_{T,load}$, we now recalculate V_{OL}, again according to (5.38).

$$V_{OL} = 0.186 \text{ V}$$

$$V_{T,load} = -2.95 \text{ V}$$

At this point, we can stop the iteration process since the threshold voltage of the load device has not changed in the two significant digits after the decimal point. Continuing the iteration would not produce a perceptible improvement of V_{OL}.

The calculation of V_{IL} involves simultaneous solution of (5.33), (5.39), and (5.41), using numerical iterations. When the input voltage is equal to V_{IL}, we expect that the output voltage is slightly lower than the output high voltage, V_{OH}. As a first-order approximation, assume that $V_{out} = V_{OH} = 5$ V for $V_{in} = V_{IL}$. Then, the threshold voltage of the load device can be estimated as $V_{T,load}(V_{out} = 5\,V) = -2.36$ V. Substituting this value into (5.41) gives V_{IL} as a function of the output voltage V_{out}.

$$V_{IL}(V_{out}) = V_{T0} + \frac{k_{load}}{k_{driver}} \cdot [V_{out} - V_{DD} + |V_{T,load}(V_{out})|]$$

$$= 1 + \left(\frac{1}{6}\right)(V_{out} - 5 + 2.36) = 0.167 V_{out} + 0.56$$

This expression can be rearranged as

$$V_{out} = 6V_{IL} - 3.35$$

Now, substitute this into the KCL equation (5.39) to obtain the following quadratic equation for V_{IL}:

$$\frac{k_{driver}}{2} \cdot (V_{IL} - V_{T0})^2 = \frac{k_{load}}{2} \cdot [2|V_{T,load}(V_{out})| \cdot (V_{DD} - 6V_{IL} + 3.35)$$

$$- (V_{DD} - 6V_{IL} + 3.35)^2]$$

$$2 \cdot (V_{IL} - 1)^2 = \frac{1}{3} \cdot [2 \cdot 2.36 \cdot (5 - 6V_{IL} + 3.35) - (5 - 6V_{IL} + 3.35)^2]$$

The solution of this second-order equation yields two possible values for V_{IL}.

$$V_{IL} = \begin{cases} 0.98 \text{ V} \\ \underline{1.36 \text{ V}} \end{cases}$$

Note that V_{IL} must be larger than the threshold voltage V_{T0} of the driver transistor, hence, $V_{IL} = 1.36$ V is the physically correct solution. The output voltage level at this point can also be found as

$$V_{out} = 6 \cdot 1.36 - 3.35 = 4.81 \text{ V}$$

which significantly improves our initial assumption of $V_{out} = 5$ V. At this point, the threshold voltage of the load transistor must be recalculated, in order to update its value. Substituting $V_{out} = 4.81$ V into (5.33) yields $V_{T,load} = -2.38$ V. We observe that this value is only slightly higher (by 20 mV) than the threshold voltage value used in the previous calculations. For practical purposes, we can terminate the numerical iteration at this stage and accept $V_{IL} = 1.36$ V as a fairly accurate estimate.

To calculate V_{IH}, we first have to find the numerical value of $(dV_{T,load}/V_{out})$ using (5.45). When the input voltage is equal to V_{IH}, the output voltage is expected to be relatively low. As a first-order approximation, assume that the output voltage level is $V_{out} = V_{OL} \approx 0.2$ V when $V_{in} = V_{IH}$. The threshold voltage of the load device can also be estimated as $V_{T,load}(V_{out} = 0.2 \text{ V}) = -2.95$ V. Thus,

$$\frac{dV_{T,load}}{dV_{out}} = \frac{\gamma}{2\sqrt{|2\phi_F| + V_{out}}} = \frac{0.4}{2\sqrt{0.6 + 0.2}} = 0.22$$

This value can now be used in (5.44) to find V_{IH} as a function of the output voltage.

$$V_{IH}(V_{out}) = V_{T0} + 2V_{out} + \frac{k_{load}}{k_{driver}} \cdot [-V_{T,load}(V_{out})] \cdot \left(\frac{dV_{T,load}}{dV_{out}}\right)$$

$$= 1 + 2V_{out} + \left(\frac{1}{6}\right) \cdot 2.95 \cdot 0.22 = 2V_{out} + 1.1$$

This expression is rearranged as:

$$V_{out} = 0.5 \, V_{IH} - 0.55$$

Next, substitute V_{out} in the KCL equation (5.42), to obtain

$$2 \cdot [2 \cdot (V_{IH} - 1) \cdot (0.5V_{IH} - 0.55) - (0.5V_{IH} - 0.55)^2] = \frac{1}{3} \cdot (2.95)^2$$

The solution of this simple quadratic equation yields two values for V_{IH}.

$$V_{IH} = \begin{cases} -0.35 \text{ V} \\ \underline{2.43 \text{ V}} \end{cases}$$

where $V_{IH} = 2.43$ V is the physically correct solution. The output voltage level at this point is calculated as

$$V_{out} = 0.5 \cdot 2.43 - 0.55 = 0.67 \text{ V}$$

With this updated output voltage value, we can now reevaluate the load threshold voltage as $V_{T,load}(V_{out} = 0.67 \text{ V}) = -2.9$ V, and the $(dV_{T,load}/V_{out})$ value as

$$\frac{dV_{T,load}}{dV_{out}} = 0.18$$

Note both of these values are fairly close to those used at the beginning of this iteration process. Repeating the iterative calculation will provide only a marginal improvement of accuracy, thus, we may accept $V_{IH} = 2.43$ V as a good estimate.

In conclusion, the noise margins for high signal levels and for low signal levels can be found as follows:

$$NM_H = V_{OH} - V_{IH} = 2.57 \text{ V}$$

$$NM_L = V_{IL} - V_{OL} = 1.17 \text{ V}$$

5.4 CMOS Inverter

All of the inverter circuits considered so far had the general circuit structure shown in Fig. 5.3, consisting of an enhancement-type nMOS driver transistor and a load device which can be a resistor, an enhancement-type nMOS transistor, or a depletion-type

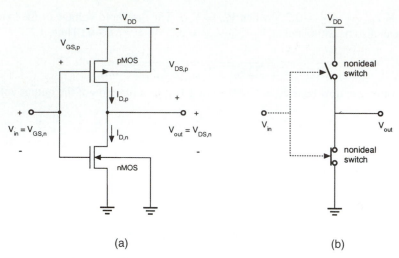

Figure 5.16   (a) CMOS inverter circuit. (b) Simplified view of the CMOS inverter, consisting of two complementary nonideal switches.

nMOS transistor acting as a nonlinear resistor. In this general configuration, the input signal is always applied to the gate of the driver transistor, and the operation of the inverter is controlled primarily by switching the driver. Now, we will turn our attention to a radically different inverter structure, which consists of an enhancement-type nMOS transistor and an enhancement-type pMOS transistor, operating in complementary mode (Fig. 5.16). This configuration is called Complementary MOS (CMOS). The circuit topology is complementary push-pull in the sense that for high input, the nMOS transistor drives (pulls down) the output node while the pMOS transistor acts as the load, and for low input the pMOS transistor drives (pulls up) the output node while the nMOS transistor acts as the load. Consequently, both devices contribute equally to the circuit operation characteristics.

The CMOS inverter has two important advantages over the other inverter configurations. The first and perhaps the most important advantage is that the steady-state power dissipation of the CMOS inverter circuit is virtually negligible, except for small power dissipation due to leakage currents. However, as mentioned earlier, the trend of increasing subthreshold leakage currents in deep sub-micron technologies causes great design challenges. In all other inverter structures examined so far, a nonzero steady-state current is drawn from the power source when the driver transistor is turned on, which results in a significant DC power consumption. The other advantages of the CMOS configuration are that the voltage transfer characteristic (VTC) exhibits a full output voltage swing between 0 V and V_{DD}, and that the VTC transition is usually very sharp. Thus, the VTC of the CMOS inverter resembles that of an ideal inverter.

Since nMOS and pMOS transistors must be fabricated on the same chip side-by-side, the CMOS process is more complex than the standard nMOS-only process. In particular, the CMOS process must provide an n-type substrate for the pMOS transistors and a p-type substrate for the nMOS transistors. This can be achieved by building either n-type *tubs* (wells) on a p-type wafer, or by building p-type tubs on an n-type wafer (cf. Chapter 2). In addition, the close proximity of an nMOS and a pMOS transistor may lead to the formation of two parasitic bipolar transistors, causing a *latch-up* condition. In order to prevent this undesirable effect, additional *guard rings* must be built around the nMOS and the pMOS transistors as well (cf. Chapter 13). The increased process complexity of CMOS fabrication may be considered as the price being paid for the improvements achieved in power consumption and noise margins.

Circuit Operation

In Fig. 5.16, note that the input voltage is connected to the gate terminals of both the nMOS and the pMOS transistors. Thus, both transistors are driven directly by the input signal, V_{in}. The substrate of the nMOS transistor is connected to the ground, while the substrate of the pMOS transistor is connected to the power supply voltage, V_{DD}, in order to reverse-bias the source and drain junctions. Since $V_{SB} = 0$ for both devices, there will be no substrate-bias effect for either device. It can be seen from the circuit diagram in Fig. 5.16 that

$$V_{GS,n} = V_{in}$$
$$V_{DS,n} = V_{out}$$

(5.51)

and also,

$$V_{GS,p} = -(V_{DD} - V_{in})$$
$$V_{DS,p} = -(V_{DD} - V_{out})$$

(5.52)

We will start our analysis by considering two simple cases. When the input voltage is smaller than the nMOS threshold voltage, i.e., when $V_{in} < V_{T0,n}$, the nMOS transistor is cut-off. At the same time, the pMOS transistor is on, operating in the linear region. Since the drain currents of both transistors are approximately equal to zero (except for small leakage currents), i.e.,

$$I_{D,n} = I_{D,p} = 0$$

(5.53)

the drain-to-source voltage of the pMOS transistor is also equal to zero, and the output voltage V_{OH} is equal to the power supply voltage.

$$V_{out} = V_{OH} = V_{DD}$$

(5.54)

On the other hand, when the input voltage exceeds $(V_{DD} + V_{T0,p})$, the pMOS transistor is turned off. In this case, the nMOS transistor is operating in the linear region, but its drain-to-source voltage is equal to zero because condition (5.53) is satisfied. Consequently, the output voltage of the circuit is

$$V_{out} = V_{OL} = 0$$

(5.55)

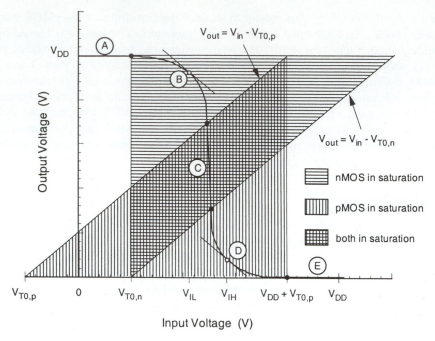

Figure 5.17 Operating regions of the nMOS and the pMOS transistors.

Next, we examine the operating modes of the nMOS and the pMOS transistors as functions of the input and output voltages. The nMOS transistor operates in *saturation* if $V_{in} > V_{T0,n}$ *and* if the following condition is satisfied.

$$V_{DS,n} \geq V_{GS,n} - V_{T0,n} \quad \Leftrightarrow \quad V_{out} \geq V_{in} - V_{T0,n} \tag{5.56}$$

The pMOS transistor operates in *saturation* if $V_{in} < (V_{DD} + V_{T0,p})$, *and* if:

$$V_{DS,p} \leq V_{GS,p} - V_{T0,p} \quad \Leftrightarrow \quad V_{out} \leq V_{in} - V_{T0,p} \tag{5.57}$$

Both of these conditions for device saturation are illustrated graphically as shaded areas on the $V_{out} - V_{in}$ plane in Fig. 5.17. A typical CMOS inverter voltage transfer characteristic is also superimposed for easy reference. Here, we identify five distinct regions, labeled A through E, each corresponding to a different set of operating conditions. The table below lists these regions and the corresponding critical input and output voltage levels.

Region	V_{in}	V_{out}	nMOS	pMOS
A	$< V_{T0,n}$	V_{OH}	cut-off	linear
B	V_{IL}	high $\approx V_{OH}$	saturation	linear
C	V_{th}	V_{th}	saturation	saturation
D	V_{IH}	low $\approx V_{OL}$	linear	saturation
E	$> (V_{DD} + V_{T0,p})$	V_{OL}	linear	cut-off

In Region A, where $V_{in} < V_{T0,n}$, the nMOS transistor is cut-off and the output voltage is equal to $V_{OH} = V_{DD}$. As the input voltage is increased beyond $V_{T0,n}$ (into Region B), the nMOS transistor starts conducting in saturation mode and the output voltage begins to decrease. Also note that the critical voltage V_{IL} which corresponds to $(dV_{out}/dV_{in}) = -1$ is located within Region B. As the output voltage further decreases, the pMOS transistor enters saturation at the boundary of Region C. It is seen from Fig. 5.17 that the inverter threshold voltage, where $V_{in} = V_{out}$, is located in Region C. When the output voltage V_{out} falls below $(V_{in} - V_{T0,n})$, the nMOS transistor starts to operate in linear mode. This corresponds to Region D in Fig. 5.17, where the critical voltage point V_{IH} with $(dV_{out}/dV_{in}) = -1$ is also located. Finally, in Region E, with the input voltage $V_{in} > (V_{DD} + V_{T0,p})$, the pMOS transistor is cut-off, and the output voltage is $V_{OL} = 0$.

In a simplistic analogy, the nMOS and the pMOS transistors can be seen as nearly ideal switches—controlled by the input voltage—that connect the output node to the power supply voltage or to the ground potential, depending on the input voltage level. The qualitative overview of circuit operation, illustrated in Fig. 5.17 and discussed so far, also highlights the complementary nature of the CMOS inverter. The most significant feature of this circuit is that the current drawn from the power supply in both of these steady-state operating points, i.e., in Region A and in Region E, is nearly equal to zero. The only current that flows in either case is the very small leakage current of the reverse-biased source and drain junctions. The CMOS inverter can *drive* any load, such as interconnect capacitance or fan-out logic gates which are connected to its output node, either by supplying current to the load, or by sinking current from the load.

The steady-state input-output voltage characteristics of the CMOS inverter can be better visualized by considering the interaction of individual nMOS and pMOS transistor characteristics in the current-voltage space. We already know that the drain current $I_{D,n}$ of the nMOS transistor is a function of the voltages $V_{GS,n}$ and $V_{DS,n}$. Hence, the nMOS drain current is also a function of the inverter input and output voltages V_{in} and V_{out}, according to (5.51).

$$I_{D,n} = f(V_{in}, V_{out})$$

This two-variable function, which is essentially described by the current equations (3.54) through (3.56), can be represented as a *surface* in the three-dimensional current-voltage space. Figure 5.18 shows this $I_{D,n}(V_{in}, V_{out})$ surface for the nMOS transistor.

Similarly, the drain current $I_{D,p}$ of the pMOS transistor is also a function of the inverter input and output voltages V_{in} and V_{out}, according to (5.52).

$$I_{D,p} = f(V_{in}, V_{out})$$

This two-variable function, described by the current equations (3.57) through (3.59), can be represented as another surface in the three-dimensional current-voltage space. Figure 5.19 shows the corresponding $I_{D,p}(V_{in}, V_{out})$ surface for the pMOS transistor.

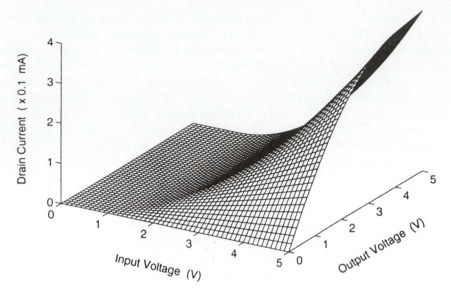

Figure 5.18   Current-voltage surface representing the nMOS transistor characteristics.

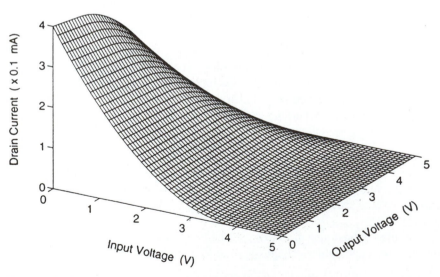

Figure 5.19   Current-voltage surface representing the pMOS transistor characteristics.

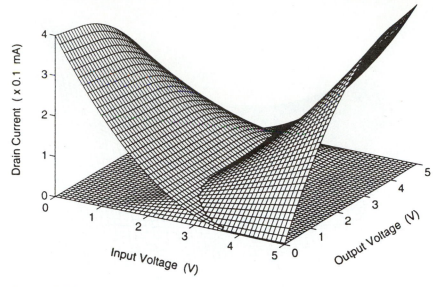

Figure 5.20 Intersection of the current-voltage surfaces shown in Figs. 5.18 and 5.19.

Remember that in a CMOS inverter operating in steady-state, the drain current of the nMOS transistor is always equal to the drain current of the pMOS transistor, according to KCL.

$$I_{D,n} = I_{D,p}$$

Thus, the *intersection* of the two current-voltage surfaces shown in Figs. 5.18 and 5.19 will give the operating curve of the CMOS inverter circuit in the three-dimensional current-voltage space. The intersection of the two characteristic surfaces is shown in Fig. 5.20. The intersecting surfaces are shown from a different viewing angle in Fig. 5.21, with the intersection curve highlighted in bold.

It is clear that the vertical projection of the intersection curve on the $V_{in} - V_{out}$ plane produces the typical CMOS inverter voltage transfer characteristic already shown in Fig. 5.17. Similarly, the horizontal projection of the intersection curve on the $I_D - V_{in}$ plane gives the steady-state current drawn by the inverter from the power supply voltage as a function of the input voltage. In the following, we will present an in-depth analysis of the CMOS inverter static characteristics, by calculating the critical voltage points on the VTC. It has already been established that $V_{OH} = V_{DD}$ and $V_{OL} = 0$ for this inverter; thus, we will devote our attention to V_{IL}, V_{IH} and the inverter switching threshold, V_{th}.

Calculation of V_{IL}

By definition, the slope of the VTC is equal to (-1), i.e., $dV_{out}/dV_{in} = -1$ when the input voltage is $V_{in} = V_{IL}$. Note that in this case, the nMOS transistor operates in

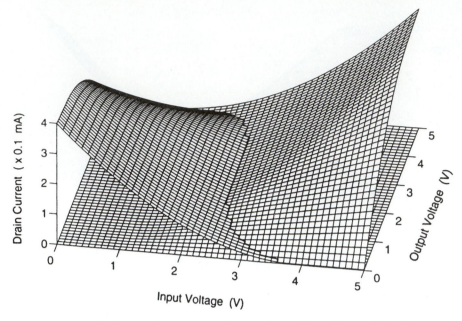

Figure 5.21   The intersecting current-voltage surfaces shown from a different viewing angle. Notice that projection of the intersection curve on the voltage plane gives the VTC.

saturation while the pMOS transistor operates in the linear region. From $I_{D,n} = I_{D,p}$, we obtain the following current equation:

$$\frac{k_n}{2} \cdot (V_{GS,n} - V_{T0,n})^2 = \frac{k_p}{2} \cdot \left[2 \cdot (V_{GS,p} - V_{T0,p}) \cdot V_{DS,p} - V_{DS,p}^2\right] \qquad (5.58)$$

Using equations (5.51) and (5.52), this expression can be rewritten as

$$\frac{k_n}{2} \cdot (V_{in} - V_{T0,n})^2$$
$$= \frac{k_p}{2} \cdot \left[2 \cdot (V_{in} - V_{DD} - V_{T0,p}) \cdot (V_{out} - V_{DD}) - (V_{out} - V_{DD})^2\right] \qquad (5.59)$$

To satisfy the derivative condition at V_{IL}, we differentiate both sides of (5.59) with respect to V_{in}.

$$k_n \cdot (V_{in} - V_{T0,n}) = k_p \cdot \left[(V_{in} - V_{DD} - V_{T0,p}) \cdot \left(\frac{dV_{out}}{dV_{in}}\right) + (V_{out} - V_{DD})\right.$$
$$\left. - (V_{out} - V_{DD}) \cdot \left(\frac{dV_{out}}{dV_{in}}\right)\right] \qquad (5.60)$$

Substituting $V_{in} = V_{IL}$ and $(dV_{out}/dV_{in}) = -1$ in (5.60), we obtain

$$k_n \cdot (V_{IL} - V_{T0,n}) = k_p \cdot (2V_{out} - V_{IL} + V_{T0,p} - V_{DD}) \qquad (5.61)$$

The critical voltage V_{IL} can now be found as a function of the output voltage V_{out}, as follows:

$$V_{IL} = \frac{2V_{out} + V_{T0,p} - V_{DD} + k_R V_{T0,n}}{1 + k_R} \quad (5.62)$$

where k_R is defined as

$$k_R = \frac{k_n}{k_p}$$

This equation must be solved together with the KCL equation (5.59) to obtain the numerical value of V_{IL} and the corresponding output voltage, V_{out}. Note that the solution is fairly straightforward and does not require numerical iterations as in the previous cases, since none of the transistors is subject to substrate-bias effects.

Calculation of V_{IH}

When the input voltage is equal to V_{IH}, the nMOS transistor operates in the linear region, and the pMOS transistor operates in saturation. Applying KCL to the output node, we obtain

$$\frac{k_n}{2} \cdot \left[2 \cdot (V_{GS,n} - V_{T0,n}) \cdot V_{DS,n} - V_{DS,n}^2 \right] = \frac{k_p}{2} \cdot (V_{GS,p} - V_{T0,p})^2 \quad (5.63)$$

Using equations (5.51) and (5.52), this expression can be rewritten as

$$\frac{k_n}{2} \cdot \left[2 \cdot (V_{in} - V_{T0,n}) \cdot V_{out} - V_{out}^2 \right] = \frac{k_p}{2} \cdot (V_{in} - V_{DD} - V_{T0,p})^2 \quad (5.64)$$

Now, differentiate both sides of (5.64) with respect to V_{in}.

$$k_n \cdot \left[(V_{in} - V_{T0,n}) \cdot \left(\frac{dV_{out}}{dV_{in}} \right) + V_{out} - V_{out} \cdot \left(\frac{dV_{out}}{dV_{in}} \right) \right]$$
$$= k_p \cdot (V_{in} - V_{DD} - V_{T0,p}) \quad (5.65)$$

Substituting $V_{in} = V_{IH}$ and $(dV_{out}/dV_{in}) = -1$ in (5.65), we obtain

$$k_n \cdot (-V_{IH} + V_{T0,n} + 2V_{out}) = k_p \cdot (V_{IH} - V_{DD} - V_{T0,p}) \quad (5.66)$$

The critical voltage V_{IH} can now be found as a function of V_{out} as follows:

$$V_{IH} = \frac{V_{DD} + V_{T0,p} + k_R \cdot (2V_{out} + V_{T0,n})}{1 + k_R} \quad (5.67)$$

Again, this equation must be solved simultaneously with the KCL equation (5.64) to obtain the numerical values of V_{IH} and V_{out}.

Calculation of V_{th}

The inverter threshold voltage is defined as $V_{th} = V_{in} = V_{out}$. Since the CMOS inverter exhibits large noise margins and a very sharp VTC transition, the inverter threshold voltage emerges as an important parameter characterizing the DC performance of the inverter. For $V_{in} = V_{out}$, both transistors are expected to be in saturation mode; hence, we can write the following KCL equation.

$$\frac{k_n}{2} \cdot (V_{GS,n} - V_{T0,n})^2 = \frac{k_p}{2} \cdot (V_{GS,p} - V_{T0,p})^2 \quad (5.68)$$

Replacing $V_{GS,n}$ and $V_{GS,p}$ in (5.68) according to (5.51) and (5.52), we obtain

$$\frac{k_n}{2} \cdot (V_{in} - V_{T0,n})^2 = \frac{k_p}{2} \cdot (V_{in} - V_{DD} - V_{T0,p})^2 \tag{5.69}$$

The correct solution for V_{in} for this equation is

$$V_{in} \cdot \left(1 + \sqrt{\frac{k_p}{k_n}}\right) = V_{T0,n} + \sqrt{\frac{k_p}{k_n}} \cdot (V_{DD} + V_{T0,p}) \tag{5.70}$$

Finally, the inverter threshold (switching threshold) voltage V_{th} is found as

$$V_{th} = \frac{V_{T0,n} + \sqrt{\dfrac{1}{k_R}} \cdot (V_{DD} + V_{T0,p})}{\left(1 + \sqrt{\dfrac{1}{k_R}}\right)} \tag{5.71}$$

Note that the inverter threshold voltage is defined as $V_{th} = V_{in} = V_{out}$. When the input voltage is equal to V_{th}, however, we find that the output voltage can actually attain any value between $(V_{th} - V_{T0,n})$ and $(V_{th} - V_{T0,p})$, without violating the voltage conditions used in this analysis. This is due to the fact that the VTC segment corresponding to Region C in Fig. 5.17 becomes completely vertical if the channel-length modulation effect is neglected, i.e., if $\lambda = 0$. In more realistic cases with $\lambda > 0$, the VTC segment in Region C exhibits a finite, but very large, slope. Figure 5.22 shows the variation of the inversion (switching) threshold voltage V_{th} as a

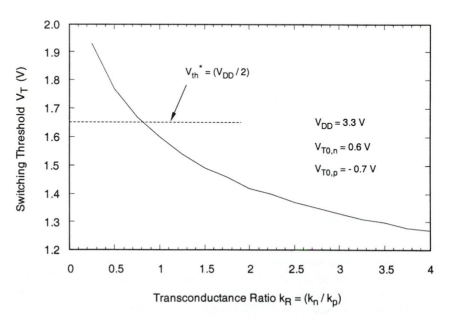

Figure 5.22   Variation of the inversion threshold voltage as a function of k_R.

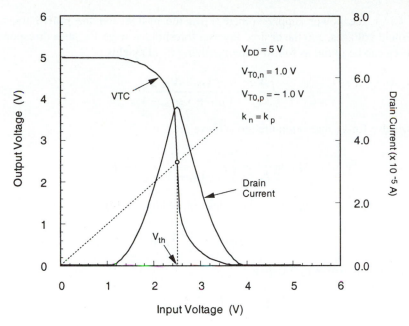

Figure 5.23 Typical VTC and the power supply current of a CMOS inverter circuit.

function of the transconductance ratio k_R, and for fixed values of V_{DD}, $V_{T0,n}$ and $V_{T0,p}$.

It has already been established that the CMOS inverter does not draw any significant current from the power source, except for small leakage and subthreshold currents, when the input voltage is either smaller than $V_{T0,n}$ or larger than $(V_{DD} + V_{T0,p})$. The nMOS and the pMOS transistors conduct a nonzero current, on the other hand, during low-to-high and high-to-low transitions, i.e., in Regions B, C, and D. It can be shown that the current being drawn from the power source during transition reaches its peak value when $V_{in} = V_{th}$. In other words, the maximum current is drawn when both transistors are operating in saturation mode. Figure 5.23 shows the voltage transfer characteristic of a typical CMOS inverter circuit and the power supply current, as a function of the input voltage.

Design of CMOS Inverters

The inverter threshold voltage V_{th} was identified as one of the most important parameters that characterize the steady-state input-output behavior of the CMOS inverter circuit. The CMOS inverter can, by virtue of its complementary push-pull operating mode, provide a full output voltage swing between 0 and V_{DD}, and therefore, the noise margins are relatively wide. Thus, the problem of designing a CMOS inverter can be reduced to setting the inverter threshold to a desired voltage value.

Given the power supply voltage V_{DD}, the nMOS and the pMOS transistor threshold voltages, and the desired inverter threshold voltage V_{th}, the corresponding ratio k_R can be found as follows. Reorganizing (5.71) yields

$$\sqrt{\frac{1}{k_R}} = \frac{V_{th} - V_{T0,n}}{V_{DD} + V_{T0,p} - V_{th}} \tag{5.72}$$

Now solve for k_R that is required to achieve the given V_{th}.

$$k_R = \frac{k_n}{k_p} = \left(\frac{V_{DD} + V_{T0,p} - V_{th}}{V_{th} - V_{T0,n}}\right)^2 \tag{5.73}$$

Recall that the switching threshold voltage of an *ideal* inverter is defined as

$$V_{th,ideal} = \frac{1}{2} \cdot V_{DD} \tag{5.74}$$

Substituting (5.74) in (5.73) gives

$$\left(\frac{k_n}{k_p}\right)_{ideal} = \left(\frac{0.5V_{DD} + V_{T0,p}}{0.5V_{DD} - V_{T0,n}}\right)^2 \tag{5.75}$$

for a near-ideal CMOS VTC that satisfies the condition (5.74). Since the operations of the nMOS and the pMOS transistors of the CMOS inverter are fully complementary, we can achieve completely symmetric input-output characteristics by setting the threshold voltages as $V_{T0} = V_{T0,n} = |V_{T0,p}|$. This reduces (5.75) to:

$$\left(\frac{k_n}{k_p}\right)_{\substack{symmetric \\ inverter}} = 1 \tag{5.76}$$

Note that the ratio k_R is defined as

$$\frac{k_n}{k_p} = \frac{\mu_n \, C_{ox} \cdot \left(\dfrac{W}{L}\right)_n}{\mu_p \, C_{ox} \cdot \left(\dfrac{W}{L}\right)_p} = \frac{\mu_n \cdot \left(\dfrac{W}{L}\right)_n}{\mu_p \cdot \left(\dfrac{W}{L}\right)_p} \tag{5.77}$$

assuming that the gate oxide thickness t_{ox}, and hence, the gate oxide capacitance C_{ox} have the same value for both nMOS and pMOS transistors. The unity-ratio condition (5.76) for the ideal symmetric inverter requires that

$$\frac{\left(\dfrac{W}{L}\right)_n}{\left(\dfrac{W}{L}\right)_p} = \frac{\mu_p}{\mu_n} \approx \frac{230 \text{ cm}^2/\text{V} \cdot \text{s}}{580 \text{ cm}^2/\text{V} \cdot \text{s}} \tag{5.78}$$

Hence,

$$\left(\frac{W}{L}\right)_p \approx 2.5 \left(\frac{W}{L}\right)_n \tag{5.79}$$

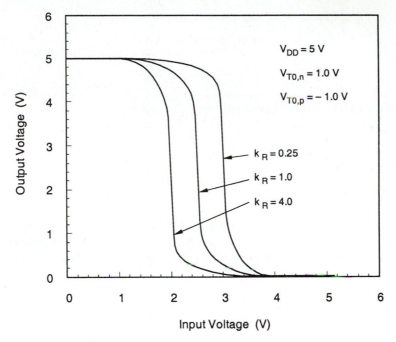

Figure 5.24 Voltage transfer characteristics of three CMOS inverters, with different nMOS-to-pMOS ratios.

It should be noted that the numerical values used in (5.78) for electron and hole mobilities are *typical* values, and that exact μ_n and μ_p values will vary with surface doping concentration of the substrate and the tub. The VTCs of three CMOS inverter circuits with different k_R ratios are shown in Fig. 5.24. It can be seen clearly that the inverter threshold voltage V_{th} shifts to lower values with increasing k_R ratio.

For a symmetric CMOS inverter with $V_{T0,n} = |V_{T0,p}|$ and $k_R = 1$, the critical voltage V_{IL} can be found, using (5.62), as follows:

$$V_{IL} = \frac{1}{8} \cdot (3V_{DD} + 2V_{T0,n}) \tag{5.80}$$

Also, the critical voltage V_{IH} is found as

$$V_{IH} = \frac{1}{8} \cdot (5V_{DD} - 2V_{T0,n}) \tag{5.81}$$

Note that the sum of V_{IL} and V_{IH} is always equal to V_{DD} in a symmetric inverter.

$$V_{IL} + V_{IH} = V_{DD} \tag{5.82}$$

The noise margins NM_L and NM_H for this symmetric CMOS inverter are now calculated using (5.3) and (5.4).

$$NM_L = V_{IL} - V_{OL} = V_{IL}$$
$$NM_H = V_{OH} - V_{IH} = V_{DD} - V_{IH} \tag{5.83}$$

which are equal to each other, and also to V_{IL}.

$$NM_L = NM_H = V_{IL} \tag{5.84}$$

EXAMPLE 5.4

Consider a CMOS inverter circuit with the following parameters:

$V_{DD} = 3.3$ V

$V_{T0,n} = 0.6$ V

$V_{T0,p} = -0.7$ V

$k_n = 200 \ \mu\text{A/V}^2$

$k_p = 80 \ \mu\text{A/V}^2$

Calculate the noise margins of the circuit. Notice that the CMOS inverter being considered here has $k_R = 2.5$ and $V_{T0,n} \neq |V_{T0,p}|$; hence, it is not a symmetric inverter.

First, the output low voltage V_{OL} and the output high voltage V_{OH} are found, using (5.54) and (5.55), as $V_{OL} = 0$ and $V_{OH} = 3.3$ V. To calculate V_{IL} in terms of the output voltage, we use (5.62).

$$V_{IL} = \frac{2V_{out} + V_{T0,p} - V_{DD} + k_R V_{T0,n}}{1 + k_R}$$

$$= \frac{2V_{out} - 0.7 - 3.3 + 1.5}{1 + 2.5} = 0.57V_{out} - 0.71$$

Now substitute this expression into the KCL equation (5.59).

$$2.5(0.57V_{out} - 0.71 - 0.6)^2 = 2(0.57V_{out} - 0.71 - 3.3 + 0.7)(V_{out} - 3.3)$$
$$- (V_{out} - 3.3)^2$$

This expression yields a second-order polynomial in V_{out}, as follows:

$$0.66V_{out}^2 + 0.05V_{out} - 6.65 = 0$$

Only one root of this quadratic equation corresponds to a physically correct solution for V_{out} (i.e., $V_{out} > 0$).

$$V_{out} = 3.14 \text{ V}$$

From this value, we can calculate the critical voltage V_{IL} as:

$$V_{IL} = 0.57 \cdot 3.14 - 0.71 = \underline{1.08 \text{ V}}$$

To calculate V_{IH} in terms of the output voltage, use (5.67):

$$V_{IH} = \frac{V_{DD} + V_{T0,p} + k_R \cdot (2V_{out} + V_{T0,n})}{1 + k_R}$$

$$= \frac{3.3 - 0.7 + 2.5(2V_{out} + 0.6)}{1 + 2.5} = 1.43V_{out} + 1.17$$

Next, substitute this expression into the KCL equation (5.64) to obtain a second-order polynomial in V_{out}.

$$2.5\big[2(1.43V_{out} + 1.17 - 0.6)V_{out} - V_{out}^2\big] = (1.43V_{out} - 1.43)^2$$

$$2.61V_{out}^2 + 6.94V_{out} - 2.04 = 0$$

Again, only one root of this quadratic equation corresponds to the physically correct solution for V_{out} at this operating point, i.e., when $V_{in} = V_{IH}$.

$$V_{out} = 0.27 \text{ V}$$

From this value, we can calculate the critical voltage V_{IH} as:

$$V_{IH} = 1.43 \cdot 0.37 + 1.17 = \underline{\underline{1.55 \text{ V}}}$$

Finally, we find the noise margins for low voltage levels and for high voltage levels using (5.3) and (5.4).

$$NM_L = V_{IL} - V_{OL} = 1.08 \text{ V}$$

$$NM_H = V_{OH} - V_{IH} = 1.75 \text{ V}$$

Supply Voltage Scaling in CMOS Inverters

In the following, we will briefly examine the effects of supply voltage scaling, i.e., reduction of V_{DD}, upon the static voltage transfer characteristics of CMOS inverters. The overall power dissipation of any digital circuit is a strong function of the supply voltage V_{DD}. With the growing trend for reducing the power dissipation in large-scale integrated systems and especially in portable applications, reduction (or scaling) of the power supply voltage emerges as one of the most widely practiced measures for low-power design. While such reduction is usually very effective, several important issues must also be addressed so that the system performance is not sacrificed. In this context, it is quite relevant to explore the influence of supply voltage scaling upon the VTC of simple CMOS inverter circuits.

The expressions we have developed in this section for V_{IL}, V_{IH}, and V_{th} indeed show that the static characteristics of the CMOS inverter allow significant variation of the supply voltage without affecting the functionality of the basic inverter. Neglecting second-order effects such as subthreshold conduction, it can be seen that the CMOS inverter will continue to operate correctly with a supply voltage which is as low as the following limit value.

$$V_{DD}^{min} = V_{T0,n} + |V_{T0,p}| \tag{5.85}$$

This means that correct inverter operation will be sustained if at least one of the transistors remains in conduction, for any given input voltage. Figure 5.25 shows the voltage transfer characteristics of a CMOS inverter, obtained with different supply voltage levels. The exact shape of the VTC near the limit value is essentially determined by subthreshold conduction properties of the nMOS and pMOS transistors, yet it is clear that the circuit operates as an inverter over a large range of supply voltages levels.

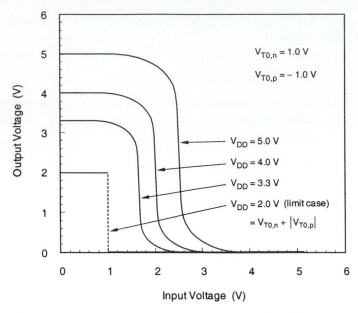

Figure 5.25   Voltage transfer characteristics of a CMOS inverter, obtained with different power supply voltage levels.

It is interesting to note that the CMOS inverter will continue to operate, albeit somewhat differently, even beyond the limit described in (5.85). If the power supply voltage is reduced *below* the sum of the two threshold voltages, the VTC will contain a region in which none of the transistors is conductiong. The output voltage level within this region is then determined by the *previous state* of the output, since the previous output level is always preserved as stored charge at the output node. Thus, the VTC exhibits a *hysteresis* behavior for very low supply voltage levels, which is illustrated in Fig. 5.26.

Power and Area Considerations

Since the CMOS inverter does not draw any significant current from the power source in both of its steady-state operating points ($V_{out} = V_{OH}$ and $V_{out} = V_{OL}$), the DC power dissipation of this circuit is almost negligible. The drain current that flows through the nMOS and the pMOS transistors in both cases is essentially limited to the reverse leakage current of the source and drain pn-junctions, and in short-channel MOSFETs, the relatively small subthreshold current. This unique property of the CMOS inverter was already identified as one of the most important advantages of this configuration. In many applications requiring a low overall power consumption, CMOS is preferred over other circuit alternatives for this reason. It must be noted, however, that the CMOS inverter does conduct a significant amount of current during a *switching event,* i.e., when the output voltage changes from a low to high state, or

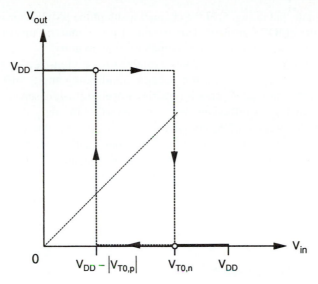

Figure 5.26 Voltage transfer characteristic of a CMOS inverter, operated with a supply voltage which is lower than the limit given in (5.85).

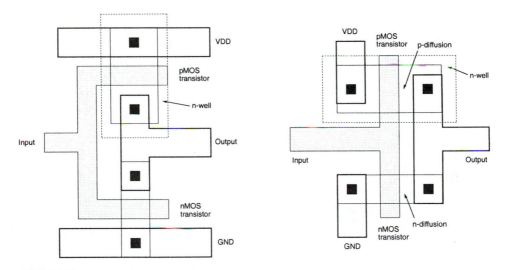

Figure 5.27 Two sample layouts of CMOS inverter circuits (for p-type substrate).

from a high to low state. The detailed calculation of this *dynamic power dissipation* will be examined in Chapters 6 and 11.

Figure 5.27 shows two layout examples for the simple CMOS inverter circuit. In both cases, it is assumed that the circuit is being built on a *p*-type wafer, which also provides the substrate for the nMOS transistor. The pMOS transistor, on the other hand, must be placed in an n-well (dotted lines), which becomes the substrate for this device. Also note that in Fig. 5.27 the channel width of the pMOS transistor is larger than that of the nMOS transistor. This is typical for symmetric inverter configurations, in which the k_R ratio is set approximately equal to unity.

Compared to other inverter layouts examined in previous sections, the CMOS inverters shown in Fig. 5.27 do not occupy significantly more area. The added complexity of the fabrication process (creating n-well diffusion, separate p-type and n-type source and drain diffusions, etc.) appears to be the only drawback for the inverter example. Because of the complementary nature of this circuit configuration, however, CMOS random logic circuits require significantly more transistors for the same function than their nMOS counterparts. Consequently, CMOS logic circuits tend to occupy more area than comparable nMOS logic circuits, which apparently affects the integration density of pure-CMOS logic. The actual integration density of nMOS logic, on the other hand, is limited by power dissipation and heat generation problems.

Exercise Problems

5.1   Design a resistive-load inverter with R $= 1\,k\Omega$, such that $V_{OL} = 0.6\,V$. The enhancement-type nMOS driver transistor has the following parameters:

$$V_{DD} = 5.0\,V$$
$$V_{T0} = 1.0\,V$$
$$\gamma = 0.2\,V^{1/2}$$
$$\lambda = 0$$
$$\mu_n C_{ox} = 22.0\,\mu A/V^2$$

   a.   Determine the required aspect ratio, W/L.

   b.   Determine V_{IL} and V_{IH}.

   c.   Determine noise margins NM_L and NM_H.

5.2   Layout of resistive-load inverter:

   a.   Draw the layout of the resistive-load inverter designed in Problem 5.1 using a polysilicon resistor with sheet resistivity of 25 Ω/square and the minimum feature size of 2 μm. It should be noted that L stands for the effective channel length which is related to the mask channel length as $L = L_M + d - 2L_D$, where we assume d (process error) $= 0$ and $L_D = 0.25\,\mu$m. To save chip area, use minimum sizes for W and L. Also, the circuit area can be reduced by using the folded layout (snake pattern) of the resistor.

   b.   Perform circuit extraction to obtain SPICE input list from the layout.

   c.   Run SPICE simulation of the circuit to obtain the DC voltage transfer characteristic (VTC) curve. Plot the VTC and check whether the calculated values in Problem 5.1 match the SPICE simulation results.

5.3   Refer to the CMOS fabrication process described in Chapter 2. Draw cross-sections of the following device along the lines A-A' and B-B'.

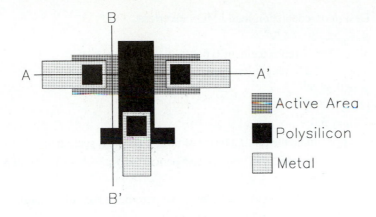

5.4 Consider the following nMOS inverter circuit which consists of two enhancement-type nMOS transistors, with the parameters:

$V_{T0} = 0.8$ V

$\mu_n C_{ox} = 45.0 \ \mu A/V^2$

$\gamma = 0.38 \ V^{1/2}$

$|2\phi_F| = 0.6$ V

$\lambda = 0$

$V_{DD} = 5.0$ V

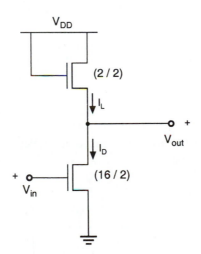

a. Calculate V_{OH} and V_{OL} values. Note that the substrate-bias effect of the load device must be taken into consideration.

b. Interpret the results in terms of noise margins and static (DC) power dissipation.

c. Calculate the steady-state current which is drawn from the DC power supply when the input is a logic "1", i.e., when $V_{in} = V_{OH}$.

5.5   Design of a depletion-load nMOS inverter:

$\mu_n C_{ox} = 30\,\mu\text{A/V}^2$

$V_{T0} = 0.8$ V (enhancement-type)

$V_{T0} = -2.8$ V (depletion-type)

$\gamma = 0.38\,\text{V}^{1/2}$

$|2\phi_F| = 0.6$ V

$V_{DD} = 5.0$ V

   a. Determine the (W/L) ratios of both transistors such that:

      (i) the static (DC) power dissipation for $V_{in} = V_{OH}$ is 250 mW, *and*

      (ii) $V_{OL} = 0.3$ V.

   b. Calculate V_{IL} and V_{IH} values, and determine the noise margins.

   c. Plot the VTC of the inverter circuit.

5.6   Consider a CMOS inverter with the following parameters:

nMOS	$V_{T0,n} = 0.6$ V	$\mu_n C_{ox} = 60\,\mu\text{A/V}^2$	$(W/L)_n = 8$
pMOS	$V_{T0,p} = -0.7$ V	$\mu_p C_{ox} = 25\,\mu\text{A/V}^2$	$(W/L)_p = 12$

Calculate the noise margins and the switching threshold (V_{th}) of this circuit. The power supply voltage is $V_{DD} = 3.3\,\text{V}$.

5.7   Design of a CMOS inverter circuit:

Use the same device parameters as in Problem 5.6. The power supply voltage is $V_{DD} = 3.3$ V. The channel length of both transistors is $L_n = L_p = 0.8\,\mu\text{m}$.

   a. Determine the (W_n/W_p) ratio so that the switching (inversion) threshold voltage of the circuit is $V_{th} = 1.4$ V.

   b. The CMOS fabrication process used to manufacture this inverter allows a variation of the $V_{T0,n}$ value by $\pm 15\%$ around its nominal value, and a variation of the $V_{T0,p}$ value by $\pm 20\%$ around its nominal value. Assuming that all other parameters (such as μ_n, μ_p, C_{ox}, W_n, W_p) always retain their nominal values, find the upper and lower limits of the switching threshold voltage (V_{th}) of this circuit.

5.8   Consider the CMOS inverter designed in Problem 5.7, with the following circuit configuration:

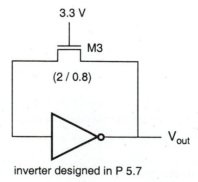

inverter designed in P 5.7

a. Calculate the output voltage level V_{out}.

b. Determine if the process-related variation of $V_{T0,n}$ of M3 has any influence upon the output voltage V_{out}.

c. Calculate the *total current* being drawn from the power supply source, and determine its variation due to process-related threshold-voltage variations.

5.9 Consider a CMOS inverter, with the following device parameters:

nMOS	$V_{T0,n} = 0.6$ V	$\mu_n C_{ox} = 60\ \mu\text{A/V}^2$
pMOS	$V_{T0,p} = -0.8$ V	$\mu_p C_{ox} = 20\ \mu\text{A/V}^2$

Also: $V_{DD} = 3$ V

$\quad\quad \lambda = 0$

a. Determine the (W/L) ratios of the nMOS and the pMOS transistor such that the switching threshold is $V_{th} = 1.5$ V.

b. Plot the VTC of the CMOS inverter using SPICE.

c. Determine the VTC of the inverter for $\lambda = 0.05$ V^{-1} and $\lambda = 0.1$ V^{-1}.

d. Discuss how the noise margins are influenced by nonzero λ value. Note that transistors with very short channel lengths (manufactured with sub-micron design rules) tend to have larger λ values than long-channel transistors.

5.10 Consider the CMOS inverter designed in Problem 5.9 above, with $\lambda = 0.1$ V^{-1}. Now consider a cascade connection of four *identical* inverters, as shown below.

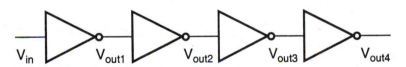

a. If the input voltage is $V_{in} = 1.55$ V, find V_{out1}, V_{out2}, V_{out3} and V_{out4}. (Note that this requires solving KCL equations for each subsequent stage, using the nonzero λ value).

b. How many stages are necessary to restore a *true* logic output level?

c. Verify your result with SPICE simulation.

6 CHAPTER

MOS Inverters: Switching Characteristics and Interconnect Effects

In this chapter, we will investigate the dynamic (time-domain) behavior of the inverter circuits. The switching characteristics of digital integrated circuits and, in particular, of inverter circuits, essentially determine the overall operating speed of digital systems. As already shown in the design example presented in Chapter 1, the transient performance requirements of a digital system are usually among the most important design specifications that must be met by the circuit designer. Therefore, the switching speed of the circuit must be estimated and optimized very early in the design phase.

The closed-form delay expressions will be derived under the assumption of pulse excitation, for lumped load capacitances. While exact circuit simulation (SPICE) usually provides the most accurate estimation of the time-domain behavior of complex circuits, the delay expressions presented here can also be used in many cases to provide a quick and accurate approximation of the switching characteristics. In the following, specific emphasis will be given to the switching behavior of CMOS inverter.

6.1 Introduction

Consider the cascade connection of two CMOS inverter circuits shown in Fig. 6.1. The parasitic capacitances associated with each MOSFET are illustrated individually. Here, the capacitances C_{gd} and C_{gs} are primarily due to gate overlap with diffusion, while C_{db} and C_{sb} are voltage-dependent junction capacitances, as discussed in Chapter 3. The capacitance component C_g is due to the thin-oxide capacitance over the gate area. In addition, we also consider the lumped interconnect capacitance C_{int}, which represents the parasitic capacitance contribution of the metal or polysilicon connection between the two inverters. It is assumed that a pulse

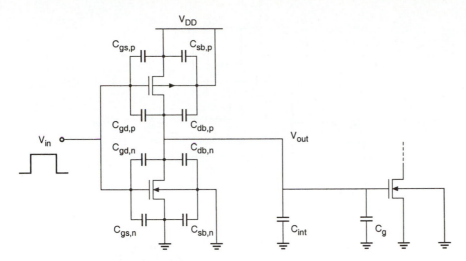

Figure 6.1   Cascaded CMOS inverter stages.

waveform is applied to the input of the first-stage inverter. We wish to analyze the time-domain behavior of the first-stage output, V_{out}.

The problem of analyzing the output voltage waveform is fairly complicated, even for this relatively simple circuit, because a number of nonlinear, voltage-dependent capacitances are involved. To simplify the problem, we first combine the capacitances seen in Fig. 6.1 into an equivalent *lumped* linear capacitance, connected between the output node of the inverter and the ground. This combined capacitance at the output node will be called the load capacitance, C_{load}.

$$C_{load} = C_{gd,n} + C_{gd,p} + C_{db,n} + C_{db,p} + C_{int} + C_g \qquad (6.1)$$

Note that some of the parasitic capacitance components shown in Fig. 6.1 do not appear in this lumped capacitance expression. In particular, $C_{sb,n}$ and $C_{sb,p}$ have no effect on the transient behavior of the circuit since the source-to-substrate voltages of both transistors are always equal to zero. The capacitances $C_{gs,n}$ and $C_{gs,p}$ are also not included in (6.1) because they are connected between the input node and the ground (or the power supply). The capacitance terms $C_{db,n}$ and $C_{db,p}$ in (6.1) are the equivalent junction capacitances calculated for a particular output voltage transition, according to (3.109) and (3.114). The reader is referred to Chapter 3 for details concerning the calculation of parasitic junction capacitances.

The first-stage CMOS inverter is shown with the single lumped output load capacitance C_{load} in Fig. 6.2. Now, the problem of analyzing the switching behavior can be handled more easily. In fact, the question of inverter transient response is reduced to finding the charge-up and charge-down times of a single capacitance which is charged and discharged through one transistor. The delay times calculated using C_{load} may slightly overestimate the actual inverter delay, but this is not considered a significant deficiency in a first-order approximation.

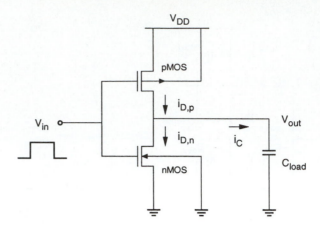

Figure 6.2 First-stage CMOS inverter with lumped output load capacitance.

6.2 Delay-Time Definitions

Before we begin the derivation of delay expressions, we will present some commonly used delay time definitions. The input and output voltage waveforms of a typical inverter circuit are shown in Fig. 6.3. The propagation delay times τ_{PHL} and τ_{PLH} determine the input-to-output signal delay during the high-to-low and low-to-high transitions of the output, respectively. By definition, τ_{PHL} is the time delay between the $V_{50\%}$-transition of the *rising* input voltage and the $V_{50\%}$-transition of the *falling* output voltage. Similarly, τ_{PLH} is defined as the time delay between the $V_{50\%}$-transition of the *falling* input voltage and the $V_{50\%}$-transition of the *rising* output voltage.

To simplify the analysis and the derivation of delay expressions, the input voltage waveform is usually assumed to be an ideal step pulse with zero rise and fall times. Under this assumption, τ_{PHL} becomes the time required for the output voltage to fall from V_{OH} to the $V_{50\%}$ level, and τ_{PLH} becomes the time required for the output voltage to rise from V_{OL} to the $V_{50\%}$ level. The voltage point $V_{50\%}$ is defined as follows.

$$V_{50\%} = V_{OL} + \frac{1}{2}(V_{OH} - V_{OL}) = \frac{1}{2}(V_{OL} + V_{OH}) \qquad (6.2)$$

Thus, the propagation delay times τ_{PHL} and τ_{PLH} are found from Fig. 6.3 as

$$\begin{aligned} \tau_{PHL} &= t_1 - t_0 \\ \tau_{PLH} &= t_3 - t_2 \end{aligned} \qquad (6.3)$$

The average propagation delay τ_P of the inverter characterizes the average time required for the input signal to propagate through the inverter.

$$\tau_P = \frac{\tau_{PHL} + \tau_{PLH}}{2} \qquad (6.4)$$

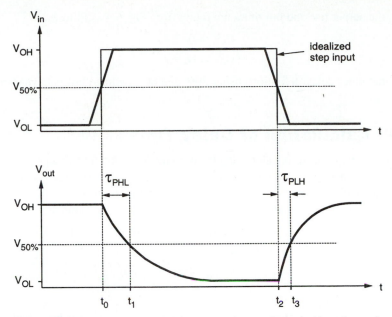

Figure 6.3 Input and output voltage waveforms of a typical inverter, and the definitions of propagation delay times. The input voltage waveform is idealized as a step pulse for simplicity.

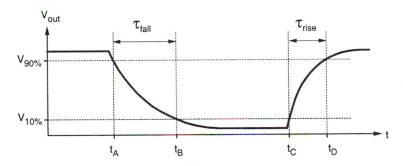

Figure 6.4 Output voltage rise and fall times.

We will refer to Fig. 6.4 for the definition of output voltage rise and fall times. The rise time τ_{rise} is defined here as the time required for the output voltage to rise from the $V_{10\%}$ level to $V_{90\%}$ level. Similarly, the fall time τ_{fall} is defined here as the time required for the output voltage to drop from the $V_{90\%}$ level to $V_{10\%}$ level. The voltage levels $V_{10\%}$ and $V_{90\%}$ are defined as

$$V_{10\%} = V_{OL} + 0.1 \cdot (V_{OH} - V_{OL}) \tag{6.5}$$

$$V_{90\%} = V_{OL} + 0.9 \cdot (V_{OH} - V_{OL}) \tag{6.6}$$

Thus, the output rise and fall times are found from Fig. 6.4 as follows.

$$\tau_{fall} = t_B - t_A$$
$$\tau_{rise} = t_D - t_C \tag{6.7}$$

Note that other delay definitions using 20% and 80% voltage levels have also been used.

6.3   Calculation of Delay Times

The simplest approach for calculating the propagation delay times τ_{PHL} and τ_{PLH} is based on estimating the average capacitance current during charge down and charge up, respectively. If the capacitance current during an output transition is approximated by a constant average current I_{avg}, the delay times are found as

$$\tau_{PHL} = \frac{C_{load} \cdot \Delta V_{HL}}{I_{avg,HL}} = \frac{C_{load} \cdot (V_{OH} - V_{50\%})}{I_{avg,HL}} \tag{6.8}$$

$$\tau_{PLH} = \frac{C_{load} \cdot \Delta V_{LH}}{I_{avg,LH}} = \frac{C_{load} \cdot (V_{50\%} - V_{OL})}{I_{avg,LH}} \tag{6.9}$$

Note that the average current during high-to-low transition can be calculated by using the current values at the beginning and the end of the transition.

$$I_{avg,HL} = \frac{1}{2} \left[i_C \left(V_{in} = V_{OH}, V_{out} = V_{OH} \right) + i_C (V_{in} = V_{OH}, V_{out} = V_{50\%}) \right] \tag{6.10}$$

Similarly, the average capacitance current during low-to-high transition is

$$I_{avg,LH} = \frac{1}{2} \left[i_C \left(V_{in} = V_{OL}, V_{out} = V_{50\%} \right) + i_C (V_{in} = V_{OL}, V_{out} = V_{OL}) \right] \tag{6.11}$$

While the average-current method is relatively simple and requires minimal calculation, it neglects the variations of the capacitance current between the beginning and end points of the transition. Therefore, we do not expect the average-current method to provide a very accurate estimate of the delay times. Still, this approach can provide rough, first-order estimates of the charge-up and charge-down delay times.

The propagation delay times can be found more accurately by solving the state equation of the output node in the time domain. The differential equation associated with the output node is given below. Note that the capacitance current is also a function of the output voltage.

$$C_{load} \frac{dV_{out}}{dt} = i_C = i_{D,p} - i_{D,n} \tag{6.12}$$

First, we consider the rising-input case for a CMOS inverter. Initially, the output voltage is assumed to be equal to V_{OH}. When the input voltage switches from low (V_{OL}) to high (V_{OH}), the nMOS transistor is turned on and it starts to discharge the load capacitance. At the same time, the pMOS transistor is switched off; thus,

$$i_{D,p} \approx 0 \tag{6.13}$$

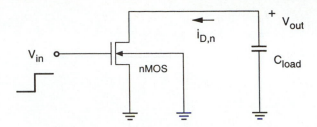

Figure 6.5 Equivalent circuit of the CMOS inverter during high-to-low output transition.

The circuit given in Fig. 6.2 can now be reduced to a single nMOS transistor and a capacitor, as shown in Fig. 6.5. The differential equation describing the discharge event is then

$$C_{load}\frac{dV_{out}}{dt} = -i_{D,n} \tag{6.14}$$

Note that in other types of inverter circuits, such as the resistive-load inverter or the depletion-load inverter, the load device continues to conduct a nonzero current when the input is switched from low to high. However, the load current is usually negligible in comparison to the driver current. Therefore, (6.14) can be used to calculate the charge-down time not only in CMOS inverters, but also in almost all common types of inverter circuits.

The input and output voltage waveforms during this high-to-low transition are illustrated in Fig. 6.6. When the nMOS transistor starts conducting, it initially operates in the saturation region. When the output voltage falls below $(V_{DD} - V_{T,n})$, the nMOS transistor starts to conduct in the linear region. These two operating regions are also shown in Fig. 6.6. First, consider the nMOS transistor operating in saturation.

$$i_{D,n} = \frac{k_n}{2}(V_{in} - V_{T,n})^2$$

$$= \frac{k_n}{2}(V_{OH} - V_{T,n})^2, \qquad \text{for} \quad V_{OH} - V_{T,n} < V_{out} \le V_{OH} \tag{6.15}$$

Since the saturation current is practically independent of the output voltage (neglecting channel-length modulation), the solution of (6.14) in the time interval between t_0 and t_1' can be found as

$$\int_{t=t_0}^{t=t_1'} dt = -C_{load}\int_{V_{out}=V_{OH}}^{V_{out}=V_{OH}-V_{T,n}} \left(\frac{1}{i_{D,n}}\right) dV_{out}$$

$$= -\frac{2C_{load}}{k_n(V_{OH} - V_{T,n})^2}\int_{V_{out}=V_{OH}}^{V_{out}=V_{OH}-V_{T,n}} dV_{out} \tag{6.16}$$

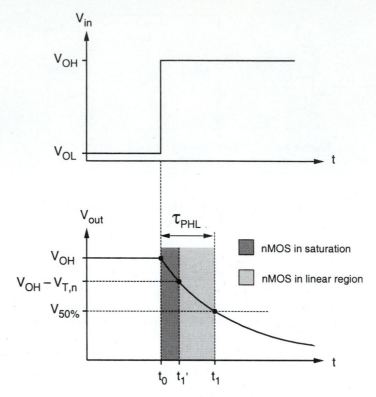

Figure 6.6 Input and output voltage waveforms during high-to-low transition.

Evaluating this simple integral yields

$$t_1' - t_0 = \frac{2C_{load}V_{T,n}}{k_n(V_{OH} - V_{T,n})^2} \tag{6.17}$$

At $t = t_1'$, the output voltage will be equal to $(V_{DD} - V_{T,n})$ and the transistor will be at the saturation-linear region boundary. Next, consider the nMOS transistor operating in the linear region.

$$i_{D,n} = \frac{k_n}{2} \left[2(V_{in} - V_{T,n})V_{out} - V_{out}^2 \right]$$

$$= \frac{k_n}{2} \left[2(V_{OH} - V_{T,n})V_{out} - V_{out}^2 \right], \qquad \text{for} \quad V_{out} \leq V_{OH} - V_{T,n} \tag{6.18}$$

The solution of (6.14) in the time interval between t_1' and t_1 can be found as

$$\int_{t=t_1'}^{t=t_1} dt = -C_{load} \int_{V_{out}=V_{OH}-V_{T,n}}^{V_{out}=V_{50\%}} \left(\frac{1}{i_{D,n}} \right) dV_{out}$$

$$= -2\,C_{load} \int_{V_{out}=V_{OH}-V_{T,n}}^{V_{out}=V_{50\%}} \left(\frac{1}{k_n \left[2(V_{OH} - V_{T,n})V_{out} - V_{out}^2 \right]} \right) dV_{out} \tag{6.19}$$

Evaluating this integral yields

$$t_1 - t_1' = -\frac{2C_{load}}{k_n} \frac{1}{2(V_{OH} - V_{T,n})} \ln\left(\frac{V_{out}}{2(V_{OH} - V_{T,n}) - V_{out}}\right)\Bigg|_{V_{out}=V_{OH}-V_{T,n}}^{V_{out}=V_{50\%}} \tag{6.20}$$

$$t_1 - t_1' = \frac{C_{load}}{k_n(V_{OH} - V_{T,n})} \ln\left(\frac{2(V_{OH} - V_{T,n}) - V_{50\%}}{V_{50\%}}\right) \tag{6.21}$$

Finally, the propagation delay time for high-to-low output transition (τ_{PHL}) can be found by combining (6.17) and (6.21):

$$\tau_{PHL} = \frac{C_{load}}{k_n(V_{OH} - V_{T,n})} \left[\frac{2V_{T,n}}{V_{OH} - V_{T,n}} + \ln\left(\frac{4(V_{OH} - V_{T,n})}{V_{OH} + V_{OL}} - 1\right)\right] \tag{6.22a}$$

For $V_{OH} = V_{DD}$ and $V_{OL} = 0$, as is the case for the CMOS inverter, (6.22a) becomes:

$$\tau_{PHL} = \frac{C_{load}}{k_n(V_{DD} - V_{T,n})} \left[\frac{2V_{T,n}}{V_{DD} - V_{T,n}} + \ln\left(\frac{4(V_{DD} - V_{T,n})}{V_{DD}} - 1\right)\right] \tag{6.22b}$$

EXAMPLE 6.1

Consider the CMOS inverter circuit shown in Fig. 6.2, with $V_{DD} = 3.3$ V. The I–V characteristics of the nMOS transistor are specified as follows: when $V_{GS} = 3.3$ V, the drain current reaches its saturation level $I_{sat} = 2$ mA for $V_{DS} \geq 2.5$ V. Assume that the input signal applied to the gate is a step pulse that switches instantaneously from 0 V to 3.3 V. Using the data above, calculate the delay time necessary for the output to fall from its initial value of 3.3 V to 1.65 V, assuming an output load capacitance of 300 fF.

For the solution, consider the simplified pull-down circuit shown in Fig. 6.5. We will assume that the nMOS transistor operates in saturation from $t = 0$ to $t = t_1' = t_{sat}$, and that it will operate in the linear region from $t = t_1' = t_{sat}$ to $t = t_2 = t_{delay}$. We can also deduce from the I–V characteristics that $V_{T,n} = 0.8$ V, since the nMOS transistor enters saturation when $V_{DS} \geq V_{GS} - V_{T,n}$. The voltage V_{GS} is equal to 3.3 V for $t \geq 0$.

The current equation for the saturation region can be written as

$$C\frac{dV_{out}}{dt} = -I_D = -I_{sat} = -\frac{1}{2}k_n(V_{OH} - V_{T,n})^2$$

We can calculate the amount of time in which the nMOS transistor operates in saturation (t_{sat}), by integrating this equation.

$$\int_{t=0}^{t=t_{sat}} dt = -\int_{V_{out}=3.3}^{V_{out}=2.5} \frac{C}{I_D} dV_{out}$$

$$t_{sat} = \frac{V_{T,n}C}{I_{sat}} = \frac{0.8\,\text{V} \cdot 300\,\text{fF}}{2\,\text{mA}} = 120\,\text{[ps]}$$

The transconductance k_n of the nMOS transistor can be found as follows:

$$k_n = \frac{2I_{sat}}{(V_{OH} - V_{T,n})^2} = \frac{2 \times 2 \times 10^{-3}}{(3.3 - 0.8)^2} = 0.640 \times 10^{-3} \; [\text{A/V}^2]$$

Now, the current equation for the linear operating region is

$$C\frac{dV_{out}}{dt} = -I_D = -\frac{1}{2}k_n\left[2(V_{OH} - V_{T,n})V_{out} - V_{out}^2\right]$$

Integrating this differential equation between the two voltage boundary conditions yields the time in which the nMOS transistor operates in the linear region during this transition.

$$\int_{t=t_{sat}}^{t=t_{delay}} dt = -2C \int_{V_{out}=2.5}^{V_{out}=1.65} \frac{dV_{out}}{k_n\left[2(V_{OH} - V_{T,n})V_{out} - V_{out}^2\right]}$$

$$t_{delay} - t_{sat} = -\frac{C}{k_n}\frac{1}{(V_{OH} - V_{T,n})} \ln\left(\frac{V_{out}}{2(V_{OH} - V_{T,n}) - V_{out}}\right)\Bigg|_{V_{out}=2.5}^{V_{out}=1.65}$$

$$= \frac{C}{k_n}\frac{1}{(V_{OH} - V_{T,n})}$$

$$\times \left[\ln\left(\frac{2(V_{OH} - V_{T,n}) - V_{1.65}}{V_{1.65}}\right) - \ln\left(\frac{2(V_{OH} - V_{T,n}) - V_{2.5}}{V_{2.5}}\right)\right]$$

$$= \frac{0.3 \times 10^{-12}}{0.640 \times 10^{-3}}\frac{1}{(3.3 - 0.8)}\left[\ln\left(\frac{5 - 1.65}{1.65}\right) - \ln\left(\frac{5 - 2.5}{2.5}\right)\right]$$

$$= 133 \; [\text{ps}]$$

Thus, the total delay time is found to be

$$t_{delay} = 120 + 133 = 253 \; [\text{ps}]$$

Note that t_{delay} corresponds to the propagation delay time τ_{PHL} for falling output.

The average-current method presented earlier in this Section can also be used to estimate the propagation delay times as well as the rise and fall times of inverter circuits. This simple approach can in some cases produce fairly accurate first-order results, as presented in the following example.

EXAMPLE 6.2

For the CMOS inverter shown in Fig. 6.2 with a power supply voltage of $V_{DD} = 5$ V, determine the fall time τ_{fall}, which is defined as the time elapsed between the time point at which $V_{out} = V_{90\%} = 4.5$ V and the time point at which $V_{out} = V_{10\%} = 0.5$ V. Use both the average-current method and the differential equation method for calculating τ_{fall}. The output load capacitance is 1 pF. The nMOS transistor parameters are given as

$$\mu_n C_{ox} = 20\,\mu\text{A/V}^2$$
$$(W/L)_n = 10$$
$$V_{T,n} = 1.0\,\text{V}$$

Using a simple expression similar to (6.10), we can determine the average capacitor current during the charge-down event described earlier.

$$I_{avg} = \frac{1}{2}[I(V_{in} = 5\,\text{V}, V_{out} = 4.5\,\text{V}) + I(V_{in} = 5\,\text{V}, V_{out} = 0.5\,\text{V})]$$

$$= \frac{1}{2}\left[\frac{1}{2}k_n(V_{in} - V_{T,n})^2 + \frac{1}{2}k_n\left(2(V_{in} - V_{T,n})V_{out} - V_{out}^2\right)\right]$$

$$= \frac{1}{2} \cdot \frac{1}{2} \cdot 20 \times 10^{-6} \cdot 10[(5-1)^2 + (2(5-1)0.5 - 0.5^2)] = 0.9875\,[\text{mA}]$$

The fall time is then found as

$$\tau_{fall} = \frac{C \cdot \Delta V}{I_{avg}} = \frac{1 \times 10^{-12}(4.5 - 0.5)}{0.9875 \times 10^{-3}} = 4.05 \times 10^{-9}\,[\text{s}] = 4.05\,[\text{ns}]$$

Now, we will recalculate the fall time using the differential equation approach. The nMOS transistor operates in the saturation region for $4.0\,\text{V} \le V_{out} \le 4.5\,\text{V}$. Writing the current equation for the saturation region, we obtain

$$C\frac{dV_{out}}{dt} = -\frac{1}{2}k_n(V_{in} - V_{T,n})^2, \qquad \text{where} \quad k_n = \mu_n C_{ox}\left(\frac{W}{L}\right)_n$$

$$\frac{dV_{out}}{dt} = \frac{-20 \times 10^{-6} \cdot 10 \cdot (5-1)^2}{2 \cdot 1 \times 10^{-12}} = -1.6 \times 10^9\,[\text{V/s}]$$

Integrating this simple expression yields the time during which the nMOS transistor operates in saturation.

$$\int_{t=0}^{t=t_{sat}} dt = -\frac{1}{1.6 \times 10^9}\int_{V_{out}=4.5}^{V_{out}=4} dV_{out}$$

$$t_{sat} = \frac{0.5}{1.6 \times 10^9} = 0.3125 \times 10^{-9}\,[\text{s}] = 0.3125\,[\text{ns}]$$

The nMOS transistor operates in the linear region for $0.5\,\text{V} \le V_{out} \le 4.0\,\text{V}$. The current equation for this operating region is written as follows:

$$C\frac{dV_{out}}{dt} = -\frac{1}{2}k_n\left[2(V_{in} - V_{T,n})V_{out} - V_{out}^2\right]$$

Integrating this equation, we obtain the delay component during which the nMOS transistor operates in the linear region.

$$\int_{t=t_{sat}}^{t=t_{delay}} dt = -2C \int_{V_{out}=4}^{V_{out}=0.5} \frac{dV_{out}}{k_n \left[2(V_{in} - V_{T,n})V_{out} - V_{out}^2 \right]}$$

$$\tau_{fall} - t_{sat} = \frac{C}{k_n} \frac{1}{(V_{in} - V_{T,n})}$$

$$\times \left[\ln \left(\frac{2(V_{in} - V_{T,n}) - V_{0.5}}{V_{0.5}} \right) - \ln \left(\frac{2(V_{in} - V_{T,n}) - V_{4.0}}{V_{4.0}} \right) \right]$$

$$= \frac{1 \times 10^{-12}}{20 \times 10^{-6} \cdot 10 \cdot 4} \left[\ln \left(\frac{8 - 0.5}{0.5} \right) - \ln \left(\frac{8 - 4}{4} \right) \right]$$

$$= 3.385 \times 10^{-9} \, [\text{s}] = 3.385 \, [\text{ns}]$$

Thus, the fall time of the CMOS inverter is found as follows:

$$\tau_{fall} = 3.6975 \, [\text{ns}]$$

In a CMOS inverter, the charge-up event of the output load capacitance for falling input transition is completely analogous to the charge-down event for rising input. When the input voltage switches from high (V_{OH}) to low (V_{OL}), the nMOS transistor is cut off, and the load capacitance is being charged up through the pMOS transistor. Following a very similar derivation procedure, the propagation delay time τ_{PLH} can be found as

$$\tau_{PLH} = \frac{C_{load}}{k_p(V_{OH} - V_{OL} - |V_{T,p}|)}$$

$$\times \left[\frac{2|V_{T,p}|}{V_{OH} - V_{OL} - |V_{T,p}|} + \ln \left(\frac{2(V_{OH} - V_{OL} - |V_{T,p}|)}{V_{OH} - V_{50\%}} - 1 \right) \right] \quad (6.23\text{a})$$

For $V_{OH} = V_{DD}$ and $V_{OL} = 0$, (6.23a) becomes

$$\tau_{PLH} = \frac{C_{load}}{k_p(V_{DD} - |V_{T,p}|)} \left[\frac{2|V_{T,p}|}{V_{DD} - |V_{T,p}|} + \ln \left(\frac{4(V_{DD} - |V_{T,p}|)}{V_{DD}} - 1 \right) \right] \quad (6.23\text{b})$$

Comparing the delay expression (6.23b) with (6.22b), we can see that the sufficient conditions for *balanced* propagation delays, i.e., for $\tau_{PHL} = \tau_{PLH}$, in a CMOS inverter are:

$$V_{T,n} = |V_{T,p}| \quad \text{and} \quad k_n = k_p \ (\text{or} \ W_p/W_n = \mu_n/\mu_p)$$

The calculation of τ_{PLH} in different types of inverters depends on the load device and its operation, yet the analysis procedure is very similar to the CMOS case. We will consider the nMOS depletion-load inverter case as an example in the following. When the input voltage switches from high to low, the enhancement-type nMOS driver transistor is turned off. The output load capacitance is then being charged up through the depletion-type load transistor. The differential equation describing this event is

$$C_{load}\frac{dV_{out}}{dt} = i_{D,load}(V_{out}) \tag{6.24}$$

Note that the load device is initially in saturation, and enters the linear region when the output voltage rises above $(V_{DD} + V_{T,load})$, where $V_{T,load} < 0$.

$$i_{D,load} = \frac{k_{n,load}}{2}(|V_{T,load}|)^2, \qquad \text{for} \quad V_{out} \leq V_{DD} - |V_{T,load}| \tag{6.25}$$

$$i_{D,load} = \frac{k_{n,load}}{2}[2|V_{T,load}|(V_{DD} - V_{out}) - (V_{DD} - V_{out})^2]$$
$$\text{for} \quad V_{out} > V_{DD} - |V_{T,load}| \tag{6.26}$$

The delay time τ_{PLH} can be found as follows:

$$\tau_{PLH} = C_{load}\left[\int_{V_{out}=V_{OL}}^{V_{out}=V_{DD}-|V_{T,load}|}\left(\frac{dV_{out}}{i_{D,load}(sat)}\right)\right.$$
$$\left. + \int_{V_{out}=V_{DD}-|V_{T,load}|}^{V_{out}=V_{50\%}}\left(\frac{dV_{out}}{i_{D,load}(linear)}\right)\right] \tag{6.27}$$

$$\tau_{PLH} = \frac{C_{load}}{k_{n,load}|V_{T,load}|}\left[\frac{2(V_{DD} - |V_{T,load}| - V_{OL})}{|V_{T,load}|}\right.$$
$$\left. + \ln\left(\frac{2|V_{T,load}| - (V_{DD} - V_{50\%})}{V_{DD} - V_{50\%}}\right)\right] \tag{6.28}$$

All of the delay time derivations in this section were made under the simplifying assumption that the input signal waveform is a step pulse with zero rise and fall times. Now, we consider the case where the input voltage waveform is not an ideal (step) pulse waveform, but has finite rise and fall times, τ_r and τ_f. The exact calculation of the output voltage delay times is more complicated under this more realistic assumption, since both the nMOS transistor and the pMOS transistor conduct current during the charge-up and charge-down events. To simplify the estimation of the actual propagation delays, we can utilize the propagation delay times calculated under the step-input assumption, using the following empirical expressions:

$$\tau_{PHL}(actual) = \sqrt{\tau_{PHL}^2(step\ input) + \left(\frac{\tau_r}{2}\right)^2} \tag{6.29}$$

$$\tau_{PLH}(actual) = \sqrt{\tau_{PLH}^2(step\ input) + \left(\frac{\tau_f}{2}\right)^2} \tag{6.30}$$

Here, τ_{PHL} (*step input*) and τ_{PLH} (*step input*) denote the propagation delay time values calculated assuming a step pulse input waveform at the input, i.e., using (6.22b) and (6.23b). While the expressions given before are purely empirical, they provide a simple estimation of how much the propagation delays are *increased* as a result of nonzero input rise and fall times.

Another important issue to be considered here is the fact that the delay expressions in the previous pages were derived using the simple current-voltage relationships originally developed for long-channel transistors. As explained in Chapter 3, these current expressions based on the gradual channel approximation can still be used for sub-micron MOS transistors with proper parameter adjustments; therefore, the delay analysis presented in this section remains largely valid for small-geometry devices as well. Yet it should be noted that the current driving capability of sub-micron transistors is significantly reduced as a result of channel velocity saturation; a small-geometry transistor cannot be expected to have the same maximum charge/discharge current as a long-channel transistor with the same (W/L)-ratio.

In particular, the saturation current of a deep-sub-micron nMOS transistor is no longer a quadratic function of $(V_{GS} - V_T)$, but is rather described as

$$I_{sat} = \kappa W_n (V_{GS} - V_T) \tag{6.31}$$

where κ is a function of the carrier saturation velocity, the channel length, and the *degree* of velocity saturation in the channel. Assuming that the discharge current of the nMOS transistor between $V_{out} = V_{DD}$ and $V_{out} = V_{50\%}$ can be approximated by its saturation current (in most cases, this rough assumption can be used to produce a first-order estimate of propagation delay times with an error of about 10%) and using the average current method, we obtain

$$\tau_{PHL} \approx \frac{C_{load} V_{50\%}}{I_{sat}} = \frac{C_{load}(V_{DD}/2)}{\kappa W_n (V_{DD} - V_T)} \tag{6.32}$$

Note that in this case, the propagation delay has only a weak dependence on the power supply voltage. Better estimates of propagation delay can be obtained by calculating the charge/discharge current using an accurate short-channel MOSFET model (such as the Sakurai-Newton current model) and applying the average-current method.

6.4   Inverter Design with Delay Constraints

The design of CMOS inverters and, in general, of CMOS logic circuits, based on timing (delay) specifications is one of the most fundamental issues in digital circuit design which ultimately determine the overall performance of complex systems. In most cases, the delay constraints should be considered together with other design constraints such as noise margins, logic (inversion) threshold, silicon area, and power dissipation. Thus, the design process usually involves the balancing of mostly conflicting requirements for optimum overall performance. In the following, we will consider some of the fundamental aspects of CMOS inverter design based on timing constraints.

The delay expressions developed in the previous section will form the basis of our design approach. The goal is to determine the channel dimensions (W_n, W_p) of nMOS and pMOS transistors which satisfy certain timing requirements. In the most

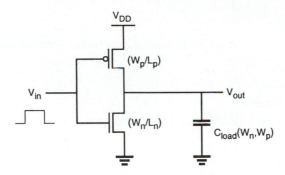

Figure 6.7 General circuit structure considered in the inverter design problem.

general case, we must take into account that the combined load capacitance of the inverter is also a function of the transistor sizes (Fig. 6.7). The load capacitance C_{load} in (6.1) consists of *intrinsic* components (parasitic drain capacitances which depend on transistor dimensions) and *extrinsic* components (interconnect/wiring capacitance and fan-out capacitances which are usually independent of transistor dimensions of the inverter under consideration).

If C_{load} mainly consists of extrinsic components, and if this overall load capacitance can be estimated accurately and independently of the transistor dimensions, then the problem of inverter design can be reduced to a rather straightforward application of the delay equations derived in Section 6.3. Given a required (*target*) delay value of τ_{PHL}^*, the (W/L)-ratio of the nMOS transistor can be found as

$$\left(\frac{W_n}{L_n}\right) = \frac{C_{load}}{\tau_{PHL}^* \mu_n C_{ox}(V_{DD} - V_{T,n})}\left[\frac{2V_{T,n}}{V_{DD} - V_{T,n}} + \ln\left(\frac{4(V_{DD} - V_{T,n})}{V_{DD}} - 1\right)\right]$$

(6.33)

Similarly, the (W/L)-ratio of the pMOS transistor to satisfy a given target value of τ_{PLH}^* can be calculated as

$$\left(\frac{W_p}{L_p}\right) = \frac{C_{load}}{\tau_{PLH}^* \mu_p C_{ox}(V_{DD} - |V_{T,p}|)}\left[\frac{2|V_{T,p}|}{V_{DD} - |V_{T,p}|}\right.$$
$$\left. + \ln\left(\frac{4(V_{DD} - |V_{T,p}|)}{V_{DD}} - 1\right)\right]$$
(6.34)

Transistor dimensions to satisfy other timing constraints (e.g., rise and fall times) can be determined using a similar approach. In most cases, the transistor sizes found from delay requirements must also meet other design criteria such as noise margins and logic inversion threshold. In all cases, the transistor dimensions should be chosen so that all delay times remain *below* the required target value. The design approach is demonstrated in the following example.

EXAMPLE 6.3

A company has access to a CMOS fabrication process with the device parameters listed below.

$$\mu_n C_{ox} = 120 \, \mu\text{A/V}^2$$

$$\mu_p C_{ox} = 60 \, \mu\text{A/V}^2$$

$$L = 0.6 \, \mu\text{m for both nMOS and pMOS devices}$$

$$V_{T0,n} = 0.8 \, \text{V}$$

$$V_{T0,p} = -1.0 \, \text{V}$$

$$W_{min} = 1.2 \, \mu\text{m}$$

Design a CMOS inverter by determining the channel widths W_n and W_p of the nMOS and pMOS transistors, to meet the following performance specifications.

- $V_{th} = 1.5$ V for $V_{DD} = 3$ V,
- Propagation delay times $\tau^*_{PHL} \leq 0.2$ ns and $\tau^*_{PLH} \leq 0.15$ ns,
- A falling delay of 0.35 ns for an output transition from 2 V to 0.5 V, assuming a combined output load capacitance of 300 fF and ideal step input.

We start our design by satisfying the time delay constraints. First, the *minimum* (W/L) ratios of the nMOS and pMOS transistors which are dictated by the propagation delay constraints can be found using (6.33) and (6.34), as follows.

$$\left(\frac{W_n}{L_n}\right) = \frac{C_{load}}{\tau^*_{PHL}\mu_n C_{ox}(V_{DD} - V_{T,n})}\left[\frac{2V_{T,n}}{V_{DD} - V_{T,n}} + \ln\left(\frac{4(V_{DD} - V_{T,n})}{V_{DD}} - 1\right)\right]$$

$$= \frac{300 \times 10^{-15}}{0.2 \times 10^{-9} \cdot 120 \times 10^{-6} \cdot (3 - 0.8)}\left[\frac{2 \cdot 0.8}{3 - 0.8} + \ln\left(\frac{4(3 - 0.8)}{3} - 1\right)\right]$$

$$= 7.9$$

$$\left(\frac{W_p}{L_p}\right) = \frac{C_{load}}{\tau^*_{PLH}\mu_p C_{ox}(V_{DD} - |V_{T,p}|)}$$

$$\times \left[\frac{2|V_{T,p}|}{V_{DD} - |V_{T,p}|} + \ln\left(\frac{4(V_{DD} - |V_{T,p}|)}{V_{DD}} - 1\right)\right]$$

$$= \frac{300 \times 10^{-15}}{0.15 \times 10^{-9} \cdot 60 \times 10^{-6} \cdot (3 - 1)}\left[\frac{2 \cdot 1}{3 - 1} + \ln\left(\frac{4(3 - 1)}{3} - 1\right)\right]$$

$$= 25.2$$

During the falling output transition (from 2 V to 0.5 V), the nMOS transistor of the CMOS inverter will operate entirely in the linear region. The current equation of the nMOS transistor in this region is

$$C_{load}\frac{dV_{out}}{dt} = -\frac{1}{2}\mu_n C_{ox}\frac{W_n}{L_n}\left[2(V_{OH} - V_{T0,n})V_{out} - V_{out}^2\right]$$

By integrating this expression, we obtain the following relationship.

$$t_{delay} = 0.35 \times 10^{-9} = -2C_{load} \int_{V_{out}=2}^{V_{out}=0.5} \frac{dV_{out}}{\mu_n C_{ox} \frac{W_n}{L_n} \left[2(V_{OH} - V_{T0,n})V_{out} - V_{out}^2 \right]}$$

$$t_{delay} = \frac{-C_{load}}{\mu_n C_{ox} \frac{W_n}{L_n}} \frac{1}{(V_{OH} - V_{T0,n})} \ln \left(\frac{V_{out}}{2(V_{OH} - V_{T0,n}) - V_{out}} \right) \Bigg|_2^{0.5}$$

$$t_{delay} = \frac{-C_{load}}{\mu_n C_{ox} \left(\frac{W_n}{L_n} \right)} \frac{1}{(3 - 0.8)} \left[\ln \left(\frac{0.5}{2(3 - 0.8) - 0.5} \right) - \ln \left(\frac{2}{2(3 - 0.8) - 2} \right) \right]$$

$$0.35 \times 10^{-9} = \frac{-300 \times 10^{-15}}{120 \times 10^{-6} \left(\frac{W_n}{L_n} \right) 2.2} [-2.054 + 0.182]$$

Now we solve this equation for the nMOS transistor (W/L) ratio:

$$\left(\frac{W_n}{L_n} \right) = 6.1$$

Notice that this ratio is *smaller* than the (W/L)-ratio found from the propagation delay constraint. Thus, we take the larger ratio which will satisfy both timing constraints, and determine the size of the nMOS transistor as $W_n = 4.7 \, \mu$m, for the given $L_n = 0.6 \, \mu$m. Next, the logic threshold constraint of $V_{th} = 1.5$ V will help determine the pMOS transistor dimensions. Using (5.87) for the logic threshold voltage of the CMOS inverter,

$$V_{th} = \frac{V_{T0,n} + \sqrt{\frac{1}{k_R}(V_{DD} + V_{T0,p})}}{1 + \sqrt{\frac{1}{k_R}}} = 1.5$$

we find that the ratio k_R which satisfies this design constraint is equal to 0.51. This value can now be used to calculate the (W/L)-ratio of the pMOS transistor, as follows.

$$k_R = \frac{\mu_n C_{ox} \left(\frac{W_n}{L_n} \right)}{\mu_p C_{ox} \left(\frac{W_p}{L_p} \right)} = \frac{120 \times 10^{-6} (7.9)}{60 \times 10^{-6} \left(\frac{W_p}{L_p} \right)} = 0.51$$

$$\left(\frac{W_p}{L_p} \right) = 31$$

Note that this ratio is *larger* than the one found from the propagation delay constraint earlier. Since the larger ratio will satisfy both the timing constraint and the V_{th}-constraint, we determine the pMOS transistor size as $W_p = 18.6 \, \mu$m, for the given $L_p = 0.6 \, \mu$m.

In the previous discussion, we have assumed that the combined output load capacitance C_{load} is mainly dominated by its extrinsic components, and hence, that it is not very sensitive to device dimensions. This assumption allowed us to treat C_{load} as a constant design parameter. Yet in most cases, we have to take into account that the intrinsic components of C_{load} are *increasing* functions of the device dimensions, W_n and W_p. The expression (6.1) describing the components of the output load capacitance becomes

$$C_{load} = C_{gd,n}(W_n) + C_{gd,p}(W_p)$$
$$+ C_{db,n}(W_n) + C_{db,p}(W_p)$$
$$+ C_{int} + C_g$$
$$= f(W_n, W_p) \tag{6.35}$$

Note that the fan-out capacitance C_g is also a function of the device dimensions in the next-stage gate(s). In the following discussion, the fan-out capacitance C_g will be treated as a constant parameter which is independent of the device dimensions of the driving gate. It is clear from (6.35) that the determination of device dimensions to satisfy delay constraints will not be as straightforward as suggested by the earlier equations, (6.33) and (6.34). Any effort to increase the channel width of nMOS and pMOS transistors in order to reduce delay will inevitably increase the intrinsic components of the load capacitance.

To gain some insight into the transistor sizing problem under delay constraints and to analyze the interaction of various design parameters, we will consider the simplified mask layout of a CMOS inverter in Fig. 6.8. Here, the diffusion areas of both nMOS and pMOS transistors have a simple rectangular geometry, and the drain region length D_{drain} is assumed to be same for both devices. The relatively small gate-to-drain capacitances $C_{gd,n}$ and $C_{gd,p}$ will be neglected in the following analysis. The drain regions of nMOS and pMOS transistors are highlighted in Fig. 6.8. Using the junction capacitance expressions (3.109) and (3.114) derived in Chapter 3, the drain parasitic capacitances can be found as

$$C_{db,n} = W_n D_{drain} C_{j0,n} K_{eq,n} + 2(W_n + D_{drain}) C_{jsw,n} K_{eq,n} \tag{6.36}$$

$$C_{db,p} = W_p D_{drain} C_{j0,p} K_{eq,p} + 2(W_p + D_{drain}) C_{jsw,p} K_{eq,p} \tag{6.37}$$

where $C_{j0,n}$ and $C_{j0,p}$ denote the zero-bias junction capacitances for n-type and p-type diffusion regions, $C_{j0sw,n}$ and $C_{j0sw,p}$ denote the zero-bias sidewall junction capacitances, and $K_{eq,n}$ and $K_{eq,n}$ denote the voltage equivalence factors. The combined output load capacitance then becomes

$$C_{load} = (W_n C_{j0,n} K_{eq,n} + W_p C_{j0,p} K_{eq,p}) D_{drain}$$
$$+ 2(W_n + D_{drain}) C_{jsw,n} K_{eq,n}$$
$$+ 2(W_p + D_{drain}) C_{jsw,p} K_{eq,p}$$
$$+ C_{int} + C_g \tag{6.38}$$

Thus, the total capacitive load of the inverter can now be expressed as

$$C_{load} = \alpha_0 + \alpha_n W_n + \alpha_p W_p \tag{6.39}$$

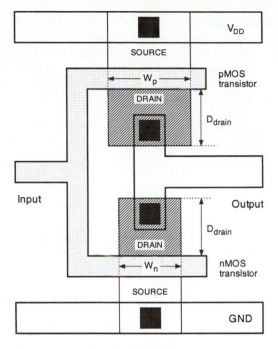

Figure 6.8 Simplified CMOS inverter mask layout used for delay analysis.

where

$$\alpha_0 = 2D_{drain}(C_{jsw,n}K_{eq,n} + C_{jsw,p}K_{eq,p}) + C_{int} + C_g \qquad (6.40)$$

$$\alpha_n = K_{eq,n}(C_{j0,n}D_{drain} + 2C_{jsw,n}) \qquad (6.41)$$

$$\alpha_p = K_{eq,p}(C_{j0,p}D_{drain} + 2C_{jsw,p}) \qquad (6.42)$$

Using (6.39), the propagation delay expressions (6.22b) and (6.23b) for falling and rising output transitions are rewritten as

$$\tau_{PHL} = \left(\frac{\alpha_0 + \alpha_n W_n + \alpha_p W_p}{W_n}\right) \times \left(\frac{L_n}{\mu_n C_{ox}(V_{DD} - V_{T,n})}\right)$$

$$\times \left[\frac{2V_{T,n}}{V_{DD} - V_{T,n}} + \ln\left(\frac{4(V_{DD} - V_{T,n})}{V_{DD}} - 1\right)\right] \qquad (6.43)$$

$$\tau_{PLH} = \left(\frac{\alpha_0 + \alpha_n W_n + \alpha_p W_p}{W_p}\right) \times \left(\frac{L_p}{\mu_p C_{ox}(V_{DD} - |V_{T,p}|)}\right)$$

$$\times \left[\frac{2|V_{T,p}|}{V_{DD} - |V_{T,p}|} + \ln\left(\frac{4(V_{DD} - |V_{T,p}|)}{V_{DD}} - 1\right)\right] \qquad (6.44)$$

Note that the channel lengths L_n and L_p are usually fixed and equal to each other. Also, the ratio between the channel widths W_n and W_p is usually dictated by other design constraints such as noise margins and the logic inversion threshold. Let this transistor *aspect ratio* be defined as

$$R(\text{aspect ratio}) \equiv \frac{W_p}{W_n} \tag{6.45}$$

Now, the propagation delay τ_{PHL} for the falling output transition can be expressed as a function of the single design parameter W_n, and τ_{PLH} can be expressed as a function of the single design parameter W_p.

$$\tau_{PHL} = \Gamma_n \left(\frac{\alpha_0 + (\alpha_n + R\alpha_p)W_n}{W_n} \right) \tag{6.46a}$$

$$\tau_{PLH} = \Gamma_p \left(\frac{\alpha_0 + \left(\dfrac{\alpha_n}{R} + \alpha_p \right) W_p}{W_p} \right) \tag{6.46b}$$

where Γ_n and Γ_p are defined as

$$\Gamma_n = \left(\frac{L_n}{\mu_n C_{ox}(V_{DD} - V_{T,n})} \right) \times \left[\frac{2V_{T,n}}{V_{DD} - V_{T,n}} + \ln \left(\frac{4(V_{DD} - V_{T,n})}{V_{DD}} - 1 \right) \right] \tag{6.47a}$$

$$\Gamma_p = \left(\frac{L_p}{\mu_p C_{ox}(V_{DD} - |V_{T,p}|)} \right) \times \left[\frac{2|V_{T,p}|}{V_{DD} - |V_{T,p}|} + \ln \left(\frac{4(V_{DD} - |V_{T,p}|)}{V_{DD}} - 1 \right) \right] \tag{6.47b}$$

Given *target* delay values τ_{PHL}^* and τ_{PLH}^*, the minimum channel widths of the nMOS transistor and the pMOS transistor which satisfy these delay constraints can be calculated from (6.46a) and (6.46b), by solving for W_n and W_p, respectively.

The important conclusion to be drawn from (6.46a) and (6.46b) is that there exists an inherent limitation to switching speed in CMOS inverters, due to drain parasitic capacitances. It can be seen that *increasing* W_n and W_p to reduce the propagation delay times will have a diminishing influence upon delay beyond certain values, and the delay values will asymptotically approach a *limit value* for large W_n and W_p. From (6.46a) and (6.46b), the limit-delay values can be found as

$$\tau_{PHL}^{limit} = \Gamma_n(\alpha_n + R\alpha_p) \tag{6.48a}$$

$$\tau_{PLH}^{limit} = \Gamma_p \left(\frac{\alpha_n}{R} + \alpha_p \right) \tag{6.48b}$$

The propagation delay times of a CMOS inverter cannot be reduced beyond these limit values, which are dictated by technology-related parameters such as doping densities, minimum channel length, and minimum layout design rules (e.g., D_{drain}). Also, note that the propagation delay limit is *independent* of the extrinsic capacitance components, C_{int} and C_g. How fast this asymptotic limit is approached in a specific case depends on the *ratio* of intrinsic and extrinsic capacitance components of C_{load}. If the extrinsic components dominate the total load capacitance, then delay reduction can be achieved for wider range of W_n and W_p. If, on the other hand, the intrinsic capacitance component is dominant, then the speed limit is reached for smaller values of W_n and W_p.

EXAMPLE 6.4

To illustrate some of the fundamental issues discussed in this section, we consider the design of a CMOS inverter using the $0.8\,\mu$m technology parameters given in Chapter 4. The power supply voltage is $V_{DD} = 3.3$ V; the extrinsic capacitance component of the load (which consists of interconnect capacitance and next-stage fan-in capacitance) is 100 fF. The channel lengths of nMOS and pMOS transistors are $L_n = L_p = 0.8\,\mu$m. The transistor aspect ratio is chosen as $R = (W_p/W_n) = 2.75$.

In this example, a number of CMOS inverter circuits with various transistor widths were designed to drive the *same* extrinsic capacitance. The actual mask layout of each inverter was drawn, the parasitic capacitances were calculated from the layout (parasitic extraction), and the extracted circuit file was simulated using SPICE in order to determine the transient response and the propagation delay of each inverter. The simulated output voltage waveforms obtained for five different inverter designs are shown below.

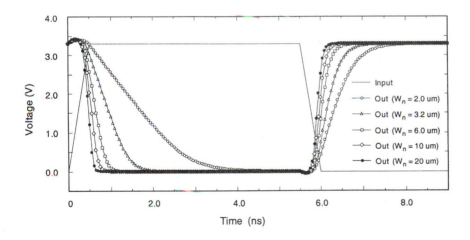

As expected, the inverter circuit with the smallest transistor dimensions ($W_n = 2\,\mu$m, $W_p = 5.5\,\mu$m) has the largest propagation delay. The delay is reduced by increasing the channel widths of both the nMOS and the pMOS devices. Initially, the amount of delay reduction can be quite significant; for example, the propagation delay τ_{PHL} is reduced by about 50% when the transistor widths are increased to $W_n = 3.2\,\mu$m and $W_p = 8.8\,\mu$m. Yet the rate of delay reduction gradually diminishes when the transistor widths are further increased, and the delays approach limit values which are described by (6.48). For example, the effect of a 100% width increase from $W_n = 10\,\mu$m to $W_n = 20\,\mu$m is almost negligible, due to the increased drain parasitic capacitances of both transistors, as explained in the previous discussion.

In the following figure, the falling-output propagation delay τ_{PHL} (obtained from SPICE simulation) is plotted as a function of the nMOS channel width. The delay asymptotically approaches a limit value of about 0.2 ns, which is mainly determined by technology-specific parameters, independent of the extrinsic capacitance component.

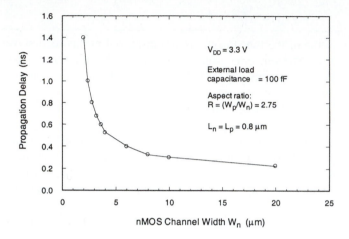

In addition to the fact that the influence of device sizing upon propagation delay is inherently limited by the parasitic capacitances, the overall silicon area occupied by the circuit should also be considered. In fact, the increase in silicon area can be viewed as a design trade-off for delay reduction, since the circuit speed improvements are typically obtained at the expense of increased transistor dimensions. In our example, the circuit area is proportional to W_n and W_p, since the other transistor dimensions are simply kept constant while the channel widths are increased to reduce the delay. Based on the simulation results given above, it can be argued that increasing W_n beyond about 4–5 μm will result in a waste of valuable silicon area, since the obtainable delay reduction is very small beyond that point.

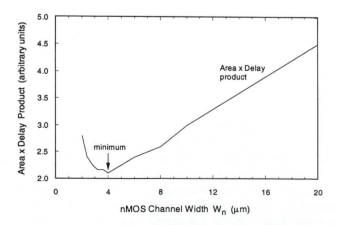

A practical measure used for quantifying *design quality* is the (Area × Delay) product, which takes into account the silicon-area cost of transistor sizing for delay reduction. While the propagation delay asymptotically approaches a limit value for increasing channel widths, the (Area × Delay) product exhibits a clear minimum around $W_n = 4\,\mu$m, indicating the optimum choice both in terms of speed and overall silicon area.

Note that the analytical discussions of CMOS inverter design issues presented in this section were mainly based on the conventional (long-channel) current-voltage model of MOS transistors. In small-geometry (submicron) transistors, other effects such as carrier velocity saturation should be included in the analysis of the switching speed.

Nevertheless, the intrinsic delay of state-of-the-art submicron inverters can typically be on the order of a few tens of picoseconds, still allowing very high switching speeds (at least theoretically). In fact, the speed limitations of modern sub-micron logic circuits mainly stem from the constraints imposed by interconnect parasitics, rather than the intrinsic delays of individual gates. This issue will be examined in the following sections.

CMOS Ring Oscillator Circuit

The following circuit example illustrates some of the basic notions associated with the switching characteristics of inverters, which were introduced in the preceding sections. At the same time, this example will provide us with a simple demonstration of astable behavior in digital circuits.

Consider the cascade connection of three identical CMOS inverters, as shown in Fig. 6.9, where the output node of the third inverter is connected to the input node of the first inverter. As such, the three inverters form a voltage feedback loop. It can be found by simple inspection that this circuit does not have a stable operating point. The only DC operating point, at which the input and output voltages of all inverters are equal to the logic threshold V_{th}, is inherently unstable in the sense that any disturbance in node voltages would make the circuit drift away from the DC operating

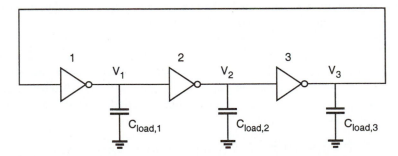

Figure 6.9 Three-stage ring oscillator circuit consisting of identical inverters.

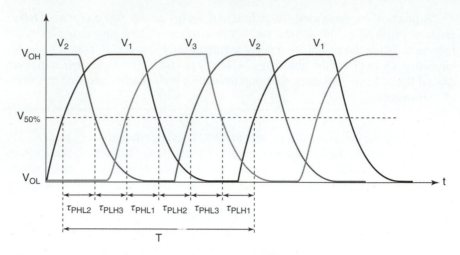

Figure 6.10 Typical voltage waveforms of the three inverters shown in Fig. 6.9.

point. In fact, a closed-loop cascade connection of any *odd* number of inverters will display astable behavior; i.e., such a circuit will oscillate once any of the inverter input or output voltages deviate from the unstable operating point, V_{th}. Therefore, the circuit is called a *ring oscillator*. A detailed analysis of closed-loop cascade circuits consisting of identical inverters will be presented in Chapter 8. Here, a qualitative understanding of the circuit behavior will be sought.

Figure 6.10 shows the typical output voltage waveforms of the three inverters during oscillation. As the output voltage V_1 of the first inverter stage rises from V_{OL} to V_{OH}, it triggers the second inverter output V_2 to fall, from V_{OH} to V_{OL}. Note that the difference between the $V_{50\%}$-crossing times of V_1 and V_2 is the signal propagation delay τ_{PHL2}, of the second inverter. As the output voltage V_2 of the second inverter falls, it triggers the output voltage V_3 of the third inverter to rise from V_{OL} to V_{OH}. Again, the difference between the $V_{50\%}$-crossing times of V_2 and V_3 is the signal propagation delay τ_{PLH3}, of the third inverter. It can be seen from Fig. 6.10 that each inverter triggers the next inverter in the cascade connection, and the last inverter again triggers the first, thus sustaining the oscillation.

In this three-stage circuit, the oscillation period T of any of the inverter output voltages can be expressed as the sum of six propagation delay times (Fig. 6.10). Since the three inverters in the closed-loop cascade connection are assumed to be identical, and since the output load capacitances are equal to each other ($C_{load1} = C_{load2} = C_{load3}$), we can also express the oscillation period T in terms of the average propagation delay τ_P, as follows:

$$
\begin{aligned}
T &= \tau_{PHL1} + \tau_{PLH1} + \tau_{PHL2} + \tau_{PLH2} + \tau_{PHL3} + \tau_{PLH3} \\
&= 2\,\tau_P + 2\,\tau_P + 2\,\tau_P \\
&= 3 \cdot 2\,\tau_P = 6\,\tau_P
\end{aligned}
\tag{6.49}
$$

Generalizing this relationship for an arbitrary odd number (n) of cascade-connected inverters, we obtain

$$f = \frac{1}{T} = \frac{1}{2 \cdot n \cdot \tau_P} \qquad (6.50)$$

Thus, the oscillation frequency (f) is found to be a very simple function of the average propagation delay of an inverter stage. This relationship can also be utilized to *measure* the average propagation delay of a typical inverter with *minimum* capacitive loading, simply by fabricating a ring oscillator circuit consisting of n identical inverters, and by accurately determining its oscillation frequency. From (6.50) we obtain

$$\tau_P = \frac{1}{2 \cdot n \cdot f} \qquad (6.51)$$

Typically, the number n is made much larger than just 3 or 5, in order to keep the oscillation frequency of the circuit within an easily measurable range. The ring oscillator frequency measurements are routinely utilized to characterize a particular design and/or a new fabrication process. Also, the ring oscillator circuit can be used as a very simple pulse generator, where the output waveform is utilized as a simple master clock signal generated on-chip. However, for higher accuracy and stability of the oscillation frequency, an off-chip crystal oscillator is usually preferred.

6.5 Estimation of Interconnect Parasitics

The classical approach for determining the switching speed of a logic gate is based on the assumption that the loads are mainly capacitive and lumped. In the previous sections, we have examined the relatively simple delay models for inverters with purely capacitive load at the output node, which can be used to estimate the transient behavior of the circuit once the load is determined. The conventional delay estimation approaches seek to classify three main components of the output load, all of which are assumed to be purely capacitive, as: (i) internal parasitic capacitances of the transistors, (ii) interconnect (line) capacitances, and (iii) input capacitances of the fan-out gates. Of these three components, the load conditions imposed by the interconnection lines present serious problems, especially in submicron circuits.

Figure 6.11 shows a simple situation where an inverter is driving three other inverters, linked by interconnection lines of different length and geometry. If the load from each interconnection line can be approximated by a lumped capacitance, then the total load seen by the primary inverter is simply the sum of all capacitive components described above. In most cases, however, the load conditions imposed by the interconnection line are far from being simple. The line, itself a three-dimensional structure in metal and/or polysilicon, usually has a non-negligible resistance in addition to its capacitance. The (length/width) ratio of the wire usually dictates that the parameters are *distributed,* making the interconnect a true transmission line. Also, an interconnect is rarely isolated from other influences. In realistic conditions, the interconnection line is in very close proximity to a number of other lines, either on the

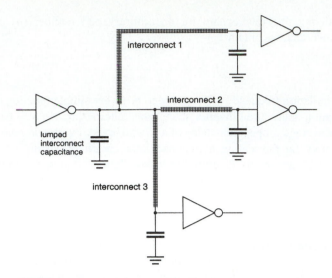

Figure 6.11   An inverter driving three other inverters over interconnection lines.

same level or on different levels. The capacitive/inductive coupling and the signal interference between neighboring lines should also be taken into consideration for an accurate estimation of delay.

In general, if the time of flight across the interconnection line (as determined by the speed of light) is much shorter than the signal rise/fall times, then the wire can be modeled as a capacitive load, or as a lumped or distributed RC network. If the interconnection lines are sufficiently long and the rise times of the signal waveforms are comparable to the time of flight across the line, then the *inductance* also becomes important, and the interconnection lines must be modeled as *transmission lines*. The following is a simple rule of thumb which can be used to determine when to use transmission-line models.

$$\tau_{rise}(\tau_{fall}) < 2.5 \times \left(\frac{l}{v}\right) \qquad \Rightarrow \quad \{\text{transmission-line modeling}\}$$

$$2.5 \times \left(\frac{l}{v}\right) < \tau_{rise}(\tau_{fall}) < 5 \times \left(\frac{l}{v}\right) \quad \Rightarrow \quad \left\{ \begin{array}{l} \text{either transmission-line} \\ \text{or lumped modeling} \end{array} \right\} \qquad (6.52)$$

$$\tau_{rise}(\tau_{fall}) > 5 \times \left(\frac{l}{v}\right) \qquad \Rightarrow \quad \{\text{lumped modeling}\}$$

Here, l is the interconnect line length, and v is the propagation speed. Note that transmission line analysis gives the correct result irrespective of the rise/fall time and the interconnect length; yet the same result can be obtained with the same accuracy using lumped approximation when rise/fall times are sufficiently large. For example, the longest wire on a VLSI chip may be about 2 cm. The time of flight of a signal across this wire, assuming $\varepsilon_r = 4$, is approximately 133 ps, which is shorter than typical on-chip signal rise/fall times. Thus, either a capacitive or an RC model

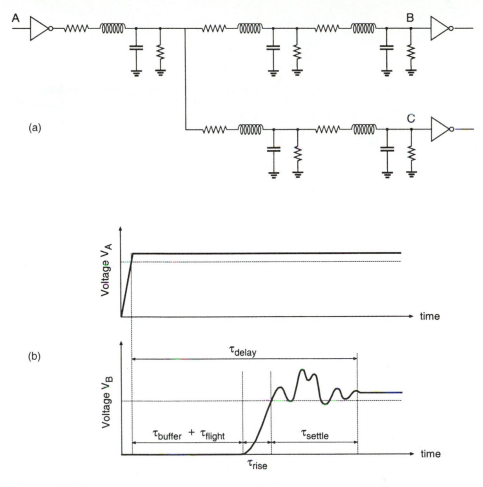

Figure 6.12　(a) An RLCG interconnection tree. (b) Typical signal waveforms at the nodes A and B, showing the signal delay and the various delay components.

is adequate for the wire. On the other hand, the time of flight across a 10 cm multichip module (MCM) interconnect in an alumina substrate is approximately 1 ns, which is of the same order of magnitude as the rise times of signals generated by some drivers. In this case, the interconnection lines should be modeled by taking into consideration the RLCG (resistance, inductance, capacitance, and conductance) parasitics as shown in Fig. 6.12. Note that the signal integrity can be significantly degraded especially when the output impedance of the driver is significantly lower than the characteristic impedance of the transmission line.

The transmission-line effects have not been a serious concern in CMOS VLSI chips until recently, since the gate delays due to capacitive load components dominated the line delay in most cases. But as the fabrication technologies move to finer sub-micron design rules, the intrinsic gate delays tend to decrease significantly. In

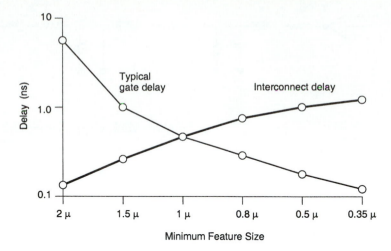

Figure 6.13 Interconnect delay dominates gate delay in sub-micron CMOS technologies.

contrast, the overall chip size and the worst-case line length on a chip tend to increase, mainly due to increasing chip complexity; thus, the importance of interconnect delay increases in sub-micron technologies. In addition, as the widths of metal lines shrink while thicknesses increase, the transmission line effects and signal coupling between neighboring lines become even more pronounced.

This fact is illustrated in Fig. 6.13, where typical intrinsic gate delay and interconnect delay are plotted qualitatively, for different technologies. It can be seen that for sub-micron technologies, the interconnect delay starts to dominate the gate delay. In order to deal with the implications and to optimize a system for speed, chip designers must have reliable and efficient means for (i) estimating the interconnect parasitics in a large chip, and (ii) simulating the transient effects.

Having witnessed that the interconnection delay becomes a dominant factor in sub-micron VLSI chips, we need to know which interconnections in a large chip may cause serious problems in terms of delay. The hierarchical structure of most VLSI designs offers some insight on this question. In a chip consisting of several functional modules, each module contains a relatively large number of local connections between its functional blocks, logic gates, and transistors. Since these intra-module connections are usually made over short distances, their influence on speed can be simulated easily with conventional models. Yet there are also a fair amount of longer connections between the modules on a chip, the so-called inter-module connections. It is usually these inter-module connections which should be scrutinized in the early design phases for possible timing problems. Figure 6.14 shows the typical statistical distribution of wire lengths on a chip, normalized for the chip diagonal length. The distribution plot clearly exhibits two distinct peaks, one for the relatively shorter intra-module connections, and the other for the longer inter-module connections. Also note that a small number of interconnections may be very long, typically longer

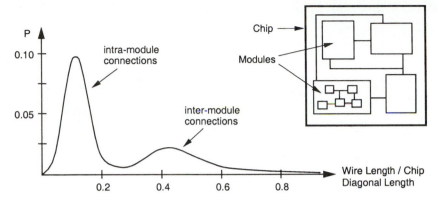

Figure 6.14 Statistical distribution of interconnection length on a typical chip.

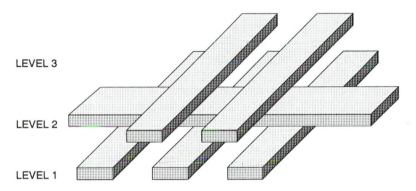

Figure 6.15 An example of six interconnect lines running on three different levels.

than the chip diagonal length. These lines are usually required for global signal bus connections, and for clock distribution networks. Although their numbers are relatively small, these long interconnections are usually the most problematic ones.

Interconnect Capacitance Estimation

In a large-scale integrated circuit, the parasitic interconnect capacitances are among the most difficult parameters to estimate accurately. Each interconnection line (wire) is a three-dimensional structure in metal and/or polysilicon, with significant variations of shape, thickness, and vertical distance from the ground plane (substrate). Also, each interconnect line is typically surrounded by a number of other lines, either on the same level or on different levels. Figure 6.15 shows a simplified view of six interconnections on three different levels, running in close proximity of each other. The accurate estimation of the parasitic capacitances of these wires with respect to the ground plane, as well as with respect to each other, is a complicated task.

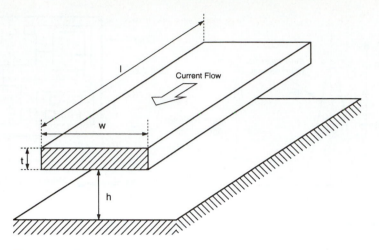

Figure 6.16   Interconnect segment running parallel to the surface, which is used for parasitic resistance and capacitance estimations.

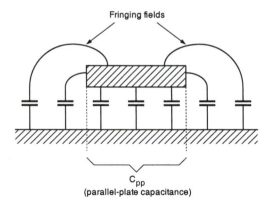

Figure 6.17   Influence of fringing electric fields upon the parasitic wire capacitance.

First, consider the section of a single interconnect which is shown in Fig. 6.16. It is assumed that this wire segment has a length of (l) in the current direction, a width of (w) and a thickness of (t). Moreover, we assume that the interconnect segment runs parallel to the chip surface and is separated from the ground plane by a dielectric (oxide) layer of height (h). Now, the correct estimation of the parasitic capacitance with respect to ground is an important issue. Using the basic geometry given in Fig. 6.16, one can calculate the parallel-plate capacitance C_{pp} of the interconnect segment. However, in interconnect lines where the wire thickness (t) is comparable in magnitude to the ground-plane distance (h), *fringing electric fields* significantly increase the total parasitic capacitance (Fig. 6.17).

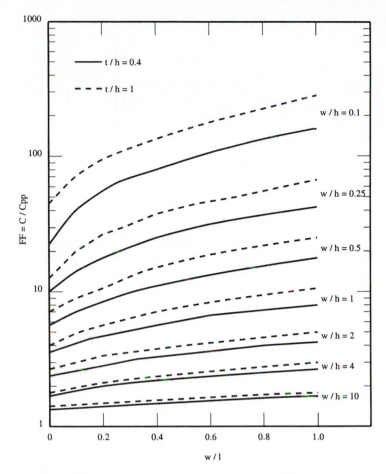

Figure 6.18 Variation of the fringing-field factor with the interconnect geometry.

Figure 6.18 shows the variation of the fringing-field factor $FF = C_{total}/C_{pp}$, as a function of (t/h), (w/h), and (w/l). It can be seen that the influence of fringing fields increases with the decreasing (w/h) ratio, and that the fringing-field capacitance can be as much as 10–20 times larger than the parallel-plate capacitance. It was mentioned earlier that the sub-micron fabrication technologies allow the width of the metal lines to be decreased rather significantly, but the thickness of the line must be preserved in order to ensure structural integrity. This situation, which involves narrow metal lines with a considerable vertical thickness, makes these interconnection lines especially vulnerable to fringing field effects.

A set of simple formulas developed by Yuan and Trick in the early 1980s can be used to estimate the capacitance of the interconnect structures in which fringing fields complicate the parasitic capacitance calculation. The following two cases are considered for two different ranges of width line width (w).

$$C = \varepsilon \left[\frac{\left(w - \frac{t}{2}\right)}{h} + \frac{2\pi}{\ln\left(1 + \frac{2h}{t} + \sqrt{\frac{2h}{t}\left(\frac{2h}{t} + 2\right)}\right)} \right] \quad \text{for} \quad w \geq \frac{t}{2} \quad (6.53)$$

$$C = \varepsilon \left[\frac{w}{h} + \frac{\pi\left(1 - 0.0543 \cdot \frac{t}{2h}\right)}{\ln\left(1 + \frac{2h}{t} + \sqrt{\frac{2h}{t}\left(\frac{2h}{t} + 2\right)}\right)} + 1.47 \right] \quad \text{for} \quad w < \frac{t}{2} \quad (6.54)$$

These formulas provide accurate approximation of the parasitic capacitance values to within 10% error, even for very small values of (t/h). Figure 6.19 shows a different view of the line capacitance as a function of (w/h) and (t/h). The linear dash-dotted line in this plot represents the corresponding parallel-plate capacitance, and the other two curves represent the actual capacitance, taking into account the fringing-field effects. We see that the actual wire capacitance decreases as its width is reduced with respect to its thickness; yet the capacitance levels off at approximately 1 pF/cm, when the wire width is approximately equal to insulator thickness.

Now consider the more realistic case where the interconnection line is not completely isolated from the surrounding structures, but is coupled with other lines

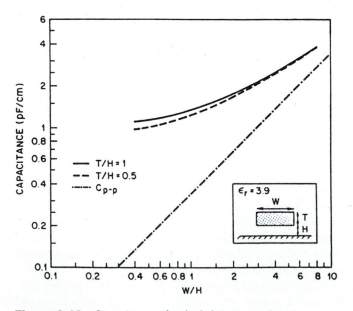

Figure 6.19 Capacitance of a single interconnect, as a function of (w/h) and (t/h).

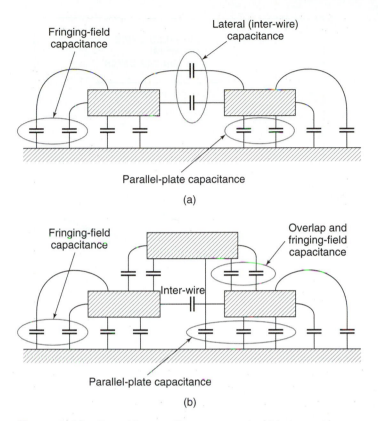

Figure 6.20   Capacitive coupling components, (a) between two
parallel lines running on the same level, and (b) between three parallel
lines running on two different levels.

running in parallel. In this case, the total parasitic capacitance of the line is not only
increased by the fringing-field effects, but also by the capacitive coupling between the
lines. The capacitance components associated with parallel interconnection lines (in
two different configurations) are depicted in Fig. 6.20. Note that the capacitive coupling between neighboring lines is increased when the thickness of the wire is comparable to its width. This coupling between the interconnect lines is mainly responsible for *signal crosstalk,* where transitions in one line can cause noise in the other lines.
Figure 6.21 shows the capacitance of a line which is coupled with two other lines on
both sides (at the same level), separated by the minimum design rule. Especially if
both of the neighboring lines are biased at ground potential, the total parasitic capacitance of the interconnect running in the middle (with respect to the ground plane) can
be more than 20 times as large as the simple parallel-plate capacitance.

Figure 6.22 shows the cross-section view of a double-metal CMOS structure,
where the individual parasitic capacitances between the layers are also indicated. The
cross-section does not show a MOSFET, but just a portion of a diffusion region
over which some metal lines may pass. The inter-layer capacitances between Metal-2
and Metal-1, Metal-1 and Polysilicon, and Metal-2 and Polysilicon are labeled as

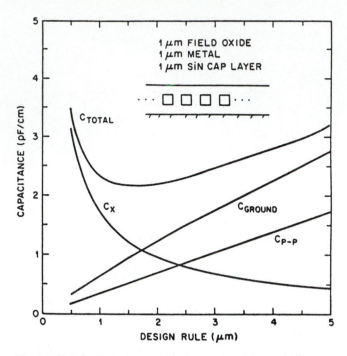

Figure 6.21   Capacitance of an interconnect line which is coupled with two other parallel lines on both sides, as a function of the minimum distance between the lines. C_{TOTAL} indicates the combined capacitance of the line, while C_{GROUND} and C_X indicate the capacitance to the ground plane and the lateral (inter-line) capacitance, respectively. The pure parallel plate capacitance is also shown as a reference.

C_{m2m1}, C_{m1p}, and C_{m2p}, respectively. The other parasitic capacitance components are defined with respect to the substrate. If the metal line passes over an active region, the oxide thickness underneath is smaller (because of the active area window), and consequently, the capacitance is larger. These special cases are labeled as C_{m1a} and C_{m2a}. Otherwise, the thick field oxide layer results in a smaller capacitance value.

The vertical thickness values of the different layers in a typical 0.8-micron CMOS technology are given in Table 6.1, to serve as an example. Notice that especially for the Metal-2 layer, the minimum (w/t) ratio can be as low as 1.6, which would lead to a significant increase of the fringing field capacitance components. Also, the (t/h) ratio for the Metal-1 layer is approximately equal to 1. Table 6.2 contains the capacitance values between the various layers shown in Fig. 6.22, for the same 0.8-micron CMOS process. The perimeter values are to be used to calculate the fringing field capacitances.

For the estimation of interconnect capacitances in a complicated three-dimensional structure, the exact geometry must be taken into account for every portion of the wire. Yet this requires an excessive amount of computation in a large

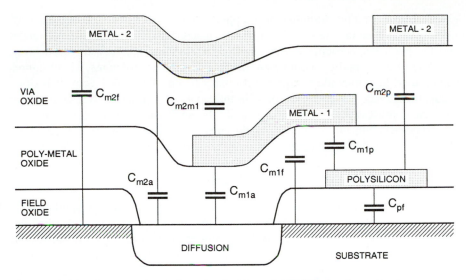

Figure 6.22 Cross-sectional view of a double-metal CMOS structure, showing capacitances between various layers.

Table 6.1 Thickness values of different layers in a typical 0.8 micron CMOS process

Field oxide thickness	$0.52~\mu$m	
Gate oxide thickness	16.0 nm	$(= 0.016~\mu$m$)$
Polysilicon thickness	$0.35~\mu$m	(minimum width $0.8~\mu$m)
Poly-metal oxide thickness	$0.65~\mu$m	
Metal-1 thickness	$0.60~\mu$m	(minimum width $1.4~\mu$m)
Via oxide thickness	$1.00~\mu$m	
Metal-2 thickness	$1.00~\mu$m	(minimum width $1.4~\mu$m)
n^+ junction depth	$0.40~\mu$m	
p^+ junction depth	$0.40~\mu$m	
n-well junction depth	$3.50~\mu$m	

Table 6.2 Parasitic capacitance values between various layers, for a typical double-metal 0.8 micron CMOS technology

Poly over field oxide	C_{pf}	Area	0.066 fF/μm^2
		Perimeter	0.046 fF/μm
Metal-1 over field oxide	C_{m1f}	Area	0.030 fF/μm^2
		Perimeter	0.044 fF/μm
Metal-2 over field oxide	C_{m2f}	Area	0.016 fF/μm^2
		Perimeter	0.042 fF/μm
Metal-1 over Poly	C_{m1p}	Area	0.053 fF/μm^2
		Perimeter	0.051 fF/μm
Metal-2 over Poly	C_{m2p}	Area	0.021 fF/μm^2
		Perimeter	0.045 fF/μm
Metal-2 over Metal-1	C_{m2m1}	Area	0.035 fF/μm^2
		Perimeter	0.051 fF/μm

circuit, even if simple formulas are applied for the calculation of capacitances. Usually, chip manufacturers supply the area capacitance (parallel-plate capacitance) and the perimeter capacitance (fringing-field capacitance) figures for each layer, which are backed up by measurement of capacitance test structures. These figures can be used to extract the parasitic capacitances from the mask layout. It is often prudent to include test structures on chip that enable the designer to independently calibrate a process to a set of design tools. In some cases where the entire chip performance is influenced by the parasitic capacitance or coupling of a specific line, accurate 3-D simulation of interconnect parasitic effects is the only reliable solution.

Interconnect Resistance Estimation

The parasitic resistance of a metal or polysilicon line can also have a significant influence on the signal propagation delay over that line. The resistance of a line depends on the type of material used (e.g., polysilicon, aluminum, or gold), the dimensions of the line and finally, the number and locations of the contacts on that line. Consider again the interconnection line shown in Fig. 6.16. The total resistance in the indicated current direction can be found as

$$R_{wire} = \rho \cdot \frac{l}{w \cdot t} = R_{sheet}\left(\frac{l}{w}\right) \tag{6.55}$$

where (ρ) represents the characteristic resistivity of the interconnect material, and R_{sheet} represents the sheet resistivity of the line, in (Ω/square).

$$R_{sheet} = \frac{\rho}{t} \tag{6.56}$$

For a typical polysilicon layer, the sheet resistivity is between 20–40 Ω/square, whereas the sheet resistivity of silicided polysilicon (polycide) is about 2–4 Ω/square. The sheet resistance of aluminum is usually much smaller, approximately 0.1 Ω/square. Typical metal-poly and metal-diffusion contact resistance values are between 20–30 Ω, while typical via resistance is about 0.3 Ω.

Using (6.55), we can estimate the total parasitic resistance of a wire segment based on its geometry. In most short-distance aluminum and polycide interconnects, the amount of parasitic wire resistance is usually negligible. On the other hand, the effects of the parasitic resistance must be taken into account for longer wire segments. As a first-order approximation in simulations, the total lumped resistance may be assumed to be connected in series with the total lumped capacitance of the wire. Yet in most cases, more accurate approaches are needed to estimate the interconnect delay, as we will examine in detail next.

6.6 Calculation of Interconnect Delay

RC Delay Models

As we have already discussed in the previous Section, an interconnect line can be modeled as a lumped *RC* network if the time of flight across the interconnection line is significantly shorter than the signal rise/fall times. This is usually the case in most

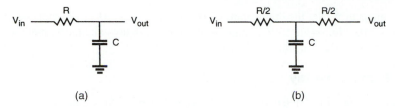

Figure 6.23 (a) Simple lumped RC model of an interconnect line, where R and C represent the total line resistance and capacitance, respectively. (b) The T-model of the same line.

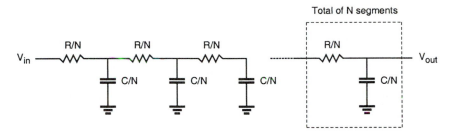

Figure 6.24 Distributed RC ladder network model consisting of N equal segments.

on-chip interconnects; thus, we will mainly concentrate on the calculation of delay in RC networks, in the following.

The simplest model which can be used to represent the resistive and capacitive parasitics of the interconnect line consists of one lumped resistance and one lumped capacitance (Fig. 6.23(a)). Assuming that the capacitance is discharged initially, and assuming that the input signal is a rising step pulse at time $t = 0$, the output voltage waveform of this simple RC circuit is found as

$$V_{out}(t) = V_{DD}\left(1 - exp\left(-\frac{t}{RC}\right)\right) \qquad (6.57)$$

The rising output voltage reaches the 50%-point at $t = \tau_{PLH}$; thus, we have

$$V_{50\%} = V_{DD}\left(1 - exp\left(-\frac{\tau_{PLH}}{RC}\right)\right) \qquad (6.58)$$

and the propagation delay for the simple lumped RC network is found as

$$\tau_{PLH} \approx 0.69\, RC \qquad (6.59)$$

Unfortunately, this simple lumped RC network model provides a very rough approximation of the actual transient behavior of the interconnect line. The accuracy of the simple lumped RC model can be significantly improved by dividing the total line resistance into two equal parts (the T-model), as shown in Fig. 6.23b. The transient behavior of an interconnect line can be more accurately represented using the RC ladder network, shown in Fig. 6.24. Here, each RC-segment consists of a series resistance (R/N), and a capacitance (C/N) connected between the node and the ground. It can be expected that the accuracy of this model increases with increasing N, where the transient behavior approaches that of a *distributed RC* line for very large values of N. Yet the delay analysis of such RC networks of higher complexity requires either full-scale SPICE simulation, or other delay calculation methods such as the Elmore delay formula.

The Elmore Delay

Consider a general RC tree network, as shown in Fig. 6.25. Note that (i) there are no resistor loops in this circuit, (ii) all of the capacitors in an RC tree are connected between a node and the ground, and (iii) there is one input node in the circuit. Also notice that there is a unique resistive path, from the input node to any other node in the circuit.

Inspecting the general topology of this RC tree network, we can make the following path definitions:

■   Let P_i denote the unique path from the input node to node i, $i = 1, 2, 3, \ldots, N$.

■   Let $P_{ij} = P_i \cap P_j$ denote the portion of the path between the input and the node i, which is *common* to the path between the input and node j.

Assuming that the input signal is a step pulse at time $t = 0$, the Elmore delay at node i of this RC tree is given by the following expression.

$$\tau_{Di} = \sum_{j=1}^{N} C_j \sum_{\substack{for \ all \\ k \in P_{ij}}} R_k \tag{6.60}$$

Calculation of the Elmore delay is equivalent to deriving the first-order time constant (first moment of the impulse response) of this circuit. Note that although this delay is still an *approximation* for the actual signal propagation delay from the input node to node i, it provides a fairly simple and accurate means of predicting the behavior of the RC line. The procedure to calculate the delay at any node in the circuit is very

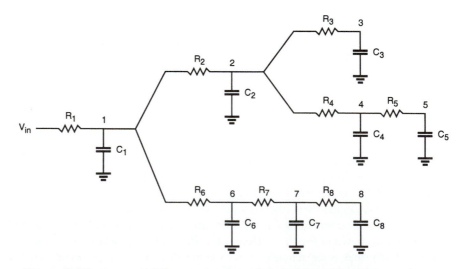

Figure 6.25   A general RC tree network consisting of several branches.

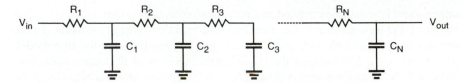

Figure 6.26 Simple *RC* ladder network consisting of one branch.

straightforward. For example, the Elmore delay at node 7 can be found according to (6.60) as

$$\tau_{D7} = R_1 C_1 + R_1 C_2 + R_1 C_3 + R_1 C_4 + R_1 C_5 + (R_1 + R_6) C_6$$
$$+ (R_1 + R_6 + R_7) C_7 + (R_1 + R_6 + R_7) C_8 \tag{6.61}$$

Similarly, the Elmore delay at node 5 can be calculated as

$$\tau_{D7} = R_1 C_1 + (R_1 + R_2) C_2 + (R_1 + R_2) C_3 + (R_1 + R_2 + R_4) C_4$$
$$+ (R_1 + R_2 + R_4 + R_5) C_5 + R_1 C_6 + R_1 C_7 + R_1 C_8 \tag{6.62}$$

As a special case of the general *RC* tree network, consider now the simple *RC* ladder network shown in Fig. 6.26. Here, the entire network consists of one single branch, and the Elmore delay from the input to the output (node *N*) is found according to (6.60) as

$$\tau_{DN} = \sum_{j=1}^{N} C_j \sum_{k=1}^{j} R_k \tag{6.63}$$

If we further assume a *uniform RC* ladder network, consisting of identical elements (R/N) and (C/N) as shown in Fig. 6.24, then the Elmore delay from the input to the output node becomes

$$\tau_{DN} = \sum_{j=1}^{N} \left(\frac{C}{N}\right) \sum_{k=1}^{j} \left(\frac{R}{N}\right)$$
$$= \left(\frac{C}{N}\right) \left(\frac{R}{N}\right) \left(\frac{N(N+1)}{2}\right) = RC \left(\frac{N+1}{2N}\right) \tag{6.64}$$

For very large *N* (distributed *RC* line behavior), this delay expression reduces to

$$\tau_{DN} = \frac{RC}{2} \qquad \text{for} \quad N \to \infty \tag{6.65}$$

Thus, we see that the propagation delay of a distributed *RC* line is considerably *smaller* than that of a lumped *RC* network shown in (6.59).

If the length of the interconnection line is sufficiently large and the rise/fall times of the signal waveforms are comparable to the time of flight across the line, then the interconnect line must be modeled as a transmission line, according to (6.52). While simple closed form solutions of transmission line equations are not available, a number of circuit simulators such as SPICE support lossless transmission line models which can be used to estimate the time-domain behavior of the interconnect line. Alternatively, we can use approximate solutions for the time-domain behavior of the transmission line, which are based on simplifying assumptions.

EXAMPLE 6.5

In this example, we will examine the signal propagation delay across a long polysilicon interconnect line, and we will compare various simulation models that can be used to represent the transient behavior of the interconnect. First, consider a uniform polysilicon line with a length of $1000\,\mu$m and a width of $4\,\mu$m. Assuming a sheet resistance value of 30 Ω/square, the total lumped resistance of the line can be found using (6.55), as

$$R_{lumped} = R_{sheet} \times (\text{\# of squares})$$

$$= 30\,(\Omega/\text{square}) \times \left(\frac{1000\,\mu\text{m}}{4\,\mu\text{m}}\right) = 7.5\,\text{k}\Omega$$

To calculate the total capacitance associated with this interconnect line, we have to consider both the parallel-plate capacitance component and the fringing-field component. Using the unit capacitance values given in Table 6.2, we obtain

$$C_{parallel\text{-}plate} = (\text{unit area capacitance}) \times (\text{area})$$

$$= 0.066\,\text{fF}/\mu\text{m}^2 \times (1000\,\mu\text{m} \times 4\,\mu\text{m}) = 264\,\text{fF}$$

$$C_{fringe} = (\text{unit length capacitance}) \times (\text{perimeter})$$

$$= 0.046\,\text{fF}/\mu\text{m} \times (1000\,\mu\text{m} + 1000\,\mu\text{m} + 4\,\mu\text{m} + 4\,\mu\text{m}) = 92\,\text{fF}$$

$$C_{lumped\_total} = C_{parallel\text{-}plate} + C_{fringe} = 356\ \text{fF}$$

The simplest model which can be used to represent the resistive and capacitive parasitics of the interconnect line is the lumped RC network model shown in Fig. 6.23(a). Assuming a step input pulse, the propagation delay time of the lumped RC network can be found using (6.59), as $\tau_{PLH} = \tau_{PHL} = 1.9$ ns. To improve the accuracy of our delay estimations, we will also use the T-model shown in Fig. 6.23(b), which is obtained by dividing the total lumped resistance into two equal parts. Notice that the single capacitance component of the T-model is equal to the total lumped capacitance value found above. Finally, we will consider using a distributed RC ladder network, consisting of 10 segments, to represent the interconnect line for higher accuracy. In this case, each segment of the RC network (Fig. 6.24) will have a resistance component of 750 Ω and a capacitance component of 35.6 fF. The simulated output voltage waveforms (obtained by using the lumped RC network model and the distributed RC ladder network model in SPICE simulations) are shown next.

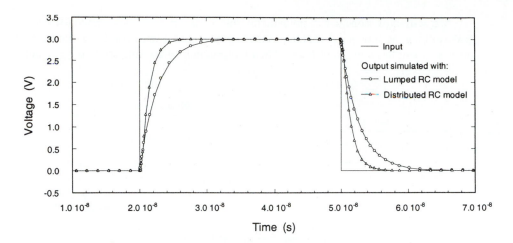

Notice that there is a significant difference between the transient response characteristics of the simplistic lumped *RC* model and the more realistic distributed *RC* ladder network model. The lumped *RC* representation of the interconnect line causes an *overestimation* of the propagation delay time.

The simulated output voltage waveforms (on the rising edge of the signal) are shown in greater detail below. It can be seen that the propagation delay of the lumped *RC* network is about 1.9 ns, while the propagation delay of the 10-segment *RC* ladder network is about 1.1 ns. Also notice that the relatively simple T-model, which only consists of one lumped capacitor and two lumped resistors, yields a significantly more accurate transient response compared to the lumped *RC* network model. Thus, the T-model should clearly be the choice when a simple transient estimation is required, especially in cases where simulation time constraints may restrict the use of a more complex, multi-stage *RC* ladder network to represent the exact transient behavior of the interconnection line. This observation will be reinforced in the next example, where we consider the behavior of a nonuniform polysilicon line.

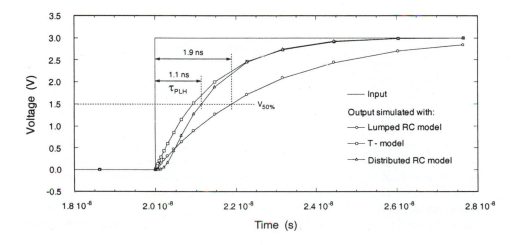

We will now consider a polysilicon line consisting of two segments, each 500 μm long, as shown below. One of the segments has a width of 2 μm, while the other segment has a width of 6 μm. The two terminals of the line are labeled A and B, respectively.

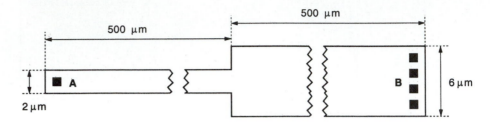

polysilicon interconnect line

The resistance of each segment, as well as the total lumped resistance of the entire interconnect line, can be found using (6.55), as follows.

$$R_{lumped\_1} = 30\,(\Omega/\text{square}) \times \left(\frac{500\,\mu\text{m}}{2\,\mu\text{m}}\right) = 7.5\,\text{k}\Omega$$

$$R_{lumped\_2} = 30\,(\Omega/\text{square}) \times \left(\frac{500\,\mu\text{m}}{6\,\mu\text{m}}\right) = 2.5\,\text{k}\Omega$$

$$R_{lumped\_total} = R_{lumped\_1} + R_{lumped\_2} = 10\,\text{k}\Omega$$

The parasitic capacitances associated with this line are calculated as

$$C_{parallel\text{-}plate\_1} = 0.066\,\text{fF}/\mu\text{m}^2 \times (500\,\mu\text{m} \times 2\,\mu\text{m}) = 66\,\text{fF}$$
$$C_{parallel\text{-}plate\_2} = 0.066\,\text{fF}/\mu\text{m}^2 \times (500\,\mu\text{m} \times 6\,\mu\text{m}) = 198\,\text{fF}$$
$$C_{fringe\_1} \approx C_{fringe\_2} = 46\,\text{fF}$$
$$C_{lumped\_total} = C_{parallel\text{-}plate\_1} + C_{parallel\text{-}plate\_2} + C_{fringe\_1} + C_{fringe\_2} = 356\,\text{fF}$$

Notice that the lumped resistance and capacitance values of this nonuniform interconnect line are quite similar to those of the uniform interconnect segment examined earlier; in fact, the total lumped capacitance value is exactly the same. Yet the fact that the geometry of this line is not uniform will have significant consequences upon its transient behavior, especially when considering signal transmission from A to B, and from B to A.

To represent the nonuniform geometry of the line, the 10-segment RC ladder network model used for SPICE simulations is constructed of two sections, each of which consists of five identical RC segments. The five RC segments corresponding to the narrower portion of the wire have a resistance component of 1.5 kΩ and a capacitance component of 22.4 fF each, while the five RC segments corresponding to the wider portion of the wire have a resistance component of 500 Ω and a capacitance component of 48.8 fF each. The simulated output voltage waveforms (obtained by using the lumped RC network model and the distributed RC ladder network model) are shown below. In the RC ladder network, the transient behavior is simulated for signal transmission from A to B, as well as for signal transmission from B to A.

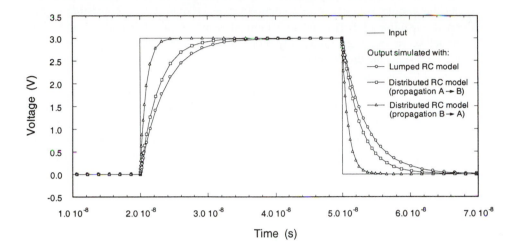

The simulated output voltage waveforms show that the lumped RC model line again causes an *overestimation* of the propagation delay time compared to the distributed RC ladder network representation. In addition, the transient behavior of the ladder network now depends on the *direction* of signal flow, and there are significant differences between the propagation delay times from A to B and from B to A. This directional dependence of propagation delay times can not be observed in the simple lumped RC model. The simulated output voltage waveforms (rising edge of the signal, signal propagation from B to A) are shown in greater detail below.

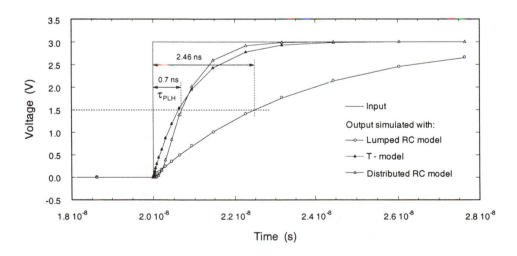

It can be seen that the propagation delay of the lumped RC network is about 2.5 ns (regardless of signal flow direction), while the propagation delay of the 10-segment RC ladder network is about 0.7 ns (for signal flow from B to A). In this case, the difference between the delay times is significantly more pronounced due to the nonuniform distribution of parasitics. As in the previous example, we also construct the simple T-model of this nonuniform interconnect line, consisting of one lumped capacitor (356 fF) and two lumped resistors (2.5 kΩ and 7.5 kΩ). The T-model again yields a significantly more accurate transient response, and it correctly represents the directional dependence of propagation delay times which is due to the nonuniform geometry of the line. This example also shows the importance of proper wire sizing in particular for the width. For instance, doubling the wire width given the same length would reduce the delay. The inclusion of coupling capacitances is becoming increasingly more important for accurate analysis of delay and signal integrity. For such analysis, SPICE simulation is much more needed. For signal integrity (SI) and prevention of serious crosstalks, designers utilize capacitive shielding around critical signal lines using ground or power lines.

6.7 Switching Power Dissipation of CMOS Inverters

It was shown in Chapter 5 that the static power dissipation of the CMOS inverter is quite negligible. During switching events where the output load capacitance is alternatingly charged up and charged down, on the other hand, the CMOS inverter inevitably dissipates power. In the following section, we will derive the expressions for the dynamic power consumption of the CMOS inverter.

Consider the simple CMOS inverter circuit shown in Fig. 6.27. We will assume that the input voltage is an ideal step waveform with negligible rise and fall times.

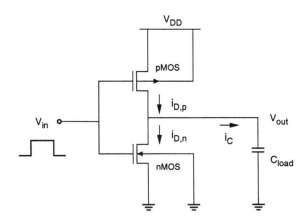

Figure 6.27 CMOS inverter used in the dynamic power-dissipation analysis.

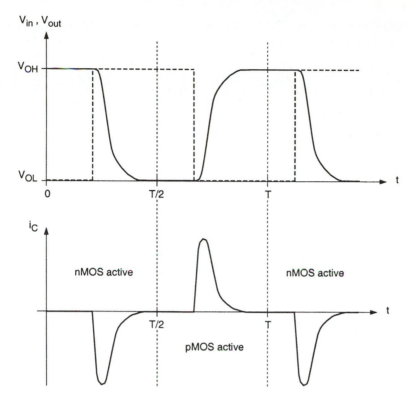

Figure 6.28 Typical input and output voltage waveforms and the capacitor current waveform during switching of the CMOS inverter.

Typical input and output voltage waveforms and the expected load capacitor current waveform are shown in Fig. 6.28. When the input voltage switches from low to high, the pMOS transistor in the circuit is turned off, and the nMOS transistor starts conducting. During this phase, the output load capacitance C_{load} is being discharged through the nMOS transistor. Thus, the capacitor current equals the instantaneous drain current of the nMOS transistor. When the input voltage switches from high to low, the nMOS transistor in the circuit is turned off, and the pMOS transistor starts conducting. During this phase, the output load capacitance C_{load} is being charged up through the pMOS transistor; therefore, the capacitor current equals the instantaneous drain current of the pMOS transistor.

Assuming periodic input and output waveforms, the average power dissipated by any device over one period can be found as follows:

$$P_{avg} = \frac{1}{T} \int_0^T v(t) \cdot i(t)\, dt \qquad (6.66)$$

Since during switching, the nMOS transistor and the pMOS transistor in a CMOS inverter conduct current for one-half period each, the average power dissipation of the CMOS inverter can be calculated as the power required to charge up and charge down the output load capacitance.

$$P_{avg} = \frac{1}{T} \left[\int_0^{T/2} V_{out} \left(-C_{load} \frac{dV_{out}}{dt} \right) dt + \int_{T/2}^{T} (V_{DD} - V_{out}) \left(C_{load} \frac{dV_{out}}{dt} \right) dt \right]$$

(6.67)

Evaluating the integrals in (6.67), we obtain

$$P_{avg} = \frac{1}{T} \left[\left(-C_{load} \frac{V_{out}^2}{2} \right) \Big|_0^{T/2} + \left(V_{DD} \cdot V_{out} \cdot C_{load} - \frac{1}{2} C_{load} V_{out}^2 \right) \Big|_{T/2}^{T} \right]$$

(6.68)

$$P_{avg} = \frac{1}{T} C_{load} V_{DD}^2$$

(6.69)

Noting that $f = 1/T$, this expression can also be written as:

$$P_{avg} = C_{load} \cdot V_{DD}^2 \cdot f$$

(6.70)

It is clear that the average power dissipation of the CMOS inverter is proportional to the switching frequency f. Therefore, the low-power advantage of CMOS circuits becomes less prominent in high-speed operation, where the switching frequency is high. Also note that the average power dissipation is independent of all transistor characteristics and transistor sizes. Consequently, the switching delay times have no relevance to the amount of power consumption during the switching events. The reason for this is that the switching power is solely dissipated for charging and discharging the output capacitance from V_{OL} to V_{OH}, and vice versa.

For this reason, the switching power expression derived for the CMOS inverter also applies to all general CMOS circuits, as shown in Fig. 6.29. A general CMOS logic circuit consists of an nMOS logic block between the output node and the ground, and a pMOS logic block between the output and V_{DD}. As in the simple CMOS inverter case, either the pMOS block *or* the nMOS block can conduct depending on the input voltage combination, but mostly not both at the same time, although slowly changing input signals can cause simultaneous conductions. Generally speaking, switching power is dissipated mainly for charging and discharging the output capacitance.

To summarize, if the total parasitic capacitance in the circuit can be lumped at the output node with reasonable accuracy, if the output voltage swing is between 0 and V_{DD}, and if the input voltage waveforms are assumed to be ideal step inputs, the average switching power expression (6.70) will hold for any CMOS logic circuit, when the leakage power is neglected.

Note that under realistic conditions, when the input voltage waveform deviates from ideal step input and has nonzero rise and fall times, for example, both the nMOS and the pMOS transistor will simultaneously conduct a certain amount of current during the switching event. This is called the short-circuit current, since in this case, the two transistors temporarily form a conducting path between the V_{DD} and the ground. The additional power dissipation, which is due to the short-circuit current, cannot be

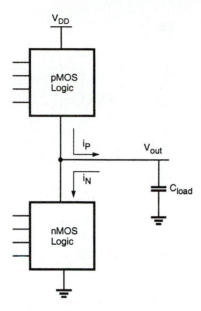

Figure 6.29   Generalized CMOS
logic circuit.

predicted by the power-dissipation formula (6.70), since the short-circuit current is not being utilized to charge or discharge the output load capacitor. We must be aware that this additional power-dissipation term can be quite significant under some non-ideal conditions. If the load capacitance is increased, on the other hand, the short-circuit dissipation term usually becomes negligible in comparison to the power dissipation which is due to the charging/discharging of capacitances.

Power Meter Simulation

In the following, we present a simple circuit simulation approach which can be used to estimate the average power dissipation of arbitrary circuits (including the effects of short circuit and leakage currents), under realistic operating conditions. According to (6.66), the average power dissipation of any device or circuit which is driven by a periodic input waveform can be found by integrating the product of its instantaneous terminal voltage and its instantaneous terminal current over one period. If we have to determine the amount of P_{avg} drawn from the power supply over one period, the problem is reduced to finding only the time-average of the power supply current, since the power supply voltage is a constant.

Using a simple simulation model called the *power meter,* we can estimate the average power dissipation of an arbitrary device or circuit driven by a periodic input, with transient circuit simulation. Consider the circuit structure shown in Fig. 6.30, in which a zero-volt independent voltage source is connected in series with the power supply voltage source V_{DD} of the device or circuit in question. Consequently, the

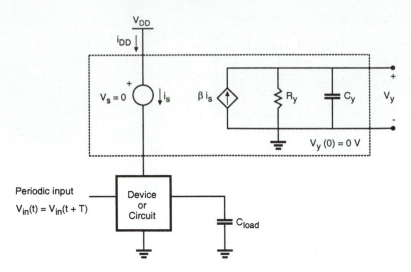

Figure 6.30   The power meter circuit used for the simulation of average dynamic power dissipation of an arbitrary device or circuit.

instantaneous power supply current $i_{DD}(t)$ which is being drawn by the circuit will also pass through the zero-volt voltage source, $i_s(t) = i_{DD}(t)$.

The power meter circuit consists of three elements: a linear current-controlled current source, a capacitor, and a resistor, all connected in parallel. The current equation for the common node of the power meter circuit can be written as follows:

$$C_y \frac{dV_y}{dt} = \beta i_s - \frac{V_y}{R_y} \tag{6.71}$$

The initial condition of the node voltage V_y is set as $V_y(0) = 0\,\text{V}$. Then, the time-domain solution of $V_y(t)$ can be found by integrating (6.71).

$$V_y(t) = \frac{\beta}{C_y} \int_0^t exp\left(-\frac{t-\tau}{R_y C_y}\right) i_{DD}(\tau)\,d\tau \tag{6.72}$$

Assuming $R_y C_y \gg T$, the voltage value $V_y(T)$ at the end of one period can be approximated as follows.

$$V_y(T) \approx \frac{\beta}{C_y} \int_0^T i_{DD}(\tau)\,d\tau \tag{6.73}$$

If the constant coefficient value of the current-controlled current source is set to be

$$\beta = V_{DD} \frac{C_y}{T} \tag{6.74}$$

the voltage value $V_y(T)$ at the end of one period will be found by transient simulation as:

$$V_y(T) = V_{DD} \cdot \frac{1}{T} \int_0^T i_{DD}(\tau)\,d\tau \tag{6.75}$$

Note that the right-hand side of (6.75) corresponds to the average power drawn from the power supply source over one period. Thus, the value of the node voltage V_y at $t = T$ gives the average power dissipation.

The power meter circuit shown in Fig. 6.30 can be easily simulated using a conventional circuit simulation program such as SPICE, and it enables us to accurately estimate the average power dissipation of any circuit with arbitrary complexity. Also note that the power meter circuit inherently takes into account the additional power dissipation due to the short-circuit currents, which may arise because of nonideal input conditions. In the following example, we present a sample SPICE simulation of the power meter for estimating the dynamic power dissipation of a CMOS inverter circuit.

EXAMPLE 6.6

Consider the simple CMOS inverter circuit shown in Fig. 6.27. We will assume that the circuit is being driven by a square-wave input signal with period $T = 20$ ns, and that the total output load capacitance is equal to 1 pF. The power supply voltage is 5 V. Using the average dynamic power-dissipation formula (6.70) derived earlier, we can calculate the expected power dissipation to be $P_{avg} = 1.25$ mW.

Now, the circuit with an attached power meter will be simulated using SPICE. The corresponding circuit input file is listed here for reference. The controlled current source coefficient is calculated as 0.025, according to (6.74). The resistance and capacitance values R_y and C_y are chosen as $100 \text{k}\Omega$ and 100 pF to satisfy the condition $R_y C_y \gg T$.

```
Power meter simulation:
mn 3 2 0 0 nmod w=10u l=1u
mp 3 2 4 1 pmod w=20u l=1u
vdd 1 0 5
vtstp 1 4 0
.model nmod nmos(vto=1 kp=20u)
.model pmod pmos(vto=-1 kp=10u)
vin 2 0 pulse(0 5 8n 2n 2n 8n 20n)
cl 3 0 1p
fp 0 9 vtstp 0.025
rp 9 0 100k
cp 9 0 100p
.tran 1n 60n uic
.print tran v(3) v(2)
.print tran i(vtstp)
.print tran v(9)
.end
```

The simulation results are plotted on the following page. It can be seen that here, the significant power supply current is being drawn from the voltage source V_{DD} only during the charge-up phase of the output capacitor. The power meter output voltage by the end of the first period corresponds to exactly 1.25 mW, as expected.

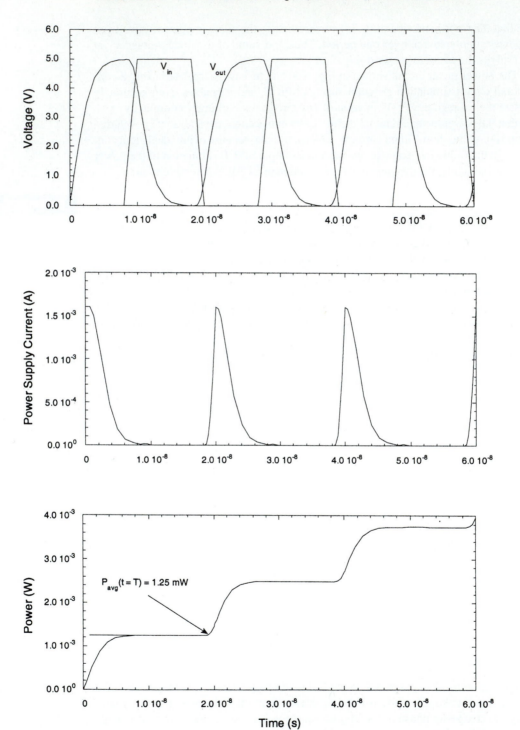

Power-Delay Product

The power-delay product (PDP) is a fundamental parameter which is often used for measuring the quality and the performance of a CMOS process and gate design. As a physical quantity, the power-delay product can be interpreted as the average *energy* required for a gate to switch its output voltage from low to high and from high to low. We have already seen that in a CMOS logic gate, energy is dissipated (i) by the pMOS network while the output load capacitance C_{load} is being charged up from 0 to V_{DD}, and (ii) by the nMOS network while the output load capacitance is being charged down from V_{DD} to 0. Following a simple analysis procedure which is very similar to the one used for deriving the average dynamic power dissipation (6.69) in CMOS logic gates and ignoring the short-circuit and leakage currents, the amount of energy required to switch the output can be found as

$$PDP = C_{load}V_{DD}^2 \qquad (6.76)$$

The energy described by (6.76) is mainly dissipated as heat when the nMOS and pMOS transistors conduct current during switching. Thus, from a design point-of-view, it is desirable to minimize the power-delay product. Since the PDP is a function of the output load capacitance and the power supply voltage, the designer should try to keep both C_{load} and V_{DD} as small as possible when designing a CMOS logic gate. The power-delay product is also defined as

$$PDP = 2P_{avg}^* \tau_P \qquad (6.77)$$

where P_{avg}^* is the average switching power dissipation at *maximum* operating frequency and τ_P is the average propagation delay, as defined in (6.4). The factor of 2 in (6.77) accounts for *two* transitions of the output, from low to high and from high to low. Using (6.69) and (6.4), this expression can be rewritten as

$$
\begin{aligned}
PDP &= 2\left(C_{load}V_{DD}^2 f_{max}\right)\tau_P \\
&= 2\left[C_{load}V_{DD}^2 \left(\frac{1}{\tau_{PHL}+\tau_{PLH}}\right)\right]\left(\frac{\tau_{PHL}+\tau_{PLH}}{2}\right) \qquad (6.78) \\
&= C_{load}V_{DD}^2
\end{aligned}
$$

which is found to be identical to (6.76). Note that calculating PDP with the generic definition of P_{avg} (6.70) may result in a *misleading interpretation* that the amount of energy required per switching event is a function of the operating frequency. Thus design engineers often use energy-delay product for performance comparison.

Appendix

Super Buffer Design

The term *super buffer* has been used to describe a chain of inverters designed to drive a large capacitive load with minimal signal propagation delay time. To reduce delay time, it is necessary for the buffer circuit to provide quickly a large amount of pull-up or pull-down current to charge or discharge the load capacitor. One seemingly obvious method would be to use large pMOS and nMOS transistors in the inverter driving the load capacitor. However, such a large buffer has a large input capacitance, which in turn creates a large load for the previous stage. Then an alert designer would suggest increasing the transistor sizes in the previous stage. If so, then what about the sizing of the transistors in the stage prior to the previous stage? Thus the effect of the large load can be propagated to many gates preceding the last-stage driver, and indeed such fine tuning of transistors is practiced in custom design. An alternative method of handling a large capacitive load is to use a super buffer between a logic gate facing the large load and the load itself as shown in Fig. A.1.

Now a major objective of super buffer design becomes:

> *Given the load capacitance faced by a logic gate, design a scaled chain*
> *of N inverters such that the delay time between the logic gate and the*
> *load capacitance node is minimized.*

To solve this problem, let us first introduce an equivalent inverter for the logic gate (NAND2 in this case). For simplicity, it is assumed that the pull-up and pull-down delays of the first-stage inverter driving an identical inverter are the same, say τ_0. The next design task is to determine the following:

- the number of stages, N
- the optimal scale factor, α

To determine these quantities, the following observations can be made under uniform α-scaling of inverters from one stage to the next in the super buffer shown in Fig. A.2.

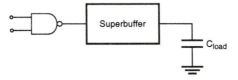

Figure A.1 Using a super buffer circuit to drive a large capacitive load.

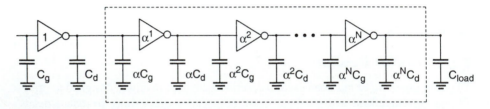

Figure A.2 Scaled super buffer circuit consisting of *N* inverter stages.

For the super buffer, the following observations can be made:

- C_g denotes the input capacitance of the first stage inverter
- C_d denotes the drain capacitance of the first stage inverter
- the inverters in the chain are scaled up by a factor of α per stage
- $C_{load} = \alpha^{N+1} C_g$ (A.1)
- all inverters have identical delay of $\tau_0(C_d + \alpha C_g)/(C_d + C_g)$ (A.2)

where τ_0 represents the per-gate delay in the ring oscillator circuit with load capacitance $(C_d + C_g)$. Thus the total delay time from the input terminal to the load capacitance node becomes

$$\tau_{total} = (N+1)\tau_0 \left(\frac{C_d + \alpha C_g}{C_d + C_g} \right) \tag{A.3}$$

There are two unknowns in this equation. To solve for these unknowns, consider the relationship between α and N in (A.1), i.e.,

$$(N+1) = \frac{\ln \left(\dfrac{C_{load}}{C_g} \right)}{\ln \alpha} \tag{A.4}$$

By combining (A.3) and (A.4), the following delay relationship can be derived.

$$\tau_{total} = \frac{\ln \left(\dfrac{C_{load}}{C_g} \right)}{\ln \alpha} \tau_0 \left(\frac{C_d + \alpha C_g}{C_d + C_g} \right) \tag{A.5}$$

To minimize the delay, we set the derivative of (A.5) with respect to α equal to zero and solve for α.

$$\frac{\partial \tau_{total}}{\partial \alpha} = \tau_0 \ln \left(\frac{C_{load}}{C_g} \right) \left[-\frac{\frac{1}{\alpha}}{(\ln \alpha)^2} \left(\frac{C_d + \alpha C_g}{C_d + C_g} \right) + \frac{1}{\ln \alpha} \left(\frac{C_g}{C_d + C_g} \right) \right] = 0$$

$$\tag{A.6}$$

Solving for α in (A.6) we obtain the following condition for the optimal scale factor.

$$\alpha(\ln \alpha - 1) = \frac{C_d}{C_g} \tag{A.7}$$

A special case of the above equation occurs when the drain capacitance is neglected; i.e., $C_d = 0$. In that case, the optimal scale factor becomes the natural number $e = 2.718$. However, in reality the drain parasitics cannot be ignored and hence, (A.6) should be considered instead.

Exercise Problems

6.1   Does the inverter with a lower V_{OL} always have the shorter high-to-low switching time? Justify your answer.

6.2   Consider switching delays for 1 pF in a 10-kΩ resistive-load inverter circuit, where

$$\mu_n \cdot C_{ox} = 25 \, \mu\text{A/V}^2$$
$$W/L = 10$$
$$V_{T0} = 1.0 \, \text{V}$$

a.   Find τ_{PHL} (50% high-to-low transition delay) by using the average-current method. Assume that the input signal is an ideal rectangular pulse switching between 0 and 5 V with zero rise/fall times. You will have to calculate V_{OL} to solve this problem.

b.   By using an appropriate differential equation and the proper initial voltage across the capacitor (when the input voltage is at V_{OH}) which is V_{OL} and not 0 V, calculate τ_{PLH}. Use the same input voltage as in part (a).

6.3   Consider a CMOS ring oscillator consisting of an odd number (n) of identical inverters connected in a ring configuration as shown in Fig. 6.7. The layout of the ring oscillator is such that the interconnection (wiring) parasitics can be assumed to be zero. Therefore, the delay of each stage is the same and the average gate delay is called the intrinsic delay (τ_P) as long as identical gates are used. The ring oscillator circuit is often used to quote the circuit speed of a particular technology using the ring oscillator frequency (f).

a.   Derive an expression for the intrinsic delay (τ_P) in terms of the number of stages n.

b.   Show that τ_P is independent of the transistor sizes, i.e., it remains the same when all the gates are scaled uniformly up or down.

6.4   Suppose that the resistive-load inverter examined in Exercise 6.2 is connected to a load capacitance of 1.0 pF which is initially discharged. The gate of the nMOS transistor is driven by a rectangular pulse which changes from high to low at $t = 0$. As a result, the nMOS transistor begins to charge up the capacitor. Solve the following two parts by using the differential equations and not by using the average-current methods.

a.   Determine the 50% low-to-high delay time, which is defined as the time difference between 50% points of input and output waveforms when the output waveform switches from low to high.

b.   Determine the 50% high-to-low delay time, which is defined as the time difference between 50% points of input and output waveforms when the output waveform switches from high to low when the capacitor is initially charged to 5.0 V.

6.5 A resistive-load inverter with $R_L = 50$ kW has the following device parameters:

$V_{TN}(V_{SB} = 0) = 1.0$ V
$\gamma = 0.5$ V$^{1/2}$
$t_{ox} = 0.05\ \mu$m
$\mu_n = 500$ cm^2/V·s
$W = 10\ \mu$m and $L = 1.0\ \mu$m

A ring oscillator is formed by connecting nine identical inverters in a closed loop. We are interested in finding the resulting oscillation frequency.

a. Find the delay times of the inverter for an ideal step pulse whose voltage swing is between V_{OL} and V_{OH}, i.e., τ_{PHL} and τ_{PLH}. It should be noted that the loading capacitance of each inverter is strictly due to the drain parasitic capacitance and the gate capacitance of the following stage. For simplicity, neglect the drain parasitics and assume C_{load} is equal to gate capacitance.

b. The rise and fall times are defined between 10% and 90% of the full voltage swing. But for simplicity, we will assume that $\tau_{fall} = 2 \cdot \tau_{PHL}$ and $\tau_{rise} = 2 \cdot \tau_{PLH}$.

   Estimate the actual propagation delays τ_{PHL} and τ_{PLH} by using the rise and fall times of the inverter and the ideal delays found in (a).

c. Find the oscillation frequency from the information in part (b).

6.6 The layout of a CMOS inverter is shown in Fig. P6.6. This inverter is driving another inverter, which is identical to the one shown below except that the

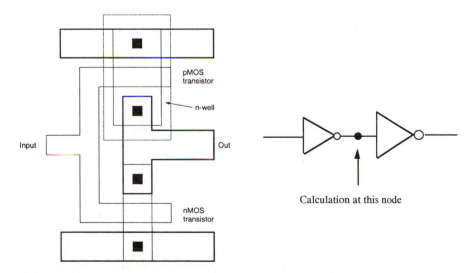

Figure P6.6

transistor widths are three times larger. Calculate τ_{PLH} and τ_{PHL}. Assume that the interconnect capacitance is negligible. The parameters are given as:

$V_{TP} = -1.0$ V $V_{TN} = 1.0$ V
$k_n' = 40\,\mu A/V^2$ $k_p' = 20\,\mu A/V^2$
$C_{ox} = 69\,nF/cm^2$ $C_{jsw} = 2.2\,pF/cm$
$C_{j0} = 7\,nF/cm^2$ $\phi_0 = 0.86$ V
$L_D = 1\,\mu m$ $L_{mask} = 5\,\mu m$

The source and drain region length is $Y = 12\,\mu m$ and the channel width is $W = 10\,\mu m$ for both transistors.

6.7 For an nMOS depletion-load inverter circuit, calculate the propagation delay times τ_{PLH} and τ_{PHL} assuming that:

- the inverter is driving an identical gate (fan-out $= 1$)
- the interconnect capacitance is negligible
- the lateral diffusion for both transistors is $L_D = 0.25\,\mu m$
- $L_{mask} = 2\,\mu m$, $W_{mask} = 10\,\mu m$, $Y = 10\,\mu m$ for the depletion-type nMOS
- $L_{mask} = 2\,\mu m$, $W_{mask} = 15\,\mu m$, $Y = 10\,\mu m$ for the enhancement-type nMOS

Use the device parameters given in Example 5.3 and the following information for calculating junction capacitances.

$t_{ox} = 0.1\,\mu m$
$x_j = 1.0\,\mu m$
$N_A = 10^{16}\,cm^{-3}$
$N_D = 10^{19}\,cm^{-3}$
$N_A(sw) = 10^{17}\,cm^{-3}$

6.8 Consider a CMOS inverter, with the following device parameters:

nMOS $V_{T0,n} = 0.8$ V $\mu_n C_{ox} = 50\,\mu A/V^2$
pMOS $V_{T0,p} = -1.0$ V $\mu_p C_{ox} = 20\,\mu A/V^2$

The power supply voltage is $V_{DD} = 5$ V. Both transistors have a channel length of $L_n = L_p = 1\,\mu m$. The total output load capacitance of this circuit is $C_{out} = 2\,pF$, which is independent of transistor dimensions.

a. Determine the channel width of the nMOS and the pMOS transistors such that the switching threshold voltage is equal to 2.2 V and the output rise time is $\tau_{rise} = 5$ ns.

b. Calculate the average propagation delay time τ_p for the circuit designed in (a).

c. How do the switching threshold V_{th} and the delay times change if the power supply voltage is dropped from 5 V to 3.3 V. Provide an interpretation of the results.

6.9 Consider a CMOS inverter with the same process parameters as in Problem 6.8. The switching threshold is designed to be equal to 2.4 V.

A simplified expression of the total output load capacitance is given as:

$$C_{out} = 500 \text{ fF} + C_{db,n} + C_{db,p}$$

Furthermore, we know that the drain-to-substrate parasitic capacitances of the nMOS and the pMOS transistors are functions of the channel width. A set of simplified capacitance expressions are given below.

$$C_{db,n} = (100 + 9 \, W_n)\text{fF}$$

$$C_{db,p} = (80 + 7 \, W_p)\text{fF}$$

where W_n and W_p are expressed in μm.

a. Determine the channel width of both transistors such that the propagation delay τ_{PHL} is smaller than 0.825 ns.

b. Assume now that the CMOS inverter has been designed with $(W/L)_n = 6$ and $(W/L)_p = 15$, and that the total output load capacitance is 250 fF. Calculate the output rise time and fall time using the average current method.

6.10 Consider a CMOS inverter with the following parameters:

$$V_{T0,n} = 1.0 \text{ V} \qquad \mu_n C_{ox} = 45 \; \mu\text{A/V}^2 \qquad (W/L)_n = 10$$

$$V_{T0,p} = -1.2 \text{ V} \qquad \mu_p C_{ox} = 25 \; \mu\text{A/V}^2 \qquad (W/L)_p = 20$$

The power supply voltage is 5 V, and the output load capacitance is 1.5 pF.

a. Calculate the rise time and the fall time of the output signal using

(i) exact method (differential equations)

(ii) average current method

b. Determine the *maximum* frequency of a periodic square-wave input signal so that the output voltage can still exhibit a full logic swing from 0 V to 5 V in each cycle.

c. Calculate the dynamic power dissipation at this frequency.

d. Assume that the output load capacitance is mainly dominated by fixed fan-out components (which are independent of W_n and W_p). We want to re-design the inverter so that the propagation delay times are reduced by 25%. Determine the required channel dimensions of the nMOS and the pMOS transistors. How does this re-design influence the switching (inversion) threshold?

7 CHAPTER

Combinational MOS Logic Circuits

7.1 Introduction

Combinational logic circuits, or gates, which perform Boolean operations on multiple input variables and determine the outputs as Boolean functions of the inputs, are the basic building blocks of all digital systems. In this chapter, we will examine the static and dynamic characteristics of various combinational MOS logic circuits. It will be seen that many of the basic principles used in the design and analysis of MOS inverters in Chapters 5 and 6 can be directly applied to combinational logic circuits as well.

The first major class of combinational logic circuits to be presented in this chapter is the nMOS depletion-load gates. Our purpose for including nMOS depletion-load circuits here is mainly pedagogical, to emphasize the *load* concept, which is still being widely used in many areas in digital circuit design. We will examine simple circuit configurations such as two-input NAND and NOR gates and then expand our analysis to more general cases of multiple-input circuit structures. Next, the CMOS logic circuits will be presented in a similar fashion. We will stress the similarities and differences between the nMOS depletion-load logic and CMOS logic circuits and point out the advantages of CMOS gates with examples. The design of complex logic gates, which allows the realization of complex Boolean functions of multiple variables, will be examined in detail. Finally, we will devote the last section to CMOS transmission gates and to transmission gate (TG) logic circuits.

In its most general form, a combinational logic circuit, or gate, performing a Boolean function can be represented as a multiple-input, single-output system, as depicted in Fig. 7.1. All input variables are represented by node voltages, referenced to the ground potential. Using *positive logic convention*, the Boolean (or logic) value of "1" can be represented by a high voltage of V_{DD}, and the Boolean (or logic) value of "0" can be represented by a low voltage of 0. The output node is loaded with a capacitance C_L, which represents the combined parasitic device capacitances in the

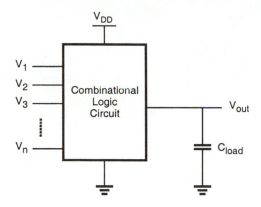

Figure 7.1 Generic combinational logic circuit (gate).

circuit and the interconnect capacitance components *seen* by the output node. This output load capacitance certainly plays a very significant role in the dynamic operation of the logic gate.

As in the simple inverter case, the voltage transfer characteristic (VTC) of a combinational logic gate provides valuable information on the DC operating performance of the circuit. Critical voltage points such as V_{OL} or V_{th} are considered to be important design parameters for combinational logic circuits. Other design parameters and concerns include the dynamic (transient) response characteristics of the circuit, the silicon area occupied by the circuit, and the amount of static and dynamic power dissipation.

7.2 MOS Logic Circuits with Depletion nMOS Loads

Two-Input NOR Gate

The first circuit to be examined in this section is the two-input NOR gate. The circuit diagram, the logic symbol, and the corresponding truth table of the gate are given in Fig. 7.2. The Boolean OR operation is performed by the parallel connection of the two enhancement-type nMOS driver transistors. If the input voltage V_A *or* the input voltage V_B is equal to the logic-high level, the corresponding driver transistor turns on and provides a conducting path between the output node and the ground. Hence, the output voltage becomes low. In this case, the circuit operates like a depletion-load inverter with respect to its static behavior. A similar result is achieved when both V_A and V_B are high, in which case two parallel conducting paths are created between the output node and the ground. If, on the other hand, both V_A and V_B are low, both driver transistors remain cut-off. The output node voltage is pulled to a logic-high level by the depletion-type nMOS load transistor.

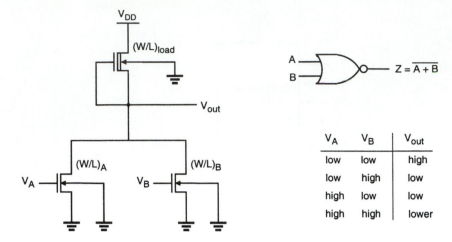

Figure 7.2 A two-input depletion-load NOR gate, its logic symbol, and the corresponding truth table. Note that the substrates of all transistors are connected to ground.

The DC analysis of the circuit can be simplified significantly by considering the structural similarities between this circuit and the simple nMOS depletion-load inverter. In the following, the calculation of output low and output high voltages will be examined.

Calculation of V_{OH}

When both input voltages V_A and V_B are lower than the corresponding driver threshold voltage, the driver transistors are turned off and should conduct no drain current (other than the subthreshold leakage currents that will be neglected here for simplicity). Consequently, the load device, which operates in the linear region, should also have zero drain current. In particular, its linear region current equation becomes

$$I_{D,load} = \frac{k_{n,load}}{2} \cdot \left[2|V_{T,load}(V_{OH})| \cdot (V_{DD} - V_{OH}) - (V_{DD} - V_{OH})^2 \right] = 0 \quad (7.1)$$

The solution of this equation gives $V_{OH} = V_{DD}$.

Calculation of V_{OL}

To calculate the output low voltage V_{OL}, we must consider three different cases, i.e., three different input voltage combinations, which produce a conducting path from the output node to the ground. These cases are

(i) $V_A = V_{OH}$ $V_B = V_{OL}$

(ii) $V_A = V_{OL}$ $V_B = V_{OH}$

(iii) $V_A = V_{OH}$ $V_B = V_{OH}$

For the first two cases, (i) and (ii), the NOR circuit reduces to a simple nMOS depletion-load inverter. Assuming that the threshold voltages of the two enhancement-type driver transistors are identical ($V_{T0,A} = V_{T0,B} = V_{T0}$), the driver-to-load ratio of the corresponding inverter can be found as follows. In case (i), where the driver transistor A is on, the ratio is

$$k_R = \frac{k_{driver,A}}{k_{load}} = \frac{k'_{n,driver}\left(\dfrac{W}{L}\right)_A}{k'_{n,load}\left(\dfrac{W}{L}\right)_{load}} \tag{7.2}$$

In case (ii), where the driver transistor B is on, the ratio is

$$k_R = \frac{k_{driver,B}}{k_{load}} = \frac{k'_{n,driver}\left(\dfrac{W}{L}\right)_B}{k'_{n,load}\left(\dfrac{W}{L}\right)_{load}} \tag{7.3}$$

The output low voltage level V_{OL} in both cases is found by using (5.38), as follows:

$$V_{OL} = V_{OH} - V_{T0} - \sqrt{(V_{OH} - V_{T0})^2 - \left(\frac{k_{load}}{k_{driver}}\right) \cdot |V_{T,load}(V_{OL})|^2} \tag{7.4}$$

Note that if the (W/L) ratios of both drivers are identical, i.e., $(W/L)_A = (W/L)_B$, the output low voltage (V_{OL}) values calculated for case (i) and case (ii) will be identical.

In case (iii), where both driver transistors are turned on, the saturated load current is the sum of the two linear-mode driver currents.

$$I_{D,load} = I_{D,driverA} + I_{D,driverB} \tag{7.5}$$

$$\frac{k_{load}}{2}|V_{T,load}(V_{OL})|^2 = \frac{k_{driver,A}}{2}\left[2(V_A - V_{T0})V_{OL} - V_{OL}^2\right]$$

$$+ \frac{k_{driver,B}}{2}\left[2(V_B - V_{T0})V_{OL} - V_{OL}^2\right] \tag{7.6}$$

Since the gate voltages of both driver transistors are equal ($V_A = V_B = V_{OH}$), we can devise an *equivalent* driver-to-load ratio for the NOR structure:

$$k_R = \frac{k_{driver,A} + k_{driver,B}}{k_{load}} = \frac{k'_{n,driver}\left[\left(\dfrac{W}{L}\right)_A + \left(\dfrac{W}{L}\right)_B\right]}{k'_{n,load}\left(\dfrac{W}{L}\right)_{load}} \tag{7.7}$$

Thus, the NOR gate with both of its inputs tied to a logic-high voltage is replaced with an nMOS depletion-load inverter circuit with the driver-to-load ratio given by

(7.7). The output voltage level in this case is

$$V_{OL} = V_{OH} - V_{T0} - \sqrt{(V_{OH} - V_{T0})^2 - \left(\frac{k_{load}}{k_{driver,A} + k_{driver,B}}\right) \cdot |V_{T,load}(V_{OL})|^2}$$

(7.8)

Note that the V_{OL} given by (7.8) is *lower* than the V_{OL} values calculated for case (i) and for case (ii), when only one input is logic-high. We conclude that the worst-case condition from the static operation viewpoint, i.e., the highest possible V_{OL} value, is observed in case (i) or (ii).

This result also suggests a simple design strategy for NOR gates. Usually, we have to achieve a certain maximum V_{OL} for the worst case, i.e., when only one input is high. Thus, we assume that one input (either V_A or V_B) is logic-high and determine the driver-to-load ratio of the resulting inverter using (7.4). Then set

$$k_{driver,A} = k_{driver,B} = k_R k_{load}$$

(7.9)

This design choice yields two identical driver transistors, which guarantee the required value of V_{OL} in the worst case. When both inputs are logic-high, the output voltage is even lower than the required maximum V_{OL}, thus the design constraint is satisfied.

EXERCISE 7.1

Consider the depletion-load nMOS NOR2 gate shown in Fig. 7.2, with the following parameters: $\mu_n C_{ox} = 25\ \mu A/V^2$, $V_{T0,driver} = 1.0\ V$, $V_{T0,load} = -3.0\ V$, $\gamma = 0.4\ V^{1/2}$, and $|2\phi_F| = 0.6\ V$. The transistor dimensions are given as $(W/L)_A = 2$, $(W/L)_B = 4$, and $(W/L)_{load} = 1/3$. The power supply voltage is $V_{DD} = 5\ V$.

Calculate the output voltage levels for all four valid input voltage combinations.

Generalized NOR Structure with Multiple Inputs

At this point, we can expand our analysis to generalized n-input NOR gates, which consist of n parallel driver transistors, as shown in Fig. 7.3. Note that the combined current I_D in this circuit is supplied by the driver transistors which are turned on, i.e., transistors which have gate voltages higher than the threshold voltage V_{T0}.

The combined pull-down current can then be expressed as follows:

$$I_D = \sum_{k(on)} I_{D,k} = \begin{cases} \sum_{k(on)} \frac{\mu_n C_{ox}}{2} \left(\frac{W}{L}\right)_k \left[2(V_{GS,k} - V_{T0})V_{out} - V_{out}^2\right] & \text{linear} \\ \\ \sum_{k(on)} \frac{\mu_n C_{ox}}{2} \left(\frac{W}{L}\right)_k (V_{GS,k} - V_{T0})^2 & \text{saturation} \end{cases}$$

(7.10)

Assuming that the input voltages of all driver transistors are identical,

$$V_{GS,k} = V_{GS} \qquad \text{for} \quad k = 1, 2, \ldots, n$$

(7.11)

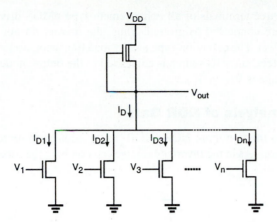

Figure 7.3 Generalized n-input NOR gate.

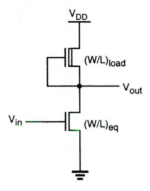

Figure 7.4 Equivalent inverter circuit corresponding to the n-input NOR gate.

the pull-down current expression can be rewritten as

$$
I_D = \begin{cases}
\dfrac{\mu_n C_{ox}}{2} \left(\displaystyle\sum_{k(on)} \left(\dfrac{W}{L}\right)_k \right) \left[2(V_{GS} - V_{T0})V_{out} - V_{out}^2 \right] & linear \\[3ex]
\dfrac{\mu_n C_{ox}}{2} \left(\displaystyle\sum_{k(on)} \left(\dfrac{W}{L}\right)_k \right) (V_{GS} - V_{T0})^2 & saturation
\end{cases}
\tag{7.12}
$$

Thus, the multiple-input NOR gate can also be reduced to an equivalent inverter, shown in Fig. 7.4, for static analysis. The (W/L) ratio of the driver transistor here is

$$
\left(\frac{W}{L}\right)_{equivalent} = \sum_{k(on)} \left(\frac{W}{L}\right)_k
\tag{7.13}
$$

Note that the source terminals of all enhancement-type nMOS driver transistors in the NOR gate are connected to ground. Thus, the drivers do not experience any substrate-bias effect. The depletion-type nMOS load transistor, however, is subject to substrate-bias effect, since its source is connected to the output node, and its source-to-substrate voltage is $V_{SB} = V_{out}$.

Transient Analysis of NOR Gate

Figure 7.5 shows the two-input NOR (NOR2) gate with all of its relevant parasitic device capacitances. As in the inverter case, we can combine the capacitances seen in

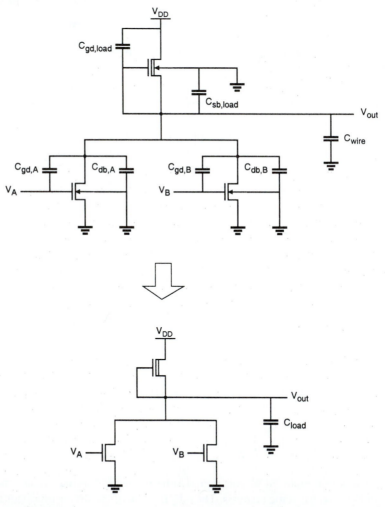

Figure 7.5 Parasitic device capacitances in the NOR2 gate and the lumped equivalent load capacitance. The gate-to-source capacitances of the driver transistors are included in the load of the previous stages driving the inputs A and B.

Fig. 7.5 into one lumped capacitance, connected between the output node and the ground. The value of this combined load capacitance, C_{load}, can be found as

$$C_{load} = C_{gd,A} + C_{gd,B} + C_{gd,load} + C_{db,A} + C_{db,B} + C_{sb,load} + C_{wire} \qquad (7.14)$$

Note that the output load capacitance given in (7.14) is valid for simultaneous as well as for single-input switching, i.e., the load capacitance C_{load} will be present at the output node even if only one input is active and all other inputs are low. This fact must be taken into account in calculations using the inverter equivalent of the NOR gate. The load capacitance at the output node of the *equivalent* inverter corresponding to a NOR gate is always *larger* than the total lumped load capacitance of an actual inverter with the same dimensions. Hence, while the static (DC) behaviors of the NOR gate and the equivalent inverter are essentially identical in this case, the actual transient response of the NOR gate will be slower than that of the equivalent inverter.

Two-Input NAND Gate

Next, we will examine the two-input NAND (NAND2) gate. The circuit diagram, the logic symbol, and the corresponding truth table of the gate are given in Fig. 7.6. The Boolean AND operation is performed by the series connection of the two enhancement-type nMOS driver transistors. There is a conducting path between the output node and the ground only if the input voltage V_A *and* the input voltage V_B are equal to logic-high, i.e., only if both of the series-connected drivers are turned on. In this case, the output voltage will be low, which is the complemented result of the AND operation. Otherwise, either one or both of the driver transistors will be off, and the output voltage will be pulled to a logic-high level by the depletion-type nMOS load transistor.

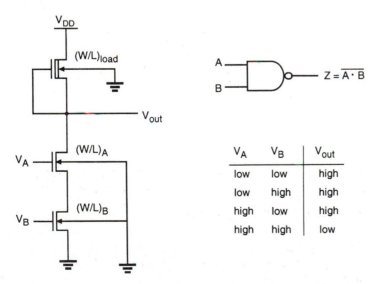

V_A	V_B	V_{out}
low	low	high
low	high	high
high	low	high
high	high	low

Figure 7.6 A two-input depletion-load NAND gate, its logic symbol, and the corresponding truth table. Notice the substrate-bias effect for all nMOS transistors except one.

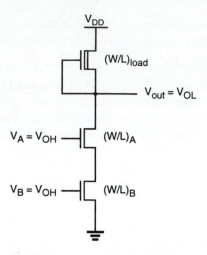

Figure 7.7 The NAND2 gate with both of its inputs at logic-high level.

Figure 7.6 shows that all transistors except the one next to the ground are subject to substrate-bias effect, since their source voltages are larger than zero. We have to consider this fact in detailed calculations. For all of the three input combinations which produce a logic-high output voltage, the corresponding V_{OH} value can easily be found as $V_{OH} = V_{DD}$. The calculation of the logic-low voltage V_{OL}, on the other hand, requires a closer investigation.

Consider the NAND2 gate with both of its inputs equal to V_{OH}, as shown in Fig. 7.7. It can easily be seen that the drain currents of all transistors in the circuit are equal to each other.

$$I_{D,load} = I_{D,driverA} = I_{D,driverB} \tag{7.15}$$

$$\frac{k_{load}}{2}|V_{T,load}(V_{OL})|^2 = \frac{k_{driver,A}}{2}\left[2(V_{GS,A} - V_{T,A})V_{DS,A} - V_{DS,A}^2\right]$$

$$= \frac{k_{driver,B}}{2}\left[2(V_{GS,B} - V_{T,B})V_{DS,B} - V_{DS,B}^2\right] \tag{7.16}$$

The gate-to-source voltages of both driver transistors can be assumed to be approximately equal to V_{OH}. Also, we may neglect, for simplicity, the substrate-bias effect for driver transistor A, and assume $V_{T,A} = V_{T,B} = V_{T0}$, since the source-to-substrate voltage of driver A is relatively low. The drain-to-source voltages of both driver transistors can then be solved from (7.16) as

$$V_{DS,A} = V_{OH} - V_{T0} - \sqrt{(V_{OH} - V_{T0})^2 - \left(\frac{k_{load}}{k_{driver,A}}\right)\cdot|V_{T,load}(V_{OL})|^2} \tag{7.17}$$

$$V_{DS,B} = V_{OH} - V_{T0} - \sqrt{(V_{OH} - V_{T0})^2 - \left(\frac{k_{load}}{k_{driver,B}}\right)\cdot|V_{T,load}(V_{OL})|^2} \tag{7.18}$$

Let the two driver transistors be identical; i.e., $k_{driver,A} = k_{driver,B} = k_{driver}$. Noting that the output voltage V_{OL} is equal to the sum of the drain-to-source voltages of both drivers, we obtain

$$V_{OL} \approx 2\left(V_{OH} - V_{T0} - \sqrt{(V_{OH} - V_{T0})^2 - \left(\frac{k_{load}}{k_{driver}}\right) \cdot |V_{T,load}(V_{OL})|^2}\right) \quad (7.19)$$

The following analysis gives a better and more accurate view of the operation of two series-connected driver transistors. Consider the two identical enhancement-type nMOS transistors with their gate terminals connected. At this point, the only simplifying assumption will be $V_{T,A} = V_{T,B} = V_{T0}$. When both driver transistors are in the linear region, the drain currents can be written as

$$I_{D,A} = \frac{k_{driver}}{2}\left[2(V_{GS,A} - V_{T0})V_{DS,A} - V_{DS,A}^2\right] \quad (7.20)$$

$$I_{D,B} = \frac{k_{driver}}{2}\left[2(V_{GS,B} - V_{T0})V_{DS,B} - V_{DS,B}^2\right] \quad (7.21)$$

Since $I_{D,A} = I_{D,B}$, this current can also be expressed as

$$I_D = I_{D,A} = I_{D,B} = \frac{I_{D,A} + I_{D,B}}{2} \quad (7.22)$$

Using $V_{GS,A} = V_{GS,B} - V_{DS,B}$, (7.22) yields

$$I_D = \frac{k_{driver}}{4}[2(V_{GS,B} - V_{T0})(V_{DS,A} + V_{DS,B}) - (V_{DS,A} + V_{DS,B})^2] \quad (7.23)$$

Now let $V_{GS} = V_{GS,B}$ and $V_{DS} = V_{DS,A} + V_{DS,B}$. The drain-current expression can then be written as follows.

$$I_D = \frac{k_{driver}}{4}[2(V_{GS} - V_{T0})V_{DS} - V_{DS}^2] \quad (7.24)$$

Thus, two nMOS transistors connected in series and with the same gate voltage behave like *one* nMOS transistor with $k_{eq} = 0.5k_{driver}$.

Generalized NAND Structure with Multiple Inputs

At this point, we expand our analysis to generalized n-input NAND gates, which consist of n series-connected driver transistors, as shown in Fig. 7.8. Neglecting the substrate-bias effect, and assuming that the threshold voltages of all transistors are equal to V_{T0}, the driver current I_D in the linear region can be derived as in Eq. (7.25) whereas I_D in saturation is taken as its extension.

$$I_D = \frac{\mu_n C_{ox}}{2}\left(\frac{1}{\sum_{k(on)}\frac{1}{\left(\frac{W}{L}\right)_k}}\right) \cdot \begin{cases} [2(V_{in} - V_{T0})V_{out} - V_{out}^2] & linear \\ (V_{in} - V_{T0})^2 & saturation \end{cases} \quad (7.25)$$

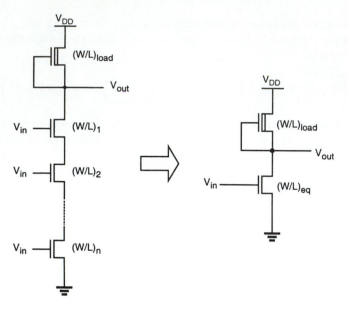

Figure 7.8   The generalized NAND structure and its inverter equivalent.

Hence, the (W/L) ratio of the equivalent driver transistor is

$$\left(\frac{W}{L}\right)_{equivalent} = \frac{1}{\displaystyle\sum_{k(on)} \frac{1}{\left(\dfrac{W}{L}\right)_k}} \qquad (7.26)$$

If the series-connected transistors are identical, i.e., $(W/L)_1 = (W/L)_2 = \cdots = (W/L)$, the width-to-length ratio of the equivalent transistor becomes

$$\left(\frac{W}{L}\right)_{equivalent} = \frac{1}{n}\left(\frac{W}{L}\right) \qquad (7.27)$$

The NAND design strategy which emerges from this analysis is summarized as follows, for an *n*-input NAND. First, we determine the (W/L) ratios for an *equivalent inverter* that satisfies the required V_{OL} value. This gives us the driver transistor ratio $(W/L)_{driver}$ and the load transistor ratio $(W/L)_{load}$. Then, we set the (W/L) ratios of all NAND driver transistors as $(W/L)_1 = (W/L)_2 = \cdots = n(W/L)_{driver}$. This guarantees that the series structure consisting of n driver transistors has an equivalent (W/L) ratio of $(W/L)_{driver}$ when all inputs are logic-high.

For a two-input NAND gate, this means that each driver transistor must have a (W/L) ratio twice that of the equivalent inverter driver. If the area occupied by the depletion-type load transistor is negligible, the resulting NAND2 structure will occupy approximately four times the area occupied by the equivalent inverter which has the same static characteristics. Note that large (W/L) ratios will also result in large parasitic capacitances.

Transient Analysis of NAND Gate

Figure 7.9 shows a NAND2 gate with all parasitic device capacitances. As in the inverter case, we can combine the capacitances seen in Fig. 7.9 into one lumped capacitance, connected between the output node and the ground. The value of the lumped capacitance C_{load}, however, depends on the input voltage conditions.

Assume, for example, that the input V_A is equal to V_{OH} and the other input V_B is switching from V_{OH} to V_{OL}. In this case, both the output voltage V_{out} and the internal node voltage V_x will rise, resulting in

$$C_{load} = C_{gd,load} + C_{gd,A} + C_{gd,B} + C_{gs,A} + C_{db,A}$$
$$+ C_{db,B} + C_{sb,A} + C_{sb,load} + C_{wire} \tag{7.28}$$

Note that this approximation fully reflects the internal node capacitances into the lumped output capacitance C_{load}, and thus, is quite conservative. In reality, only a fraction of the internal node capacitance is reflected into C_{load}. The use of overly

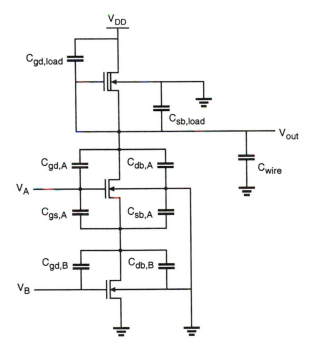

Figure 7.9 Parasitic device capacitances in the NAND2 gate.

conservative capacitance values in the design will lead to a conservative speed estimate, and force the designer to increase transistor sizes to counter the expected large delays.

Now consider another case where V_B is equal to V_{OH} and V_A switches from V_{OH} to V_{OL}. In this case, the output voltage V_{out} will rise, but the internal node voltage V_x will remain low because the bottom driver transistor is on. Thus, the lumped output capacitance is

$$C_{load} = C_{gd,load} + C_{gd,A} + C_{db,A} + C_{sb,load} + C_{wire} \qquad (7.29)$$

It should be noted that the load capacitance in this case is smaller than the load capacitance found in the previous case. Thus, it is expected that the high-to-low switching delay from signal B connected to the bottom transistor is larger than the high-to-low switching delay from signal A connected to the top transistor.

EXAMPLE 7.1

A depletion-load nMOS NAND2 gate is simulated with SPICE for two different input switching events. The SPICE input file of the circuit is listed in the following. Note that the total capacitance between the intermediate node X and the ground is assumed to be half of the total capacitance appearing between the output node and the ground.

```
NAND2 circuit delay analysis
m1 3 1 0 0 mn w=5u l=1u
m2 4 2 3 0 mn w=5u l=1u
m3 5 4 4 0 mnd w=1u l=3u
cl 4 0 0.1p
cp 3 0 0.05p
vdd 5 0 dc 5.0
* case 1 (upper input switching from high to low)
vin1 2 0 dc pulse (5.0 0.0 1ns 1ns 2ns 40ns 50ns)
vin2 1 0 dc 5.0
* case 2 (lower input switching from high to low)
* vin1 2 0 dc 5.0
* vin2 1 0 dc pulse (5.0 0.0 1ns 1ns 2ns 40ns 50ns)
.model mn nmos (vto=1.0 kp=25u gamma=0.4)
.model mnd nmos (vto=-3.0 kp=25u gamma=0.4)
.tran 0.1ns 40ns
.print tran v(1) v(2) v(4)
.end
```

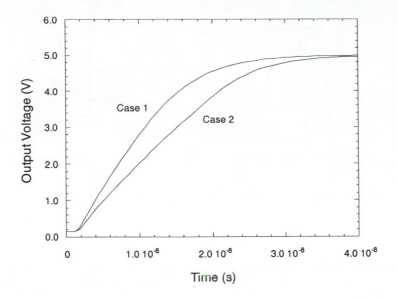

The simulated transient response of the NAND2 gate for both cases is plotted against time above. The time delay difference between the two cases is clearly visible. In fact, the propagation delay time in Case 2 is about 30% larger than that in Case 1, which proves that the input switching order has a significant influence on speed.

7.3 CMOS Logic Circuits

CMOS NOR2 (Two-Input NOR) Gate

The design and analysis of CMOS combinational logic circuits can be based on the basic principles developed for the nMOS depletion-load logic circuits in the previous section. Figure 7.10 shows the circuit diagram of a two-input CMOS NOR gate. Note that the circuit consists of a parallel-connected n-net and a series-connected complementary p-net. The input voltages V_A and V_B are applied to the gates of one nMOS and one pMOS transistor.

The complementary nature of the operation can be summarized as follows: When either one or both inputs are high, i.e., when the n-net creates a conducting path between the output node and the ground, the p-net is cut-off. On the other hand, if both input voltages are low, i.e., the n-net is cut-off, then the p-net creates a conducting path between the output node and the supply voltage V_{DD}. Thus, the dual or complementary circuit structure allows that, for any given input combination, the output is connected either to V_{DD} or to ground via a low-resistance path. A DC current path between the V_{DD} and ground is not established for any of the input combinations. This results in the fully complementary operation mode already examined for the simple CMOS inverter circuit.

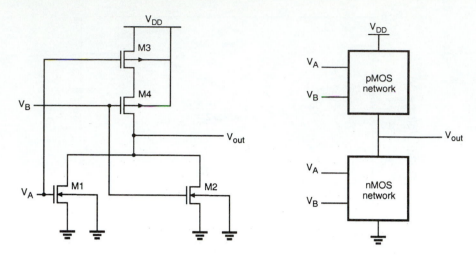

Figure 7.10   A CMOS NOR2 gate and its complementary operation: Either the nMOS network (n-net) is on and the pMOS network is off, or the pMOS network (p-net) is on and the nMOS network is off.

The output voltage of the CMOS NOR2 gate will attain a logic-low voltage of $V_{OL} = 0$ and a logic-high voltage of $V_{OH} = V_{DD}$. For circuit design purposes, the switching threshold voltage V_{th} of the CMOS gate emerges as an important design criterion. We start our analysis of the switching threshold by assuming that both input voltages switch simultaneously, i.e., $V_A = V_B$. Furthermore, it is assumed that the device sizes in each block are identical, $(W/L)_{n,A} = (W/L)_{n,B}$ and $(W/L)_{p,A} = (W/L)_{p,B}$, and the substrate-bias effect for the pMOS transistors is neglected for simplicity.

By definition, the output voltage is equal to the input voltage at the switching threshold.

$$V_A = V_B = V_{out} = V_{th} \tag{7.30}$$

It is obvious that the two parallel nMOS transistors are saturated at this point, because $V_{GS} = V_{DS}$. The combined drain current of the two nMOS transistors is

$$I_D = k_n(V_{th} - V_{T,n})^2 \tag{7.31}$$

Thus, we obtain the first equation for the switching threshold V_{th}.

$$V_{th} = V_{T,n} + \sqrt{\frac{I_D}{k_n}} \tag{7.32}$$

Examination of the p-net in Fig. 7.10 shows that the pMOS transistor M3 operates in the linear region, while the other pMOS transistor, M4, is in saturation for

$V_{in} = V_{out}$. Thus,

$$I_{D3} = \frac{k_p}{2}\left[2(V_{DD} - V_{th} - |V_{T,p}|)V_{SD3} - V_{SD3}^2\right] \qquad (7.33)$$

$$I_{D4} = \frac{k_p}{2}(V_{DD} - V_{th} - |V_{T,p}| - V_{SD3})^2 \qquad (7.34)$$

The drain currents of both pMOS transistors are identical, i.e., $I_{D3} = I_{D4} = I_D$. Thus,

$$V_{DD} - V_{th} - |V_{T,p}| = 2\sqrt{\frac{I_D}{k_p}} \qquad (7.35)$$

This yields the second equation of the switching threshold voltage V_{th}. Combining (7.32) and (7.35), we obtain

$$V_{th}(\text{NOR2}) = \frac{V_{T,n} + \frac{1}{2}\sqrt{\frac{k_p}{k_n}}(V_{DD} - |V_{T,p}|)}{1 + \frac{1}{2}\sqrt{\frac{k_p}{k_n}}} \qquad (7.36)$$

Now compare this expression with the switching threshold voltage of the CMOS inverter, which was derived in Chapter 5.

$$V_{th}(\text{INR}) = \frac{V_{T,n} + \sqrt{\frac{k_p}{k_n}}(V_{DD} - |V_{T,p}|)}{1 + \sqrt{\frac{k_p}{k_n}}} \qquad (7.37)$$

If $k_n = k_p$ and $V_{T,n} = |V_{T,p}|$, the switching threshold of the CMOS inverter is equal to $V_{DD}/2$. Using the same parameters, the switching threshold of the NOR2 gate is

$$V_{th}(\text{NOR2}) = \frac{V_{DD} + V_{T,n}}{3} \qquad (7.38)$$

which is not equal to $V_{DD}/2$. For example, when $V_{DD} = 5$ V and $V_{T,n} = |V_{T,p}| = 1$ V, the switching threshold voltages of the NOR2 gate and the inverter are

$$V_{th}(\text{NOR2}) = 2 \text{ V}$$

$$V_{th}(\text{INR}) = 2.5 \text{ V}$$

The switching threshold voltage of the NOR2 gate can also be obtained by using the equivalent-inverter approach. When both inputs are identical, the parallel-connected nMOS transistors can be represented by a single nMOS transistor with $2k_n$. Similarly, the series-connected pMOS transistors are represented by a single pMOS transistor with $k_p/2$. The resulting equivalent CMOS inverter is shown in Fig. 7.11.

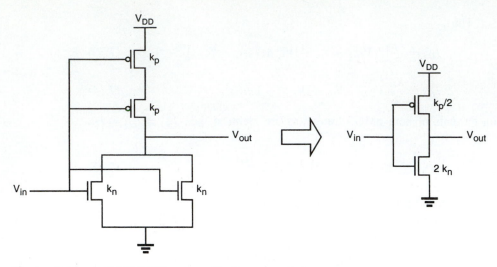

Figure 7.11 A CMOS NOR2 gate and its inverter equivalent.

Using the inverter switching threshold expression (7.37) for the equivalent inverter circuit, we obtain

$$V_{th}(\text{NOR2}) = \frac{V_{T,n} + \sqrt{\dfrac{k_p}{4k_n}}(V_{DD} - |V_{T,p}|)}{1 + \sqrt{\dfrac{k_p}{4k_n}}} \qquad (7.39)$$

which is identical to (7.36).

From (7.36), we can easily derive simple design guidelines for the NOR2 gate. For example, in order to achieve a switching threshold voltage of $V_{DD}/2$ for simultaneous switching, we have to set $V_{T,n} = |V_{T,p}|$ and $k_p = 4k_n$.

Figure 7.12 shows the CMOS NOR2 gate with the parasitic device capacitances, the inverter equivalent, and the corresponding lumped output load capacitance. In the worst case, the total lumped load capacitance is assumed to be equal to the sum of all internal parasitic device capacitances seen in Fig. 7.12.

CMOS NAND2 (Two-Input NAND) Gate

Figure 7.13 shows a two-input CMOS NAND (NAND2) gate. The operating principle of this circuit is the exact dual of the CMOS NOR2 operation examined earlier. The n-net consisting of two series-connected nMOS transistors creates a conducting path between the output node and the ground only if both input voltages are logic-high, i.e., are equal to V_{OH}. In this case, both of the parallel-connected pMOS transistors in the p-net will be off. For all other input combinations, either one or both

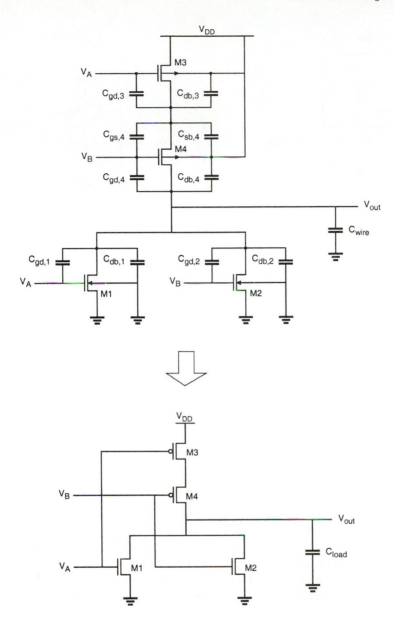

Figure 7.12 Parasitic device capacitances of the CMOS NOR2 circuit and the simplified equivalent with the lumped output load capacitance.

of the pMOS transistors will be turned on, while the n-net is cut-off, thus creating a current path between the output node and the power supply voltage.

Using an analysis similar to the one developed for the NOR2 gate, we can easily calculate the switching threshold for the CMOS NAND2 gate. Again, we will assume that the device sizes in each block are identical, with $(W/L)_{n,A} = (W/L)_{n,B}$ and

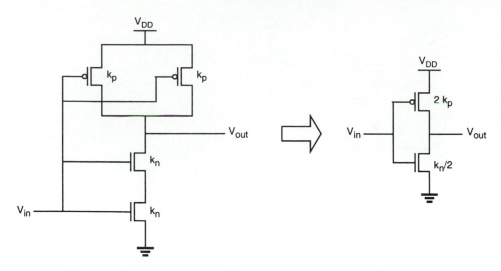

Figure 7.13   A CMOS NAND2 gate and its inverter equivalent.

$(W/L)_{p,A} = (W/L)_{p,B}$. The switching threshold for this gate is then found as

$$V_{th}(\text{NAND2}) = \frac{V_{T,n} + 2\sqrt{\dfrac{k_p}{k_n}}(V_{DD} - |V_{T,p}|)}{1 + 2\sqrt{\dfrac{k_p}{k_n}}} \tag{7.40}$$

As we can see from (7.40), a switching threshold voltage of $V_{DD}/2$ (for simultaneous switching) is achieved by setting $V_{T,n} = |V_{T,p}|$ and $k_n = 4k_p$ in the NAND2.

At this point, we can state the following observation about the area requirements of CMOS combinational logic gates. In comparison with equivalent nMOS depletion-load logic, the total number of transistors in CMOS gates is about twice the number of transistors in nMOS gates ($2n$ vs. $(n + 1)$ for n inputs). The silicon area occupied by the CMOS gate, however, is not necessarily twice the area occupied by the nMOS depletion-load gate, since a significant portion of the silicon area must be reserved for signal routing and contacts in both cases. Thus, the area disadvantage of CMOS logic may actually be smaller than the simple transistor count suggests.

Layout of Simple CMOS Logic Gates

In the following, we will examine simplified layout examples for CMOS NOR2 and NAND2 gates. Figure 7.14 shows a sample layout of a CMOS NOR2 gate, using single-layer metal and single-layer polysilicon.

In this example, the p-type diffusion area for pMOS transistors and the n-type diffusion area for nMOS transistors are aligned to allow simple routing of the gate signals via two parallel polysilicon lines running vertically. Figure 7.15 shows the

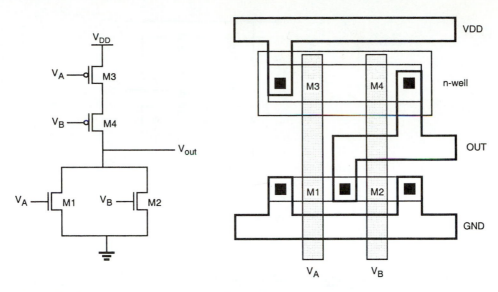

Figure 7.14 Sample layout of the CMOS NOR2 gate.

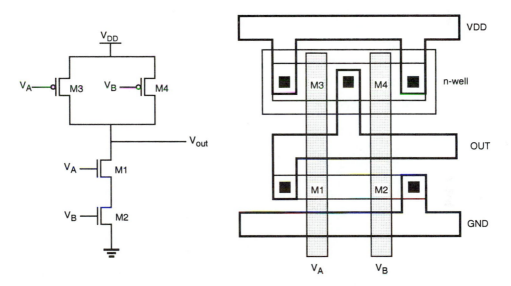

Figure 7.15 Sample layout of the CMOS NAND2 gate.

layout of a CMOS NAND2 gate, using the same basic layout principles as in the NOR2 layout example. This layout style allows the so-called feedthrough of signals in standard cell layouts. For signal routing among multiple standard cells, the feedthrough feature provides area savings in signal routing, for without it, as in the case of nMOS standard cell, separate feedthrough cells need to be inserted just for signal routing purpose.

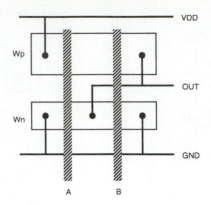

Figure 7.16 Stick-diagram layout of the CMOS NOR2 gate.

Finally, Fig. 7.16 shows a simplified (stick diagram) view of the CMOS NOR2 gate layout given in Fig. 7.14. Here, the diffusion areas are depicted by rectangles, the metal connections and contacts are represented by solid lines and circles, respectively, and the polysilicon columns are represented by cross-hatched strips. The stick-diagram layout does not carry any information on the actual geometry relations of the individual features, but it conveys valuable information on the relative placement of the transistors and their interconnections.

7.4 Complex Logic Circuits

To realize other complex Boolean functions of multiple input variables, the basic circuit structures and design principles developed for simple NOR and NAND gates in the previous sections can easily be extended to complex logic gates. The ability to realize complex logic functions using a small number of transistors is one of the most attractive features of nMOS and CMOS logic circuits.

Consider the following Boolean function as an example.

$$Z = \overline{A(D + E) + BC} \tag{7.41}$$

The nMOS depletion-load complex logic gate that is used to realize this function is shown in Fig. 7.17. Inspection of the circuit topology reveals the simple design principle of the pull-down network:

- OR operations are performed by parallel-connected drivers.
- AND operations are performed by series-connected drivers.
- Inversion is provided by the nature of MOS circuit operation.

The design principles stated here for individual inputs and corresponding driver transistors can also be extended to circuit sub-blocks, so that Boolean OR and

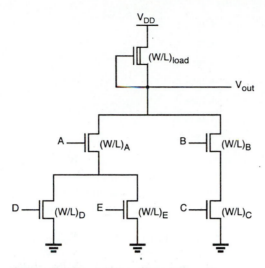

Figure 7.17 nMOS complex logic gate realizing the Boolean function given in (7.41).

AND operations can be performed in a nested circuit structure. Thus, we obtain a circuit topology which consists of series- and parallel-connected branches, as shown above.

In Fig. 7.17, the left nMOS driver branch consisting of three driver transistors is used to perform the logic function A(D + E), while the right-hand side branch performs the function BC. By connecting the two branches in parallel, and by placing the load transistor between the output node and the power supply voltage V_{DD}, we obtain the complex function given in (7.41). Each input variable is assigned to only one driver.

For the analysis and design of complex logic gates, we can employ the equivalent-inverter approach already used for the simpler NOR and NAND gates. It can be shown for the circuit in Fig. 7.17 that, if all input variables are logic-high, the equivalent-driver (W/L) ratio of the pull-down network consisting of five nMOS transistors is

$$\left(\frac{W}{L}\right)_{equivalent} = \frac{1}{\dfrac{1}{\left(\dfrac{W}{L}\right)_B} + \dfrac{1}{\left(\dfrac{W}{L}\right)_C}} + \frac{1}{\dfrac{1}{\left(\dfrac{W}{L}\right)_A} + \dfrac{1}{\left(\dfrac{W}{L}\right)_D + \left(\dfrac{W}{L}\right)_E}} \qquad (7.42)$$

For calculating the logic-low voltage level V_{OL}, we have to consider various cases, since the value of V_{OL} actually depends on the number *and* the configuration of the conducting nMOS transistors in each case. All possible configurations are tabulated on the next page. Each configuration is assigned a class number which reflects the total resistance of the current path from V_{out} node to ground.

A-D	Class 1
A-E	Class 1
B-C	Class 1
A-D-E	Class 2
A-D-B-C	Class 3
A-E-B-C	Class 3
A-D-E-B-C	Class 4

Assuming that all driver transistors have the same (W/L) ratio, a Class 1 path such as (B-C) has the highest series resistance, followed by Class 2, Class 3, etc. Consequently, the logic-low voltage levels corresponding to each class have the following order, where each subscript numeral represents the class number.

$$V_{OL1} > V_{OL2} > V_{OL3} > V_{OL4} \qquad (7.43)$$

The design of complex logic gates is based on the same ideas as the design of NOR and NAND gates. We usually start by specifying a maximum V_{OL} value. The design objective is to determine the driver and load transistor sizes so that the complex logic gate achieves the specified V_{OL} value even in the worst case. The given V_{OL} value first allows us to find the $(W/L)_{load}$ and $(W/L)_{driver}$ ratios for an equivalent inverter. Next, we have to identify all worst-case (Class 1) paths in the circuit, and determine the transistor sizes in these worst-case paths such that each Class 1 path has the equivalent driver ratio of $(W/L)_{driver}$.

In this example, this design strategy yields the following ratios for the three worst-case paths.

$$\left(\frac{W}{L}\right)_A = \left(\frac{W}{L}\right)_D = 2\left(\frac{W}{L}\right)_{driver}$$

$$\left(\frac{W}{L}\right)_A = \left(\frac{W}{L}\right)_E = 2\left(\frac{W}{L}\right)_{driver} \qquad (7.44)$$

$$\left(\frac{W}{L}\right)_B = \left(\frac{W}{L}\right)_C = 2\left(\frac{W}{L}\right)_{driver}$$

The transistor sizes found above guarantee that, for all other input combinations, the logic-low output voltage level will be less than the specified V_{OL}.

Complex CMOS Logic Gates

The realization of the n-net, or pull-down network, is based on the same basic design principles examined earlier. The pMOS pull-up network, on the other hand, must be the dual network of the n-net. This means that all parallel connections in the nMOS pull-down network will correspond to a series connection in the pMOS pull-up network, and all series connections in the pull-down network correspond to a parallel connection in the pull-up network.

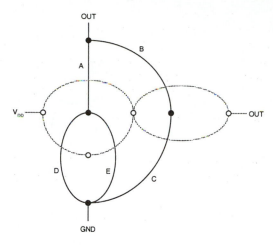

Figure 7.18 Construction of the dual pull-up graph from the pull-down graph, using the dual-graph concept.

Figure 7.18 shows the simple construction of the dual p-net (pull-up) graph from the n-net (pull-down) graph. Each driver transistor in the pull-down network is represented by an edge, and each node is represented by a vertex in the pull-down graph. Next, a new vertex is created within each confined area in the pull-down graph, and neighboring vertices are connected by edges which cross each edge in the pull-down graph only once. This new graph represents the pull-up network. The resulting CMOS complex logic gate is shown in Fig. 7.19.

Layout of Complex CMOS Logic Gates

Now, we will investigate the problem of constructing a minimum-area layout for the complex CMOS logic gate. Figure 7.20 shows the stick-diagram layout of a "first attempt," using an arbitrary ordering of the polysilicon gate columns. Note that in this case, the separation between the polysilicon columns must allow for one diffusion-to-diffusion separation and two metal-to-diffusion contacts in between. This certainly consumes a considerable amount of extra silicon area.

If we can minimize the number of diffusion-area breaks both for nMOS and for pMOS transistors, the separation between the polysilicon gate columns can be made smaller, which will reduce the overall horizontal dimension and, hence, the circuit layout area. The number of diffusion breaks can be minimized by changing the *ordering* of the polysilicon columns.

A simple method for finding the optimum gate ordering is the Euler-path approach: find a Euler path in the pull-down graph and a Euler path in the pull-up graph with identical ordering of input labels; i.e., find a common Euler path for both graphs. The Euler path is defined as an uninterrupted path that traverses each edge (branch) of the graph exactly once. Figure 7.21 shows the construction of a common Euler path for both graphs in our example.

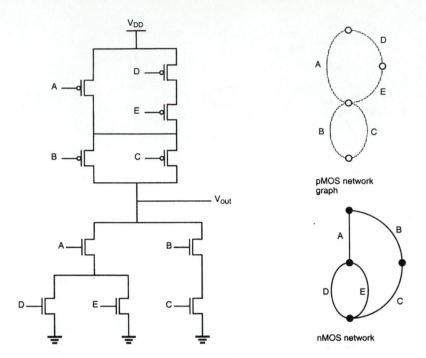

pMOS network
graph

nMOS network

Figure 7.19 A complex CMOS logic gate realizing the Boolean function (7.41).

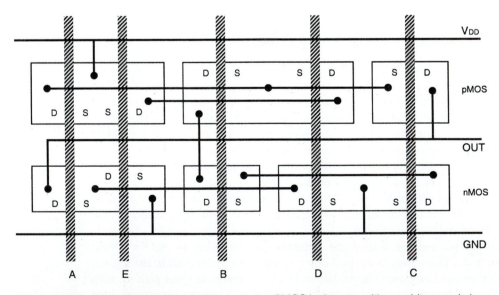

Figure 7.20 Stick-diagram layout of the complex CMOS logic gate, with an arbitrary ordering of the polysilicon gate columns.

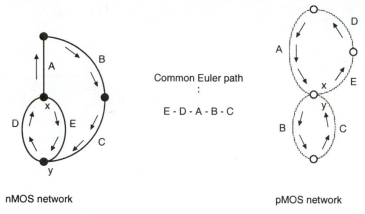

Common Euler path
:

E - D - A - B - C

nMOS network   pMOS network

Figure 7.21 Finding a common Euler path in both graphs for n-net and p-net provides a gate ordering that minimizes the number of diffusion breaks and, thus, minimizes the logic-gate layout area. In both cases, the Euler path starts at (x) and ends at (y).

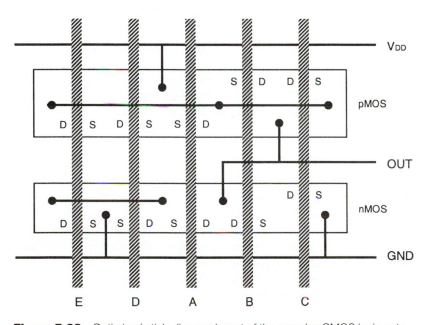

Figure 7.22 Optimized stick-diagram layout of the complex CMOS logic gate.

It is seen that there is a common sequence (E-D-A-B-C) in both graphs, i.e., a Euler path. The polysilicon gate columns can be arranged according to this sequence, which results in uninterrupted p-type and n-type diffusion areas. The stick diagram of the new layout is shown in Fig. 7.22. In this case, the polysilicon column separation Δd has to allow for only one metal-to-diffusion contact. The advantages of this new layout are more compact (smaller) layout area, simple routing of signals, and consequently, less parasitic capacitance.

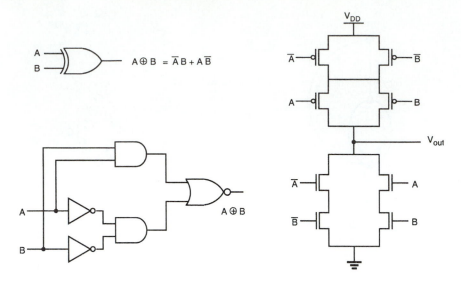

Figure 7.23   Full-CMOS implementation of the XOR function.

As a further example of complex CMOS gates, the full-CMOS implementation of the exclusive-OR (XOR) function is shown in Fig. 7.23. Note that two additional inverters are also needed to obtain the inverse of both input variables (A and B). With these inverters, the CMOS XOR circuit in Fig. 7.23 requires a total of 12 transistors. Other CMOS realizations of the XOR gate that can be implemented with fewer transistors will be examined later.

AOI and OAI Gates

While theoretically there are no strict limitations on the topology of the pull-down and the corresponding pull-up networks in a complex CMOS logic gate, we may recognize two important circuit categories as subsets of the general complex CMOS gate topology. These are the AND-OR-INVERT (AOI) gates and the OR-AND-INVERT (OAI) gates. The AOI gate, as its name suggests, enables the sum-of-products realization of a Boolean function in one logic stage (Fig. 7.24). The pull-down net of the AOI gate consists of parallel branches of series-connected nMOS driver transistors. The corresponding p-type pull-up network can simply be found using the dual-graph concept.

The OAI gate, on the other hand, enables the product-of-sums realization of a Boolean function in one logic stage (Fig. 7.25). The pull-down net of the OAI gate consists of series branches of parallel-connected nMOS driver transistors, while the corresponding p-type pull-up network can be found using the dual-graph concept.

Pseudo-nMOS Gates

The large area requirements of complex CMOS gates present a problem in high-density designs, since two complementary transistors, one nMOS and one pMOS, are needed for every input. One possible approach to reduce the number of transistors is

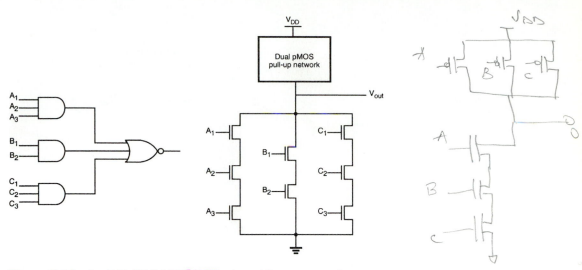

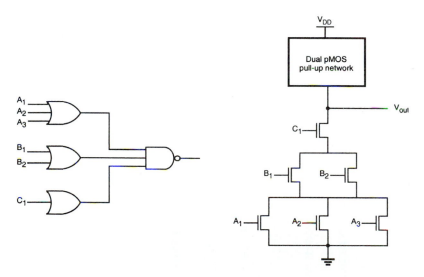

Figure 7.24 An AND-OR-INVERT (AOI) gate and the corresponding pull-down net.

Figure 7.25 An OR-AND-INVERT (OAI) gate, and the corresponding pull-down net.

to use a single pMOS transistor, with its gate terminal connected to ground, as the *load* device (Fig. 7.26). With this simple pull-up arrangement, the complex gate can be implemented with much fewer transistors. The similarities of pseudo-nMOS gates to depletion-load nMOS logic gates are obvious.

The most significant disadvantage of using a pseudo-nMOS gate instead of a full-CMOS gate is the nonzero static power dissipation, since the always-on pMOS load device conducts a steady-state current when the output voltage is lower than

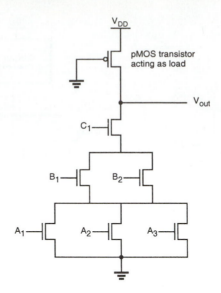

Figure 7.26 The pseudo-nMOS implementation of the OAI gate in Fig. 7.25.

V_{DD}. Also, the value of V_{OL} and the noise margins are now determined by the *ratio* of the pMOS load transconductance to the pull-down or driver transconductance.

EXAMPLE 7.2

The simplified layout of a CMOS complex logic circuit is given below. Draw the corresponding circuit diagram, and find an equivalent CMOS inverter circuit for simultaneous switching of all inputs, assuming that $(W/L)_p = 15$ for all pMOS transistors and $(W/L)_n = 10$ for all nMOS transistors.

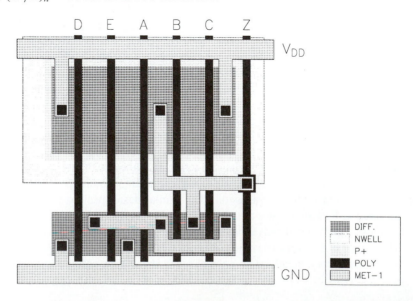

The circuit diagram can be found from the layout by inspection:

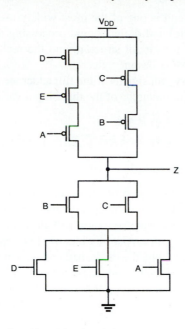

The Boolean function realized by this circuit is

$$Z = \overline{(D + E + A)(B + C)}$$

The equivalent (W/L) ratios of the nMOS network and the pMOS network are determined by using the series-parallel equivalency rules discussed earlier in this chapter, as follows.

$$\left(\frac{W}{L}\right)_{n,eq} = \cfrac{1}{\cfrac{1}{\left(\dfrac{W}{L}\right)_D + \left(\dfrac{W}{L}\right)_E + \left(\dfrac{W}{L}\right)_A} + \cfrac{1}{\left(\dfrac{W}{L}\right)_B + \left(\dfrac{W}{L}\right)_C}}$$

$$= \cfrac{1}{\cfrac{1}{30} + \cfrac{1}{20}} = 12$$

$$\left(\frac{W}{L}\right)_{p,eq} = \cfrac{1}{\cfrac{1}{\left(\dfrac{W}{L}\right)_D} + \cfrac{1}{\left(\dfrac{W}{L}\right)_E} + \cfrac{1}{\left(\dfrac{W}{L}\right)_A}} + \cfrac{1}{\cfrac{1}{\left(\dfrac{W}{L}\right)_B} + \cfrac{1}{\left(\dfrac{W}{L}\right)_C}}$$

$$= \cfrac{1}{\cfrac{1}{15} + \cfrac{1}{15} + \cfrac{1}{15}} + \cfrac{1}{\cfrac{1}{15} + \cfrac{1}{15}} = 12.5$$

CMOS Full-Adder Circuit

The one-bit full adder circuit is one of the most widely used building blocks in all data processing (arithmetic) and digital signal processing architectures. In the following, we will examine the circuit structure and the realization of the full adder using the conventional CMOS design style.

The sum_out and carry_out signals of the full adder are defined as the following two combinational Boolean functions of the three input variables, A, B, and C.

$$\text{sum\_out} = A \oplus B \oplus C$$
$$= ABC + A\bar{B}\bar{C} + \bar{A}\bar{B}C + \bar{A}\bar{C}B$$
$$\text{carry\_out} = AB + AC + BC$$

A gate-level realization of these two functions is shown in Fig. 7.27. Note that instead of realizing the two functions independently, we use the carry_out signal to generate the sum output. This implementation will ultimately reduce the circuit complexity and, hence, save chip area. Also, we identify two separate sub-networks consisting of several gates (highlighted with dashed boxes) which will be utilized for the transistor-level realization of the full-adder circuit.

The transistor-level design of the CMOS full-adder circuit is shown in Fig. 7.28. Note that the circuit contains a total of 14 nMOS and 14 pMOS transistors, together with the two CMOS inverters which are used to generate the outputs.

The mask layout of the full-adder circuit, which has been designed using the simple layout optimization strategy described earlier in this section, is shown in Fig. 7.29. Note, however, that all nMOS and pMOS transistors in this layout have the same (W/L) ratio. In order to optimize the transient (time-domain) performance of the circuit, it is usually necessary to adjust the transistor dimensions individually, as already shown in Chapter 6. A performance-optimized and more compact mask layout of the

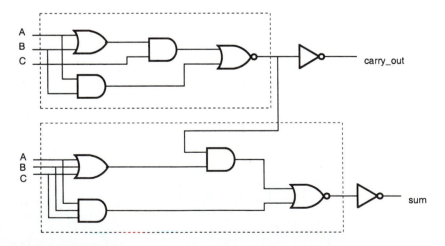

Figure 7.27 Gate-level schematic of the one-bit full-adder circuit.

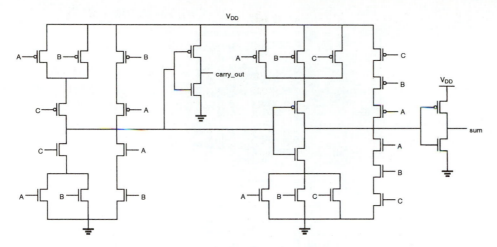

Figure 7.28 Transistor-level schematic of the one-bit full-adder circuit.

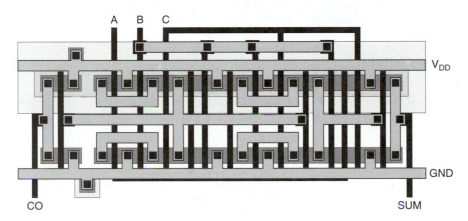

Figure 7.29 Mask layout of the CMOS full-adder circuit using minimum-size transistors.

same CMOS full-adder circuit is shown in Fig. 7.30. The simulated input and output voltage waveforms of the one-bit CMOS full adder are shown in Fig. 7.31. Please refer to the detailed design example that was presented in Chapter 1 for further design information.

The full-adder circuit presented here can be used as the basic building block of a general n-bit binary adder, which accepts two n-bit binary numbers as input and produces the binary sum at the output. The simplest such adder can be constructed by a cascade-connection of full adders, where each adder stage performs a two-bit addition, produces the corresponding sum bit, and passes the carry output on to the next stage. Hence, this cascade-connected adder configuration is called the carry ripple adder (Fig. 7.32). The overall speed of the carry ripple adder is obviously limited by

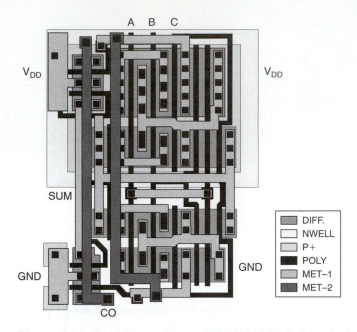

Figure 7.30 Mask layout of the optimized CMOS full adder circuit.

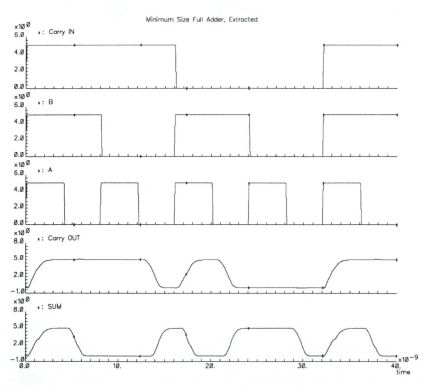

Figure 7.31 Simulated input and output waveforms of the CMOS full-adder circuit.

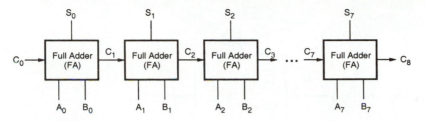

Figure 7.32 Block diagram of a carry ripple adder chain consisting of full adders.

the delay of the carry bits rippling through the carry chain; therefore, a fast carry_out response becomes essential for the overall performance of the adder chain. Such a signal path is called a (timing) critical path.

7.5 CMOS Transmission Gates (Pass Gates)

In this section, we will examine a simple switch circuit called the CMOS transmission gate (TG) or pass gate, and present a new class of logic circuits which use the TGs as their basic building blocks. As shown in Fig. 7.33, the CMOS transmission gate consists of one nMOS and one pMOS transistor, connected in parallel. The gate voltages applied to these two transistors are also set to be complementary signals. As such, the CMOS TG operates as a bidirectional switch between the nodes A and B which is controlled by signal C.

If the control signal C is logic-high, i.e., equal to V_{DD}, then both ransistors are turned on and provide a low-resistance current path between the nodes A and B. If,

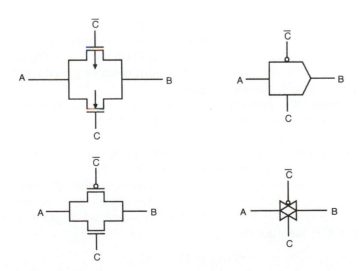

Figure 7.33 Four different representations of the CMOS transmission gate (TG).

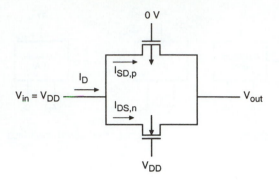

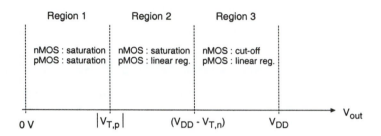

Figure 7.34   Bias conditions and operating regions of the CMOS transmission gate, shown as functions of the output voltage.

on the other hand, the control signal C is low, then both transistors will be off, and the path between the nodes A and B will be an open circuit. This condition is also called the high-impedance state.

Note that the substrate terminal of the nMOS transistor is connected to ground and the substrate terminal of the pMOS transistor is connected to V_{DD}. Thus, we must take into account the substrate-bias effect for both transistors, depending on the bias conditions. Figure 7.33 also shows three other commonly used symbolic representations of the CMOS transmission gate.

For a detailed DC analysis of the CMOS transmission gate, we will consider the following bias condition, shown in Fig. 7.34. The input node (A) is connected to a constant logic-high voltage, $V_{in} = V_{DD}$. The control signal is also logic-high, thus ensuring that both transistors are turned on. The output node (B) may be connected to a capacitor, which represents capacitive loading of the subsequent logic stages driven by the transmission gate. We will now investigate the input-output current-voltage relationship of the CMOS TG as a function of the output voltage V_{out}.

It can be seen from Fig. 7.34 that the drain-to-source and the gate-to-source voltages of the nMOS transistor are

$$V_{DS,n} = V_{DD} - V_{out}$$
$$V_{GS,n} = V_{DD} - V_{out}$$

(7.45)

Thus, the nMOS transistor will be turned off for $V_{out} > V_{DD} - V_{T,n}$ and will operate in the saturation mode for $V_{out} < V_{DD} - V_{T,n}$. The V_{DS} and V_{GS} voltages of the pMOS transistor are

$$V_{DS,p} = V_{out} - V_{DD}$$
$$V_{GS,p} = -V_{DD}$$

(7.46)

Consequently, the pMOS transistor is in saturation for $V_{out} < |V_{T,p}|$, and it operates in the linear region for $V_{out} > |V_{T,p}|$. Note that, unlike the nMOS transistor, the pMOS transistor remains turned on, regardless of the output voltage level V_{out}.

This analysis has shown that we can identify three operating regions for the CMOS transmission gate, depending on the output voltage level. These operating regions are depicted in Fig. 7.34 as functions of V_{out}. The total current flowing through the transmission gate is the sum of the nMOS drain current and the pMOS drain current.

$$I_D = I_{DS,n} + I_{SD,p}$$

(7.47)

At this point, we may devise an *equivalent resistance* for each transistor in this structure, as follows.

$$R_{eq,n} = \frac{V_{DD} - V_{out}}{I_{DS,n}}$$

$$R_{eq,p} = \frac{V_{DD} - V_{out}}{I_{SD,p}}$$

(7.48)

The total equivalent resistance of the CMOS TG will then be the parallel equivalent of these two resistances, $R_{eq,n}$ and $R_{eq,p}$. Now, we will calculate the equivalent resistance values for the three operating regions of the transmission gate.

Region 1

Here, the output voltage is smaller than the absolute value of the pMOS transistor threshold voltage, i.e., $V_{out} < |V_{T,p}|$. According to Fig. 7.34, both transistors are in saturation. We obtain the equivalent resistance of both devices as

$$R_{eq,n} = \frac{2(V_{DD} - V_{out})}{k_n(V_{DD} - V_{out} - V_{T,n})^2}$$

(7.49)

$$R_{eq,p} = \frac{2(V_{DD} - V_{out})}{k_p(V_{DD} - |V_{T,p}|)^2}$$

(7.50)

Note that the source-to-substrate voltage of the nMOS transistor is equal to the output voltage V_{out}, while the source-to-substrate voltage of the pMOS transistor is equal to zero. Thus, we have to take into account the substrate-bias effect for the nMOS transistor in our calculations.

Region 2

In this region, $|V_{T,p}| < V_{out} < (V_{DD} - V_{T,n})$. Thus, the pMOS transistor now operates in the linear region, while the nMOS transistor continues to operate in saturation.

$$R_{eq,n} = \frac{2(V_{DD} - V_{out})}{k_n(V_{DD} - V_{out} - V_{T,n})^2} \tag{7.51}$$

$$R_{eq,p} = \frac{2(V_{DD} - V_{out})}{k_p\left[2(V_{DD} - |V_{T,p}|)(V_{DD} - V_{out}) - (V_{DD} - V_{out})^2\right]}$$

$$= \frac{2}{k_p\left[2(V_{DD} - |V_{T,p}|) - (V_{DD} - V_{out})\right]} \tag{7.52}$$

Region 3

Here, the output voltage is $V_{out} > (V_{DD} - V_{T,n})$. Consequently, the nMOS transistor will be turned off, which results in an open-circuit equivalent. The pMOS transistor will continue to operate in the linear region.

$$R_{eq,p} = \frac{2}{k_p\left[2(V_{DD} - |V_{T,p}|) - (V_{DD} - V_{out})\right]} \tag{7.53}$$

Combining the equivalent resistance values found for the three operating regions, we can now plot the total resistance of the CMOS transmission gate as a function of the output voltage V_{out}, as shown in Fig. 7.35.

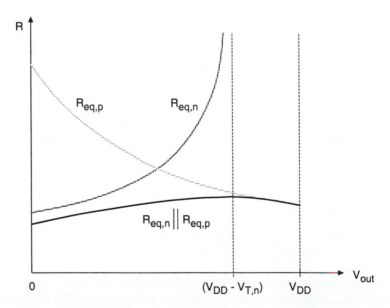

Figure 7.35   Equivalent resistance of the CMOS transmission gate, plotted as a function of the output voltage.

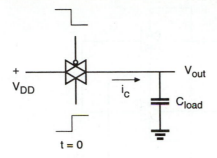

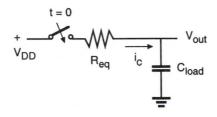

Figure 7.36 Replacing the CMOS TG with its resistor equivalent for transient analysis.

It can be seen that the total equivalent resistance of the TG remains relatively constant, i.e., its value is almost independent of the output voltage, whereas the individual equivalent resistances of both the nMOS and the pMOS transistors are strongly dependent on V_{out}. This property of the CMOS TG is naturally quite desirable. A CMOS pass gate which is turned on by a logic-high control signal can be replaced by its simple equivalent resistance for dynamic analysis, as shown in Fig. 7.36.

The implementation of CMOS transmission gates in logic circuit design usually results in compact circuit structures which may even require a smaller number of transistors than their standard CMOS counterparts. Note that the control signal *and* its complement must be available simultaneously for TG applications. Figure 7.37 shows a two-input multiplexer (MUX) circuit consisting of two CMOS transmission gates. The operation of the multiplexer can be understood quite easily: If the control input S is logic-high, then the bottom TG will conduct, and the output will be equal to the input B. If the control signal is low, the bottom TG will turn off and the top TG will connect the input A to the output node.

Figure 7.38 shows an eight-transistor implementation of the logic XOR function, using two CMOS TGs and two CMOS inverters. The same function can also be implemented using only six transistors, as shown in Fig. 7.39.

Using the generalized multiplexer approach, each Boolean function can be realized with a TG logic circuit. As an example, Fig. 7.40(a) shows the TG logic implementation of a three-variable Boolean function. Note that the three input variables

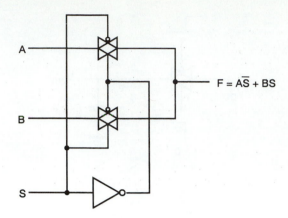

Figure 7.37 Two-input multiplexer circuit
implemented using two CMOS TGs.

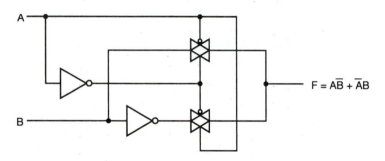

Figure 7.38 Eight-transistor CMOS TG implementation of the XOR
function.

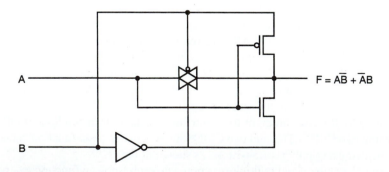

Figure 7.39 Six-transistor CMOS TG implementation of the XOR
function.

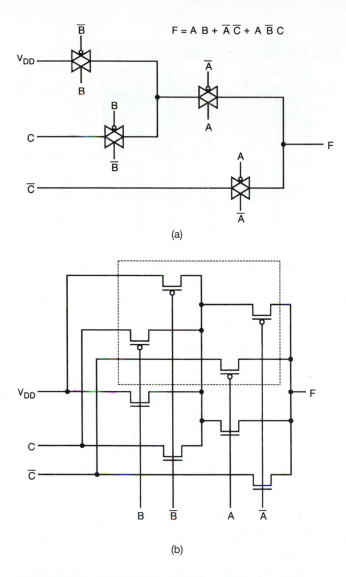

$$F = A\,B + \overline{A}\,\overline{C} + A\,\overline{B}\,C$$

(a)

(b)

Figure 7.40 (a) CMOS TG realization of a three-variable Boolean function. (b) All pMOS transistors can be placed into one n-well to save area.

and their inverses must be used to control the CMOS transmission gates. Including the three inverters not shown here, the TG implementation requires a total of 14 transistors. An important point in TG logic design is that a conducting TG network (*low-impedance* path) should always be provided between the output node and one of the inputs, for all possible input combinations. This is to make sure that the output node with its capacitive load is never left in a *high-impedance* state.

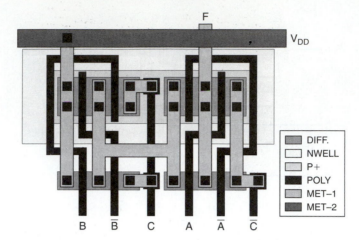

Figure 7.41 Mask layout of the CMOS TG circuit shown in Fig. 7.40.

If each CMOS transmission gate in TG logic circuits is realized with a full nMOS-pMOS pair, the disjoint n-well structures of the pMOS transistors and the diffusion contacts may cause a significant overall area increase. In an attempt to reduce the silicon area occupied by TG circuits, the transmission gates can be laid out as separated nMOS-pMOS pairs with all pMOS transistors placed in one single n-well, as shown in Fig. 7.40(b). However, the *routing* area required for connecting the p-type diffusion regions to input signals must be carefully considered. The layout of the TG circuit is given in Fig. 7.41.

Complementary Pass-Transistor Logic (CPL)

The complexity of full-CMOS pass-gate logic circuits can be reduced dramatically by adopting another circuit concept, called Complementary Pass-transistor Logic (CPL). The main idea behind CPL is to use a purely nMOS pass-transistor network for the logic operations, instead of a CMOS TG network. All inputs are applied in complementary form; i.e., every input signal and its inverse must be provided. The circuit also produces complementary outputs, to be used by subsequent CPL stages. Thus, the CPL circuit essentially consists of complementary inputs, an nMOS pass transistor logic network to generate complementary outputs, and CMOS output inverters to restore the output signals. The circuit diagrams of a CPL NOR2 and a CPL NAND2 are shown in Fig. 7.42.

The elimination of pMOS transistors from the pass-gate network significantly reduces the parasitic capacitances associated with each node in the circuit, thus, the operation speed is typically higher compared to a full-CMOS counterpart. But the improvement in transient characteristics comes at a price of increased process complexity. In CPL circuits, the threshold voltages of the nMOS transistors in the pass-gate network must be reduced to about 0 V through threshold-adjustment

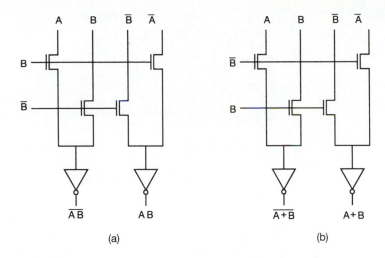

(a)   (b)

Figure 7.42 Circuit diagram of (a) CPL NAND2 gate and (b) CPL NOR2 gate.

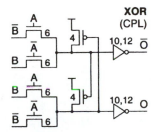

Figure 7.43 Circuit diagram of a CPL-based XOR gate.

implants, in order to eliminate the threshold-voltage drop. This, on the other hand, reduces the overall noise immunity and makes the transistors more susceptible to subthreshold conduction in the off-mode. Also note that the CPL design style is highly modular, a wide range of functions can be realized by using the same basic pass-transistor structures.

Regarding the transistor count, CPL circuits do not always offer a marked advantage over conventional CMOS. The NAND2 and NOR2 circuits shown in Fig. 7.42 each consist of 8 transistors. XOR and XNOR functions realized with CPL have a similar complexity (i.e., transistor count) as conventional CMOS realizations. The same observation is true for the realization of full adders with CPL. The circuit diagram of a CPL-based XOR gate is shown in Fig. 7.43. Here, the cross-coupled pMOS pull-up transistors are used to speed up the output response. Transistor widths are given in λ-units. Figure 7.44 shows the circuit diagram of a CPL full-adder circuit consisting of 32 transistors. The mask layout of this CPL circuit is given in Fig. 7.45.

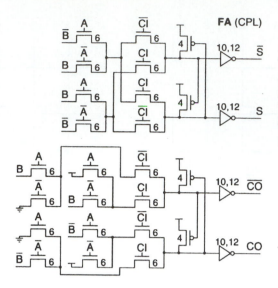

Figure 7.44   Circuit diagram of a CPL full adder.

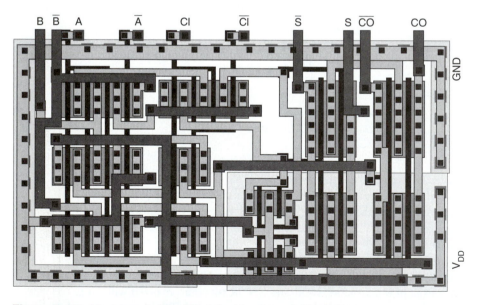

Figure 7.45   Mask layout of the CPL full adder shown in Fig. 7.44.

Exercise Problems

7.1 A CMOS circuit was designed based on company XYZ's 3-μm design rules, as shown in Fig. P7.1 with $W_N = 7\ \mu$m and $W_P = 15\ \mu$m.

a. Determine the circuit configuration and draw the circuit diagram.

b. For simple hand analysis, make the following assumptions:

   i)   Wiring parasitic capacitances and resistances are negligible.

   ii)   Device parameters are

	nMOS	pMOS
V_{T0}	1.0 V	−1.0 V
t_{ox}	500 Å	500 Å
k'	20 μA/V^2	10 μA/V^2
X_j	0.5 μm	0.5 μm
L_D	0.5 μm	0.5 μm

   iii) The total capacitance at node I is 0.6 pF.

   iv) An ideal step-pulse signal is applied to the CK terminal such that

$$V_{CK} = 5\ \text{V}, \quad t < 0$$
$$V_{CK} = 0\ \text{V}, \quad 0 \le t < T$$
$$V_{CK} = 5\ \text{V}, \quad t \ge T_w$$
$$V_{DD} = 5\ \text{V},$$

   v)   At $t = 0$, the node voltage at I is zero.

   vi) The input voltages at A_1, B_1, and B_2 are zero for $0 \le t \le T_W$.

Find the minimum T_W that allows V_I to reach 2.5 V.

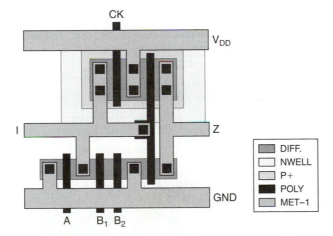

Figure P7.1

7.2 Calculate the equivalent W/L of the two nMOSTs with W_1/L and W_2/L connected in series. For simplicity, neglect the body effect, i.e., the threshold voltages of individual transistors are constant and do not depend on the source voltages. Although this is not true in reality, such an assumption is necessary for simple analysis with a reasonably good approximation.

7.3 Analytical expressions for V_{th}(logic) have been derived in Chapter 7 for the CMOS NOR2 gate. Now consider the CMOS NAND2 gate for the following cases and use $k_p = k_n = 100\ \mu A/V^2$:

- two inputs switching simultaneously
- top nMOS switching while the bottom nMOS's gate is tied to V_{DD}
- top nMOS gate is tied to V_{DD} and the gate input of the bottom nMOS is changing

a.   Derive an analytical expression for V_{th} corresponding to the first case. Also find the V_{th} value for the first case for $V_{DD} = 5$ V when the magnitudes of the threshold voltages are 1 V with $\gamma = 0$.

b.   Determine V_{th} for all three cases by using SPICE.

c.   For $C_{load} = 0.2$ pF, calculate 50% delays (low-to-high and high-to-low propagation delays) for an ideal pulse input signal for each of the three cases by assuming that C_{load} includes all of the internal parasitic capacitances. Verify the results using SPICE.

7.4 Write down the SPICE input description for transistor connections, source and drain parasitics in terms of areas, and perimeters for the layout shown in Example 7.2. Neglect the wiring capacitances in the polysilicon and metal runners. Default model names to be used for pMOS and nMOS are MODP and MODN. Assume $L = 1\ \mu m$ and $y = 10\ \mu m$ for all transistors.

7.5 For the gate shown in Fig. P7.5,

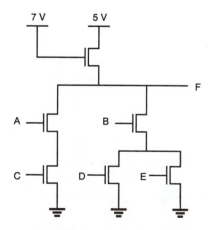

Figure P7.5

- Pull-up transistor ratio is 5/5
- Pull-down transistor ratios are 100/5
- $V_{T0} = 1.0$ V
- $\gamma = 0.4$ V$^{1/2}$
- $|2\phi_F| = 0.6$ V

a. Identify the worst-case input combination(s) for V_{OL}.

b. Calculate the worst-case value of V_{OL}. (Assume that all pull-down transistors have the same body bias and, initially, that $V_{OL} \approx 5\% \, V_{DD}$.)

7.6 A store has one express register and three regular ones. It is the store policy that the express register be open only when two or more of the other registers are busy. Assume that the Boolean variables A, B, and C reflect the status of each of the regular registers (1 busy, 0 idle). Design the logic circuit, with A, B, and C as inputs and F as output, to automatically notify the manager (by setting $F = 1$) to open the express register. Present two solutions, the first using only NAND gates, the second using only NOR gates.

7.7 Calculate $V_{OL}, V_{OH}, V_{IL}, V_{IH}, NM_L$, and NM_H for a two-input NOR gate fabricated with CMOS technology.

$(W/L)_p = 4$

$(W/L)_n = 1$

$V_{Tn} = 0.7$ V

$V_{Tp} = -0.7$ V

$\mu_n C_{ox} = 40 \; \mu$A/V^2

$\mu_p C_{ox} = 20 \; \mu$A/V^2

$V_{DD} = 5$ V

Compare your answers with SPICE simulation results.

7.8 Use a layout editor (e.g., Magic) to design a two-input CMOS NAND gate. All devices have $W = 10 \; \mu$m. The n-channel transistors have $L_{eff} = 1 \; \mu$m and the p-channel transistors have $L_{eff} = 2 \; \mu$m. Calculate the drawn channel lengths by assuming that $L_D = 0.25 \; \mu$m. Use the design rule checker to avoid rule violations. Finally, have the layout editor perform parasitic capacitance extraction.

7.9 Assume that the 2-input NAND gate in Problem 7.8 is driving a 0.1 pF load. Use hand calculations to estimate t_{PLH} and t_{PHL}. Do not forget to add in the parasitic capacitances extracted from your layout! Check your answer with SPICE. Use:

$k'_n = 20 \; \mu$A/V^2

$k'_p = 10 \; \mu$A/V^2

$V_{Tn} = |V_{Tp}| = 1.0$ V

7.10 Consider the logic circuit shown in Fig. P7.10, with
$V_{TO}(enhancement) = 1$ V, $V_{TO}(depletion) = -3$ V, and $\gamma = 0$.

a. Determine the logic function F.

b. Calculate W_L/L_L such that V_{OL} does not exceed 0.4 V.

c. Qualitatively, would W_L/L_L increase or decrease if the same conditions
in (b) are to be achieved but $\gamma = 0.4$ $V^{1/2}$?

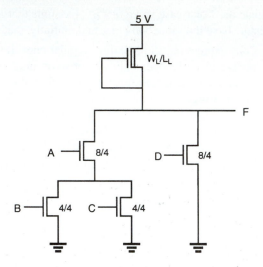

Figure P7.10

7.11 Consider the circuit shown in Fig. P7.11.

a. Determine the logic function F.

b. Design a circuit to implement the same logic function, but using NOR
gates. Draw a transistor-level schematic and use nMOS enhancement-
depletion (E-D) technology.

c. Design a circuit to implement the same logic function, but use an AOI
(AND-OR-INVERT) gate. Draw a transistor-level schematic and use
CMOS technology.

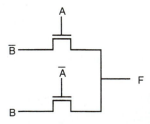

Figure P7.11

7.12 The enhancement-type MOS transistors have the following parameters:

$V_{DD} = 5$ V

$|V_{T0}| = 1.0$ V for both nMOS and pMOS transistor

$\lambda = 0.0$ V^{-1}

$\mu_p C_{ox} = 20$ μA/V^2

$\mu_n C_{ox} = 50$ μA/V^2

For a CMOS complex gate OAI432 with $(W/L)_p = 30$ and $(W/L)_n = 40$,

a. Calculate the W/L sizes of an equivalent inverter with the weakest pull-down and pull-up. Such an inverter can be used to calculate worst-case pull-up and pull-down delays, with proper incorporation of parasitic capacitances at internal nodes into the total load capacitance. In this problem, you are asked to calculate only $(W/L)_{worst-case}$ for both p-channel and n-channel MOSFETs by neglecting the parasitic capacitances.

b. Do the layout of OAI432 with minimal diffusion breaks to reduce the number of polysilicon column pitches. With proper ordering of polysilicon gate columns, the number of diffusion breaks can be minimized. One way of achieving such a goal is to find a Euler path common to both p-channel and n-channel nets using graph models. Symbolic layout that shows source and drain connections is sufficient to answer this problem.

7.13 Consider a fully complementary CMOS transmission gate with its input terminal tied to ground (0 V) while the other non-gate terminal is tied to a 1 pF load capacitor initially charged to 5 V. Use the V_{T0}, k'_p, and k'_n values in Problem 7.12. At $t = 0$, both transistors are fully turned on by clock signals to start the discharge of the capacitor.

a. Plot the effective resistance of this transmission gate as a function of capacitor voltage when $(W/L)_p = 50$ and $(W/L)_n = 40$. From the plot find the average value of the resistance. Then calculate the RC delay for the capacitor voltage to change from 5 V to 2.5 V. This can be found by solving the RC-circuit differential equation.

b. Verify your answer to part (a) by using SPICE simulation. The source/drain parasitic capacitances can be neglected.

A   B   C

V_{DD}

SUM

GND

CO

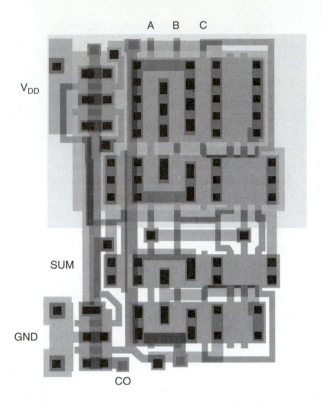

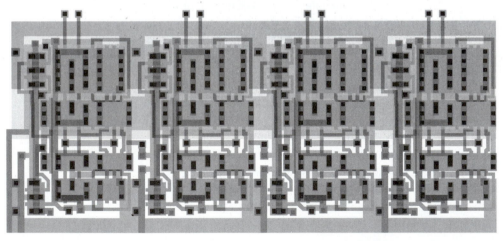

Plate 9 (Top) Layout of a CMOS full adder (FA) circuit. (Bottom) Layout of a 4-bit ripple-carry adder, consisting of four FA cells.

CHAPTER 8

Sequential MOS Logic Circuits

8.1 Introduction

In all of the combinational logic circuits examined in Chapter 7, if we neglect the propagation delay time, the output levels at any given time point are directly determined as Boolean functions of the input variables applied at that time. Thus, the combinational circuits lack the capability of storing any previous events, or displaying an output behavior which is dependent upon the previously applied inputs. Circuits of this type are also classified as non-regenerative circuits, since there is no feedback relationship between the output and the input.

The other major class of logic circuits is called sequential circuits, in which the output is determined by the current inputs as well as the previously applied input variables. Figure 8.1(a) shows a sequential circuit consisting of a combinational circuit and a memory block in the feedback loop. In most cases, the regenerative behavior of sequential circuits is due to either a direct or an indirect feedback connection between the output and the input. Regenerative operation can, under certain conditions, also be interpreted as a simple memory function. The critical components of sequential systems are the basic regenerative circuits, which can be classified into three main groups: bistable circuits, monostable circuits, and astable circuits. The general classification of non-regenerative and regenerative logic circuits is shown in Fig. 8.1.

Bistable circuits have, as their name implies, two stable states or operation modes, each of which can be attained under certain input and output conditions. Monostable circuits, on the other hand, have only one stable operating point (state). Even if the circuit experiences an external perturbation, the output eventually returns to the single stable state after a certain time period. Finally, in astable circuits, there is no stable operating point or state which the circuit can preserve for a certain time period. Consequently, the output of an astable circuit must oscillate without settling into a stable operating mode. The ring oscillator circuit examined in Chapter 6 would be a typical example of an astable regenerative circuit.

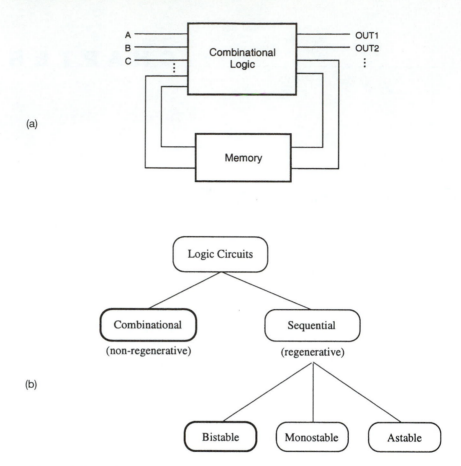

Figure 8.1   (a) Sequential circuit consisting of a combinational logic block and a memory block in the feedback loop. (b) Classification of logic circuits based on their temporal behavior.

Among these three main groups of regenerative circuit types, the bistable circuits are by far the most widely used and the most important class. All basic latch and flip-flop circuits, registers, and memory elements used in digital systems fall into this category. In the following, we will first examine the electrical behavior of the simple bistable element, and then present some of its useful applications.

8.2   Behavior of Bistable Elements

The basic bistable element to be examined in this section consists of two identical cross-coupled inverter circuits, as shown in Fig. 8.2(a). Here, the output voltage of inverter (1) is equal to the input voltage of inverter (2), i.e., $v_{o1} = v_{i2}$, and the output voltage of inverter (2) is equal to the input voltage of inverter (1), i.e., $v_{o2} = v_{i1}$. In

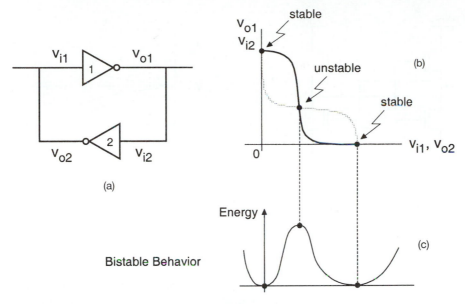

(a)

Bistable Behavior

Figure 8.2 Static behavior of the two-inverter basic bistable element: (a) Circuit schematic. (b) Intersecting voltage transfer curves of the two inverters, showing the three possible operating points. (c) Qualitative view of the potential energy levels corresponding to the three operating points.

order to investigate the static input-output behavior of both inverters, we start by plotting the voltage transfer characteristic of inverter (1) with respect to the $v_{o1} - v_{i1}$ axis pair. Notice that the input and output voltages of inverter (2) correspond to the output and input voltages of inverter (1), respectively. Consequently, we can also plot the voltage transfer characteristic of inverter (2) using the same axis pair, as shown in Fig. 8.2(b).

It can be seen that the two voltage transfer characteristics intersect at three points. Simple reasoning can help us to conclude that two of these operating points are stable, as indicated in Fig. 8.2(b). If the circuit is initially operating at one of these two stable points, it will preserve this state unless it is forced externally to change its operating point. Note that the gain of each inverter circuit, i.e., the slope of the respective voltage transfer curves, is smaller than unity at the two stable operating points. Thus, in order to change the state by moving the operating point from one stable point to the other, a sufficiently large external voltage perturbation must be applied so that the voltage gain of the inverter loop becomes larger than unity.

On the other hand, the voltage gains of both inverters are larger than unity at the third operating point. Consequently, even if the circuit is biased at this point initially, a small voltage perturbation at the input of any of the inverters will be amplified, causing the operating point to move to one of the stable operating points. This leads to the conclusion that the third operating point is unstable. The circuit has two stable operating points, hence, it is called bistable.

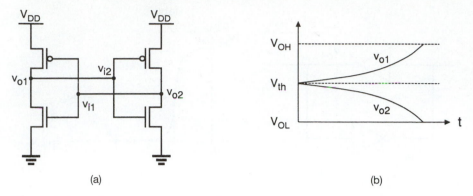

Figure 8.3   (a) Circuit diagram of a CMOS bistable element. (b) One possibility for the expected time-domain behavior of the output voltages, if the circuit is initially set at its unstable operating point.

The bistable behavior of the cross-coupled inverter circuit can also be visualized qualitatively by examining the total potential energy level at each of the three possible operating points (Fig. 8.2(c)). It is seen that the potential energy is at its minimum at two of the three operating points, since the voltage gains of both inverters are equal to zero. By contrast, the energy attains a maximum at the operating point at which the voltage gains of both inverters are maximum. Thus, the circuit has two stable operating points corresponding to the two energy minima, and one unstable operating point corresponding to the potential energy maximum.

Figure 8.3(a) shows the circuit diagram of a CMOS two-inverter bistable element. Note that at the unstable operating point of this circuit, all four transistors are in saturation, resulting in maximum loop gain for the circuit. If the initial operating condition is set at this point, any small voltage perturbation will cause significant changes in the operating modes of the transistors. Thus, we expect the output voltages of the two inverters to diverge and eventually settle at V_{OH} and V_{OL}, respectively, as illustrated in Fig. 8.3(b). The direction in which each output voltage diverges is determined by the initial perturbation polarity. In the following, we will examine this event more closely using a small-signal analysis approach.

Consider the bistable circuit shown in Fig. 8.4, which is initially operating at $v_{o1} = v_{o2} = V_{th}$, i.e., at the unstable operating point. For our analysis, we will assume that the input (gate) capacitance C_g of each inverter is much larger than its output (drain) capacitance C_d; i.e., $C_g \gg C_d$. The small-signal drain current supplied by each inverter (1 and 2) can be expressed, in terms of the small-signal gate voltage of that inverter, as follows. Note that the drain current of each inverter is also equal to the gate current of the other inverter.

$$i_{g1} = i_{d2} = g_m v_{g2}$$
$$i_{g2} = i_{d1} = g_m v_{g1}$$

(8.1)

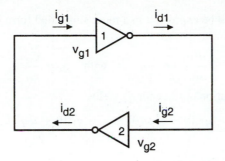

Figure 8.4 Small-signal input and output currents of the inverters.

Here, g_m represents the small-signal transconductance of the inverter. The gate voltages of both inverters can be expressed in terms of the gate charges, q_1 and q_2.

$$v_{g1} = \frac{q_1}{C_g} \qquad v_{g2} = \frac{q_2}{C_g} \tag{8.2}$$

Note that the small-signal gate current of each inverter can be written as a function of the time derivative of its small-signal gate voltage, as follows.

$$i_{g1} = C_g \frac{dv_{g1}}{dt}$$

$$i_{g2} = C_g \frac{dv_{g2}}{dt} \tag{8.3}$$

Combining (8.1) with (8.3), we obtain:

$$g_m v_{g2} = C_g \frac{dv_{g1}}{dt} \tag{8.4}$$

$$g_m v_{g1} = C_g \frac{dv_{g2}}{dt} \tag{8.5}$$

Expressing the gate voltages in terms of the gate charges, these two differential equations can also be written as

$$\frac{g_m}{C_g} q_2 = \frac{dq_1}{dt} \tag{8.6}$$

$$\frac{g_m}{C_g} q_1 = \frac{dq_2}{dt} \tag{8.7}$$

Both differential equations given in (8.6) and (8.7) can now be combined to yield a second-order differential equation describing the time behavior of gate charge q_1.

$$\frac{g_m}{C_g} q_1 = \frac{C_g}{g_m} \frac{d^2 q_1}{dt^2} \quad \Rightarrow \quad \frac{d^2 q_1}{dt^2} = \left(\frac{g_m}{C_g} \right)^2 q_1 \tag{8.8}$$

This equation can also be expressed in a more simplified form by using τ_0, the transit time constant.

$$\frac{d^2 q_1}{dt^2} = \frac{1}{\tau_0^2} q_1 \qquad \text{with} \quad \tau_0 = \frac{C_g}{g_m} \tag{8.9}$$

The time-domain solution of (8.9) for q_1 yields

$$q_1(t) = \frac{q_1(0) - \tau_0 q_1'(0)}{2} e^{-\frac{t}{\tau_0}} + \frac{q_1(0) + \tau_0 q_1'(0)}{2} e^{+\frac{t}{\tau_0}} \tag{8.10}$$

where the initial condition is given as:

$$q_1(0) = C_g \cdot v_{g1}(0) \tag{8.11}$$

Note that $v_{g1} = v_{o2}$ and $v_{g2} = v_{o1}$. Replacing the gate charge of both inverters with the corresponding output-voltage variables, we obtain,

$$v_{o2}(t) = \frac{1}{2}(v_{o2}(0) - \tau_0 v_{o2}'(0))e^{-\frac{t}{\tau_0}} + \frac{1}{2}(v_{o2}(0) + \tau_0 v_{o2}'(0))e^{+\frac{t}{\tau_0}} \tag{8.12}$$

$$v_{o1}(t) = \frac{1}{2}(v_{o1}(0) - \tau_0 v_{o1}'(0))e^{-\frac{t}{\tau_0}} + \frac{1}{2}(v_{o1}(0) + \tau_0 v_{o1}'(0))e^{+\frac{t}{\tau_0}} \tag{8.13}$$

For large values of t, the time-domain expressions (8.12) and (8.13) can be simplified as follows.

$$v_{o1}(t) \approx \frac{1}{2}(v_{o1}(0) + \tau_0 v_{o1}'(0))e^{+\frac{t}{\tau_0}}$$

$$v_{o2}(t) \approx \frac{1}{2}(v_{o2}(0) + \tau_0 v_{o2}'(0))e^{+\frac{t}{\tau_0}} \tag{8.14}$$

Note that the magnitude of both output voltages increases exponentially with time. Depending on the polarity of the initial small perturbations $dv_{o1}(0)$ and $dv_{o2}(0)$, the output voltages of both inverters will diverge from their initial value of V_{th} to either V_{OL} or V_{OH}. In fact, the polarity of the output-voltage perturbation dv_{o1} must always be opposite to that of dv_{o2}, because of the charge-conservation principle. Hence, the two output voltages always diverge into opposite directions, as expected. The phase-plane representation of this event, illustrated in Fig. 8.5, shows that the operating point $(v_{o1} = V_{th}, v_{o2} = V_{th})$ is unstable. The two operating points $(v_{o1} = V_{OL}, v_{o2} = V_{OH})$ and $(v_{o1} = V_{OH}, v_{o2} = V_{OL})$ can be shown to be stable, using small-signal models at the corresponding operating points.

$$\begin{aligned} v_{o1}: \quad V_{th} &\rightarrow V_{OH} \quad \text{or} \quad V_{OL} \\ v_{o2}: \quad V_{th} &\rightarrow V_{OL} \quad \text{or} \quad V_{OH} \end{aligned} \tag{8.15}$$

In addition to the time-domain analysis presented here, an interesting observation can be made for the two-inverter bistable element: While the bistable circuit is settling from its unstable operating point into one of its stable operating points, we can envision a signal traveling the loop consisting of the two cascaded inverters several times (Fig. 8.6). The time-domain behavior of the output voltage v_{o1} during

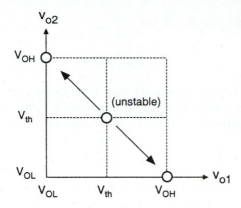

Figure 8.5 Phase-plane representation of the bistable circuit behavior.

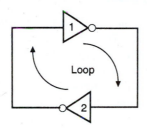

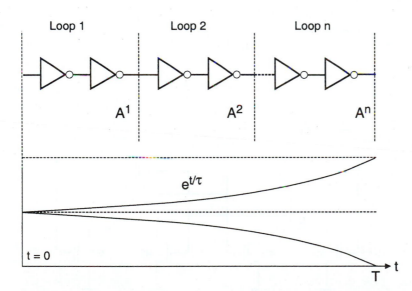

Figure 8.6 Propagation of a transient signal in the two-inverter loop during settling.

this period is approximated as follows:

$$\frac{v_{o1}(t)}{v_{o1}(0)} = e^{+\frac{t}{\tau_0}} \tag{8.16}$$

If during a time interval T, the signal travels the loop n times, then this is equivalent to the same signal propagating along a cascaded inverter chain consisting of $2n$ inverters. Expressing the loop gain (combined voltage gain of two cascaded inverters) with A, we obtain,

$$A^n = e^{+\frac{T}{\tau_0}} \tag{8.17}$$

This expression describes the time-domain behavior of the diverging process until it reaches stable points, as depicted in Fig. 8.6.

8.3   SR Latch Circuit

The bistable element consisting of two cross-coupled inverters (Fig. 8.2) has two stable operating modes, or states. The circuit preserves its state (either one of the two possible modes) as long as the power supply voltage is provided; hence, the circuit can perform a simple memory function of *holding* its state. However, the simple two-inverter circuit alone has no provision for allowing its state to be changed externally from one stable operating mode to the other. To allow such a change of state, we must add simple switches to the bistable element, which can be used to force or trigger the circuit from one operating point to the other. Figure 8.7 shows the circuit structure of the simple CMOS SR latch, which has two such triggering inputs, S (set) and R (reset). In the literature, the SR latch is also called an SR flip-flop, since two stable states can be switched back and forth. The circuit consists of two CMOS NOR2 gates. One of the input terminals of each NOR gate is used to cross-couple to

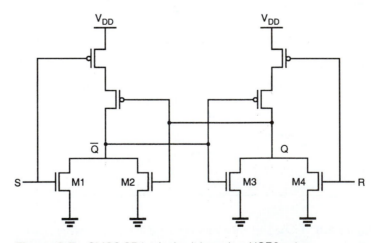

Figure 8.7   CMOS SR latch circuit based on NOR2 gates.

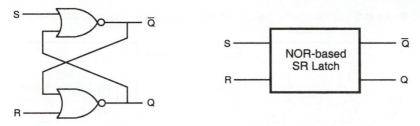

Figure 8.8 Gate-level schematic and block diagram of the NOR-based SR latch.

the output of the other NOR gate, while the second input enables triggering of the circuit.

The SR latch circuit has two complementary outputs, Q and $\overline{Q}$. By definition, the latch is said to be in its *set* state when Q is equal to logic "1" and $\overline{Q}$ is equal to logic "0." Conversely, the latch is in its *reset* state when the output Q is equal to logic "0" and $\overline{Q}$ is equal to "1." The gate-level schematic of the SR latch consisting of two NOR2 gates, and the corresponding block diagram representation, are shown in Fig. 8.8. It can easily be seen that when both input signals are equal to logic "0," the SR latch will operate exactly like the simple cross-coupled bistable element examined earlier; i.e., it will preserve (hold) either one of its two stable operating points (states) as determined by the previous inputs.

If the *set input* (S) is equal to logic "1" and the *reset input* is equal to logic "0," then the output node Q will be forced to logic "1" while the output node $\overline{Q}$ is forced to logic "0." This means that the SR latch will be *set*, regardless of its previous state. Similarly, if S is equal to "0" and R is equal to "1," then the output node Q will be forced to "0" while $\overline{Q}$ is forced to "1." Thus, with this input combination, the latch is *reset*, regardless of its previously held state. Finally, consider the case in which both of the inputs S and R are equal to logic "1." In this case, both output nodes will be forced to logic "0," which conflicts with the complementarity of Q and $\overline{Q}$. Therefore, this input combination is not permitted during normal operation and is considered to be a *not-allowed* condition. The truth table of the NOR-based SR latch is summarized in Table 8.1.

The operation of the CMOS SR latch circuit shown in Fig. 8.7 can be examined in more detail by considering the operating modes of the four nMOS transistors, M1, M2, M3, and M4. If the set input (S) is equal to V_{OH} and the reset input (R) is

Table 8.1 Truth table of the NOR-based SR latch circuit

S	R	Q_{n+1}	$\overline{Q}_{n+1}$	Operation
0	0	Q_n	$\overline{Q}_n$	hold
1	0	1	0	set
0	1	0	1	reset
1	1	0	0	not allowed

Table 8.2   Operation modes of the transistors in the NOR-based CMOS SR latch circuit

S	R	Q_{n+1}	$\overline{Q_{n+1}}$	Operation
V_{OH}	V_{OL}	V_{OH}	V_{OL}	M1 and M2 on, M3 and M4 off
V_{OL}	V_{OH}	V_{OL}	V_{OH}	M1 and M2 off, M3 and M4 on
V_{OL}	V_{OL}	V_{OH}	V_{OL}	M1 and M4 off, M2 on, or
V_{OL}	V_{OL}	V_{OL}	V_{OH}	M1 and M4 off, M3 on

equal to V_{OL}, both of the parallel-connected transistors M1 and M2 will be on. Consequently, the voltage on node $\overline{Q}$ will assume a logic-low level of $V_{OL} = 0$. At the same time, both M3 and M4 are turned off, which results in a logic-high voltage V_{OH} at node Q. If the reset input (R) is equal to V_{OH} and the set input (S) is equal to V_{OL}, the situation will be reversed (M1 and M2 turned off and M3 and M4 turned on).

When both of the input voltages are equal to V_{OL}, on the other hand, there are two possibilities. Depending on the previous state of the SR latch, either M2 *or* M3 will be on, while both of the trigger transistors M1 and M4 are off. This will generate a logic-low level of $V_{OL} = 0$ at one of the output nodes, while the complementary output node is at V_{OH}. The static operation modes and voltage levels of the NOR-based CMOS SR latch circuit are summarized in Table 8.2. For simplicity, the operating modes of the complementary pMOS transistors are not explicitly listed here.

For the transient analysis of the SR latch circuit, we have to consider an event which results in a state change, i.e., either an initially reset latch being set by applying a *set* signal, or an initially set latch being reset by applying the *reset* signal. In either case, we note that both of the output nodes undergo simultaneous voltage transitions. While one output is rising from its logic-low level to logic-high, the other output node is falling from its initial logic-high level to logic-low. Thus, an interesting problem is to estimate the amount of time required for the simultaneous switching of the two output nodes. The exact solution of this problem obviously requires the simultaneous solution of two coupled differential equations, one each for each output node. The problem can, however, be simplified considerably if we assume that the two events take place in sequence rather than simultaneously. This assumption causes an overestimation of the switching time.

To calculate the switching times for both output nodes, we first have to find the total parasitic capacitance associated with each node. Simple inspection of the circuit shows that the total lumped capacitance at each output node can be approximated as follows:

$$C_Q = C_{gb,2} + C_{gb,5} + C_{db,3} + C_{db,4} + C_{db,7} + C_{sb,7} + C_{db,8}$$
$$C_{\overline{Q}} = C_{gb,3} + C_{gb,7} + C_{db,1} + C_{db,2} + C_{db,5} + C_{sb,5} + C_{db,6}$$

(8.18)

The circuit diagram of the SR latch is shown in Fig. 8.9 together with the lumped load capacitances at the nodes Q and $\overline{Q}$. Assuming that the latch is initially reset and

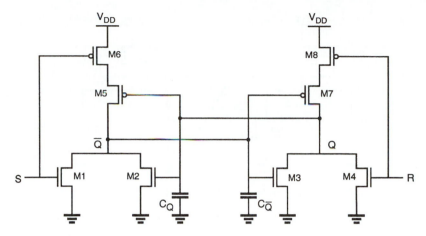

Figure 8.9 Circuit diagram of the CMOS SR latch showing the lumped load capacitances at both output nodes.

that a set operation is being performed by applying S = "1" and R = "0," the rise time associated with node Q can now be estimated as follows.

$$\tau_{rise,Q}(SR-latch) = \tau_{rise,Q}(NOR2) + \tau_{fall,\overline{Q}}(NOR2) \qquad (8.19)$$

Note that the calculation of the switching time $\tau_{rise,Q}$ requires two separate calculations for the rise and fall times of the NOR2 gates. It is obvious that by considering the two events separately, i.e., first, one of the output node voltages ($\overline{Q}$) falling from high to low due to turn-on of M1, followed by the other node voltage (Q) rising from low to high due to turn-off of M3, we are bound to overestimate the actual switching time for the SR latch. Both M2 and M4 can be assumed to be off in this process, although M2 can be turned on as Q rises, thus actually shortening the $\overline{Q}$ node fall time. This approach, however, yields a simpler first-order prediction for the time delay, as opposed to the simultaneous solution of two coupled differential equations.

The NOR-based SR latch can also be implemented by using two cross-coupled depletion-load nMOS NOR2 gates, as shown in Fig. 8.10. From the logic point of view, the operation principle of the depletion-load nMOS NOR-based SR latch is identical to that of the CMOS SR latch. In terms of power dissipation and noise margins, however, the CMOS circuit implementation offers a better alternative, since both of the CMOS NOR2 gates dissipate virtually no static power for preserving a state, and since the output voltages can exhibit a full swing between 0 and V_{DD}.

Now consider a different approach for building the basic SR latch circuit. Instead of using two NOR2 gates, we can use two NAND2 gates, as shown in Fig. 8.11. Here, one input of each NAND gate is used to cross-couple to the output of the other NAND gate, while the second input enables external triggering.

A close inspection of the NAND-based SR latch circuit reveals that in order to hold (preserve) a state, both of the external trigger inputs must be equal to logic "1."

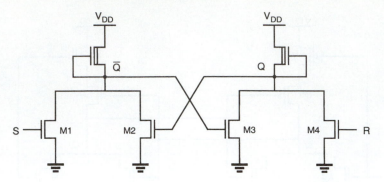

Figure 8.10   Depletion-load nMOS SR latch circuit based on NOR2 gates.

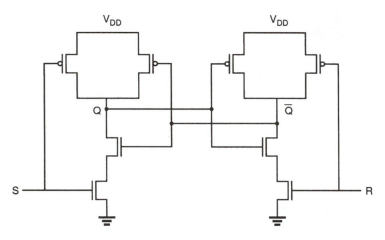

Figure 8.11   CMOS SR latch circuit based on NAND2 gates.

The operating point or the state of the circuit can be changed only by pulling the *set* input to logic zero or by pulling the *reset* input to zero. We can observe that if S is equal to "0" and R is equal to "1," the output Q attains a logic "1" value and the complementary output $\overline{Q}$ becomes logic "0." Thus, in order to *set* the NAND SR latch, a logic "0" must be applied to the *set* (S) input. Similarly, in order to *reset* the latch, a logic "0" must be applied to the *reset* (R) input. The conclusion is that the NAND-based SR latch responds to *active low* input signals, as opposed to the NOR-based SR latch, which responds to *active high* inputs. Note that if both input signals are equal to logic "0," both output nodes assume a logic-high level, which is not allowed because it violates the complementarity of the two outputs.

The gate-level schematic and the corresponding block diagram representation of the NAND-based SR latch circuit are shown in Fig. 8.12. The small circles at the S and R input terminals indicate that the circuit responds to *active low* input signals. The truth table of the NAND SR latch is also shown in the figure. The same approach

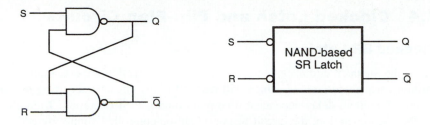

S	R	Q_{n+1}	$\overline{Q}_{n+1}$	Operation
0	0	1	1	not allowed
0	1	1	0	set
1	0	0	1	reset
1	1	Q_n	$\overline{Q}_n$	hold

Figure 8.12 Gate-level schematic and block diagram of the NAND-based SR latch.

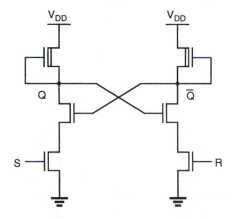

Figure 8.13 Depletion-load nMOS NAND-based SR latch circuit.

used in the timing analysis of NOR-based SR latches can be applied to NAND-based SR latches.

The NAND-based SR latch can also be implemented by using two cross-coupled depletion-load NAND2 gates, as shown in Fig. 8.13. While the operation principle is identical to that of the CMOS NAND SR latch (Fig. 8.11) from the logic point of view, the CMOS circuit implementation again offers a better alternative in terms of static power dissipation and noise margins.

8.4   Clocked Latch and Flip-Flop Circuits

Clocked SR Latch

All of the SR latch circuits examined in the previous section are essentially asynchronous sequential circuits, which will respond to the changes occurring in input signals at a circuit-delay-dependent time point during their operation. To facilitate synchronous operation, the circuit response can be controlled simply by adding a gating clock signal to the circuit, so that the outputs will respond to the input levels only during the active period of a clock pulse. For simple reference, the clock pulse will be assumed to be a periodic square waveform, which is applied simultaneously to all clocked logic gates in the system.

The gate-level schematic of a clocked NOR-based SR latch is shown in Fig. 8.14. It can be seen that if the clock (CK) is equal to logic "0," the input signals have no influence upon the circuit response. The outputs of the two AND gates will remain at logic "0," which forces the SR latch to hold its current state regardless of the S and R input signals. When the clock input goes to logic "1," the logic levels applied to the S and R inputs are permitted to reach the SR latch, and possibly change its state. Note that as in the non-clocked SR latch, the input combination S = R = "1" is not allowed in the clocked SR latch. With both inputs S and R at logic "1," the occurrence of clock pulse causes both outputs to go momentarily to zero. When the clock pulse is removed, i.e., when it becomes "0," the state of the latch is indeterminate. It can eventually settle into either state, depending on slight delay differences between the output signals.

To illustrate the operation of the clocked SR latch, a sample sequence of CK, S, and R waveforms, and the corresponding output waveform Q are shown in Fig. 8.15. Note that the circuit is strictly *level-sensitive* during active clock phases; i.e., any changes occurring in the S and R input voltages when the CK level is equal to "1" will be reflected onto the circuit outputs. Consequently, even a narrow spike or glitch occurring during an active clock phase can set or reset the latch, if the loop delay is shorter than the pulse width.

Figure 8.16 shows a CMOS implementation of the clocked NOR-based SR latch circuit, using two simple AOI gates. Notice that the AOI-based implementation of

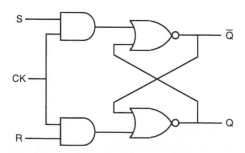

Figure 8.14 Gate-level schematic of the clocked NOR-based SR latch.

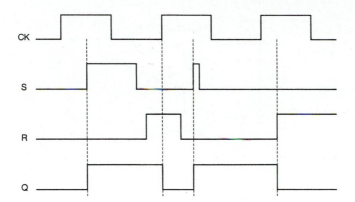

Figure 8.15 Sample input and output waveforms illustrating the operation of the clocked NOR-based SR latch circuit.

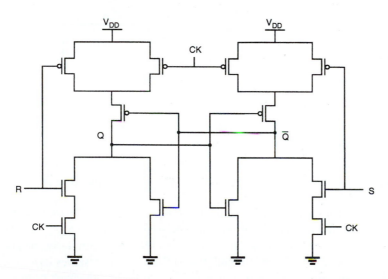

Figure 8.16 AOI-based implementation of the clocked NOR-based SR latch circuit.

the circuit results in a very small transistor count, compared with the alternative circuit realization consisting of two AND2 and two NOR2 gates.

The NAND-based SR latch can also be implemented with gating clock input, as shown in Fig. 8.17. It must be noted, however, that both input signals S and R as well as the clock signal CK are *active low* in this case. This means that changes in the input signal levels will be ignored when the clock is equal to logic "1," and that inputs will influence the outputs only when the clock is active, i.e., CK = "0." For the circuit implementation of this clocked NAND-based SR latch, we can use a simple

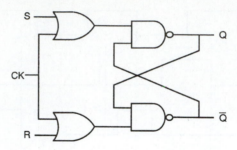

Figure 8.17 Gate-level schematic of the clocked NAND-based SR latch circuit, with active low inputs.

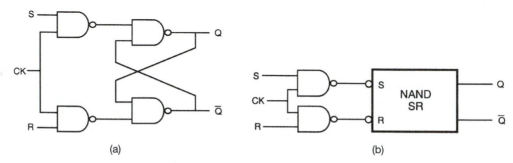

(a)   (b)

Figure 8.18 (a) Gate-level schematic of the clocked NAND-based SR latch circuit, with active high inputs. (b) Partial block diagram representation of the same circuit.

OAI structure, which is essentially analogous to the AOI-based realization of the clocked NOR SR latch circuit.

A different implementation of the clocked NAND-based SR latch is shown in Fig. 8.18. Here, both input signals and the CK signal are *active high;* i.e., the latch output Q will be set when CK = "1," S = "1," and R = "0." Similarly, the latch will be reset when CK = "1," S = "0," and R = "1." The latch preserves its state as long as the clock signal is inactive, i.e., when CK = "0." The drawback of this implementation is that the transistor count is higher than the *active low* version shown in Fig. 8.17.

Clocked JK Latch

All simple and clocked SR latch circuits examined to this point suffer from the common problem of having a not-allowed input combination; i.e., their state becomes indeterminate when both inputs S and R are activated at the same time. This problem can be overcome by adding two feedback lines from the outputs to the inputs, as shown in Fig. 8.19. The resulting circuit is called a JK latch. Figure 8.20 shows an all-NAND implementation of the JK latch with active high inputs, and the corresponding block diagram representation. The JK latch is commonly called a JK flip-flop.

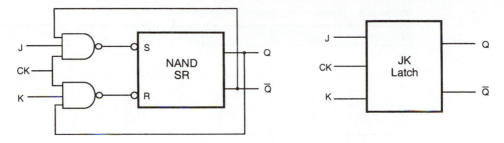

Figure 8.19 Gate-level schematic of the clocked NAND-based JK latch circuit.

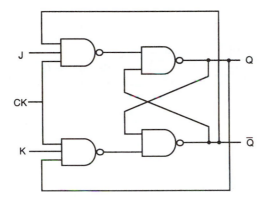

Figure 8.20 All-NAND implementation of the clocked JK latch circuit.

The J and K inputs in this circuit correspond to the set and reset inputs of the basic SR latch. When the clock is active, the latch can be set with the input combination (J = "1," K = "0"), and it can be reset with the input combination (J = "0," K = "1"). If both inputs are equal to logic "0," the latch preserves its current state. If, on the other hand, both inputs are equal to "1" during the active clock phase, the latch simply switches its state due to feedback. In other words, the JK latch does not have a not-allowed input combination. As in the other clocked latch circuits, the JK latch will hold its current state when the clock is inactive (CK = "0"). The operation of the clocked JK latch is summarized in the truth table (Table 8.3).

Figure 8.21 shows an alternative, NOR-based implementation of the clocked JK latch, and CMOS realization of this circuit. Note that the AOI-based circuit structure results in a relatively low transistor count, and consequently, a more compact circuit compared to the all-NAND realization shown in Fig. 8.20.

While there is no not-allowed input combination for the JK latch, there is still a potential problem. If both inputs are equal to logic "1" during the active phase of the clock pulse, the output of the circuit will oscillate (toggle) continuously until either

Table 8.3 Detailed truth table of the JK latch circuit

J	K	Q_n	$\overline{Q_n}$	S	R	Q_{n+1}	$\overline{Q_{n+1}}$	Operation
0	0	0	1	1	1	0	1	hold
		1	0	1	1	1	0	
0	1	0	1	1	1	0	1	reset
		1	0	1	0	0	1	
1	0	0	1	0	1	1	0	set
		1	0	1	1	1	0	
1	1	0	1	0	1	1	0	toggle
		1	0	1	0	0	1	

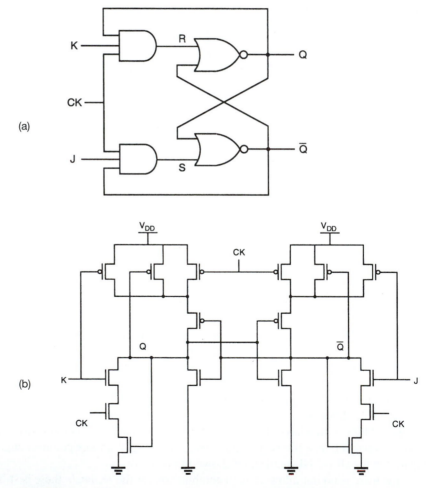

Figure 8.21 (a) Gate-level schematic of the clocked NOR-based JK latch circuit. (b) CMOS AOI realization of the JK latch.

the clock becomes inactive (goes to zero), or one of the input signals goes to zero. To prevent this undesirable timing problem, the clock pulse width must be made smaller than the input-to-output propagation delay of the JK latch circuit. This restriction dictates that the clock signal must go low before the output level has an opportunity to switch again, which prevents uncontrolled oscillation of the output. However, note that this clock constraint is difficult to implement for most practical applications.

Assuming that the clock timing constraint described above is satisfied, the output of the JK latch will toggle (change its state) only once for each clock pulse, if both inputs are equal to logic "1" (Fig. 8.22). A circuit which is operated exclusively in this mode is called a *toggle switch*.

Master-Slave Flip-Flop

Most of the timing limitations encountered in the previously examined clocked latch circuits can be prevented by using two latch stages in a cascaded configuration. The key operation principle is that the two cascaded stages are activated with opposite clock phases. This configuration is called the *master-slave flip-flop*. Our definition of flip-flop is designed to distinguish it from latches discussed previously, although they are mostly used interchangeably in the literature.

The input latch in Fig. 8.23, called the "master," is activated when the clock pulse is high. During this phase, the inputs J and K allow data to be entered into the

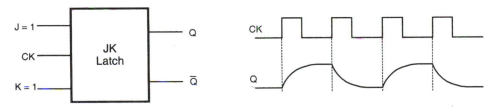

Figure 8.22 Operation of the JK latch as a toggle switch.

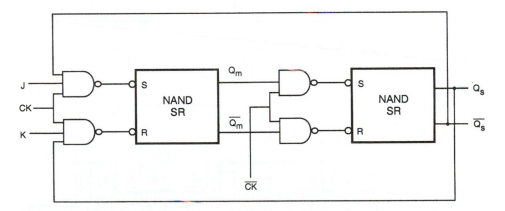

Figure 8.23 Master-slave flip-flop consisting of NAND-based JK latches.

flip-flop, and the first-stage outputs are set according to the primary inputs. When the clock pulse goes to zero, the master latch becomes inactive and the second-stage latch, called the "slave," becomes active. The output levels of the flip-flop circuit are determined during this second phase, based on the master-stage outputs set in the previous phase.

Since the master and the slave stages are effectively decoupled from each other with the opposite clocking scheme, the circuit is never *transparent;* i.e., a change occurring in the primary inputs is never reflected directly to the outputs. This very important property clearly separates the master-slave flip-flop from all of the latch circuits examined earlier in this section. Figure 8.24 shows a sample set of input and output waveforms associated with the JK master-slave flip-flop, which can help the reader to study the basic operation principles.

Because the master and the slave stages are decoupled from each other, the circuit allows for toggling when J = K = "1," but it eliminates the possibility of uncontrolled oscillations since only one stage is active at any given time. A NOR-based alternative realization for the master-slave flip-flop circuit is shown in Fig. 8.25.

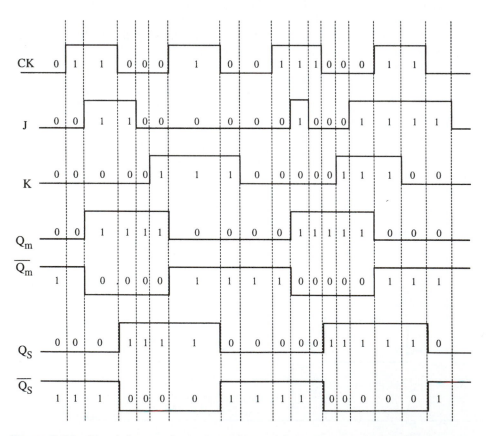

Figure 8.24 Sample input and output waveforms of the master-slave flip-flop circuit.

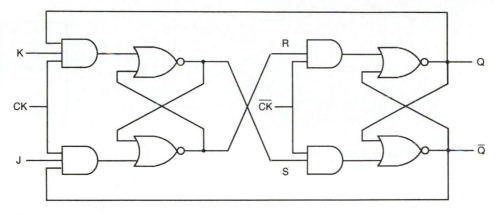

Figure 8.25 NOR-based realization of the JK master-slave flip-flop.

Figure 8.24 also shows that the master-slave flip-flop circuit examined here has the potential problem of "ones catching." When the clock pulse is high, a narrow spike or glitch in one of the inputs, for instance a glitch in the J line (or K line), may set (or reset) the master latch and thus cause an unwanted state transition, which will then be propagated into the slave stage during the following phase. This problem can be eliminated to a large extent by building an edge-triggered master-slave flip-flop, which will be examined in the following section.

8.5 CMOS D-Latch and Edge-Triggered Flip-Flop

With the widespread use of CMOS circuit techniques in digital integrated circuit design, a large selection of CMOS-based sequential circuits have also gained popularity and prominence, especially in VLSI design. Throughout this chapter, we have seen examples showing that virtually all of the latch and flip-flop circuits can be implemented with CMOS gates, and that their design is quite straightforward. However, direct CMOS implementations of conventional circuits such as the clocked JK latch or the JK master-slave flip-flop tend to require a large number of transistors.

In this section, we will see that specific versions of sequential circuits built primarily with CMOS transmission gates are generally simpler and require fewer transistors than the circuits designed with conventional structuring. As an introduction to the issue, let us first consider the simple D-latch circuit shown in Fig. 8.26. The gate-level representation of the D-latch is simply obtained by modifying the clocked NOR-based SR latch circuit. Here, the circuit has a single input D, which is directly connected to the S input of the latch. The input variable D is also inverted and connected to the R input of the latch. It can be seen from the gate-level schematic that the output Q assumes the value of the input D when the clock is active, i.e., for CK = "1." When the clock signal goes to zero, the output will simply preserve its

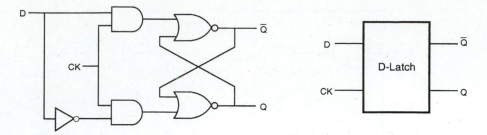

Figure 8.26 Gate-level schematic and the block diagram view of the D-latch.

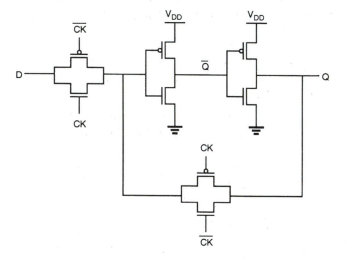

Figure 8.27 CMOS implementation of the D-latch (version 1).

state. Thus, the CK input acts as an enable signal which allows data to be accepted into the D-latch.

The D-latch finds many applications in digital circuit design, primarily for temporary storage of data or as a delay element. In the following, we will examine its simple CMOS implementation. Consider the circuit diagram given in Fig. 8.27, which shows a basic two-inverter loop and two CMOS transmission gate (TG) switches.

The TG at the input is activated by the CK signal, whereas the TG in the inverter loop is activated by the inverse of the CK signal, $\overline{CK}$. Thus, the input signal is accepted (latched) into the circuit when the clock is high, and this information is preserved as the state of the inverter loop when the clock is low. The operation of the CMOS D-latch circuit can be better visualized by replacing the CMOS transmission gates with simple switches, as shown in Fig. 8.28. A timing diagram accompanying this figure shows the time intervals during which the input and the output signals should be valid (unshaded).

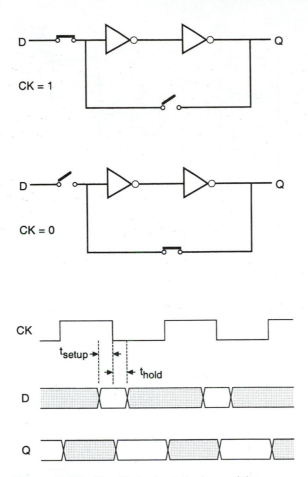

Figure 8.28   Simplified schematic view and the corresponding timing diagram of the CMOS D-latch circuit, showing the setup time and the hold time.

Note that the valid D input must be stable for a short time before (*setup time*, t_{setup}) and after (*hold time*, t_{hold}) the negative clock transition, during which the input switch opens and the loop switch closes. Once the inverter loop is completed by closing the loop switch, the output will preserve its valid level. In the D-latch design, the requirements for setup time and hold time should be met carefully. Any violation of such specifications can cause *metastability* problems which lead to seemingly chaotic transient behavior, and can result in an unpredictable state after the transitional period.

The D-latch shown in Fig. 8.27 is not an edge-triggered storage element because the output changes according to the input, i.e., the latch is transparent, while the clock is high. The transparency property makes the application of this D-latch unsuitable for counters and some data storage implementations.

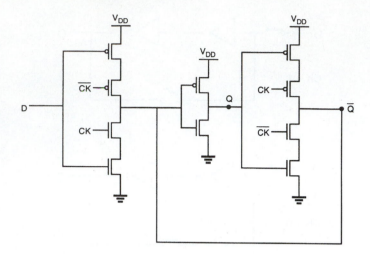

Figure 8.29   CMOS implementation of the D-latch (version 2).

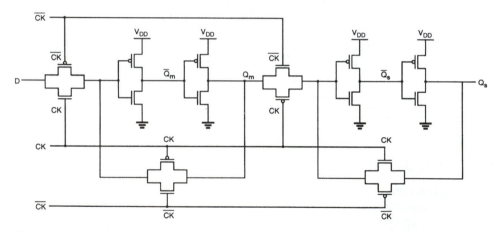

Figure 8.30   CMOS negative (falling) edge-triggered master-slave D flip-flop (DFF).

Figure 8.29 shows a different version of the CMOS D-latch. The circuit contains two tristate inverters, driven by the clock signal and its inverse. Although the circuit appears to be quite different from that shown in Fig. 8.27, the basic operation principle of the circuit is the same as that shown in Fig. 8.28. The first tri-state inverter acts as the input switch, accepting the input signal when the clock is high. At this time, the second tristate inverter is at its high-impedance state, and the output Q is following the input signal. When the clock goes low, the input buffer becomes inactive, and the second tristate inverter completes the two-inverter loop, which preserves its state until the next clock pulse.

Finally, consider the two-stage master-slave flip-flop circuit shown in Fig. 8.30, which is constructed by simply cascading two D-latch circuits. The first stage (master) is driven by the clock signal, while the second stage (slave) is driven by the inverted clock signal. Thus, the master stage is positive level-sensitive, while the slave stage is negative level-sensitive.

When the clock is high, the master stage follows the D input while the slave stage holds the previous value. When the clock changes from logic "1" to logic "0," the master latch ceases to sample the input and stores the D value at the time of the clock transition. At the same time, the slave latch becomes transparent, passing the stored master value Q_m to the output of the slave stage, Q_s. The input cannot affect the output because the master stage is disconnected from the D input. When the clock changes again from logic "0" to "1," the slave latch locks in the master latch output and the master stage starts sampling the input again. Thus, this circuit is a negative edge-triggered D flip-flop by virtue of the fact that it samples the input at the falling edge of the clock pulse.

Figure 8.31 shows the simulated input and output waveforms of the CMOS negative edge-triggered D-type flip-flop. The output of the master stage latches the applied input (D) when the clock signal is "1," and the output of the slave stage becomes valid when the clock signal drops to "0." Thus, the D-type flip-flop (DFF) essentially samples the input at every falling edge of the clock pulse.

It should be emphasized that the operation of the DFF circuit can be seriously affected if the master stage experiences a set-up time violation. This situation is illustrated in Fig. 8.32, where the input D switches from "0" to "1" immediately before the clock transition occurs (set-up time violation). As a result, the master stage fails

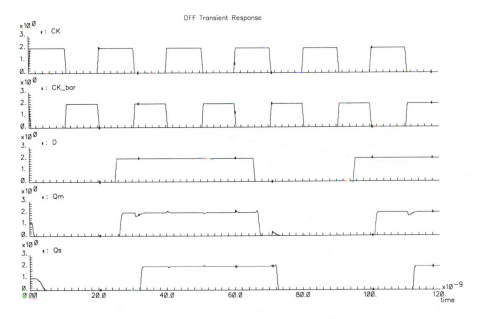

Figure 8.31   Simulated input and output waveforms of the CMOS DFF circuit in Fig. 8.30.

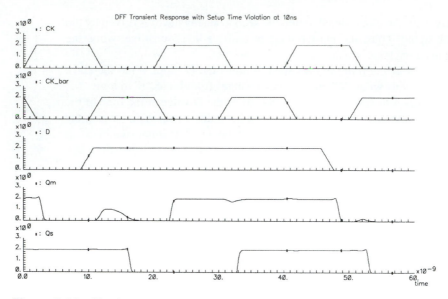

Figure 8.32 Simulated waveforms of the CMOS DFF circuit, showing a set-up time violation for the master stage input at 10 ns. The output of the master stage fails to settle at the correct level.

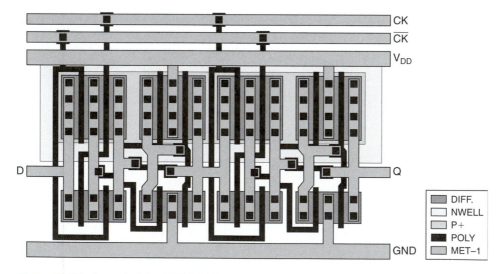

Figure 8.33 Layout of the CMOS DFF shown in Fig. 8.30.

to latch the correct value, and the slave stage produces an erroneous output. The relative timing of the input and clock signals are carefully synchronized to avoid such situations. The layout of the CMOS DFF circuit is given in Fig. 8.33.

Another implementation of edge-triggered D flip-flop is shown in Fig. 8.34, which consists of six NAND3 gates. This D flip-flop is positive edge-triggered as illustrated in the waveform chart in Fig. 8.35. Initially, all the signal values except for

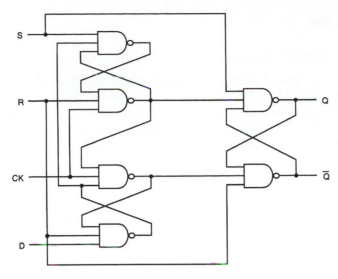

Figure 8.34 NAND3-based positive edge-triggered D flip-flop circuit.

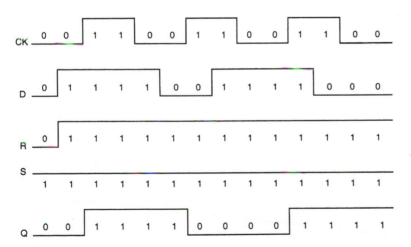

Figure 8.35 Timing diagram of the positive edge-triggered D flip-flop.

S are 0; i.e., (S, R, CK, D) = (1, 0, 0, 0), and Q = 0. In the second phase, both D and R switch to 1, i.e., (S, R, CK, D) = (1, 0, 1, 1), but no change in Q occurs and the Q value remains at 0. However, in the third phase, if CK goes to high, i.e., (S, R, CK, D) = (1, 1, 1, 1), the output of gate 2 switches to 0, which in turn sets the output of the last stage SR latch to 1. Thus, the output of this D flip-flop switches to 1 at the positive-going edge of the clock signal, CK. However, as can be observed in the ninth phase of the waveform diagram chart, the Q output is not affected by the negative-going edge of CK, nor by other signal changes.

Appendix

Schmitt Trigger Circuit

In the following, we will examine the Schmitt trigger circuit, which is a very useful regenerative circuit. The Schmitt trigger has an inverter-like voltage transfer characteristic, but with two different logic threshold voltages for increasing and for decreasing input signals. With this unique property, the circuit can be utilized for the detection of low-to-high and high-to-low switching events in noisy environments.

    The circuit diagram of a CMOS Schmitt trigger and the typical features of its voltage transfer characteristic (VTC) are show below. Also listed here is the corresponding SPICE circuit input file. In the following, we will first calculate the important points in VTC, and then compare out results with SPICE simulation.

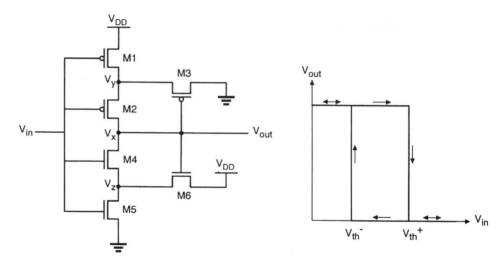

```
CMOS Schmitt Trigger DC analysis
vdd 5 0 dc 5V
vin 1 0 dc 1v
m5 2 1 0 0 mn l=1u w=1u
m4 3 1 2 0 mn l=1u w=2.5u
m6 5 3 2 0 mn l=1u w=3u
m1 4 1 5 5 mp l=1u w=1u
m2 3 1 4 5 mp l=1u w=2.5u
m3 0 3 4 5 mp l=1u w=3u
.model mn nmos vto=1 gamma=0.4 kp=2.5e-5
.model mp pmos vto=-1 gamma=0.4 kp=1.0e-5
.dc vin 0 5 0.1
.print dc v(3)
.end
```

We start our step-by-step analysis by considering a positive input sweep, i.e., assuming that the input voltage is increasing from 0 to V_{DD}.

i) At $V_{in} = 0$ V:

M1 and M2 are turned on, then

$$V_x = V_y = V_{DD} = 5 \text{ V}$$

At the same time, M4 and M5 are turned off. M3 is off; M6 is on and operates in the saturation region. Calculating the threshold voltage of M6 with $2\phi_F = -0.6$ V,

$$V_z = V_{DD} - V_{T,6} = 3.5 \text{ V}$$

ii) At $V_{in} = V_{T0,n} = 1.0$ V:

M5 starts to turn on, M4 is still off.

$$V_x = 5 \text{ V}$$

iii) At $V_{in} = 2.0$ V:

Assume M4 is off, while both M5 and M6 operate in the saturation region.

$$\frac{1}{2}k'\left(\frac{W}{L}\right)_5 (V_{in} - V_{T0,n})^2 = \frac{1}{2}k'\left(\frac{W}{L}\right)_6 (V_{DD} - V_z - V_{T,6})^2$$

$$(2-1)^2 = 3(5 - V_z - [1 + 0.4(\sqrt{0.6 + V_z} - \sqrt{0.6})])^2$$

Solving this equation for V_z, we find that there is only one physically reasonable root.

$$V_z = 2.976 \text{ V}$$

Now, we check our assumption made above; i.e., M4 is indeed turned off:

$$V_{GS,4} = 2 - 2.976 = -0.976 < V_{T0,n} = 1$$

iv) At $V_{in} = 3.5$ V:

V_z continues to decrease. Assuming M5 in linear region and M6 in saturation, we arrive at the following current equation.

$$\frac{1}{2}k'\left(\frac{W}{L}\right)_5 [2(V_{in} - V_{T0,n})V_z - V_z^2] = \frac{1}{2}k'\left(\frac{W}{L}\right)_6 (V_{DD} - V_z - V_{T,6})^2$$

$$[2(3.5 - 1.0)V_z - V_z^2] = 3(5 - V_z - [1 + 0.4(\sqrt{0.6 + V_z} - \sqrt{0.6})])^2$$

Solving this equation for V_z, we obtain $V_z = 2.2$ V. Now determine the gate-to-source voltage of M4 as

$$V_{GS,4} = 3.5 - 2.2 = 1.3 > V_{T0,n} = 1$$

It is seen that at this point, M4 is already on. Thus, the analysis above, which is based on the assumption that M4 is not conducting, can no longer be valid. At this input voltage, node x is being pulled down toward "0." This can also be seen clearly from the simulation results. We conclude that the upper logic threshold voltage V_{th}^+ is approximately equal to 3.5 V.

Next, we consider a negative input sweep, i.e., assume that the input voltage is decreasing from V_{DD} to 0.

i)   At $V_{in} = 5.0$ V:

M4 and M5 are on, so that the output voltage is $V_x = 0$ V. The pMOS transistors M1 and M2 are off, and M3 is in saturation, thus,

$$\frac{1}{2}k'\left(\frac{W}{L}\right)_3 (0 - V_y - V_{T,3})^2 = 0$$

$$V_y = -V_{T,3} = -\left[V_{T0,p} - 0.4\left(\sqrt{0.6 + V_{DD} - V_y} - \sqrt{0.6}\right)\right]$$

$$V_y = 1.5\,[\text{V}]$$

ii) At $V_{in} = 4.0$ V:

M1 is at the edge of turning on, M2 is off, and M3 is in saturation. The output voltage is still unchanged.

iii) At $V_{in} = 3.0$ V:

M1 is on and in saturation region. M3 is also in saturation, thus,

$$\frac{1}{2}k'\left(\frac{W}{L}\right)_1 (V_{in} - V_{DD} - V_{T0,p})^2 = \frac{1}{2}k'\left(\frac{W}{L}\right)_3 (0 - V_y - V_{T,3})^2$$

$$[3 - 5 - (-1)]^2 = 3\left(0 - V_y - \left[-1 - 0.4\left(\sqrt{0.6 + 5 - V_y} - \sqrt{0.6}\right)\right]\right)^2$$

The solution of this equation yields:

$$V_{in} = 2.02\ \text{V}$$

Now we determine the gate-to-source voltage of M2 as

$$V_{GS,2} = 3.0 - 2.02 = 0.98 > V_{T0,p} = -1$$

which indicates that M2 is still turned off at this point.

iv) At $V_{in} = 1.5$ V:

If M2 is still off, M1 is in the linear region, and M3 is in the saturation region:

$$\frac{1}{2}k'\left(\frac{W}{L}\right)_1 \left(2(V_{in} - V_{DD} - V_{T0,p})(V_y - V_{DD}) - (V_y - V_{DD})^2\right)$$

$$= \frac{1}{2}k'\left(\frac{W}{L}\right)_3 (0 - V_y - V_{T,3})^2$$

$$2(1.5 - 5 + 1)(V_y - 5) - (V_y - 5)^2$$
$$= 3\left(-V_y - \left[-1 - 0.4\left(\sqrt{0.6 + 5 - V_y} - \sqrt{0.6}\right)\right]\right)^2$$

Solving this quadratic equation yields

$$V_y = 2.79\ \text{V}$$

It can be shown that at this point, the pMOS transistor M2 is already turned on. Consequently, the output voltage is being pulled up to V_{DD}. We conclude that the lower logic threshold voltage V_{th}^- is approximately equal to 1.5 V.

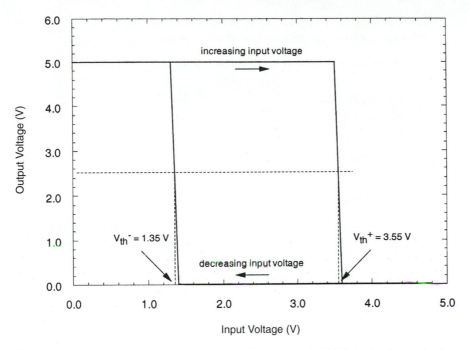

Figure 8.36 Simulated output voltage waveforms of the CMOS Schmitt trigger circuit, for increasing and for decreasing input voltage.

The SPICE simulation results are plotted in Fig. 8.36 for both increasing and decreasing input voltages. The expected hysteresis behavior and the two switching thresholds are clearly seen in the simulation results.

Exercise Problems

8.1 Figure P8.1 shows a schematic for a positive edge-triggered D flip-flop. Use a layout editor (e.g., Magic) to design a layout of the circuit. Use CMOS technology, and assume that you have an n-type substrate. On the printout of

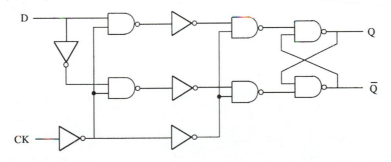

Figure P8.1

your layout, clearly indicate the location of each logic gate in Fig. P8.1. Also, calculate the parasitic capacitances of your layout.

$W_n = 4\ \mu\text{m}$ and $W_p = 8\ \mu\text{m}$ for all gates

$L_M = 2\ \mu\text{m}$

$L_D = 0.25\ \mu\text{m}$

$V_{T0,n} = 1\ \text{V}$

$V_{T0,p} = -1\ \text{V}$

$k'_n = 40\ \mu\text{A/V}^2$

$k'_p = 25\ \mu\text{A/V}^2$

$t_{ox} = 20\ \text{nm}$

8.2 For the layout in Problem 8.1, find the minimum setup time (t_{setup}) and hold time (t_{hold}) for the flip-flop using SPICE simulation. This will require you to obtain four plots.

a. A plot of the output using the minimum setup time t_{setup}

b. A plot of the output using a setup time of $0.8t_{setup}$

c. A plot of the output using the minimum hold time t_{hold}

d. A plot of the output using a hold time of $0.8t_{hold}$

8.3 We have discussed the features of the CMOS Schmitt trigger in the Appendix.

It has been pointed out that a useful application lies in the receiver circuit design to filter out noises. However, in terms of speed performance it delays the switching activity. In view of speed alone, it would be useful to reverse the switching directions. In particular, we want to have the negative-going (high-to-low transition) edge to occur at an input voltage smaller than the typical inverter's saturation voltage and also the positive-going (low-to-high transition) edge to occur at an input voltage larger than the inverter's saturation voltage. Complete the circuit connection in Fig. P8.3 to realize such a circuit block for the assembly of the following components. Justify your answer by using SPICE circuit analysis. You can use some approximation technique to simulate the circuit. For instance, the different VTC curves can be simulated by using inverters of different β ratios. To be more

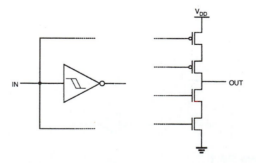

Figure P8.3

specific, for larger saturation voltage you can use an inverter with a strong
pull-up transistor, and for smaller saturation voltage use an inverter with a
strong pull-down transistor.

8.4 Consider the monostable multivibrator circuit drawn in Fig. P8.4. Calculate
the output pulse width.

$$V_T(\text{dep}) = -2 \text{ V}$$
$$V_T(\text{enh}) = 1 \text{ V}$$
$$k' = 20 \text{ } \mu\text{A/V}^2$$
$$\gamma = 0$$

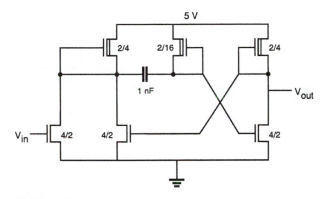

Figure P8.4

8.5 Shown in Fig. P8.5 is an nMOS Schmitt trigger. Draw the voltage transfer
characteristic. Include values for all important points on the graph. Use the
parameters in Problem 8.4 and $\lambda = 0$. W/L ratios for the transistors are given
below.

	M1	M2	M3	M4
W/L	1	0.5	10	1

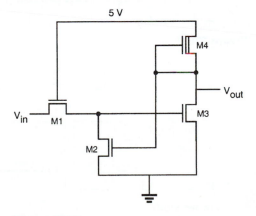

Figure P8.5

8.6 Design a circuit to implement the truth table shown in Fig. P8.6. A gate-level design is sufficient.

8.7 The circuit you have designed in Problem 8.6 is embedded in the larger circuit shown in Fig. P8.7. Complete the timing diagram for the output.

8.8 The voltage waveforms shown in Fig. P8.8 are applied to the nMOS JK master-slave flip-flop shown in Fig. 8.23. With the flip-flop initially reset, show the resulting waveforms at nodes Q_m (master flip-flop output) and Q_s (slave flip-flop output).

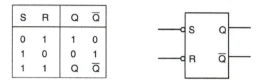

Figure P8.6

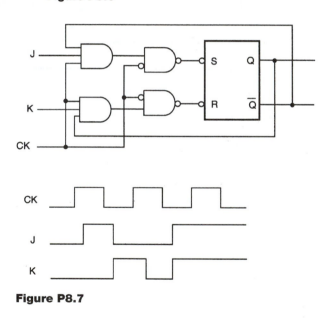

Figure P8.7

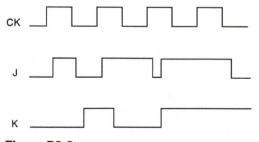

Figure P8.8

CHAPTER 9

Dynamic Logic Circuits

9.1 Introduction

A wide range of static combinational and sequential logic circuits were introduced in the previous chapters. Static logic circuits allow versatile implementation of logic functions based on static, or steady-state, behavior of simple nMOS or CMOS structures. In other words, all valid output levels in static gates are associated with steady-state operating points of the circuit in question. Hence, a typical static logic gate generates its output corresponding to the applied input voltages after a certain time delay, and it can preserve its output level (or state) as long as the power supply is provided. This approach, however, may require a large number of transistors to implement a function, and may cause a considerable time delay.

In high-density, high-performance digital implementations where reduction of circuit delay and silicon area is a major objective, *dynamic logic circuits* offer several significant advantages over static logic circuits. The operation of all dynamic logic gates depends on temporary (transient) storage of charge in parasitic node capacitances, instead of relying on steady-state circuit behavior. This operational property necessitates periodic updating of internal node voltage levels, since stored charge in a capacitor cannot be retained indefinitely. Consequently, dynamic logic circuits require periodic clock signals in order to control charge refreshing. The capability of temporarily storing a state, i.e., a voltage level, at a capacitive node allows us to implement very simple sequential circuits with memory functions. Also, the use of common clock signals throughout the system enables us to *synchronize* the operations of various circuit blocks. As a result, dynamic circuit techniques lend themselves well to synchronous logic design. Finally, the dynamic logic implementation of complex functions generally requires a smaller silicon area than does the static logic implementation. As for the power consumption which increases with the parasitic capacitances, the dynamic circuit implementation in a smaller area will, in many cases, consume less power than the static counterpart, despite its use of clock signals.

The following example presents the operation of a dynamic D-latch circuit, which essentially consists of two inverters connected in cascade. This simple circuit illustrates most of the basic operational concepts involved in dynamic circuit design.

EXAMPLE 9.1

Consider the dynamic D-latch circuit shown below. The circuit consists of two cascaded inverters and one nMOS pass transistor driving the input of the primary inverter stage.

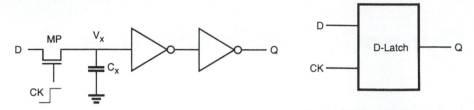

We will see that the parasitic input capacitance C_x of the primary inverter stage plays an important role in the dynamic operation of this circuit. The input pass transistor is being driven by the external periodic clock signal, as follows:

■ When the clock is high (CK = 1), the pass transistor turns on. The capacitor C_x is either charged up, or charged down through the pass transistor MP, depending on the input (D) voltage level. The output (Q) assumes the same logic level as the input.

■ When the clock is low (CK = 0), the pass transistor MP turns off, and the capacitor C_x is isolated from the input D. Since there is no current path from the intermediate node X to either V_{DD} or ground, the amount of charge stored in C_x during the previous cycle determines the output voltage level Q.

It can easily be seen that this circuit performs the function of a simple D-latch. In fact, the transistor count can be reduced by removing the last inverter stage if the latch output can be inverted. This option will be elaborated on in Section 9.2. The "hold" operation during the inactive clock cycle is accomplished by temporarily storing charge in the parasitic capacitance C_x. Correct operation of the circuit critically depends on how long a sufficient amount of charge can be retained at node X, before the output state changes due to charge leakage. Therefore, the capacitive intermediate node X is also called a *soft node*. The nature of the soft node makes the dynamic circuits more vulnerable to the so-called single-event upsets (SEUs) caused by α-particle or cosmic ray hits in integrated circuits.

In the following, we will examine the circuit operation in more detail. Assume that the dynamic D-latch circuit is being operated with a power supply voltage of $V_{DD} = 5$ V, and that the VTC of both inverters are identical, with

$$V_{OL} = 0 \text{ V}$$
$$V_{IL} = 2.1 \text{ V}$$
$$V_{IH} = 2.9 \text{ V}$$
$$V_{OH} = 5.0 \text{ V}$$

Furthermore, the threshold voltage of the pass transistor MP is given as $V_{T,n} = 0.8$ V. During the active clock phase (CK = 1), assume that the input is equal to logic "1," i.e., $V_{in} = V_{OH} = 5$ V. The pass transistor MP is conducting during this phase, and the parasitic intermediate node capacitance C_x is charged up to a logic-high level. We recall that the nMOS pass transistor is a poor conductor for logic "1," and its output voltage V_x will be *lower than* V_{OH}, by one threshold voltage: $V_x = 5.0 - 0.8 = 4.2$ V. Still, this voltage is clearly higher than the V_{IH} of the first inverter, thus, the output voltage of the first inverter will be very close to $V_{OL} = 0$ V. Consequently, the output level Q of the secondary inverter becomes a logic "1," $V_Q = V_{DD}$.

Next, the clock signal goes to zero, and the pass transistor turns off. Initially, the logic-high level at node X is preserved through charge storage in C_x. Thus, the output level Q also remains at logic "1." However, the voltage V_x eventually starts to drop from its original level of 4.2 V because of charge leakage from the soft node. It can easily be seen that in order to keep the output node Q at logic "1," the voltage level at the intermediate node X cannot be allowed to drop lower than $V_{IH} = 2.9$ V (once V_x falls *below* this level, the input of the first inverter cannot be interpreted as a logic "1"). Thus, the inactive clock phase during which the clock signal is equal to zero can, at most, be as long as it takes for the intermediate voltage V_x to drop from 4.2 V to 2.9 V, due to charge leakage. To avoid an erroneous output, the charge stored in C_x must be restored, or refreshed, to its original level before V_x reaches 2.9 V.

This example shows that the simple dynamic-charge storage principle employed in the D-latch circuit is quite feasible for preserving an output state during the inactive clock phase, assuming that the leakage currents responsible for draining the capacitance C_x are relatively small. In the following, we will examine the charge-up and charge-down events for the soft-node capacitance C_x in greater detail.

9.2 Basic Principles of Pass Transistor Circuits

The fundamental building block of nMOS dynamic logic circuits, consisting of an nMOS pass transistor driving the gate of another nMOS transistor, is shown in Fig. 9.1. As already discussed in Example 9.1, the pass transistor MP is driven by the periodic clock signal and acts as an access switch to either charge up or charge down the parasitic capacitance C_x, depending on the input signal V_{in}. Thus, the two possible operations when the clock signal is active (CK = 1) are the logic "1" transfer (charging up the capacitance C_x to a logic-high level) and the logic "0" transfer (charging down the capacitance C_x to a logic-low level). In either case, the output of the depletion-load nMOS inverter obviously assumes a logic-low or a logic-high level, depending on the voltage V_x.

Notice that the pass transistor MP provides the only current path to the intermediate capacitive node (soft node) X. When the clock signal becomes inactive (CK = 0), the pass transistor ceases to conduct and the charge stored in the parasitic

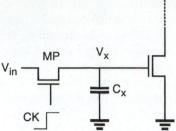

Figure 9.1 The basic building block for nMOS dynamic logic, which consists of an nMOS pass transistor driving the gate of another nMOS transistor.

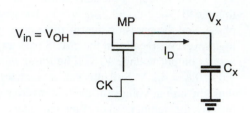

Figure 9.2 Equivalent circuit for the logic "1" transfer event.

capacitor C_x continues to determine the output level of the inverter. In the following, we will first examine the charge-up event.

Logic "1" Transfer

Assume that the soft node voltage is equal to 0 initially, i.e., $V_x(t = 0) = 0$ V. A logic "1" level is applied to the input terminal, which corresponds to $V_{in} = V_{OH} = V_{DD}$. Now, the clock signal at the gate of the pass transistor goes from 0 to V_{DD} at $t = 0$. It can be seen that the pass transistor MP starts to conduct as soon as the clock signal becomes active and that MP will operate in saturation throughout this cycle since $V_{DS} = V_{GS}$. Consequently, $V_{DS} > V_{GS} - V_{T,n}$. The circuit to be analyzed for the logic "1" transfer event can be simplified into an equivalent circuit as shown in Fig. 9.2.

The pass transistor MP operating in the saturation region starts to charge up the capacitor C_x, thus,

$$C_x \frac{dV_x}{dt} = \frac{k_n}{2}(V_{DD} - V_x - V_{T,n})^2 \tag{9.1}$$

Note that the threshold voltage of the pass transistor is actually subject to substrate bias effect and therefore, depends on the voltage level V_x. To simplify our analysis, we will neglect the substrate bias effect at this point. Integrating (9.1), we obtain

$$\int_0^t dt = \frac{2C_x}{k_n} \int_0^{V_x} \frac{dV_x}{(V_{DD} - V_x - V_{T,n})^2}$$

$$= \frac{2C_x}{k_n} \left(\frac{1}{(V_{DD} - V_x - V_{T,n})} \right) \Bigg|_0^{V_x} \tag{9.2}$$

$$t = \frac{2C_x}{k_n} \left[\left(\frac{1}{V_{DD} - V_x - V_{T,n}} \right) - \left(\frac{1}{V_{DD} - V_{T,n}} \right) \right] \tag{9.3}$$

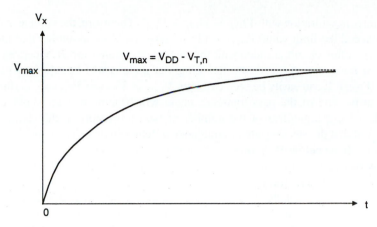

Figure 9.3 Variation of V_x as a function of time during logic "1" transfer.

This equation can be solved for $V_x(t)$, as follows.

$$V_x(t) = (V_{DD} - V_{T,n}) \frac{\left(\dfrac{k_n(V_{DD} - V_{T,n})}{2C_x}\right)t}{1 + \left(\dfrac{k_n(V_{DD} - V_{T,n})}{2C_x}\right)t} \tag{9.4}$$

The variation of the node voltage V_x according to (9.4) is plotted as a function of time in Fig. 9.3. The voltage rises from its initial value of 0 V and approaches a limit value for large t, but it cannot exceed its limit value of $V_{max} = (V_{DD} - V_{T,n})$. The pass transistor will turn off when $V_x = V_{max}$, since at this point, its gate-to-source voltage will be equal to its threshold voltage. Therefore, the voltage at node X can never attain the full power supply voltage level of V_{DD} during the logic "1" transfer. The actual value of the maximum possible voltage V_{max} at node X can be found by taking into account the substrate bias effect for MP.

$$\begin{aligned} V_{max} &= V_x|_{t \to \infty} = V_{DD} - V_{T,n} \\ &= V_{DD} - V_{T0,n} - \gamma \left(\sqrt{|2\phi_F| + V_{max}} - \sqrt{|2\phi_F|} \right) \end{aligned} \tag{9.5}$$

Thus, the voltage V_x which is obtained at node X following a logic "1" transfer can be considerably lower than V_{DD}. Also note that the rise time of the voltage V_x will be *underestimated* if the zero-bias threshold voltage V_{T0} is used in (9.3). In that case, the actual charge-up time will be longer than predicted by (9.3), because the drain current of the nMOS transistor is decreased due to the substrate bias effect.

The fact that the node voltage V_x has an upper limit of $V_{max} = (V_{DD} - V_{T,n})$ has a significant implication for circuit design. As an example, consider the following case in which a logic "1" at the input node ($V_{in} = V_{DD}$) is being transferred through a chain of cascaded pass transistors (Fig. 9.4). For simple analysis, we assume that initially all internal node voltages, V_1 through V_4, are zero. The first pass transistor

M1 operates in saturation with $V_{DS1} > V_{GS1} - V_{T,n1}$. Therefore, the voltage at node 1 cannot exceed the limit value $V_{max1} = (V_{DD} - V_{T,n1})$. Now, assuming that the pass transistors in this circuit are identical, the second pass transistor M2 operates at the *saturation boundary*. As a result, the voltage at node 2 will be equal to $V_{max2} = (V_{DD} - V_{T,n2})$. It can easily be seen that with $V_{T,n1} = V_{T,n2} = V_{T,n3} = \cdots$, the node voltage at the end of the pass transistor chain will become one threshold voltage lower than V_{DD}, regardless of the number of pass transistors in the chain. It can be observed that the steady-state internal node voltages in this circuit are always one threshold voltage below V_{DD}, regardless of the initial voltages.

Now consider a different case in which the output of each pass transistor drives the gate of another pass transistor, as depicted in Fig. 9.5.

Here, the output of the first pass transistor M1 can reach the limit $V_{max1} = (V_{DD} - V_{T,n1})$. This voltage drives the gate of the second pass transistor, which also operates in the saturation region. Its gate-to-source voltage cannot exceed $V_{Tn,2}$, hence, the upper limit for V_2 is found as $V_{max2} = V_{DD} - V_{T,n1} - V_{T,n2}$. It can be seen that in this case, each stage causes a significant loss of voltage level. The amount of voltage drop at each stage can be approximated more realistically by taking into account the corresponding substrate bias effect, which is different in all stages.

$$V_{T,n1} = V_{T0,n} - \gamma\left(\sqrt{|2\phi_F| + V_{max1}} - \sqrt{|2\phi_F|}\right)$$

$$V_{T,n2} = V_{T0,n} - \gamma\left(\sqrt{|2\phi_F| + V_{max2}} - \sqrt{|2\phi_F|}\right) \tag{9.6}$$

$$\vdots$$

The preceding analysis helped us to examine important characteristics of the logic "1" transfer event. Next, we will examine the charge-down event, which is also called a logic "0" transfer.

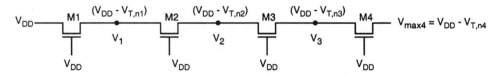

Figure 9.4 Node voltages in a pass-transistor chain during the logic "1" transfer.

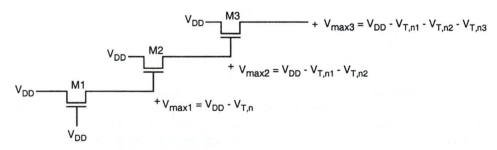

Figure 9.5 Node voltages during the logic "1" transfer, when each pass transistor is driving another pass transistor.

Logic "0" Transfer

Assume that the soft-node voltage V_x is equal to a logic "1" level initially, i.e., $V_x(t = 0) = V_{max} = (V_{DD} - V_{T,n})$. A logic "0" level is applied to the input terminal, which corresponds to $V_{in} = 0$ V. Now, the clock signal at the gate of the pass transistor goes from 0 to V_{DD} at $t = 0$. The pass transistor MP starts to conduct as soon as the clock signal becomes active, and the direction of drain current flow through MP will be opposite to that during the charge-up (logic "1" transfer) event. This means that the intermediate node X will now correspond to the drain terminal of MP and that the input node will correspond to its source terminal. With $V_{GS} = V_{DD}$ and $V_{DS} = V_{max}$, it can be seen that the pass transistor operates in the linear region throughout this cycle, since $V_{DS} < V_{GS} - V_{T,n}$.

The circuit to be analyzed for the logic "0" transfer event can be simplified into an equivalent circuit as shown in Fig. 9.6. As in the logic "1" transfer case, the depletion-load nMOS inverter does not affect this event.

The pass transistor MP operating in the linear region discharges the parasitic capacitor C_x, as follows:

$$-C_x \frac{dV_x}{dt} = \frac{k_n}{2} \left(2(V_{DD} - V_{T,n})V_x - V_x^2\right) \tag{9.7}$$

$$dt = -\frac{2C_x}{k_n} \cdot \frac{dV_x}{2(V_{DD} - V_{T,n})V_x - V_x^2} \tag{9.8}$$

Note that the source voltage of the nMOS pass transistor is equal to 0 V during this event; hence, there is no substrate bias effect for MP ($V_{T,n} = V_{T0,n}$). But the initial condition $V_x(t = 0) = (V_{DD} - V_{T,n})$ contains the threshold voltage *with* substrate bias effect, because the voltage V_x is set during the preceding logic "1" transfer event. To simplify the expressions, we will use $V_{T,n}$ in the following. Integrating both sides of (9.8) yields

$$\int_0^t dt = -\frac{2C_x}{k_n} \int_{V_{DD}-V_{T,n}}^{V_x} \left(\frac{\frac{1}{2(V_{DD} - V_{T,n})}}{2(V_{DD} - V_{T,n}) - V_x} + \frac{\frac{1}{2(V_{DD} - V_{T,n})}}{V_x} \right) dV_x \tag{9.9}$$

$$t = \frac{C_x}{k_n(V_{DD} - V_{T,n})} \left[\ln \left(\frac{2(V_{DD} - V_{T,n}) - V_x}{V_x} \right) \right] \Bigg|_{V_{DD}-V_{T,n}}^{V_x} \tag{9.10}$$

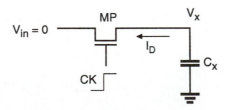

Figure 9.6 Equivalent circuit for the logic "0" transfer event.

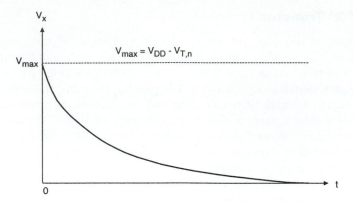

Figure 9.7 Variation of V_x as a function of time during logic "0" transfer.

Finally, the fall-time expression for the node voltage V_x can be obtained as

$$t = \frac{C_x}{k_n(V_{DD} - V_{T,n})} \ln\left(\frac{2(V_{DD} - V_{T,n}) - V_x}{V_x}\right) \tag{9.11}$$

The variation of the node voltage V_x according to (9.11) is plotted as a function of time in Fig. 9.7. It is seen that the voltage drops from its logic-high level of V_{max} to 0 V. Hence, unlike the charge-up case, the applied input voltage level (logic 0) can be transferred to the soft node without any modification during this event.

The fall time (τ_{fall}) for the soft-node voltage V_x can be calculated from (9.11) as follows. First, define the two time points $t_{90\%}$ and $t_{10\%}$ as the times at which the node voltage is equal to 0.9 V_{max} and 0.1 V_{max}, respectively. These two time points can easily be found by using (9.11).

$$t_{90\%} = \frac{C_x}{k_n(V_{DD} - V_{T,n})} \ln\left(\frac{(2 - 0.9)(V_{DD} - V_{T,n})}{0.9(V_{DD} - V_{T,n})}\right)$$

$$= \frac{C_x}{k_n(V_{DD} - V_{T,n})} \ln\left(\frac{1.1}{0.9}\right) \tag{9.12}$$

$$t_{10\%} = \frac{C_x}{k_n(V_{DD} - V_{T,n})} \ln\left(\frac{1.9}{0.1}\right) \tag{9.13}$$

The fall time of the soft-node voltage V_x is by definition the difference between $t_{10\%}$ and $t_{90\%}$, which is found as

$$\tau_{fall} = t_{10\%} - t_{90\%}$$

$$= \frac{C_x}{k_n(V_{DD} - V_{T,n})}[\ln(19) - \ln(1.22)]$$

$$= 2.74\frac{C_x}{k_n(V_{DD} - V_{T,n})} \tag{9.14}$$

Until this point, we have examined the transient charge-up and charge-down events which are responsible for logic "1" transfer and logic "0" transfer during the active clock phase, i.e., when CK = 1. Now we will turn our attention to the storage of logic levels at the soft node X during the inactive clock cycle, i.e., when CK = 0.

Charge Storage and Charge Leakage

As already discussed qualitatively in the preceding section, the preservation of a correct logic level at the soft node during the inactive clock phase depends on pre-serving a sufficient amount of charge in C_x, despite the leakage currents. To analyze the events during the inactive clock phase in more detail, consider the scenario shown in Fig. 9.8 below. We will assume that a logic-high voltage level has been transferred to the soft node during the active clock phase and that now both the input voltage V_{in} and the clock are equal to 0 V. The charge stored in C_x will gradually leak away, primarily due to the leakage currents associated with the pass transistor. The gate current of the inverter driver transistor is negligible for all practical purposes.

Figure 9.9 shows a simplified cross-section of the nMOS pass transistor, together with the lumped node capacitance C_x. We see that the leakage current responsible for

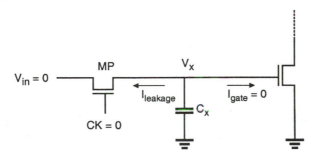

Figure 9.8 Charge leakage from the soft node.

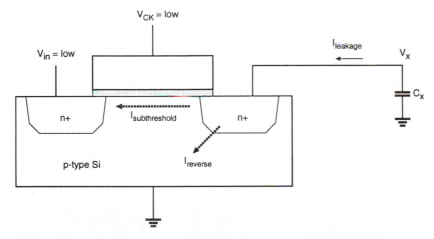

Figure 9.9 Simplified cross-section of the nMOS pass transistor, showing the leakage current components responsible for draining the soft-node capacitance C_x.

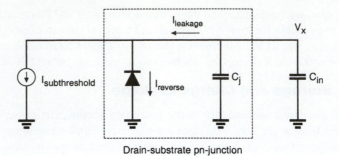

Figure 9.10 Equivalent circuit used for analyzing the charge leakage process.

draining the soft-node capacitance over time has two main components, namely, the subthreshold channel current and the reverse conduction current of the drain-substrate junction.

$$I_{leakage} = I_{subthreshold(MP)} + I_{reverse(MP)} \qquad (9.15)$$

Note that a certain portion of the total soft-node capacitance C_x is due to the reverse biased drain-substrate junction, which is also a function of the soft-node voltage V_x. Other components of C_x, which are primarily due to oxide-related parasitics, can be considered constants. In our analysis, these constant capacitance components will be represented by C_{in} (Fig. 9.10). Thus, we have to express the total charge stored in the soft node as the sum of two main components, as follows.

$$Q = Q_j(V_x) + Q_{in} \qquad \text{where} \qquad Q_{in} = C_{in} \cdot V_x$$
$$C_{in} = C_{gb} + C_{poly} + C_{metal} \qquad (9.16)$$

The total leakage current can be expressed as the time derivative of the total soft-node charge Q.

$$
\begin{aligned}
I_{leakage} &= \frac{dQ}{dt} \\[2mm]
&= \frac{dQ_j(V_x)}{dt} + \frac{dQ_{in}}{dt} \\[2mm]
&= \frac{dQ_j(V_x)}{dV_x}\frac{dV_x}{dt} + C_{in}\frac{dV_x}{dt} \qquad (9.17)
\end{aligned}
$$

where

$$\frac{dQ_j(V_x)}{dV_x} = C_j(V_x) = \frac{A \cdot C_{j0}}{\sqrt{1 + \dfrac{V_x}{\phi_0}}} = A \cdot \sqrt{\frac{q\varepsilon_{Si}N_A}{2(\phi_0 + V_x)}} \qquad (9.18)$$

according to (3.104), and

$$\phi_0 = \frac{kT}{q} \ln \left(\frac{N_D \cdot N_A}{n_i^2} \right) \tag{9.19}$$

$$C_{j0} = \sqrt{\frac{q\varepsilon_{Si} N_A N_D}{2(N_A + N_D)\phi_0}} \approx \sqrt{\frac{q\varepsilon_{Si} N_A}{2\phi_0}} \tag{9.20}$$

Also note that the subthreshold current in deep-submicron transistors can significantly exceed the reverse conduction current, especially for $V_{DS} = V_{DD}$. In long-channel transistors, the magnitude of the subthreshold current can be comparable to that of the reverse leakage current. The reverse conduction current in turn has two main components, the constant reverse saturation current I_0, and the generation current I_{gen} which originates in the depletion region and is a function of the applied bias voltage V_x.

To estimate the actual charge leakage time from the soft node, we have to solve the differential equation given in (9.17), taking into account the voltage-dependent capacitance components and the nonlinear leakage currents. For a quick estimate of the worst-case leakage behavior, on the other hand, the problem can be further simplified.

Assume that the *minimum* combined soft-node capacitance is given as

$$C_{x,min} = C_{gb} + C_{poly} + C_{metal} + C_{db,min} \tag{9.21}$$

where $C_{db,min}$ represents the minimum junction capacitance, obtained under the bias condition $V_x = V_{max}$. Now we define the *worst-case holding time* (t_{hold}) as the shortest time required for the soft-node voltage to drop from its initial logic-high value to the logic threshold voltage due to leakage. Once the soft-node voltage reaches the logic threshold, the logic stage being driven by this node will lose its previously held state.

$$t_{hold} = \frac{\Delta Q_{critical,min}}{I_{leakage,max}} \tag{9.22}$$

where

$$\Delta Q_{critical,min} = C_{x,min} \left(V_{max} - \frac{V_{DD}}{2} \right) \tag{9.23}$$

The calculation of the worst-case leakage time can be simplified with this approximation, as will be shown in the following example.

EXAMPLE 9.2

Consider the soft-node structure shown on the next page, which consists of the drain (or source, depending on current direction) terminal of the pass transistor, connected to the polysilicon gate of an nMOS driver transistor via a metal interconnect.

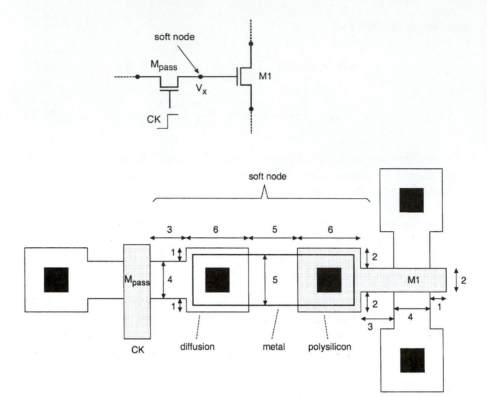

We will assume that the power supply voltage used in this circuit is $V_{DD} = 5$ V, and that the soft node has initially been charged up to its maximum voltage, V_{max}. In order to estimate the worst-case holding time, the total soft-node capacitance must be calculated first. The simplified mask layout of the structure is shown in the following. All dimensions are given in micrometers. The critical material parameters to be used in this example are listed below.

$$V_{T0} = 0.8 \text{ V}$$

$$\gamma = 0.4 \text{ V}^{1/2}$$

$$|2\phi_F| = 0.6 \text{ V}$$

$$C_{ox} = 0.065 \text{ fF/}\mu\text{m}^2$$

$$C'_{metal} = 0.036 \text{ fF/}\mu\text{m}^2$$

$$C'_{poly} = 0.055 \text{ fF/}\mu\text{m}^2$$

$$C_{j0} = 0.095 \text{ fF/}\mu\text{m}^2$$

$$C_{j0sw} = 0.2 \text{ fF/}\mu\text{m}$$

First, we calculate the oxide-related (constant) parasitic capacitance components associated with the soft node.

$$C_{gb} = C_{ox} \cdot W \cdot L_{mask}$$
$$= 0.065 \text{ fF}/\mu m^2 \cdot (4 \ \mu m \times 2 \ \mu m)$$
$$= 0.52 \text{ fF}$$
$$C_{metal} = 0.036 \text{ fF}/\mu m^2 \cdot (5 \ \mu m \times 5 \ \mu m)$$
$$= 0.90 \text{ fF}$$
$$C_{poly} = 0.055 \text{ fF}/\mu m^2 \cdot (36 \ \mu m^2 + 8 \ \mu m^2)$$
$$= 2.42 \text{ fF}$$

Now, we have to calculate the parasitic junction capacitance associated with the drain-substrate pn-junction of the pass transistor. Using the zero-bias unit capacitance values given here, we obtain

$$C_{db,max} = C_{bottom} + C_{sidewall}$$
$$= A_{bottom} \cdot C_{j0} + P_{sidewall} \cdot C_{j0sw}$$
$$= (36 \ \mu m^2 + 12 \ \mu m^2) \cdot 0.095 \text{ fF}/\mu m^2 + 30 \ \mu m \cdot 0.2 \text{ fF}/\mu m$$
$$= 4.56 \text{ fF} + 6.0 \text{ fF}$$
$$= 10.56 \text{ fF}$$

The minimum value of the drain junction capacitance is achieved when the junction is biased (in reverse) with its maximum possible voltage, V_{max}. In order to calculate the minimum capacitance value, we first find V_{max} using (9.5), as follows.

$$V_{max} = 5.0 - 0.8 - 0.4 \left(\sqrt{0.6 + V_{max}} - \sqrt{0.6} \right)$$
$$\Rightarrow V_{max} = 3.68 \text{ V}$$

Now, the minimum value of the drain junction capacitance can be calculated.

$$C_{db,min} = \frac{C_{bottom}}{\sqrt{1 + \dfrac{V_{x,max}}{\phi_0}}} + \frac{C_{sidewall}}{\sqrt{1 + \dfrac{V_{x,max}}{\phi_{0sw}}}}$$

$$= \frac{4.56 \text{ fF}}{\sqrt{1 + \dfrac{3.68}{0.88}}} + \frac{6.0 \text{ fF}}{\sqrt{1 + \dfrac{3.68}{0.95}}} = 4.71 \text{ fF}$$

The minimum value of the total soft-node capacitance is found by using (9.21).

$$C_{x,min} = C_{gb} + C_{metal} + C_{poly} + C_{db,min}$$
$$= 0.52 \text{ fF} + 0.90 \text{ fF} + 2.42 \text{ fF} + 4.71 \text{ fF}$$
$$= 8.55 \text{ fF}$$

The amount of the critical charge drop in the soft node, which will eventually cause a change of logic state, is

$$\Delta Q_{critical} = C_{x,min} \cdot \left(V_{x,max} - \frac{V_{DD}}{2} \right)$$
$$= 8.55 \text{ fF} \cdot (3.68 \text{ V} - 2.5 \text{ V})$$
$$= 10.09 \text{ fC}$$

assuming that the logic threshold voltage of the next gate is $(V_{DD}/2)$. In this example, the maximum leakage current responsible for charge depletion is given from the MOS characteristics (cf. equation (3.92) in Chapter 3) and the junction diode characteristics as

$$I_{leakage} = I_{subthreshold} + I_{reverse} = 0.85 \text{ pA}$$

Finally, we calculate the worst-case (minimum) hold time for the soft node using the expression (9.22).

$$t_{hold,min} = \frac{\Delta Q_{critical}}{I_{leakage,max}}$$
$$= \frac{10.09 \text{ fC}}{0.85 \text{ pA}} = \underline{11.87 \text{ ms}}$$

It is interesting to note that even with a very small soft-node capacitance of 8.55 fF, the worst-case hold time for this structure is relatively long, especially compared with the signal propagation delays encountered in nMOS or CMOS logic gates. This example proves the feasibility of the dynamic charge storage concept and shows that a logic state can be preserved in a soft node for a long time period when the leakage current is small.

9.3 Voltage Bootstrapping

In this section, we will briefly examine a very useful dynamic circuit technique for overcoming threshold voltage drops in digital circuits, which is called *voltage bootstrapping*. We have already seen that output voltage levels may suffer from threshold voltage drops in several circuit structures, such as pass transistor gates or enhancement-load inverters and logic gates.

Dynamic voltage bootstrapping techniques offer a simple yet effective way to overcome threshold voltage drops which occur in most situations. Consider the circuit shown in Fig. 9.11, where the voltage V_x is equal to or smaller than the power supply voltage, $V_x \leq V_{DD}$. Consequently, the enhancement-type nMOS transistor M2 will operate in saturation.

When the input voltage V_{in} is low, the maximum value that the output voltage can attain is limited by

$$V_{out}(max) = V_x - V_{T2}(V_{out}) \tag{9.24}$$

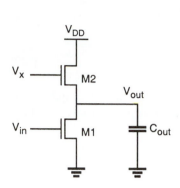

Figure 9.11 Enhancement-type circuit in which the output node is weakly driven.

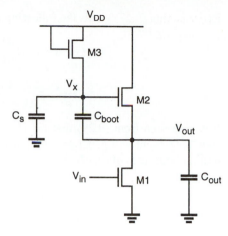

Figure 9.12 Dynamic bootstrapping arrangement to boost V_x during switching.

To overcome the threshold voltage drop and to obtain a full logic-high level (V_{DD}) at the output node, the voltage V_x must be increased. Now consider the circuit shown in Fig. 9.12, where a third transistor M3 has been added to the circuit. The two capacitors C_S and C_{boot} seen in the circuit diagram represent the capacitances which dynamically couple the voltage V_x to the ground and to the output, respectively. We will see that this circuit can produce a high V_x during switching, so that the threshold voltage drop can be overcome at the output node.

$$V_x \geq V_{DD} + V_{T2}(V_{out}) \tag{9.25}$$

Initially assume that the input voltage V_{in} is logic-high, so that M1 and M2 have a nonzero drain current and that the output voltage is low. At this point, M1 is in the linear region and M2 is in saturation. Since $I_{D3} = 0$, the initial condition for the voltage V_x can be found as

$$V_x = V_{DD} - V_{T3}(V_x) \tag{9.26}$$

Now, assume that the input switches from its logic-high level to 0 V at $t = 0$. As a result, the driver transistor M1 will turn off and the output voltage V_{out} will start to rise. This change in the output voltage level will now be coupled to V_x through the bootstrap capacitor, C_{boot}. Let i_{Cboot} represent the transient current flowing through the capacitor C_{boot} during this charge-up event, and let i_{Cs} be the current through C_S. Assuming that the two current components are approximately equal, we obtain

$$i_{Cs} \approx i_{Cboot} \Leftrightarrow C_S \frac{dV_x}{dt} \approx C_{boot} \frac{d(V_{out} - V_x)}{dt} \tag{9.27}$$

Reorganizing (9.27) yields the following equation.

$$(C_S + C_{boot})\frac{dV_x}{dt} \approx C_{boot}\frac{dV_{out}}{dt} \tag{9.28}$$

$$\frac{dV_x}{dt} \approx \frac{C_{boot}}{(C_S + C_{boot})} \cdot \frac{dV_{out}}{dt} \tag{9.29}$$

It can be seen from (9.29) that the increase in the output voltage V_{out} during this switching event will generate a proportional increase in the voltage level V_x. Integrating both sides of (9.29), we obtain

$$\int_{V_{DD}-V_{T3}}^{V_x} dV_x = \frac{C_{boot}}{(C_S + C_{boot})} \cdot \int_{V_{OL}}^{V_{DD}} dV_{out} \tag{9.30}$$

$$V_x = (V_{DD} - V_{T3}) + \frac{C_{boot}}{(C_S + C_{boot})}(V_{DD} - V_{OL}) \tag{9.31}$$

If the capacitor C_{boot} is much larger than C_S ($C_{boot} \gg C_S$), the maximum value of V_x can be approximated as

$$V_x(max) = 2V_{DD} - V_{T3} - V_{OL} \tag{9.32}$$

which proves that voltage bootstrapping can significantly boost the voltage level V_x. Now remember that in order to overcome the threshold voltage drop at the output, the minimum required voltage level V_x is

$$V_x(min) = V_{DD} + V_{T2}|_{V_{out}=V_{DD}}$$

$$= (V_{DD} - V_{T3}(V_x)) + \frac{C_{boot}}{(C_S + C_{boot})}(V_{DD} - V_{OL}) \tag{9.33}$$

This equation can be rearranged to give the required capacitance ratio, as follows.

$$\frac{C_{boot}}{(C_S + C_{boot})} = \frac{V_{T2}|_{V_{out}=V_{DD}} + V_{T3}|_{V_x}}{(V_{DD} - V_{OL})} \tag{9.34}$$

$$\frac{C_{boot}}{C_S} = \frac{V_{T2}|_{V_{out}=V_{DD}} + V_{T3}|_{V_x}}{V_{DD} - V_{OL} - V_{T2}|_{V_{out}=V_{DD}} - V_{T3}|_{V_x}} \tag{9.35}$$

Note that C_S is essentially the sum of the parasitic source-to-substrate capacitance of M3 and the gate-to-substrate capacitance of M2. To obtain a sufficiently large bootstrap capacitance C_{boot} in comparison to C_S, an extra "dummy" transistor is typically added to the circuit, as shown in Fig. 9.13.

Since its drain and source terminals are connected together, the dummy transistor simply acts as an MOS capacitor between V_x and V_{out}. Although this circuit arrangement contains two additional transistors to achieve voltage bootstrapping, the resulting circuit-performance improvement is usually well worth the extra silicon area used for the bootstrapping devices.

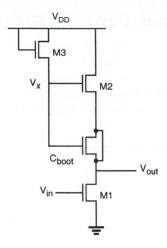

Figure 9.13 Realization of the bootstrapping capacitor with a dummy MOS device.

EXAMPLE 9.3

The transient operation of the simple bootstrap circuit shown in Fig. 9.13 is simulated using SPICE in the following. To provide the needed bootstrap capacitance C_{boot}, a dummy nMOS device with channel length $L = 5 \ \mu$m and channel width $W = 50 \ \mu$m is used. Transistor M1 has a (W/L) ratio of 2, while M2 and M3 each have a (W/L) ratio of 1.

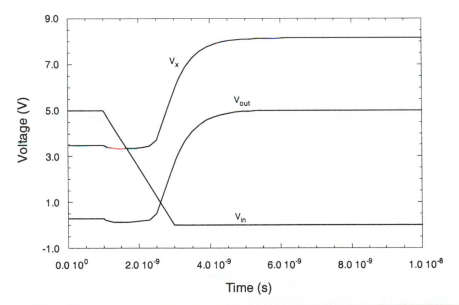

9.4 Synchronous Dynamic Circuit Techniques

Having examined the basic concepts associated with temporary storage of logic levels in capacitive circuit nodes, we now turn our attention to digital circuit design techniques which take advantage of this simple yet effective principle. In the following, we will investigate different examples of synchronous dynamic circuits implemented using depletion-load nMOS, enhancement-load nMOS, and CMOS building blocks.

Dynamic Pass Transistor Circuits

Consider the generalized view of a multi-stage synchronous circuit shown in Fig. 9.14. The circuit consists of cascaded combinational logic stages, which are interconnected through nMOS pass transistors. All inputs of each combinational logic block are driven by a single clock signal. Individual input capacitances are not shown in this figure for simplicity, but the operation of the circuit obviously depends on temporary charge storage in the parasitic input capacitances.

To drive the pass transistors in this system, two nonoverlapping clock signals, ϕ_1 and ϕ_2, are used. The nonoverlapping property of the two clock signals guarantees that at any given time point, only one of the two clock signals can be active, as illustrated in Fig. 9.15. When clock ϕ_1 is active, the input levels of Stage 1 (and also of Stage 3) are applied through the pass transistors, while the input capacitances of Stage 2 retain their previously set logic levels. During the next phase, when clock ϕ_2 is active, the input levels of Stage 2 will be applied through the pass transistors, while the input capacitances of Stage 1 and Stage 3 retain their logic levels. This allows us to incorporate the simple dynamic memory function at each stage input, and at the same time, to facilitate synchronous operation by controlling the signal flow in the circuit using the two periodic clock signals. This signal timing scheme is also called *two-phase clocking* and is one of the most widely used timing strategies.

By introducing the two-phase clocking scheme, we have not made any specific assumptions about the internal structure of the combinational logic stages. It will be seen that depletion-load nMOS, enhancement-load nMOS, or CMOS logic circuits

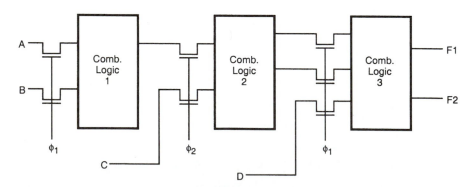

Figure 9.14 Multi-stage pass transistor logic driven by two nonoverlapping clocks.

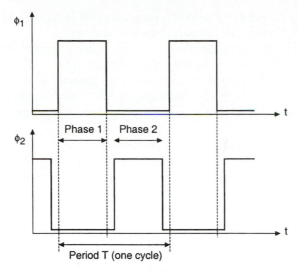

Figure 9.15 Nonoverlapping clock signals used for two-phase synchronous operation.

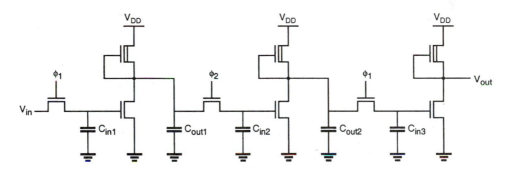

Figure 9.16 Three stages of a depletion-load nMOS dynamic shift register circuit driven with two-phase clocking.

can be used for implementing the combinational logic. Figure 9.16 shows a depletion-load dynamic shift register circuit, in which the input data are inverted once and transferred, or *shifted* into the next stage during each clock phase.

The operation of the shift register circuit is as follows. During the active phase of ϕ_1, the input voltage level V_{in} is transferred into the input capacitance C_{in1}. Thus, the valid output voltage level of the first stage is determined as the inverse of the current input during this cycle. When ϕ_2 becomes active during the next phase, the output voltage level of the first stage is transferred into the second stage input capacitance C_{in2}, and the valid output voltage level of the second stage is determined. During the active ϕ_2 phase, the first-stage input capacitance continues to retain its previous level via charge storage. When ϕ_1 becomes active again, the original data bit *written* into the register during the previous cycle is transferred into the third stage, and the first stage can now accept the next data bit.

In this circuit, the maximum clock frequency is determined by the signal propagation delay through one inverter stage. One half-period of the clock signal must be long enough to allow the input capacitance C_{in} to charge up or down, and the logic level to propagate to the output by charging C_{out}. Also notice that the logic-high input level of each inverter stage in this circuit is one threshold voltage lower than the power supply voltage level.

The same operation principle used in the simple shift register circuit can easily be extended to synchronous complex logic. Figures 9.17 and 9.18 show a two-stage circuit example implemented using depletion-load nMOS complex logic gates. In

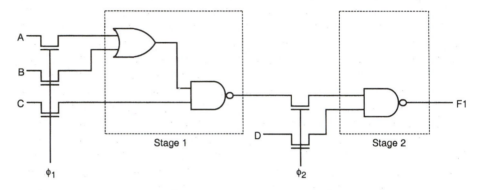

Figure 9.17   A two-stage synchronous complex logic circuit example.

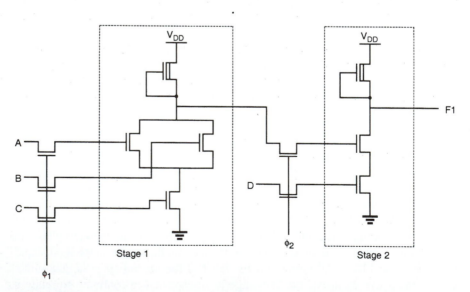

Figure 9.18   Depletion-load nMOS implementation of synchronous complex logic.

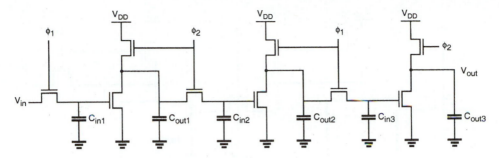

Figure 9.19   Enhancement-load dynamic shift register (ratioed logic).

a complex logic circuit such as the one shown in Fig. 9.18, we see that the signal propagation delay of each stage may be different. Thus, in order to guarantee that correct logic levels are propagated during each active clock cycle, the half-period length of the clock signal must be longer than the largest single-stage signal propagation delay found in the circuit.

Now consider a different implementation of the simple shift register circuit, using enhancement-load nMOS inverters. One important difference is that, instead of biasing the load transistors with a constant gate voltage, we apply the clock signal to the gate of the load transistor as well. It can be shown that the power dissipation and the silicon area can be reduced significantly by using this dynamic (clocked) load approach. Two variants of the dynamic enhancement-load shift register will be examined in the following, both of which are driven by two non-overlapping clock signals. Figure 9.19 shows the first implementation, where in each stage the input pass transistor and the load transistor are driven by opposite clock phases, ϕ_1 and ϕ_2.

When ϕ_1 is active, the input voltage level V_{in} is transferred into the first-stage input capacitance C_{in1} through the pass transistor. In this phase, the enhancement-type nMOS load transistor of the first-stage inverter is not active yet. During the next phase (active ϕ_2), the load transistor is turned on. Since the input logic level is still being preserved in C_{in1}, the output of the first inverter stage attains its valid logic level. At the same time, the input pass transistor of the second stage is also turned on, which allows this newly determined output level to be transferred into the input capacitance C_{in2} of the second stage. When clock ϕ_1 becomes active again, the valid output level across C_{out2} is determined, and transferred into C_{in3}. Also, a new input level can be accepted (*pipelined*) into C_{in1} during this phase.

In this circuit, the valid low-output voltage level V_{OL} of each stage is strictly determined by the driver-to-load ratio, since the output pass transistor (input pass transistor of next stage) turns on in phase with the load transistor. Therefore, this circuit arrangement is also called *ratioed dynamic logic*. The basic operation principle can obviously be extended to arbitrary complex logic, as shown in Fig. 9.20. Since the power supply current flows only when the load devices are activated by the clock signal, the overall power consumption of dynamic enhancement-load logic is generally lower than for depletion-load nMOS logic.

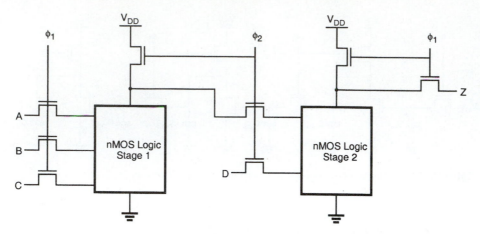

Figure 9.20 General circuit structure of ratioed synchronous dynamic logic.

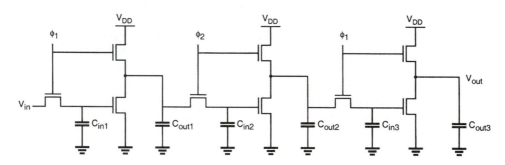

Figure 9.21 Enhancement-load dynamic shift register (ratioless logic).

Next, consider the second dynamic enhancement-load shift register implementation where, in each stage, the input pass transistor and the load transistor are driven by the same clock phase (Fig. 9.21).

When ϕ_1 is active, the input voltage level V_{in} is transferred into the first-stage input capacitance C_{in1} through the pass transistor. Note that at the same time, the enhancement-type nMOS load transistor of the first-stage inverter is active. Therefore, the output of the first inverter stage attains its valid logic level. During the next phase (active ϕ_2), the input pass transistor of the next stage is turned on, and the logic level is transferred onto the next stage. Here, we have to consider two cases, as follows.

If the output level across C_{out1} is logic-high at the end of the active ϕ_1 phase, this voltage level is transferred to C_{in2} via charge sharing over the pass transistor during the active ϕ_2 phase. Note that the logic-high level at the output node is subject to threshold voltage drop, i.e., it is one threshold voltage lower than the power supply voltage. To correctly transfer a logic-high level after charge sharing, the ratio of the capacitors (C_{out}/C_{in}) must be made large enough during circuit design.

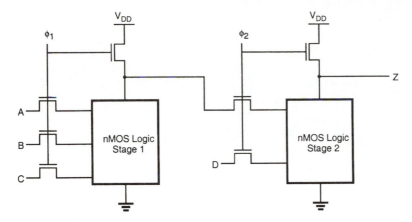

Figure 9.22 General circuit structure of ratioless synchronous dynamic logic.

If, on the other hand, the output level of the first stage is logic-low at the end of the active ϕ_1 phase, then the output capacitor C_{out1} will be completely drained to a voltage of $V_{OL} = 0$ V when ϕ_1 turns off. This can be achieved because a logic-high level is being stored in the input capacitance C_{in1} in this case, which forces the driver transistor to remain in conduction. Obviously, the logic-low level of $V_{OL} = 0$ V is also transferred into the next stage via the pass transistor during the active ϕ_2 phase.

When clock ϕ_1 becomes active again, the valid output level across C_{out2} is determined and transferred into C_{in3}. Also, a new input level can be accepted into C_{in1} during this phase. Since the valid logic-low level of $V_{OL} = 0$ V can be achieved regardless of the driver-to-load ratio, this circuit arrangement is called *ratioless dynamic logic*. The basic operation principle can be extended to arbitrary complex logic, as shown in Fig. 9.22.

9.5 Dynamic CMOS Circuit Techniques

CMOS Transmission Gate Logic

The basic two-phase synchronous logic circuit principle, in which individual logic blocks are cascaded via clock-controlled switches, can easily be adopted to CMOS structures as well. Here, static CMOS gates are used for implementing the logic blocks, and CMOS transmission gates are used for transferring the output levels of one stage to the inputs of the next stage (Fig. 9.23). Notice that each transmission gate is actually controlled by the clock signal *and* its complement. As a result, two-phase clocking in CMOS transmission gate logic requires that a total of four clock signals are generated and routed throughout the circuit.

As in the nMOS-based dynamic circuit structures, the operation of CMOS dynamic logic relies on charge storage in the parasitic input capacitances during the inactive clock cycles. To illustrate the basic operation principles, the fundamental

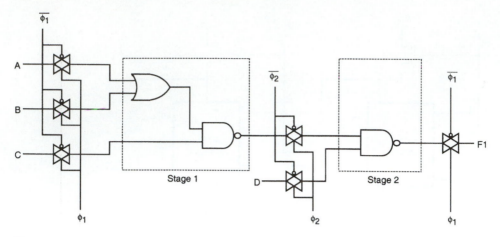

Figure 9.23 Typical example of dynamic CMOS transmission gate logic.

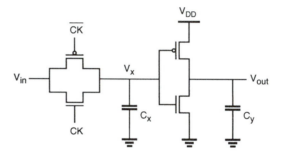

Figure 9.24   Basic building block of a CMOS transmission gate dynamic shift register.

building block of a dynamic CMOS transmission gate shift register is shown in Fig. 9.24. It consists of a CMOS inverter, which is driven by a CMOS transmission gate. During the active clock phase (CK = 1), the input voltage V_{in} is transferred onto the parasitic input capacitance C_x via the transmission gate. Note that the low on-resistance of the CMOS transmission gate usually results in a smaller transfer time compared to those for nMOS-only switches. Also, there is no threshold voltage drop across the CMOS transmission gate. When the clock signal becomes inactive, the CMOS transmission gate turns off and the voltage level across C_x can be preserved until the next cycle.

Figure 9.25 shows a single-phase CMOS shift register, which is built by cascading identical units as in Fig. 9.24 and by driving each stage alternately with the clock signal and its complement.

Ideally, the transmission gates of the odd-numbered stages would conduct during the active clock phase (when CK = 1), while the transmission gates of the even-numbered stages are off, so that the cascaded inverter stages in the chain are alternately isolated. This would ensure that inputs are permitted in alternating half

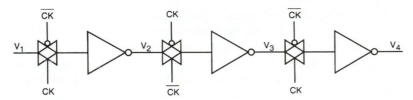

Figure 9.25 Single-phase CMOS transmission gate dynamic shift register.

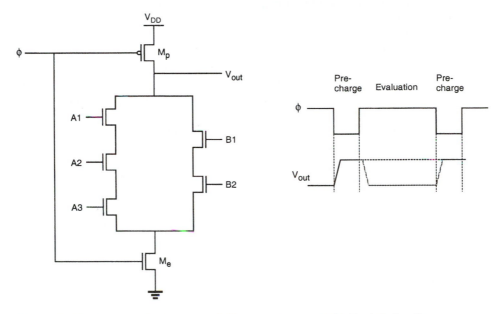

Figure 9.26 Dynamic CMOS logic gate implementing a complex Boolean function.

cycles. In practice, however, the clock signal and its complement do not constitute a truly nonoverlapping signal pair, since the clock voltage waveform has finite rise and fall times. Also, the clock skew between CK and $\overline{CK}$ may be unavoidable because one of the signals is generated by inverting the other. Therefore, true two-phase clocking with two nonoverlapping clock signals (ϕ_1 and ϕ_2) *and* their complements is usually preferred over single-phase clocking in dynamic CMOS transmission gate logic.

Dynamic CMOS Logic (Precharge-Evaluate Logic)

In the following, we will introduce a dynamic CMOS circuit technique which allows us to significantly reduce the number of transistors used to implement any logic function. The circuit operation is based on first *precharging* the output node capacitance and subsequently, *evaluating* the output level according to the applied inputs. Both of these operations are scheduled by a single clock signal, which drives one nMOS and one pMOS transistor in each dynamic stage. A dynamic CMOS logic gate which implements the function $F = (\overline{A_1 A_2 A_3 + B_1 B_2})$ is shown in Fig. 9.26.

When the clock signal is low (precharge phase), the pMOS precharge transistor M_p is conducting, while the complementary nMOS transistor M_e is off. The parasitic output capacitance of the circuit is charged up through the conducting pMOS transistor to a logic-high level of $V_{out} = V_{DD}$. The input voltages are also applied during this phase, but they have no influence yet upon the output level since M_e is turned off.

When the clock signal becomes high (evaluate phase), the precharge transistor M_p turns off and M_e turns on. The output node voltage may now remain at the logic-high level or drop to a logic low, depending on the input voltage levels. If the input signals create a conducting path between the output node and the ground, the output capacitance will discharge toward $V_{OL} = 0\,V$. The final discharged output level depends on the time span of the evaluation phase. Otherwise, V_{out} remains at V_{DD}.

The operation of the single-stage dynamic CMOS logic gate is quite straightforward. For practical multi-stage applications, however, the dynamic CMOS gate presents a significant problem. To examine this fundamental limitation, consider the two-stage cascaded structure shown in Fig. 9.27. Here, the output of the first dynamic CMOS stage drives one of the inputs of the second dynamic CMOS stage, which is assumed to be a two-input NAND gate for simplicity.

During the precharge phase, both output voltages V_{out1} and V_{out2} are pulled up by the respective pMOS precharge devices. Also, the external inputs are applied during this phase. The input variables of the first stage are assumed to be such that the output V_{out1} will drop to logic "0" during the evaluation phase. On the other hand, the external input of the second-stage NAND2 gate is assumed to be a logic "1," as shown in Fig. 9.27. When the evaluation phase begins, both output voltages V_{out1} and V_{out2} are logic-high. The output of the first stage (V_{out1}) eventually drops to its correct logic level after a certain time delay. However, since the evaluation in the second stage is done concurrently, starting with the high value of V_{out1} at the beginning of

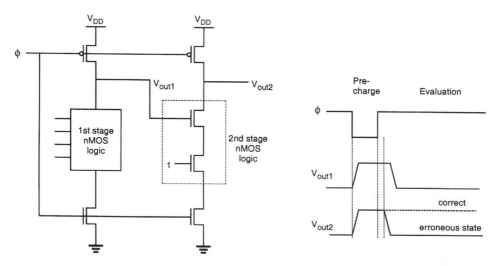

Figure 9.27   Illustration of the cascading problem in dynamic CMOS logic.

the evaluation phase, the output voltage V_{out2} at the end of the evaluation phase will be *erroneously* low. Although the first stage output subsequently assumes its correct output value once the stored charge is drained, the correction of the second-stage output is not possible.

This example illustrates that dynamic CMOS logic stages driven by the same clock signal cannot be cascaded directly. This severe limitation seems to undermine all the other advantages of dynamic CMOS logic, such as low power dissipation, large noise margins, and low transistor count. Alternative clocking schemes and circuit structures must be developed to overcome this problem. In fact, the search for viable circuit alternatives has spawned a large array of high-performance dynamic CMOS circuit techniques, some of which will be examined in the following section.

9.6 High-Performance Dynamic CMOS Circuits

The circuits presented here are variants of the basic dynamic CMOS logic gate structure. We will see that they are designed to take full advantage of the obvious benefits of dynamic operation and at the same time, to allow unrestricted cascading of multiple stages. The ultimate goal is to achieve reliable, high-speed, compact circuits using the least complicated clocking scheme possible.

Domino CMOS Logic

Consider the generalized circuit diagram of a domino CMOS logic gate shown in Fig. 9.28. A dynamic CMOS logic stage, such as the one shown in Fig. 9.26, is

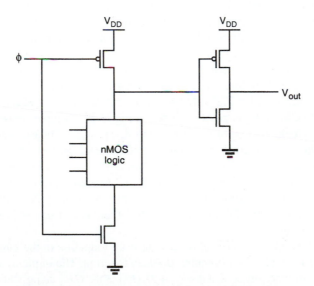

Figure 9.28 Generalized circuit diagram of a domino CMOS logic gate.

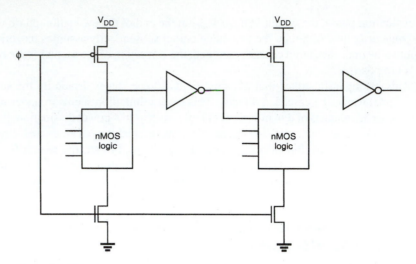

Figure 9.29   Cascaded domino CMOS logic gates.

cascaded with a static CMOS inverter stage. The addition of the inverter allows us to operate a number of such structures in cascade, as explained in the following.

During the precharge phase (when CK = 0), the output node of the dynamic CMOS stage is precharged to a high logic level, and the output of the CMOS inverter (buffer) becomes low. When the clock signal rises at the beginning of the evaluation phase, there are two possibilities: The output node of the dynamic CMOS stage is either discharged to a low level through the nMOS circuitry (1 to 0 transition), or it remains high. Consequently, the inverter output voltage can also make at most one transition during the evaluation phase, from 0 to 1. Regardless of the input voltages applied to the dynamic CMOS stage, it is not possible for the buffer output to make a 1 to 0 transition during the evaluation phase.

Remember that the problem in cascading conventional dynamic CMOS stages occurs when one or more inputs of a stage make a 1 to 0 transition *during* the evaluation phase, as illustrated in Fig. 9.27. On the other hand, if we build a system by cascading domino CMOS logic gates as shown in Fig. 9.29, all input transistors in subsequent logic blocks will be turned off during the precharge phase, since all buffer outputs are equal to 0. During the evaluation phase, each buffer output can make at most one transition (from 0 to 1), and thus each input of all subsequent logic stages can also make at most one (0 to 1) transition. In a cascade structure consisting of several such stages, the evaluation of each stage ripples the next stage evaluation, similar to a chain of dominos falling one after the other. The structure is hence called *domino CMOS logic*.

Domino CMOS logic gates allow a significant reduction in the number of transistors required to realize any complex Boolean function. The implementation of the 8-input Boolean function, $Z = AB + (C + D)(E + F) + GH$, using standard CMOS and domino CMOS, is shown in Fig. 9.30, where the reduction of circuit complexity is obvious. The distribution of the clock signal within the system is quite straightfor-

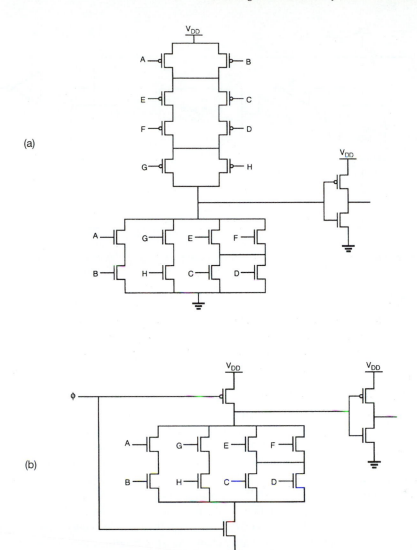

Figure 9.30 (a) An 8-input complex logic gate, realized using conventional CMOS logic and (b) domino CMOS logic.

ward, since a single clock can be used to precharge and evaluate any number of cascaded stages, as long as the signal propagation delay from the first stage to the last stage does not exceed the time span of the evaluation phase. Also, conventional static CMOS logic gates can be used together with domino CMOS gates in a cascaded configuration (Fig. 9.31). The limitation is that the number of inverting static logic stages in cascade must be even, so that the inputs of the next domino CMOS stage experience only 0 to 1 transitions during the evaluation.

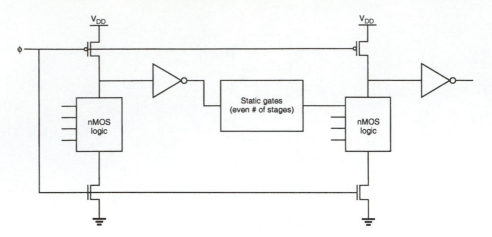

Figure 9.31   Cascading domino CMOS logic gates with static CMOS logic gates.

There are also some other limitations associated with domino CMOS logic gates. First, only non-inverting structures can be implemented using domino CMOS. If necessary, inversion must be carried out using conventional CMOS logic. Also, charge sharing between the dynamic stage output node and the intermediate nodes of the nMOS logic block during the evaluation phase may cause erroneous outputs, as will be explained in the following.

Consider the domino CMOS logic gate shown in Fig. 9.32, in which the intermediate node capacitance C_2 is comparable in size to the output node capacitance C_1. We will assume that all inputs are low initially, and that the intermediate node voltage across C_2 has an initial value of 0 V. During the precharge phase, the output node capacitance C_1 is charged up to its logic-high level of V_{DD} through the pMOS transistor. In the next phase, the clock signal becomes high and the evaluation begins. If the input signal of the uppermost nMOS transistor switches from low to high during this evaluation phase, as shown in Fig. 9.32, the charge initially stored in the output capacitance C_1 will now be shared by C_2, leading to the so-called *charge-sharing* phenomenon. The output node voltage after charge sharing becomes $V_{DD}/(1 + C_2/C_1)$. For example, if $C_1 = C_2$, the output voltage becomes $V_{DD}/2$ in the evaluation phase. Unless its logic threshold voltage is less than $V_{DD}/2$, the output voltage of the *following inverter* will then inadvertently switch high, which is a logic error. Thus, it is important to have C_2 much smaller than C_1.

Several measures can be taken in order to prevent erroneous output levels due to charge sharing in domino CMOS gates. One simple solution is to add a weak pMOS pull-up device (with a small (W/L) ratio) to the dynamic CMOS stage output, which essentially forces a high output level unless there is a strong pull-down path between the output and the ground (Fig. 9.33). It can be observed that the weak pMOS transistor will be turned on only when the precharge node voltage is kept high. Otherwise, it will be turned off as V_{out} becomes high.

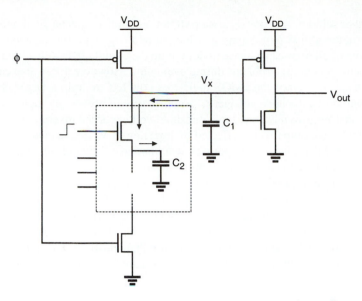

Figure 9.32 Charge sharing between the output capacitance C_1 and an intermediate node capacitance C_2 during the evaluation cycle may reduce the output voltage level.

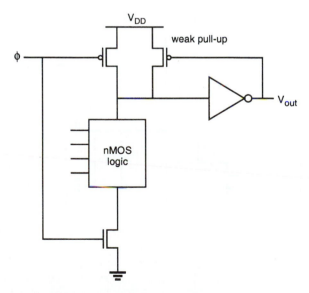

Figure 9.33 A weak pMOS pull-up device in a feedback loop can be used to prevent the loss of output voltage level due to charge sharing.

Another solution is to use separate pMOS transistors to precharge all intermediate nodes in the nMOS pull-down tree which have a large parasitic capacitance. The precharging of all high-capacitance nodes within the circuit effectively eliminates all potential charge-sharing problems during evaluation. However, it can also cause additional delay time since the nMOS logic tree now has to drain a larger charge in order to pull down the node voltage V_x. Another way of preventing logic errors due to charge sharing is to make the logic threshold voltage of the inverter smaller, such that the final stage output is not affected by lowering of V_x due to charge sharing. It should be noted that this design approach would trade off the pull-up speed (weaker pMOS transistor) for lower sensitivity to the charge-sharing problem.

The use of multiple precharge transistors also enables us to use the precharged intermediate nodes as resources for additional outputs. Thus, additional logic functions can be realized by tapping the internal nodes of the dynamic CMOS stage, as illustrated by two series-connected logic blocks in Fig. 9.34. The resulting multiple-output domino CMOS logic gate allows us to simultaneously realize several complex functions using a small number of transistors. Figure 9.35 shows the realization of

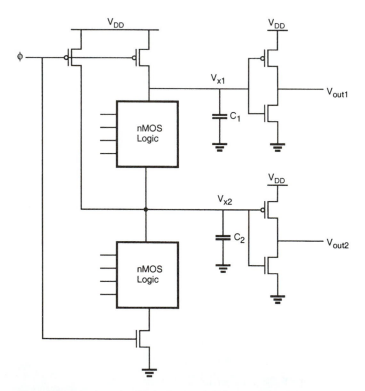

Figure 9.34 Precharging of internal nodes to prevent charge sharing also allows implementation of multiple-output domino CMOS structures.

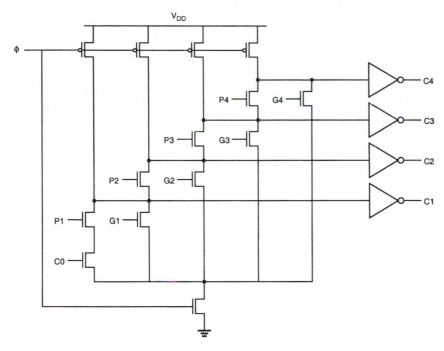

Figure 9.35 Example of a multiple-output domino CMOS gate realizing four functions.

four Boolean functions of nine variables, using a single domino CMOS logic gate. The four functions to be realized are listed in the following.

$$C_1 = G_1 + P_1 C_0$$
$$C_2 = G_2 + P_2 G_1 + P_2 P_1 C_0$$
$$C_3 = G_3 + P_3 G_2 + P_3 P_2 G_1 + P_3 P_2 P_1 C_0$$
$$C_4 = G_4 + P_4 G_3 + P_4 P_3 G_2 + P_4 P_3 P_2 G_1 + P_4 P_3 P_2 P_1 C_0$$

It can be shown that the functions C_1 through C_4 are the four carry terms to be used in a four-stage carry-lookahead adder, where the variables G_i and P_i are defined as

$$G_i = A_i \cdot B_i$$
$$P_i = A_i \oplus B_i$$

and A_i and B_i are the input bits associated with the i_{th} stage. Hence, this circuit is also known as the Manchester carry chain. The generation of the four carry terms using four separate standard CMOS logic gates or four separate single-output domino CMOS circuits, on the other hand, would require a large number of transistors and consequently a much larger silicon area. Variants of the multiple-output dynamic CMOS circuit shown in Fig. 9.35 are used widely in high-performance adder structures.

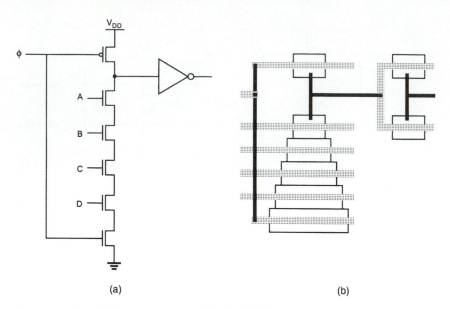

(a)   (b)

Figure 9.36   (a) Four-input domino CMOS NAND gate and (b) the corresponding stick-diagram layout to show the graded scaling of nMOS transistor sizes for improving the transient performance.

The transient performance of domino CMOS logic gates can be improved by adjusting the nMOS transistor sizes in the pull-down path, with the objective of reducing the discharge time. Shoji has shown that the best performance is obtained with a graded sizing of nMOS transistors in series structures, where the nMOS transistor closest to the output node also has the smallest (W/L) ratio. The domino CMOS circuit diagram and the corresponding stick-diagram layout of an optimized example are shown in Fig. 9.36. The fact that a graded reduction of transistor sizes from bottom to top ultimately leads to a better transient performance may seem counterintuitive. But, this effect can be explained by observing the RC delay of the combined pull-down path consisting of series-connected nMOS transistors.

Consider first the nMOS transistor closest to the output node. If the (W/L) ratio of this transistor is reduced by a certain factor, two effects occur. First, the current-driving capability will decrease, i.e., the equivalent resistance of the nMOS transistor will increase. Second, the parasitic drain capacitance associated with this transistor will decrease. If the length of the nMOS chain is sufficiently long, the increase in resistance has little influence upon the combined RC delay time, whereas a reduction of the capacitance significantly decreases the delay.

In fact, by applying Elmore's RC delay formula (see Chapter 6) to series-connected nMOS structures, one can determine if a reduction of nMOS transistor sizes will improve the transient performance. Let C_L represent the precharge-node load capacitance of the domino CMOS gate, and let C_1 represent the parasitic drain capacitance of the nMOS transistor closest to the precharge node. It is assumed that the inverter transistor sizes are fixed and that the pull-down chain contains N series-connected nMOS transistors. Shoji has shown that if the following condition

$$C_L < (N - 1)\frac{C_1}{2} \tag{9.36}$$

is satisfied, the overall delay time can be reduced by decreasing the size of the nMOS transistor closest to the output node. This result can be iteratively applied to the other transistors in the pull-down chain, which leads to graded sizing of all nMOS devices. On the other hand, if the inverter gate transistors are allowed to be optimized along with the series-connected transistors, even shorter delays can be achieved with smaller chip area.

EXAMPLE 9.4

Consider the domino CMOS NAND2 gate shown below, where $C_1 = C_2 = 0.05$ pF. First, the operation of the circuit with only one pMOS precharge transistor will be examined. Since the two capacitances C_1 and C_2 are assumed to be equal to each other, we expect that the charge-sharing phenomenon will cause erroneous output values, as explained earlier, unless specific measures are taken to prevent it.

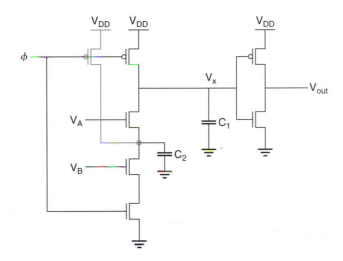

The SPICE-circuit input file of the domino CMOS NAND2 gate is listed.

```
Domino CMOS with charge sharing
vdd 10 0 dc 5V
vin 1 0 dc pulse( 5 0 1ns 0.1ns 0.1ns 10ns 22ns)
vb 4 0 dc pulse ( 0 5 35ns 0.1ns 0.1ns 11ns 22n)
va 5 0 dc pulse( 0 5 12ns 0.1ns 0.1ns 11ns 22ns)
m1 2 1 0 0 mn l=5u w=10u
m2 3 4 2 0 mn l=5u w=10u
```

```
m3 6 5 3 0 mn l=5u w=10u
m4 6 1 10 10 mp l=5u w=25u
m6 7 6 0 0 mn l=5u w=10u
m7 7 6 10 10 mp l=5u w=100u
cload 6 0 0.05p
cs 3 0 0.05p
cout 7 0 0.1p
.model mn nmos vto=1 gamma=0.4 kp=2.5e-5
.model mp pmos vto=-1 gamma=0.4 kp=1.0e-5
.tran 0.1ns 50ns
.print tran v(1) v(6) v(7) v(3)
.end
```

The transient simulation of this circuit shows that the precharge node voltage V_x drops to about 2.5 V during the evaluation phase, due to charge sharing. As a result, the inverter output voltage erroneously switches to logic-high level during the first evaluation phase.

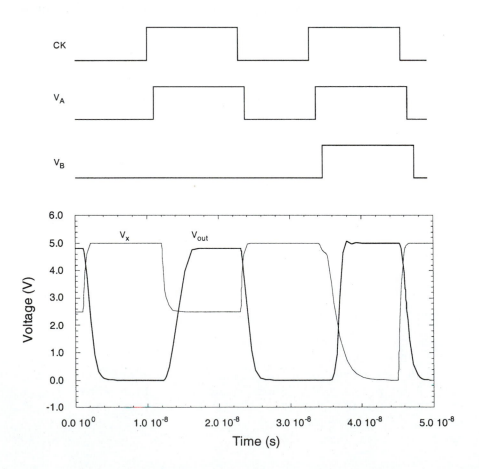

Now consider the case where an additional pMOS precharge transistor is connected between the power supply voltage V_{DD} and the intermediate node, as indicated in the circuit diagram on page 391. Both pMOS transistors conduct during the precharge phase, and charge up the node capacitances to the same voltage level. Consequently, charge sharing can no longer cause a logic error at the output node. The simulation results *with* the additional pMOS precharge transistor are plotted below, showing that the output node voltage is pulled up to logic "1" only when both inputs are equal to logic "1."

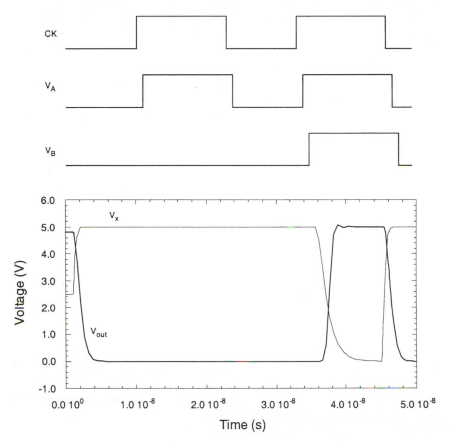

It must be emphasized that there is a speed penalty for adding another pMOS precharge transistor to the circuit. Simulation results indicate that the pull-down delay of the node voltage V_x is actually increased by about 1 ns (approx. 25 percent) as a result of the additional parasitic capacitance which is due to the precharge device.

NORA CMOS Logic (NP-Domino Logic)

In domino CMOS logic gates, all logic operations are performed by the nMOS transistors acting as pull-down networks, while the role of pMOS transistors is limited to

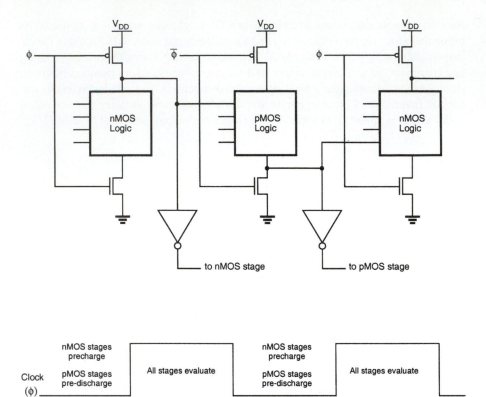

Figure 9.37 NORA CMOS logic consisting of alternating nMOS and pMOS stages, and the scheduling of precharge/evaluation phases.

precharging the dynamic nodes. As an alternative and a complement to nMOS-based domino CMOS logic, we can construct dynamic logic stages using pMOS transistors as well. Consider the circuit shown in Fig. 9.37, with alternating nMOS and pMOS logic stages.

Note that the precharge-and-evaluate timing of nMOS logic stages is accomplished by the clock signal ϕ, whereas the pMOS logic stages are controlled by the inverted clock signal, $\overline{\phi}$. The operation of the NORA CMOS circuit is as follows: When the clock signal is low, the output nodes of nMOS logic blocks are precharged to V_{DD} through the pMOS precharge transistors, whereas the output nodes of pMOS logic blocks are pre-discharged to 0 V through the nMOS discharge transistors, driven by ϕ. When the clock signal makes a low-to-high transition (note that the inverted clock signal $\overline{\phi}$ makes a high-to-low transition simultaneously), all cascaded nMOS and pMOS logic stages evaluate one after the other, much like the domino CMOS examined earlier. A simple NORA CMOS circuit example is shown in Fig. 9.38.

The advantage of NORA CMOS logic is that a static CMOS inverter is not required at the output of every dynamic logic stage. Instead, direct coupling of logic blocks is feasible by alternating nMOS and pMOS logic blocks. NORA logic is also

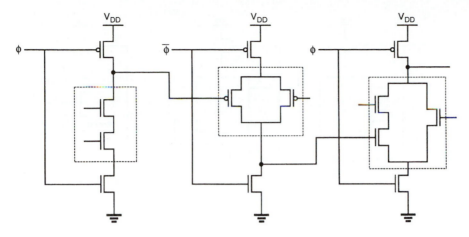

Figure 9.38 NORA CMOS logic circuit example.

compatible with domino CMOS logic. Outputs of NORA nMOS logic blocks can be inverted, and then applied to the input of a domino CMOS block, which is also driven by the clock signal ϕ. Similarly, the buffered output of a domino CMOS stage can be applied directly to the input of a NORA nMOS stage.

The second important advantage of NORA CMOS logic is that it allows pipelined system architecture. Consider the circuit shown in Fig. 9.39(a), which consists of an nMOS-pMOS logic sequence similar to the one shown in Fig. 9.37, and a clocked CMOS (C^2MOS) output buffer. It can easily be seen that all stages of this circuit perform the precharge-discharge operation when the clock is low, and all stages of the circuit evaluate output levels when the clock is high. Therefore, we will call this circuit a ϕ-section, meaning that evaluation occurs during active ϕ.

Now consider the circuit shown in Fig. 9.39(b), which is essentially the same circuit shown in Fig. 9.39(a), with the only difference being that the signals ϕ and $\bar{\phi}$ have been interchanged. In this circuit, all logic stages perform the precharge-discharge operation when the clock is high, and all stages evaluate output levels when the clock is low. Therefore, we will call this circuit a $\bar{\phi}$-section, meaning that evaluation occurs during active $\bar{\phi}$.

A pipelined system can be constructed by simply cascading alternating ϕ- and $\bar{\phi}$-sections, as shown in Fig. 9.39(c). Note that each of the sections may consist of several logic stages, and that all logic stages in one section are evaluated during the same clock cycle. When the clock is low, the ϕ-sections in the pipelined system undergo the precharge cycle, while the $\bar{\phi}$-sections undergo evaluation. When the clock signal changes from low to high, the ϕ-sections start the evaluation cycle, while the $\bar{\phi}$-sections undergo precharge. Thus, consecutive sets of input data can be processed in alternating sections of the pipelined system.

As in all dynamic CMOS structures, NORA CMOS logic gates also suffer from charge sharing and leakage. To overcome the dynamic charge sharing and soft-node

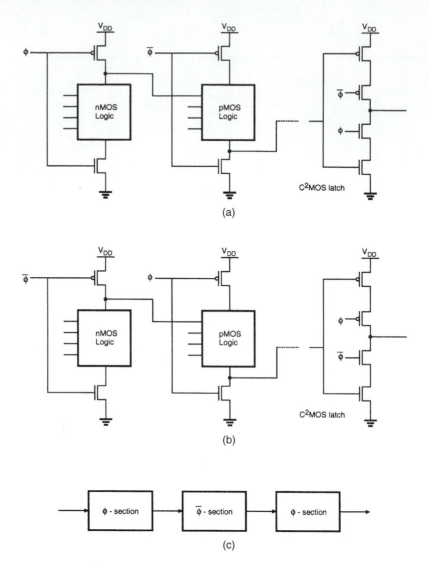

Figure 9.39   (a) NORA CMOS ϕ-section; evaluation occurs during $\phi = 1$.
(b) NORA CMOS ϕ-section; evaluation occurs during $\phi = 0$. (c) A pipelined
NORA CMOS system.

leakage problems in NORA CMOS structures, a circuit technique called Zipper
CMOS can be used.

Zipper CMOS Circuits

The basic circuit architecture of Zipper CMOS is essentially identical to NORA
CMOS, with the exception of the clock signals. The Zipper CMOS clock scheme

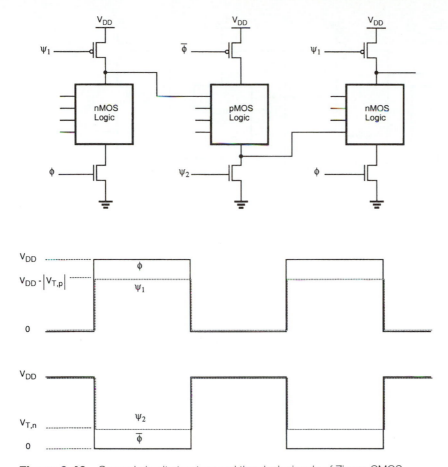

Figure 9.40 General circuit structure and the clock signals of Zipper CMOS.

requires the generation of slightly different clock signals for the precharge (discharge) transistors and for the pull-down (pull-up) transistors. In particular, the clock signals which drive the pMOS precharge and nMOS discharge transistors allow these transistors to remain in weak conduction or near cut-off during the evaluation phase, thus compensating for the charge leakage and charge-sharing problems. The generalized circuit diagram and the clock signals of the Zipper CMOS architecture are shown in Fig. 9.40.

True Single-Phase Clock (TSPC) Dynamic CMOS

The dynamic circuit technique to be presented in the following is distinctly different from the NORA CMOS circuit architecture in that it uses only one clock signal which is never inverted. Since the inverted clock signal $\overline{\phi}$ is not used anywhere in the system, no clock skew problem exists. Consequently, higher clock frequencies can be reached for dynamic pipelined operation.

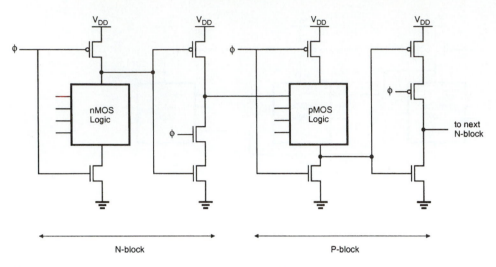

Figure 9.41   A pipelined true single-phase clock CMOS system.

Consider the circuit diagram shown in Fig. 9.41. The circuit consists of alternating stages called n-blocks and p-blocks, and each block is being driven by the same clock signal ϕ. An n-block is constructed by cascading a dynamic nMOS stage and a dynamic latch, while a p-block is constructed by cascading a dynamic pMOS stage and a dynamic latch.

When the clock signal is low, the output node of the n-block is being precharged to V_{DD} by the pMOS precharge transistor. When the clock signal switches from low to high, the logic stage output is evaluated and the output latch generates a valid output level. On the other hand, we can see by inspection that the p-block pre-discharges when the clock is high, and evaluates when the clock is low. This means that a cascade-connection of alternating n-block and p-block circuits as shown in Fig. 9.41 will allow pipelined operation using a single clock signal. Compared to NORA CMOS, we need two extra transistors per stage, but the ability to operate with a true single-phase clock signal offers very attractive possibilities from the system-design point of view.

Figure 9.42 shows the circuit diagram of a rising edge-triggered D-type flip-flop (DFF) which is constructed with the TSPC principle. The circuit consists of eleven transistors, in four simple stages. When the clock signal is low, the first stage acts as a transparent latch to receive the input signal, while the output node of the second stage is being precharged. During this cycle, the third and fourth stages simply keep the previous output state. When the clock signal switches from low to high, the first stage ceases to be transparent and the second stage starts evaluation. At the same time, the third stage becomes transparent and transmits the sampled value to the output. Note that the final stage (inverter) is only used to obtain the non-inverted output level.

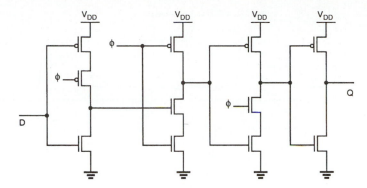

Figure 9.42 Circuit diagram of a TSPC-based rising edge-triggered DFF.

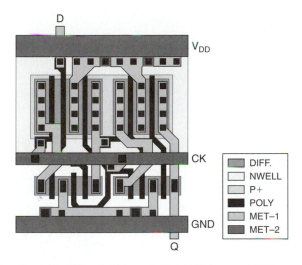

	DIFF.
	NWELL
	P+
	POLY
	MET–1
	MET–2

Figure 9.43 Mask layout of the TSPC-based DFF circuit.

The mask layout of this circuit (using 0.8 μm CMOS technology design rules) is shown in Fig. 9.43. The device dimensions are $(W/L)_n = (2\ \mu\text{m}/0.8\ \mu\text{m})$ for nMOS transistors, and $(W/L)_p = (5.6\ \mu\text{m}/0.8\ \mu\text{m})$ for pMOS transistors. The layout parasitics were determined through circuit extraction, and the extracted circuit file was simulated with SPICE to determine its performance. The simulated input and output waveforms of the TSPC-based DFF circuit are given in Fig. 9.44. It can be seen that the circuit is capable of operating with a clock frequency of 500 MHz. With their relatively simple design, low transistor count, and high speed, TSPC-based circuits offer a very favorable alternative to conventional CMOS circuits, especially in high-performance designs.

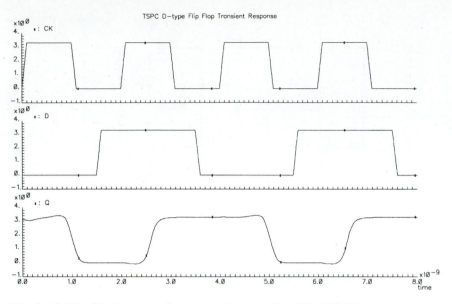

Figure 9.44 Simulation results showing the operation of the TSPC-based DFF circuit, with a clock frequency of 500 MHz.

Exercise Problems

9.1 Consider the CMOS circuit shown in Fig. P9.1 that was designed to drive a total capacitive load of $C_L = 0.2$ pF. For the n-channel devices, assume zero bias threshold voltage $V_{T0} = 1.0$ V and transconductance $k'_n = 50$ μA/V^2. For the p-channel devices, assume $V_{T0} = -1.0$ V and $k'_p = 25$ μA/V^2. Use

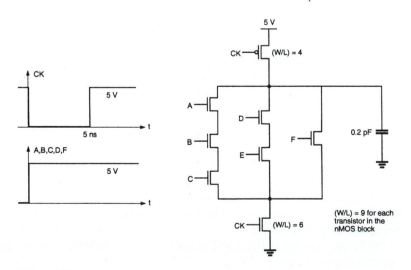

Figure P9.1

the W/L ratios for each device as shown in the figure. The initial voltage across the load capacitor C_L is 0 V. The waveform at input E is identical to 0 V for all time. For the clock CK and the rest of the inputs, the waveforms are also shown in the figure. Sketch the voltage waveform across the load capacitor C_L and provide clear marking of the 50% crossings along the time axis in nanoseconds for both rise and fall transitions. Hint: The n-channel transistor group can be approximated by a single equivalent n-channel transistor and either an average-current method or a state-equation method may be used to compute the required delay times.

9.2 In logic design, complex gates such as AOI or OAI gates are often used to combine the functions of several gates into a single gate, thus reducing the chip area and parasitics of the circuit. Let us consider an OAI432 whose logic function is $Z = \overline{(A + B + C + D)(E + F + G)(H + I)}$. Assume that only A, E, H inputs are high and other inputs are low. The device parameters are:

$C_{jsw} = 250$ pF/m
$C_{j0} = 80\ \mu\text{F/m}^2$
$C_{ox} = 350\ \mu\text{F/m}^2$
$L_D = 0.5\ \mu\text{m}$
$K_{eq} = 1.0$ for worst-case capacitance
$C_{metal} = 2.0$ pF/cm
$C_{poly} = 2.2$ pF/cm
$R_{poly} = 25\ \Omega$/sq.
$R_{metal} = 0.2\ \Omega$/sq.

Polysilicon line width $= 2\ \mu\text{m}$, metal line width $= 2\ \mu\text{m}$

$k'_n = 20\ \mu\text{A/V}^2$
$k'_p = 10\ \mu\text{A/V}^2$
$|V_{T0}| = 1.0$ V

a. Draw a full CMOS circuit diagram for this OAI432.
b. Draw a domino CMOS circuit diagram which implements Z.
c. Draw an equivalent circuit for this case by using equivalent transistor sizes with $W/L = 30/2$ (both for nMOS and pMOS).
d. By assuming that the total parasitic capacitances at the precharging node and the output nodes are 0.2 pF, calculate the delay from the end of precharging (clock signal is a rectangular pulse) to the time point at which output voltage reaches 2.5 V. For simple analysis, neglect the body effect. For approximate solution of this problem, first calculate the delay for the precharging node to fall to 2.5 V. Then use a rectangular pulse at that time point which falls from 5 V to 0 V to calculate the delay of the inverter gate. The total delay can be found by adding the two delay components. (A more accurate method is to approximate the precharging node voltage by a falling ramp function.)

9.3 Discuss the charge-sharing problems in VLSI circuits. Explain various circuit techniques used in domino CMOS circuits for solving charge-sharing problems. State as many as you know.

9.4 Bootstrapping circuits are used to increase the gate voltages of transistors such that the drain node voltage can be pulled high (despite the threshold voltage drop). The circuit shown in Fig. P9.4 can be used for such purposes and can indeed increase the node voltage at X well beyond $V_{DD} = 5$ V. Determine the maximum achievable voltage at node X using the following parameters:

$$V_{T0} = 1.0 \text{ V}$$
$$\gamma = 0.3 \text{ V}^{1/2}$$
$$|2\phi_F| = 0.6 \text{ V}$$
$$V_{OL} = 0.2 \text{ V}$$
$$C_{sub} = 12 \text{ fF}$$
$$C_{boot} = 20 \text{ fF}$$

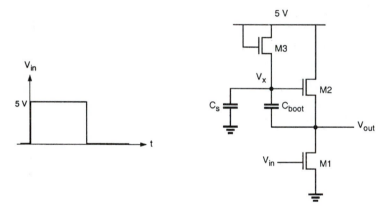

Figure P9.4

9.5 A CMOS circuit is shown in Fig. P9.5. Suppose that the precharge transistor was chosen such that node X is guaranteed to be charged to V_{DD}. All nMOS transistors have $W/L = 20$. Determine how long it takes for the node voltage at X to decrease to $0.8\ V_{DD}$ after the clock signal pulse goes to high (with zero rise time) when input voltages at A, B, D are 5 V and the input voltage at C is zero. Assume the following parameter values:

- $\gamma = 0.0 \text{ V}^{1/2}$
- $V_{T0} = 1.0 \text{ V}$
- $k_n' = 10 \text{ } \mu\text{A/V}^2$

Hint: The nMOS transistor tree between node X and the ground can be approximated by an equivalent transistor with an effective W/L.

9.6 The inputs to a domino logic gate are always LO during the precharge phase ($\phi = $ LO) and may undergo a LO-to-HI transition during the evaluation phase

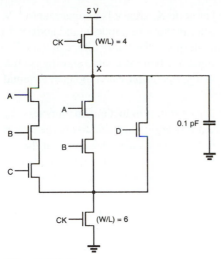

Figure P9.5

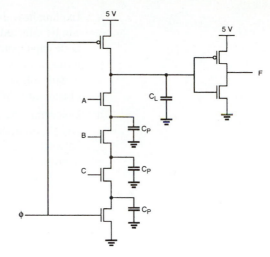

Figure P9.6

(ϕ = HI). Consider the domino 3-input AND gate shown in Fig. P9.6. If, during the evaluation phase, A = HI, B = LO, and C = LO, charge sharing will cause the voltage at the input of the inverter to drop. Given that the switching threshold of the inverter is 3 V, calculate the maximum ratio C_P/C_L necessary to ensure that charge sharing does not corrupt the value of F for the given case.

9.7 Consider the CMOS logic circuit in Fig. P9.7, which is a simple domino circuit. Node X is connected to a CMOS inverter so that the output of the inverter can be directly fed to the next stage of the domino circuit.

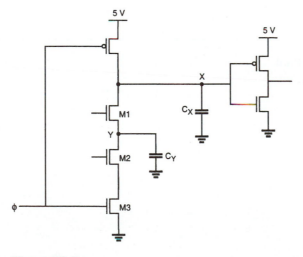

Figure P9.7

a. Explain how the voltage level at node X, after it is precharged to 5 V, can be affected by the charge sharing between node X and node Y if their node capacitances are the same. Express the final voltage at node X in terms of the initial voltage at node Y when the charge sharing is completed, following the full precharge operation when the gate terminal of transistor M2 is fixed at 0 V.

b. Determine the ratio between device transconductance parameters, k_p and k_n, of the inverter to prevent any logic error due to charge sharing between nodes X and Y under all circumstances. Assume that the magnitudes of threshold voltages in the inverter are equal to 1.0 V. The use of Level 1 transistor current equations is deemed adequate.

9.8 Consider the domino CMOS circuit shown in Fig. P9.6. Using the input voltage waveforms illustrated in Fig. P9.8 determine the output voltage waveform.

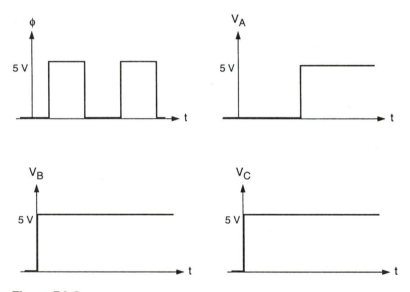

Figure P9.8

CHAPTER 10

Semiconductor Memories

Contributed by
Seung-Moon Yoo

10.1 Introduction

Semiconductor memory arrays capable of storing large quantities of digital information are essential to all digital systems. The amount of memory required in a particular system depends on the type of the application, but, in general, the number of transistors for the information (data) storage function is much larger than the number of transistors used for logic operations and other purposes. The ever-increasing demand for larger data storage capacity has driven the fabrication technology and memory development toward more compact design rules and, consequently, toward higher data storage densities. Thus, the maximum realizable data storage capacity of single-chip semiconductor memory arrays approximately doubles every two years. On-chip memory arrays have become widely used subsystems in many VLSI circuits, and commercially available single-chip read/write memory capacity has reached 1 gigabits (1 Gb). This trend toward higher memory density and large storage capacity will continue to push the leading edge of digital system design.

The area efficiency of the memory array, i.e., the number of stored data bits per unit area, is one of the key design criteria that determine the overall storage capacity and, hence, the memory *cost per bit*. Another important issue is the memory access time, i.e., the time required to store and/or retrieve a particular data bit in the memory array. The access time determines the memory *speed*, which is an important performance criterion of the memory array. Finally, the static and dynamic *power consumption* of the memory array is a significant factor to be considered in the design, because of the increasing importance of low-power applications. In the following, we will investigate different types of MOS memory arrays and discuss in detail their operations and design issues related to area, speed, and power consumption for each type.

The semiconductor memory is generally classified according to the type of data storage and data access. Read/write (R/W) memory must permit the modification (writing) of data bits stored in the memory array, as well as their retrieval (reading) on demand. The read/write memory is commonly called *Random Access Memory* (RAM), mostly due to historical reasons. Unlike sequential-access memories such as magnetic tapes, any cell can be accessed with nearly equal access time. The stored data is volatile; i.e., the stored data is lost when the power supply voltage is turned off. Based on the operation type of individual data storage cells, RAMs are classified into two main categories. *Dynamic RAMs* (DRAM) and *Static RAMs* (SRAM). The DRAM cell consists of a capacitor to store binary information, 1 (high voltage) or 0 (low voltage), and a transistor to access the capacitor. Cell information (voltage) is degraded mostly due to a junction leakage current at the storage node. Therefore, the cell data must be read and rewritten periodically (*refresh operation*) even when memory arrays are not accessed. On the other hand, the SRAM cell consists of a latch, therefore, the cell data is kept as long as the power is turned on and refresh operation is not required. Due to the advantage of low cost and high density, DRAM is widely used for the main memory in personal and mainframe computers, and engineering workstations. SRAM is mainly used for the cache memory in microprocessors, mainframe computers, engineering workstations, and memory in hand-held devices due to high speed and low power consumption.

Read-Only Memory (ROM) allows, as the name implies, only retrieval of previously stored data and does not permit modifications of the stored information contents during normal operation. ROMs are nonvolatile memories; i.e., the stored data is not lost even when the power supply is off and refresh operation is not required. Depending on the type of data storage (data write) method, ROMs are categorized as *Mask ROM,* in which data are written during chip fabrication by using a photo mask, and *Programmable ROM* (PROM), in which data are written electrically after the chip is fabricated. Depending on data erasing characteristics, PROMs are further classified into *Fuse ROM, Erasable PROM* (EPROM) and *Electrically Erasable PROM* (EEPROM). The data written by blowing the fuse electrically cannot be erased and modified in Fuse ROM. Data in EPROMs and EEPROMs can be rewritten, but the number of subsequent re-write operations is limited to 10^4–10^5. In EPROMs, ultraviolet rays that can penetrate through the crystal glass on the package are used to erase whole data in the chip simultaneously, while high electrical voltage is used to erase data in 8 bit units in EEPROMs. *Flash memory* is similar to EEPROM, where data in the block can be erased by using a high electrical voltage. A drawback of EEPROM is the slower write speed, in the order of microseconds.

Ferroelectric RAM (FRAM) utilizes the hysteresis characteristics of a ferroelectric capacitor to overcome the slow write operation of other EEPROMs. ROMs are generally used for permanent (look-up) memory in printers, fax, and game machines, and ID cards due to lower cost than RAM. Figure 10.1 and Table 10.1 show an overview and characteristic summary of the different memory types and their classifications, respectively.

Figure 10.2 shows equivalent circuits of memory cells. The DRAM cell consists of a capacitor and a switch transistor. The data are stored in the capacitors as presence and absence of charge: The presence of charge in the capacitor is considered as data "1" while the absence of charge in the capacitor as data "0". The stored charge is subject to gradual decay due to leakage current, thus *refresh* operation is required.

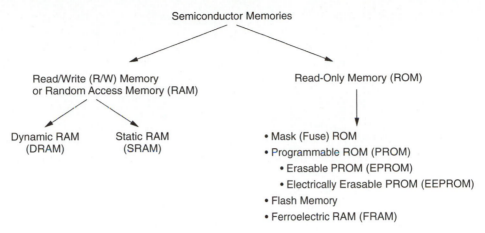

Figure 10.1   Overview of semiconductor memory types.

Table 10.1   Characteristic summary of memory devices

	Memory type					
	DRAM	**SRAM**	**UV EPROM**	**EEPROM**	**Flash**	**FRAM**
Data volatility	Yes	Yes	No	No	No	No
Data refresh operation	Required	No	No	No	No	No
Cell structure	1T-1C	6T	1T	2T	1T	1T-1C
Cell density	High	Low	High	Low	High	High
Power consumption	High	High/low	Low	Low	Low	High
Read speed (latency)	$\sim$50 ns	$\sim$10/70 ns	$\sim$50 ns	$\sim$50 ns	$\sim$50 ns	$\sim$100 ns
Write speed	$\sim$40 ns	$\sim$5/40 ns	$\sim$10 μs	$\sim$5 ms	$\sim$(10 μs$-$1 ms)	$\sim$100 ns
Endurance	High	High	High	Low	High	High
Cost	Low	High	Low	High	Low	Low
In-system writability	Yes	Yes	No	Yes	Yes	Yes
Power supply	Single	Single	Single	Multiple	Single	Single
Application example	Main memory	Cache/PDAs	Game machines	ID card	Memory card solid-state disk	Smart card, digital camera

The SRAM cell has a six-transistor latch structure to hold the state of each cell node. Since the cell data can be held indefinitely at one of the two possible states of the bistable latch as long as power supply is provided, the refresh operation is not needed in SRAMs. In Mask (Fuse) ROM, the data are programmed by a Mask pattern (blowing the fuse located at each cell, thereby severing the electrical connection of the corresponding device). Only one-time programming operation is allowed. In the EPROM and EEPROM, data can be rewritten into the cell by using ultraviolet rays or by a tunnel current, respectively. The FRAM cell has the similar structure to that of DRAM except the ferroelectric capacitor, where the cell data are modified by changing the polarization of ferroelectric material.

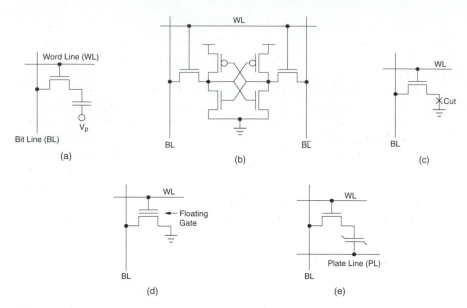

Figure 10.2 Equivalent circuits of memory cells. (a) DRAM, (b) SRAM, (c) Mask (Fuse) ROM, (d) EPROM (EEPROM), (e) FRAM.

A conceptual memory array organization is shown in Fig. 10.3. The data storage structure, or *core,* consists of individual memory cells arranged in an array of horizontal rows and vertical columns. Each *cell* is capable of storing one bit of binary information. Also, each memory cell shares a common connection with the other cells in the same row, and another common connection with the other cells in the same column. In this structure, there are 2^N rows, also called *word lines,* and 2^M columns, also called *bit lines*. Thus, the total number of memory cells in this array is $2^N \times 2^M$.

To access a particular memory cell, i.e., a particular data bit in this array, the corresponding word line *and* the corresponding bit line must be activated (selected) according to the addresses coming from the outside of the memory array. These addresses are provided by the memory controller or the processor directly. Since the signal levels at the inside and outside of the memory array are different, e.g., the Transistor-Transistor Logic (TTL) signal on the memory board and CMOS signal in the memory chip, the level of addresses is converted through the memory chip interface, called *input address buffers*. The row and column selection operations are accomplished by row and column *decoders,* respectively. The row decoder circuit selects one out of 2^N word lines according to an N-bit row address, while the column decoder circuit selects one out of 2^M bit lines according to an M-bit column address. Once a memory cell or a group of memory cells is selected in this fashion, a data read or write operation is performed on the selected single or multiple bits on a particular row. The column address serves the double duties of selecting the particular columns and routing the corresponding data content in a selected row to the output through the memory chip interface, called *data out buffers*. Typically, the data out

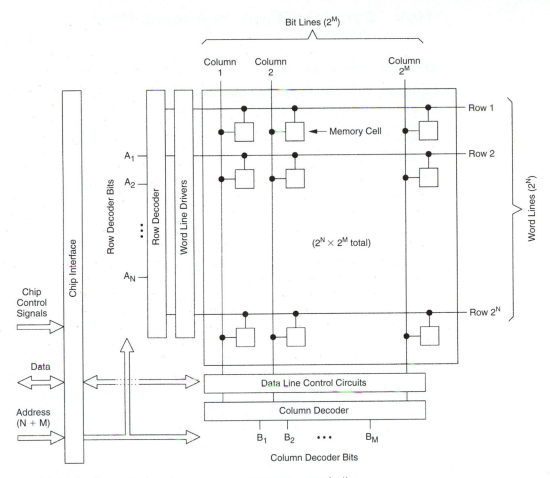

Figure 10.3 Conceptual random-access memory array organization.

buffers are required to drive relatively large current load on the board, therefore, large transistors are used. The performance of the chip interface circuit determines a major portion of the total memory speed, especially in high performance SRAMs. Other chip control signals, e.g., *Chip Select* ($\overline{CS}$), *Write Enable* ($\overline{WE}$), are also provided to activate the read or write operation of the particular memory chip out of the memory cluster on the board.

We can see from this simple discussion that individual memory cells can be accessed for data read or write operations in random order, independent of their physical locations in the memory array. Thus, the array organization can be used for *Random Access Memory* (RAM) structure. Notice that this organization can be used for both read/write memory arrays and read-only memory arrays. In the following sections, however, we will use the acronym RAM specifically for read/write memories, because it is the generally accepted abbreviation for this particular type of memory array.

10.2 Dynamic Random Access Memory (DRAM)

DRAM Configuration

Figure 10.4 shows the typical configuration of a 1 Gb DRAM device. Peripheral circuits to control DRAM operations (i.e., read, write, and refresh) and the memory chip interface (i.e., input and output buffers) are placed not at the periphery of the chip but at the border regions between memory blocks, for easy placement-and-routing and minimized signal skews. The entire memory area is divided into several blocks, where the row and column decoders are shared by neighboring blocks. Each memory block is sliced into subblocks, where each subblock has its data line control circuits. The number of cells per word and bit lines is determined by the trade-off between the chip size and performance. The chip size is usually smaller if more cells can share the same word and bit lines. However, the performance is degraded in this case since row and column address decoders must drive a larger load.

Figure 10.5 and Table 10.2 show the pin assignment of a 256 Mbit DRAM device and its function, respectively. *CLK* and *CKE* pins are used as time references for operations and data communication. $\overline{CS}$, $\overline{RAS}$, $\overline{CAS}$, and $\overline{WE}$ are pins to control the DRAM operations. Address pins are used to select the location of the memory cell. Those pins are used for both "row" and "column" addresses, using a scheme called

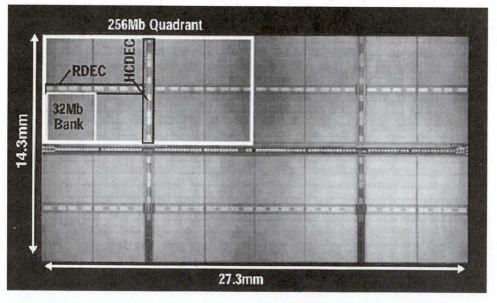

Figure 10.4 Typical configuration of DRAM chip (1 Gb DRAM).
(*Source:* Figure 2, K. Toshiaki et al., "A 390-*mm*², 16-Bank, 1-Gb DDR SDRAM with Hybrid Bitline Architecture," *IEEE J. Solid-State Circuits,* vol. 34, no. 11, pp. 1580–1588, November 1999.)

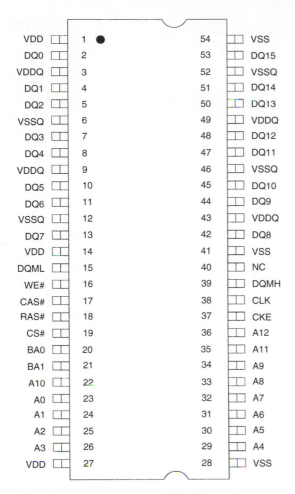

VDD	1 ●	54	VSS
DQ0	2	53	DQ15
VDDQ	3	52	VSSQ
DQ1	4	51	DQ14
DQ2	5	50	DQ13
VSSQ	6	49	VDDQ
DQ3	7	48	DQ12
DQ4	8	47	DQ11
VDDQ	9	46	VSSQ
DQ5	10	45	DQ10
DQ6	11	44	DQ9
VSSQ	12	43	VDDQ
DQ7	13	42	DQ8
VDD	14	41	VSS
DQML	15	40	NC
WE#	16	39	DQMH
CAS#	17	38	CLK
RAS#	18	37	CKE
CS#	19	36	A12
BA0	20	35	A11
BA1	21	34	A9
A10	22	33	A8
A0	23	32	A7
A1	24	31	A6
A2	25	30	A5
A3	26	29	A4
VDD	27	28	VSS

Figure 10.5 Pin assignment of 256-Mb synchronous DRAM. (The # symbol indicates signal is active "low".)

address multiplexing, which reduces the package size. Row and column addresses are captured by $\overline{RAS}$ and $\overline{CAS}$ signals.

Historical Evolution of DRAM Cell

As the continuing trend for high-density memories favors smaller memory cell sizes, the *dynamic* RAM cell with a simple structure has become a popular choice, where binary data are stored as charge in a capacitor and the presence or absence of stored charge determines the value of the stored bit. Note that the data stored as charge in a capacitor cannot be retained indefinitely, because the leakage currents eventually remove or modify the stored data. Thus, all DRAM cells require a periodic refreshing of the stored data, so that unwanted modifications due to leakage are prevented

Table 10.2   Definition and function of DRAM pins (256-Mb synchronous DRAM)

Pin name	Definition	Function
CLK	Clock input	Reference system clock for the operation and data communication
CKE	Clock Enable	Control the clock input
$\overline{CS}$	Chip Select	Activate the DRAM device from a memory cluster
$\overline{RAS}$	Row Address Strobe	Latch row address and start the cell core operation
$\overline{CAS}$	Column Address Strobe	Latch column address and start the data communication operation
$\overline{WE}$	Write Enable	Activate the write operation
A0 to A14	Address input	Select a data bit
DQ0 to DQ15	Data input and output	Communicate data with external devices
DQMU/DQML	DQ Mask for Upper (Lower) Byte	Mask byte data from the operations
V_{DD}/V_{SS}	Power pins	Power for DRAM core and peripheral circuits
$V_{DD}Q/V_{SS}Q$	Power pins	Power for DQ circuits
NC	No connection	

before they occur. The usage of a capacitor as the primary storage device generally enables the DRAM cell to be realized in a much smaller silicon area compared to the typical SRAM cell.

Figure 10.6 shows some of the steps in the historical evolution of the DRAM cells. The four-transistor cell shown in Fig. 10.6(a) is one of the earliest dynamic cells, from the 1970's. Its "write" and "read" operations are similar to those of the SRAM cell. In the "write" operation, a word line is enabled and complementary data are written from a pair of bit lines. Charge is stored at the parasitic and gate capacitances of a node connected with a high voltage bit line. Since no current path is provided to the storage nodes for restoring the charge lost due to leakage, the cell must be refreshed periodically. In the "read" operation, the voltage of a bit line is discharged to the ground through the transistor where the gate is charged with the high voltage. The read operation is non-destructive since the voltage stored at the node is maintained during the read operation.

The three-transistor DRAM cell shown in Fig. 10.6(b) was also used in the early 1970's. It utilizes a single transistor (M3) as the storage device (where the transistor is turned on or off depending on the charge stored in its gate capacitance), and one transistor each for the "read" and "write" access switches. During the "write" operation, the "write" word line is enabled and the voltage of the "write" bit line is passed onto the gate of storage device through the M1 transistor. During the "read" operation, the voltage of the "read" bit line is discharged to the ground through the M2 and M3 transistors when the gate voltage of the storage device is high. The read operation of

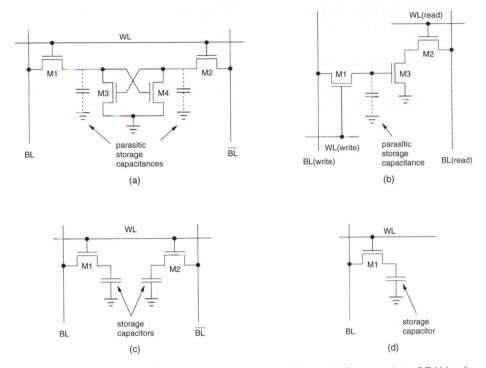

Figure 10.6 Various configurations of the dynamic RAM cell. (a) Four-transistor DRAM cell with two storage nodes. (b) Three-transistor DRAM cell with two bit lines and two word lines. (c) Two-transistor DRAM cell with two bit lines and one word line. (d) One-transistor DRAM cell with one bit line and one word line.

the three-transistor DRAM cell is also non-destructive and relatively fast, but the four lines, i.e., two bit lines and two word lines, with their additional contacts tend to increase the cell area.

The two-transistor and one-transistor DRAM cells which are shown in Fig. 10.6(c) and (d), respectively, have *explicit* storage capacitors. This means that a separate capacitor(s) must be manufactured for each storage cell, instead of relying on the gate and diffusion capacitances of the transistors for data storage. The one-transistor DRAM cell has become the industry-standard dynamic RAM cell in high-density DRAM arrays since the mid-1970's. The "read" and "write" operations of these two cells are almost the same. In the "write" operation, after the word line is enabled, the data are written into the cell through the M1 (or M2) transistor and stored at the storage capacitor. The read operation is destructive. When the charge stored in the storage cell is shared with the bit line, its charge can be changed significantly (destructed). Also, since the capacitance of the bit line is larger than that of the storage cell by about 10 times, only a small voltage difference is produced at the bit line depending on the voltage level (data) of the storage cell. Therefore, an amplifier to sense the signal difference and rewrite the data into the cell (charge restoring operation) is required for the successful "read" operation.

DRAM Cell Types

With only one transistor and one capacitor, DRAM has the smallest silicon area of the all the dynamic memory cells. However, its "read" operation is destructive. Thus, a large cell capacitance is essential to improve signal development (voltage difference) on the bit line, which limits the overall read operation as the chip operating voltage decreases. Since the voltage difference is determined by the combination of the operating voltage and the ratio of the bit line and the cell capacitance (C_s/C_{BL}), DRAM manufacturers have made a major effort to develop the technology to achieve a large capacitance cell with minimized silicon area.

The memory cell capacitance has been improved by using advanced cell structures and a high-permittivity (ϵ) dielectric material such as Ta_2O_5. A DRAM cell with a cylindrical stacked capacitor is shown in Fig. 10.7(a). The capacitor is formed on the bit line, so called a COB (Capacitor Over Bit line) structure, to increase the effective the capacitor area. A DRAM cell with a trench capacitor is shown in Fig. 10.7(b). Since the access transistor is fabricated after the capacitor is formed, this structure has the advantage of the planarization and better transistor characteristics.

Operation of Three-Transistor DRAM Cell

The circuit diagram of a typical three-transistor dynamic RAM cell is shown in Fig. 10.8 as well as the column pull-up (precharge) transistors and the column read/write circuitry. Here, the binary information is stored in the form of charge in the parasitic node capacitance C_1. The storage transistor M2 is turned on or off depending

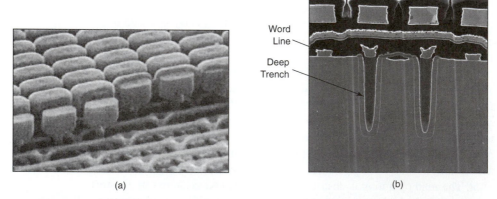

(a)   (b)

Figure 10.7   Various structures of the one-transistor DRAM cell. (a) DRAM cell with a stacked capacitor. (b) DRAM cell with a trench capacitor.
(*Source (b):* Figure 1, S. Crowder et al., "Integration of trench DRAM into a high-performance 0.18 μm logic technology with copper BEOL, IEDM (International Electron Devices Meeting) '98, pp. 1017–1020, 1998.)

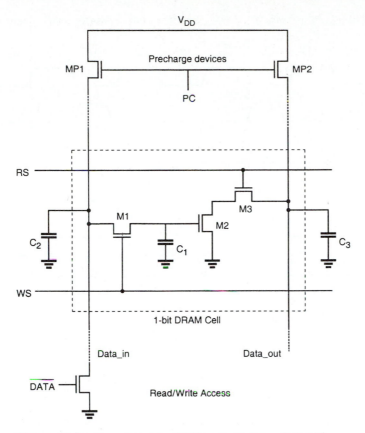

Figure 10.8   Three-transistor DRAM cell with the pull-up and read/write circuitry.

on the charge stored in C_1, and the pass transistors M1 and M3 act as access switches for data read and write operations. The cell has two separate bit lines for "data read" and "data write," and two separate word lines to control the access transistors.

The operation of the three-transistor DRAM cell and its peripheral circuitry is based on a two-phase non-overlapping clock scheme. The precharge events are driven by ϕ_1, whereas the "read" and "write" events are driven by ϕ_2. Every "data read" and "data write" operation is preceded by a precharge cycle, which is initiated with the precharge signal PC going high. During the precharge cycle, the column pull-up transistors are activated, and the corresponding column capacitances C_2 and C_3 are charged up to logic-high level. With typical enhancement type nMOS pull-up transistors ($V_{T0} \approx 1.0$ V) and a power supply voltage of 5 V, the voltage level of both columns after the precharge is approximately equal to 3.5 V.

All "data read" and "data write" operations are performed during the active ϕ_2 phase, i.e., when PC is low. Figure 10.9 depicts the typical voltage waveforms associated with the 3-T DRAM cell during a sequence of four consecutive operations:

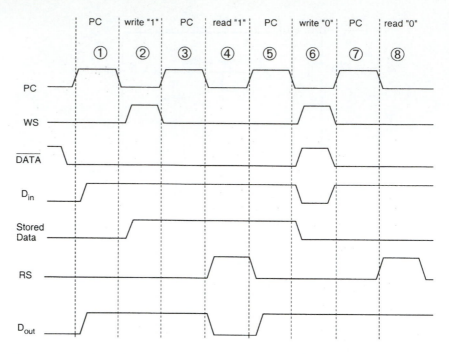

Figure 10.9 Typical voltage waveforms associated with the 3-T DRAM cell during four consecutive operations: write "1," read "1," write "0," and read "0."

write "1," read "1," write "0," and read "0." The four precharge cycles shown in Fig. 10.9 are numbered 1, 3, 5, and 7, respectively. Figure 10.10 illustrates the transient currents charging up the two columns (D_{in} and D_{out}) during a precharge cycle. The precharge cycle is effectively completed when both capacitance voltages reach their steady-state values. Note that the two column capacitances C_2 and C_3 are at least one order of magnitude larger than the internal storage capacitance C_1.

For the write "1" operation, the *inverse* data input is at the logic-low level, because the data to be written onto the DRAM cell is logic "1." Consequently, the "data write" transistor MD is turned off, and the voltage level on column D_{in} remains high. Now, the "write select" signal WS is pulled high during the active phase of ϕ_2. As a result, the write access transistor M1 is turned on. With M1 conducting, the charge on C_2 is now shared with C_1 (Fig. 10.11). Since the capacitance C_2 is very large compared to C_1, the storage node capacitance C_1 attains approximately the same logic-high level as the column capacitance C_2 at the end of the charge-sharing process.

After the write "1" operation is completed, the write access transistor M1 is turned off. With the storage capacitance C_1 charged-up to a logic-high level, transistor M2 is now conducting. In order to read this stored "1," the "read select" signal RS must be pulled high during the active phase of ϕ_2, following a precharge cycle. As the read access transistor M3 turns on, M2 and M3 create a conducting path between the "data read" column capacitance C_3 and the ground. The capacitance C_3 discharges through M2 and M3, and the falling column voltage is interpreted by the

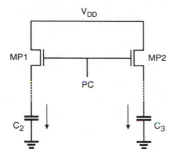

Figure 10.10 Column capacitances C_2 and C_3 are being charged-up through MP1 and MP2 during the precharge cycle.

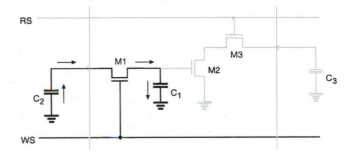

Figure 10.11 Charge sharing between C_2 and C_1 during the write "1" sequence.

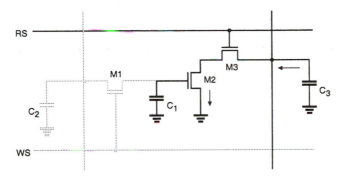

Figure 10.12 The column capacitance C_3 is discharged through the transistors M2 and M3 during the read "1" operation.

"data read" circuitry as a stored logic "1." The active portion of the DRAM cell during the read "1" cycle is shown in Fig. 10.12. Note that the 3-T DRAM cell may be read repeatedly in this fashion without disturbing the charge stored in C_1.

For the write "0" operation, the inverse data input is at the logic-high level, because the data to be written onto the DRAM cell is a logic "0." Consequently, the data write transistor is turned on, and the voltage level on column D_{in} is pulled to logic "0." Now, the "write select" signal WS is pulled high during the active phase of ϕ_2. As a result, the write access transistor M1 is turned on. The voltage level on C_2, as well as that on the storage node C_1, is pulled to logic "0" through M1 and the data write transistor, as shown in Fig. 10.13. Thus, at the end of the write "0" sequence, the storage capacitance C_1 contains a very low charge, and the transistor M2 is turned off since its gate voltage is approximately equal to zero.

In order to read this stored "0," the "read select" signal RS must be pulled high during the active phase of ϕ_2, following a precharge cycle. The read access transistor M3 turns on, but since M2 is off, there is no conducting path between the column capacitance C_3 and the ground (Fig. 10.14). Consequently, C_3 does not discharge, and the logic-high level on the D_{out} column is interpreted by the data read circuitry as a stored "0" bit.

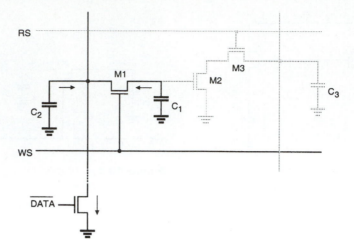

Figure 10.13 Both C_1 and C_2 are discharged via M1 and the data write transistor during the write "0" sequence.

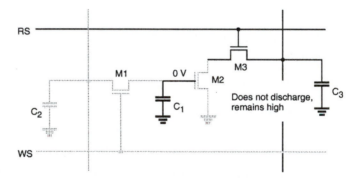

Figure 10.14 The column capacitance C_3 cannot discharge during the read "0" cycle.

As we already pointed out in the beginning of this section, the charge stored in C_1 cannot be held indefinitely, even though the "data read" operations do not significantly disturb the stored charge. The drain junction leakage current of the write access transistor M1 is the main reason for the gradual depletion of the stored charge on C_1. In order to *refresh* the data stored in the DRAM cells before they are altered due to leakage, the data must be periodically read, inverted (since the data output level reflects the inverse of the stored data), and then written back into the same cell location. This refresh operation is performed for all storage cells in the DRAM array every 2 to 4 ms. Note that all bits in one row can be refreshed at once, which significantly simplifies the procedure.

It can be seen that the three-transistor dynamic RAM cell examined here does not dissipate any static power for data storage, since there is no continuous current flow in the circuit. Also, the use of periodic precharge cycles instead of static pull-up further reduces the dynamic power dissipation. The additional peripheral circuitry required for scheduling the non-overlapping control signals and the refresh cycles does not significantly overshadow these advantages of the low-power dynamic memory.

Operation of One-Transistor DRAM Cell

Currently, the one-transistor cell is the most widely used storage structure in the DRAM industry. The circuit diagram at the one-transistor (1-T) DRAM cell consisting of one explicit storage capacitor and one access transistor is shown in Fig. 10.15(a).

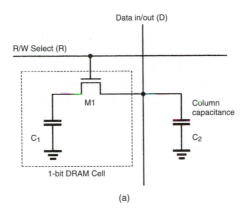

(a)

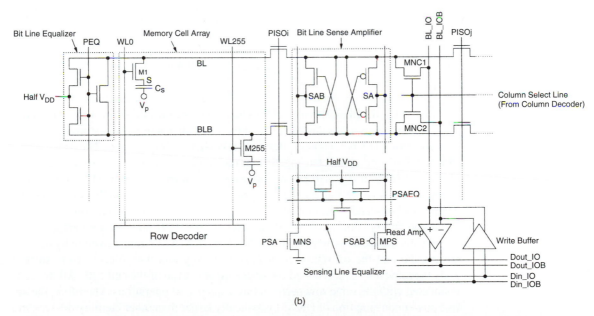

(b)

Figure 10.15 (a) Typical one-transistor (1-T) DRAM cell with its access lines. (b) Memory structure of one-transistor DRAM cell array with control circuits, where half-V_{DD} sensing, folded bit line, and shared sense amplifier architectures are incorporated.

A typical memory structure consisting of one-transistor DRAM cell array and the control circuits is shown in Fig. 10.15(b). In recent DRAM architectures, the bit lines are folded and precharged to $\frac{1}{2}V_{DD}$ to improve noise-immunity and reduce power consumption. In addition, the amplifier to sense the perturbed signal at the bit line is shared by adjacent blocks. One plate of the memory cell is biased at $\frac{1}{2}V_{DD}(V_P)$ to reduce the electric field intensity across the capacitor. The operations of one-transistor DRAM cell are "read," "write," and "refresh" operations. Before all the operations, the bit lines (*BL* and *BLB*) and sensing nodes (*SA* and *SAB*) are set to a precharge level, half-V_{DD} through bit and sensing line equalizers, respectively.

The timing diagram of the DRAM read operation for data "1" is shown in Fig. 10.16(a). Before the read operation, the signals to precharge the bit lines (*PEQ*) and bit line sense amplifier (*PSAEQ*) are disabled. The bit line amplifier is shared by two adjacent memory array blocks to reduce the chip size. Thus, the memory array select signal (*PISOi*) is set to the boosted voltage (V_{PP}). V_{PP} is an on-chip boosted voltage. Since the nMOS transistor is used as a switch, the level of V_{PP} is higher than the operating voltage (V_{DD}) plus the threshold voltage of the nMOS transistor for full charge restoration. A word line in the memory array block is selected according to the row address. The level of the word line is also higher than the operating voltage plus the threshold voltage of the cell access transistor (M1) for the same reason.

When the word line is enabled, the charge at the cell capacitor (e.g., C_S) is shared with the bit line capacitance. Since the bit line is precharged to $\frac{1}{2}V_{DD}$ a small voltage difference is developed at the bit line and the voltage of the storage node (*S*) becomes the same as that of the bit line. The resulting voltage perturbation is expressed as

$$\Delta V = \frac{C_S}{C_{BL} + C_S} \frac{V_{DD}}{2} \tag{10.1}$$

where C_S is the capacitance of the storage device and C_{BL} is the effective capacitance of the bit line including the parasitic line and junction capacitances of the cell access transistors connected with the same bit line.

Typically, ΔV is about $100 \sim 200$ mV since C_{BL} is larger than C_S by about 10 times. Note that the cell voltage has changed from V_{DD} to $\frac{1}{2}V_{DD} + \Delta V$. It means that the read operation is destructive and the cell data needs to be restored.

To sense the small signal difference, a latch amplifier is used. The sense nodes (*SA* and *SAB*) are connected together in the same block and the source transistors (*MNS* and *MPS*) are placed per memory array segment (e.g., per 64 bit lines). The control signals (*PSA* and *PSAB*) are typically activated sequentially to reduce the charge injection and short circuit current due to simultaneous activation of N and P latches in the bit line sense amplifier. Hence, the *BLB* node is discharged to the ground and the signal is amplified. The level of *BLB* and *BL* nodes eventually reaches the ground and the operating voltage, respectively and the voltage of the storage node is recovered. This is called the *restoring* operation of the cell data. All the cells connected with the same row perform this sequence of operations. Therefore, the active power consumption of DRAM is typically larger than other memory devices and is increasing for high memory density.

The data on the bit lines are read by transferring the voltage levels to the secondary data lines (*BL_IO* and *BL_IOB*). After the voltage differences on the bit lines

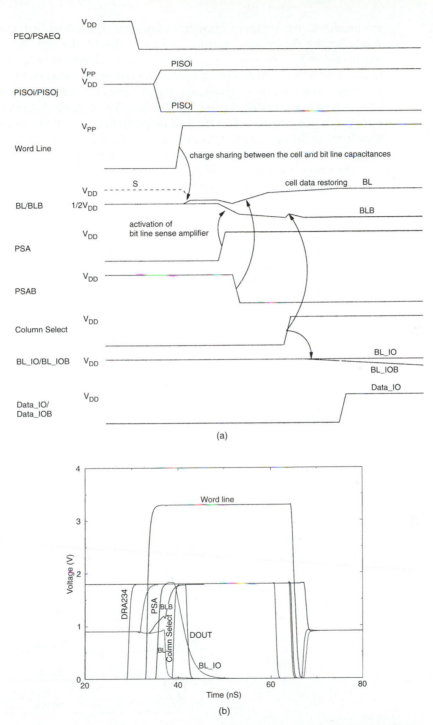

(a)

(b)

Figure 10.16 DRAM read operation. (a) Timing diagram. (b) Simulated waveforms for data "0."

are amplified, the column switch transistors are enabled by activating a column select line by the column decoder. Typically, the *BL_IO* and *BL_IOB* lines are precharged at V_{DD} or $V_{DD} - V_{tn}$, where V_{tn} is the threshold voltage of the nMOS load transistor. Since *BLB* is discharged to the ground, the level of *BL_IOB* is slowly discharged due to its large capacitance (e.g., 10 times C_{BL}) through *MNC*2. The level of *BLB* is affected slightly by the *BL_IOB* line. The voltage difference on the secondary data line is amplified by the read amplifier to the full CMOS output level and transferred to the memory interface circuit to drive the off-chip load. The simulated waveforms for the DRAM read operation is shown in Fig. 10.16(b).

The timing diagram to write data "0" into the cell is shown in Fig. 10.17. The write operation starts with reading the cell data to be modified, in a sequence that is

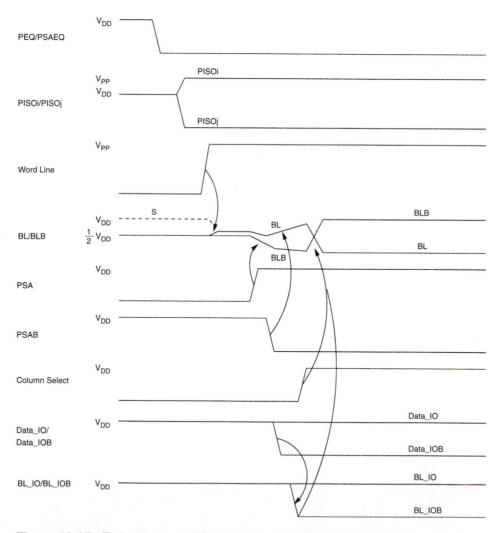

Figure 10.17 Timing diagram of DRAM write operation for data "0."

identical to the normal read operation. The data to be written into the cell is converted to CMOS level through *data input buffers* in the memory chip interface. Since the load capacitance of *BL_IO* and *BL_IOB* lines is large (e.g., $1 \sim 2$ pF) the write buffer is required to drive those lines. The column switch transistors are selected by the output of the column decoder and the levels of the bit lines and the cell data are changed. The write operation is faster than the read operation since it is performed by the strong write driver (buffer) using rail-to-rail CMOS levels.

DRAM Operation Modes

DRAM operation modes can be classified into two categories, *asynchronous* and *synchronous* modes depending on the use of a system clock. In asynchronous mode, the control of the chip and data read are done by the chip control signals such as $\overline{RAS}$ and $\overline{CAS}$ while those operations are referenced to a system clock in synchronous mode.

DRAM operation starts as $\overline{RAS}$ is pulled down. When $\overline{RAS}$ goes low, the DRAM chip activates the row circuitry such as row address buffers, row decoders, word line drivers, bit line sense amplifiers, etc. A feature of the DRAM operation is the use of *address multiplexing* scheme. That is, the address can be used as either a row or a column address by control signals, $\overline{RAS}$ and $\overline{CAS}$. The address captured at the falling edge of $\overline{RAS}$ and $\overline{CAS}$ becomes a row and a column address, respectively. The advantage of the address multiplexing scheme is to reduce the chip package size, since the package size is mainly limited by the number of pins. In typical systems, DRAM chips are used as main memory and the number of chips increases for large, high performance, and advanced systems. Therefore, since DRAM chips occupy a large portion of the system board area, small size of the package is strongly desirable.

One word line (physically multiple word lines in the chip) is selected. The length of word line (i.e., the number of cells connected to the word line) is determined by a *Refresh* cycle constraint. For example, if the refresh cycle is specified as *16K cycles per 256 ms* for a 256-Mb DRAM, it means that all DRAM cells must be refreshed by 16K row operations within 256 milliseconds. Therefore, 16K cells (256 Mb/16K) are accessed and refreshed by one row operation.

Some of data from the same word line are selected by a column address captured at the falling edge of $\overline{CAS}$. The column address generates *Column Select Line (CSL)* and connects some of bit lines to the common data lines (*BL_IO* and *BL_IOB*). With some time delay, the DRAM chip sends out data to external devices. Each $\overline{RAS}$ and $\overline{CAS}$ need precharge time (high level) to reset row and column circuitry before the new data access. The time to read the cell data from the falling edge of $\overline{RAS}$ is called t_{RAC} and memory read latency. Figure 10.18(a) shows *single bit access* in asynchronous operation, directed by different row and column addresses. Typically, the operating frequency is about 20–30 MHz.

Cell data at the same row can be read at a faster rate, about 40–60 MHz, by changing column addresses while keeping the row address active. Different *CSLs* are activated by column addresses and data can be read fast until the end cell of the same word line (*page*). This is called *page access mode* as shown in Fig. 10.18(b).

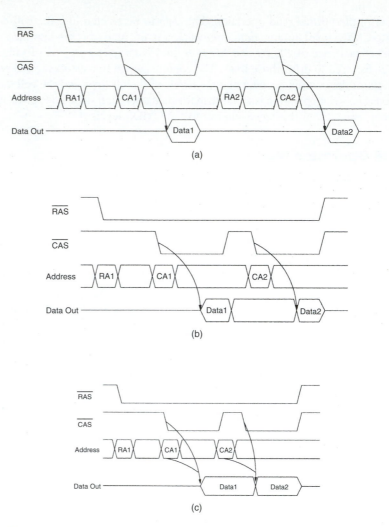

Figure 10.18 Various asynchronous DRAM read modes. (a) Single bit read. (b) Page mode read. (c) Extended Data-Out (EDO) mode read.

In page access mode, a $\overline{CAS}$ precharge time is required to reset column circuitry to capture a new column address and to guarantee a data hold time for other external systems to fetch data safely. By modifying DRAM circuitry related with the DRAM chip output interface, read frequency can be improved further. Instead of capturing the column address at the falling edge of $\overline{CAS}$, a new column address is captured at the rising edge of $\overline{CAS}$ and the read data is maintained during $\overline{CAS}$ precharge time. This is called *EDO* (*Extended Data-Out*) access mode. Due to early column address set-up, read frequency can be about 80 MHz.

In the synchronous access mode, by contrast, the data read frequency can be improved very significantly with the use of the system clock to schedule the operations.

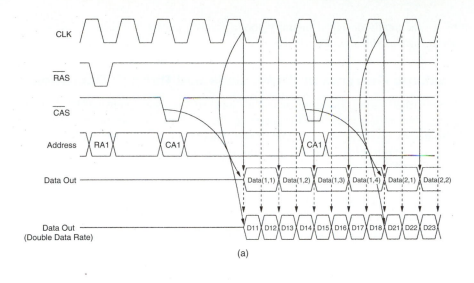

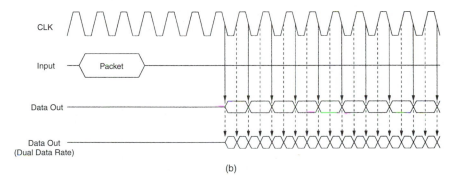

(a)

(b)

Figure 10.19 Synchronous DRAM read modes. (a) Synchronous mode read (four bit burst read). (b) Serial mode read.

All DRAM operations are done by the combination of control signals and the system clock (*CLK*) as shown in Fig. 10.19.

At the falling edge of *CLK*, control signals and addresses become active. Internal chip operations are identical but *pipelined* based on the system clock to improve data *throughput* as shown in Fig. 10.19(a). Read latency is defined as the number of the system clock cycles and typically 2 or 3. Since data is sent out with the system clock, read frequency can be about 150 – 200 MHz for conventional synchronous mode. Recently, to improve *bandwidth,* data are sent out at both edges of the system clock, resulting in an effective bandwidth increase by a factor of 2, called *DDR* (*Dual Data Rate*).

By using improved chip interface, data read frequency can be improved further to 400–800 MHz (DDR). The chip interface with small signal swing and clock recovery scheme are used to maximize the frequency. Inputs to control the chip are sent by a specifically designed memory controller as a packet and data is sent out in a serial form as shown in Fig. 10.19(b).

Leakage Currents in DRAM Cells and Refresh Operation

Figure 10.20 shows equivalent schematic and cross-sectional views of DRAM cells. For high density, the contact to connect to the cell access transistors and the bit line is shared by two adjacent cells. Different leakage mechanisms contribute to the decay of charge at the cell and total leakage current is expressed as

$$I_{leakage} = I_{sub} + I_{tunneling} + I_j + I_{cell-to-cell} \tag{10.2}$$

where I_{sub} is the leakage current through the cell access transistor, $I_{tunneling}$ the tunneling current through thin dielectric material such as Ta_2O_5, I_j the junction leakage current at the storage node, $I_{cell-to-cell}$ the leakage current across the field oxide. As

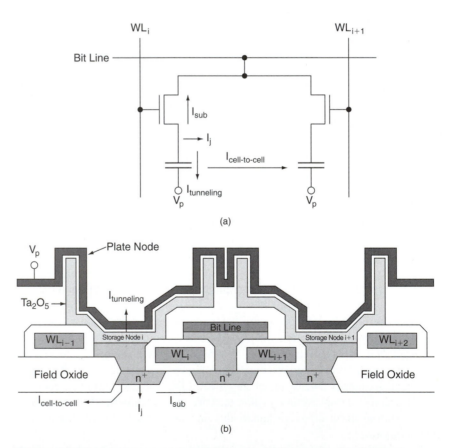

Figure 10.20 Leakage currents in DRAM cells. (a) Schematic view. (b) Cross-sectional view.

the minimum feature size is scaled-down I_j tends to take a smaller portion of the total leakage current and $I_{cell-to-cell}$ is usually negligible due to the thick field oxide as in the trench isolation technique. Since I_{sub} is a strong function of the threshold voltage, increasing V_{SB} of the access transistor is an effective and practical method to reduce this leakage component. For this purpose, an on-chip negative voltage generator, called V_{BB}, is widely used in DRAM chips. However, as the thickness of dielectric material is reduced to increase the cell capacitance, the tunneling leakage current ($I_{tunneling}$) becomes a serious issue. As already mentioned in the previous sections, a periodic refresh operation is always required to compensate for the steady leakage of charge from the storage nodes, and to ensure the conservation of data in the DRAM cells.

The refresh operation to recharge the cell capacitor is accomplished by reading and restoring operations for the cell using different control signals than those used for the normal read/write operations. A feature of all refresh operations is that the DRAM chip does not send the cell data out to the external devices. Figure 10.21 shows typical timing diagrams for refresh operations, called $ROR(\overline{RAS}$-only refresh), $CBR(\overline{CAS}$-before-$\overline{RAS})$ refresh and self refresh.

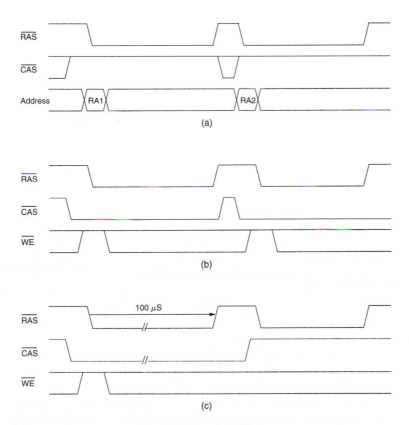

Figure 10.21 Timing diagram for various DRAM refresh operation modes. (a) ROR refresh. (b) CBR refresh. (c) Self refresh.

In the *ROR* refresh mode, the refresh address (row address) is provided by an external device. A read and restore operation is performed that is similar to the normal read operation, while the data is not sent out to the output buffers. The *CBR* refresh mode starts by lowering $\overline{CAS}$ earlier than $\overline{RAS}$. The write control signal ($\overline{WE}$) must be high. In the CBR refresh mode, the refresh address is generated by an on-chip counter. The refresh operation is performed periodically (e.g., 128 ms), as long as the DRAM chip is in operation (power on). The period for the charge of the cell to be discharged to a level where the cell data cannot be read is usually longer than the refresh interval controlled by the external devices and depends on environmental parameters such as V_{DD}, temperature, etc. Therefore the refresh period can be set differently according to the operating conditions. By keeping the *CBR* refresh timing for an extended period (e.g., 100 μs), the self refresh mode is started. Not only row addresses but also control signals to activate the row operation are generated by internal circuits, depending on how the refresh interval is programmed.

DRAM Input/Output Circuits

Since logic levels of the memory system board (e.g., TTL (0.8 V $\leftrightarrow$ 2.0 V)) and those of the memory chip (e.g., CMOS) are different, memory interface circuits are required to convert logic levels. Those circuits are called *input* and *output* buffers and are specific for different logic interfaces. Figure 10.22 and Fig. 10.23 show typical input and output buffers used in memory chips.

In an inverter-type input buffer, logic threshold voltage is determined by the ratio of pMOS and nMOS transistor sizes while a reference voltage is used in other latch- and differential amplifier-types. The inverter type has the simplest structure but its sensitivity to process, temperature and operating voltage is larger than that of other types. The latch type shows the fastest speed and lowest power consumption but it requires an additional signal (PACT) to enable the buffer. Table 10.3 compares the characteristics of three typical input buffers.

The output buffer sends the read data to external devices such as a memory controller or a processor. Unlike the input buffer, the output buffer converts a data in CMOS level to a system logic level (e.g., TTL) and needs to drive a large capacitive load. Since the outputs of several DRAM chips are usually connected to the same data bus line on board, the output buffer of any DRAM chip must be able to keep a high-impedance ($Hi - Z$) state when the chip is not selected, in order to prevent any interference of outputs from different DRAM chips. The output buffer consists of a pull-up and a pull-down driver. The pull-up driver can be built by using a pMOS transistor or an nMOS transistor, as shown in Fig. 10.23(a) and (b), respectively. During *POE* (output buffer enable signal) is low, the pull-up and pull-down drivers are turned off and the output buffer is in a tri-state. When *POE* goes high, the output buffer starts to drive the external data line depending on the read data (D).

To select a cell from a 2^{2M} memory array, M address bits are needed when the address is multiplexed. Since using a decoder with M transistors in series is practically impossible, the memory decoding scheme is composed of *pre* and *main* decoders as shown in Fig. 10.24. The timing diagram for row address decoding and

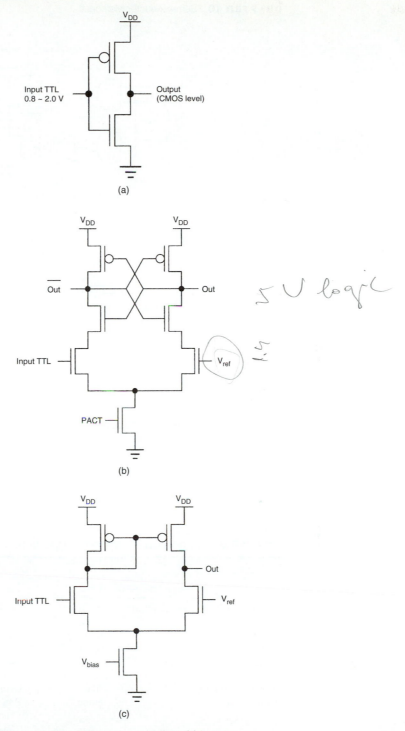

Figure 10.22 Memory input buffers. (a) Inverter type, (b) latch type, (c) differential amplifier type.

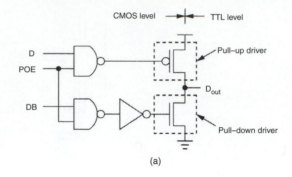

(a)

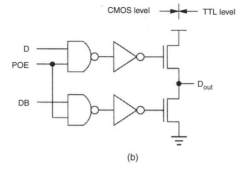

(b)

Figure 10.23   Memory output buffers. (a) pMOS pull-up and nMOS pull-down structure, (b) nMOS pull-up and nMOS pull-down structure.

Table 10.3   Characteristic comparison of input buffers

	Buffer type		
	Inverter	**Latch**	**Differential**
Logic threshold determination (V_{IH} and V_{IL})	By W_P/W_N ratio	By V_{ref}	By V_{ref}
Speed	Slow	Fastest	Fast
Standby current	Small	Smallest	Large
Sensitivity to V_{DD} and temperature	Large	Small	Small
Noise immunity	Bad	Good	Good
Constraint	None	Precharge and activation signals needed	None

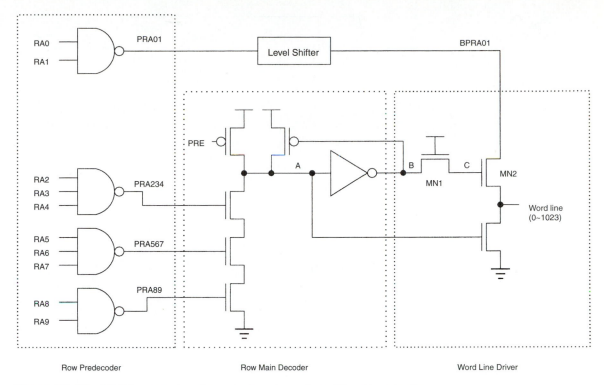

Figure 10.24   DRAM row predecoder, main decoder, and word line driver circuits.

word line selection is shown in Fig. 10.25. Address bits are grouped in 2 or 3 bits and decoded ($PRA01$, $PRA234$, $PRA567$, and $PRA89$) first in the predecoder stage. As word lines are required to have a high voltage (V_{PP}) for full data restoration, the output level of a predecoder ($PRA01$) is boosted by using a level shifter. The main decoder uses the predecoded outputs to produce the word selection signal, and a self-bootstrapped driver is placed for transferring the boosted voltage to the highly capacitive word line without signal degradation.

When the main decoder is selected, node A is discharged and becomes low. Node C is set to $V_{DD} - V_{TN}$. As the boosted signal ($BPRA01$) arrives at the drain of MN2, node C is coupled to a higher level and the boosted signal is transferred to the word line without experiencing a voltage drop. The voltage of node C is expressed as

$$V_C = V_{PP} + \Delta V = V_{DD} - V_{TN} + \frac{C_{MN2}}{C_{MN2} + C_{Cparasitic}} V_{PP} \qquad (10.3)$$

where C_{MN2} is the total effective gate and parasitic capacitance of MN2 and $C_{Cparasitic}$ is parasitic capacitance at the node C.

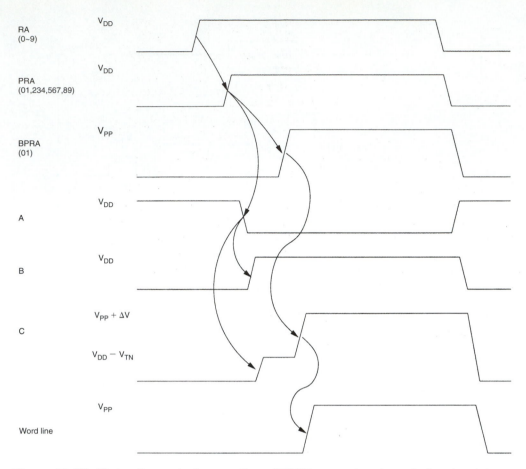

Figure 10.25 Timing diagram for the operations of DRAM row predecoder, main decoder, and word line driver circuits.

Various kinds of sense amplifiers are used to detect a signal difference on data lines (*BL_IO* and *BL_IOB*). Current-mirror differential, semilatch, and full latch types are shown in Fig. 10.26(a), (b), and (c), respectively. The current-mirror type is popular and has a good common-mode rejection ratio. But the relatively large area required for large transconductance of input transistors and the large power consumption can become limiting factors for this circuit. To improve the signal swing at the output nodes, a differential-output type is preferred as shown in Fig. 10.26(c).

The cross-coupled differential amplifier is used to achieve high speed, low power consumption, and small area penalty. In the full CMOS type, *OUT* and $\overline{OUT}$ are precharged to V_{SS} by *PSAE*. When *PSAE* goes to high, the amplifier starts to amplify a signal difference. As a node (*A* or *B*) is discharged, the other output (*B*) becomes isolated from the other input signal ($\overline{IN}$), the sensing speed can be accelerated,

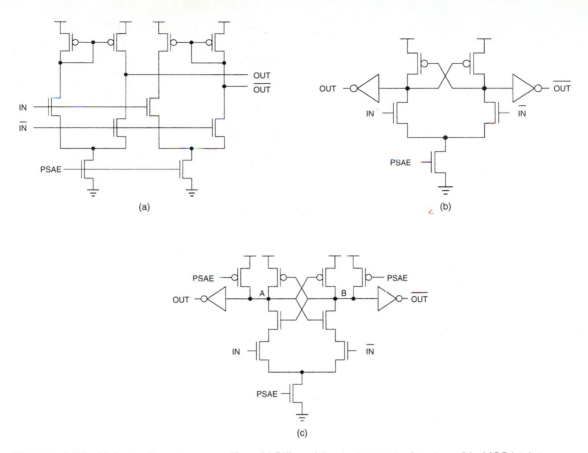

Figure 10.26 Various voltage sense amplifiers. (a) Differential-output current-mirror type, (b) pMOS latch type, (c) full CMOS latch type.

and power consumption can be small. Since there is also no static power consumption after the sensing is completed, unlike the current-mirror type, this structure can be used in low power applications. However, a precharge signal (*PSAE*) is required and output nodes are precharged properly before starting a new sensing operation. A valid signal difference must be guaranteed because its operation cannot be reversed; that is, the states of output nodes cannot be reset once they are latched until the next precharge signal arrives.

Performance of the semilatch type sense amplifier is between that of the current-mirror and that of the full CMOS latch type. The data due to invalid signal differences can be corrected with speed penalty, unlike the CMOS latch type.

Memory devices are using several on-chip voltage generators such as an internal voltage generator (V_{INT}), a half V_{DD} generator (V_{BL} and V_P), a substrate bias generator (V_{BB}), and a boosted voltage generator (V_{PP}). These on-chip voltages are used for the purpose of reducing the operating current (V_{INT}), stabilizing operations

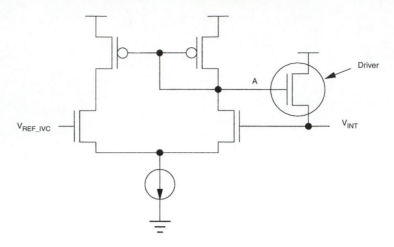

Figure 10.27   Internal voltage regulator circuit.

(V_{BL} and V_{BB}), eliminating a reliability issue (V_P) and improving performance, and programming the cell data (V_{PP}).

Lowering the operating voltage is the most effective technique to reduce the power consumption of DRAMs. But the system supply voltage has not been reduced much (e.g., the supply voltage of typical memory system is still 3.3 V while that of a 1-Gb DRAM is scaled down to 1.5 V). Therefore, an alternative method is to use a regulated voltage only for the memory chip. Figure 10.27 shows a voltage regulator circuit that consists of a level comparator and a driver. A reference voltage (V_{REF_IVC}) is used to set an internal voltage level (V_{INT}). When the internal level is lower than the reference level, the voltage of node A is raised and turns on the driver transistor to supply more current.

Most DRAM chips adopt a folded bit line structure with a half V_{DD} sensing scheme for improved noise immunity and low power consumption. That is, the bit line pairs (*BL* and *BLB*) in the same array block are used to detect the cell data and these lines are precharged at the half V_{DD} level as shown in Fig. 10.15. Since tens of thousands of cells are read (or refreshed) at every DRAM operation cycle, reducing the voltage swing of the bit line is effective to reduce power consumption. In addition, the half V_{DD} is applied to to the cell plate node (V_P) to reduce the electric field across thin dielectric material used for the cell capacitor.

Figure 10.28(a) shows a circuit for generating a half V_{DD} voltage. It is composed of a bias circuit and a driver. The transistors in the bias circuit are sized to set the voltage of node B to $V_{DD}/2$. Therefore the voltages of node A and C are $V_{DD}/2 + V_{TN}$ and $V_{DD}/2 - |V_{TP}|$, respectively and the output voltage of the driver is stabilized at $V_{DD}/2$. Since the driver transistors are weakly turned on, the static current is very small (e.g., tens of μA). The push-pull driver with large transistors is used to suppress any unexpected change at the output node quickly by turning on either transistor strongly. Simulation waveforms of the half V_{DD} generator are shown in Fig. 10.28(b).

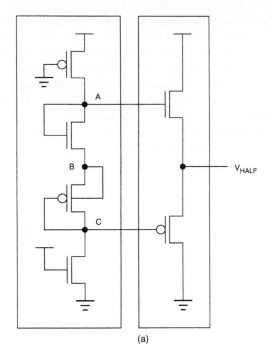

(a)

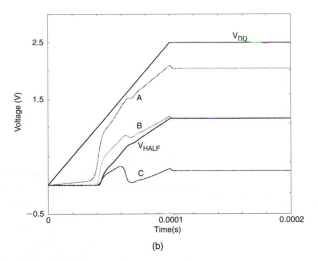

(b)

Figure 10.28 Half V_{DD} voltage generator. (a) Circuit diagram, (b) simulated output waveforms as the power supply voltage (V_{DD}) is ramped up from 0 V to 2.5 V in 100 μs.

The modern DRAM cell consists of one access transistor and one storage capacitor. The charge stored at the capacitor (typically, <100 fC) decays in time due to leakage current, thus, the stored charge must be restored before its level becomes undetectable. The characteristics of the cell play an important role in DRAM design and process since it determines the refresh cycle (during refresh operation, typically DRAM cannot be accessed) and the cell structure. The subthreshold leakage current through the access transistor is a major source of charge decay at the storage node and it is very sensitive to the change of substrate voltage. Since the charge stored at the cell is very small, the shift of threshold voltage due to voltage difference between the source and substrate voltage can cause exponential increase of leakage current and the memory cell charge can decay very quickly. To mitigate the problem, the substrate (p-type) is biased to a negative voltage level instead of V_{SS}. The change of threshold voltage for the substrate bias voltage shown in Fig. 10.29(a) is expressed as

$$\frac{\Delta V_T}{\Delta V_{SB}} \propto \sqrt{V_{SB}} \tag{10.4}$$

where V_{SB} is the difference between the source and the substrate voltage of the transistor.

The effective load capacitance of the bit line is also reduced by applying a negative substrate bias. Since it is a sum of a wire capacitance of the bit line and junction capacitances of the access transistors connected to the bit line, the effective capacitance can be smaller due to the reduced junction capacitance by increase of the junction depletion area.

The substrate bias generator is composed of a ring oscillator, a charge pump, and a level detector as shown in Fig. 10.29(b). Holes are extracted from the substrate by the charge pump to achieve a negative potential at the substrate. Figure 10.29(c) shows the behavior of signal nodes. When node A is high, the voltage of node B becomes V_{TN1}, where V_{TN1} is the threshold voltage of a MOS diode, MN1. At the

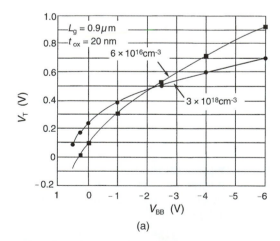

(a)

Figure 10.29 Negative substrate bias voltage (V_{BB}) generator. (a) Dependence of threshold voltage on substrate bias voltage.

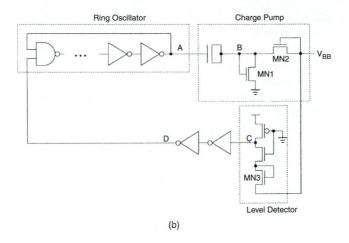

(b)

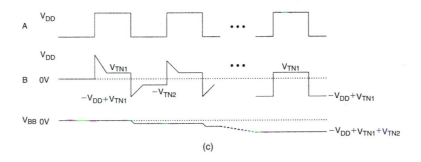

(c)

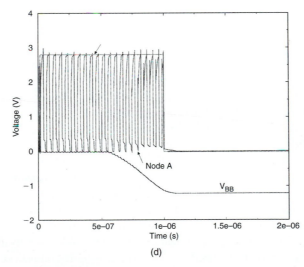

(d)

Figure 10.29 *(continued)*   (b) Circuit diagram, (c) timing diagram, (d) simulated waveforms.

falling edge of node A, node B is coupled down to $-V_{DD} + V_{TN1}$ and MN2 is turned on until node B voltage reaches $-V_{TN2}$ by extracted holes, where V_{TN2} is the threshold voltage of MN2. When node A goes high, these holes are discharged to *Ground* through MN1 and the voltage of node B becomes V_{TN1}. By repeating this operation, the substrate voltage can be lowered to $-V_{DD} + V_{TN1} + V_{TN2}$. Since the lowered substrate bias can cause the performance degradation of nMOS transistors by increasing the threshold voltage, the substrate level is set to a point where a suitable compromise between performance and leakage current reduction can be achieved. The substrate bias voltage is set by sizing transistors in the level detector. When the substrate bias is lower than a target value, MN3 is turned on more strongly, node C is pulled down, and node D becomes low, resulting in disabling the ring oscillator as shown in Fig. 10.29(d).

Contrary to the negative bias generator, an on-chip boosted voltage can be generated by supplying and reserving charge in a very large capacitor (typically, several nF). Basic operation is almost the same except for the charge pumping circuit. A circuit implementation is shown in Fig. 10.65.

10.3   Static Random Access Memory (SRAM)

As already explained in Section 10.1, read/write (R/W) memory circuits are designed to permit the modification (writing) of data bits to be stored in the memory array, as well as their retrieval (reading) on demand. The memory circuit is said to be *static* if the stored data can be retained indefinitely (as long as a sufficient power supply voltage is provided), without any need for a periodic refresh operation. We will examine the circuit structure and the operation of SRAM cells, as well as the peripheral circuits designed to read and write the data.

The data storage cell, i.e., the 1-bit memory cell in static RAM arrays, invariably consists of a simple latch circuit with two stable operating points (states). Depending on the preserved state of the two-inverter latch circuit, the data being held in the memory cell will be interpreted either as a logic "0" or as a logic "1". To access (read and write) the data contained in the memory cell via the bit line, we need at least one switch, which is controlled by the corresponding word line, i.e., the row address selection signal (Fig. 10.30(a)). Usually, two complementary access switches consisting of nMOS pass transistors are implemented to connect the 1-bit SRAM cell to the complementary bit lines (columns). This can be likened to turning the car steering wheel with both left and right hands in complementary directions.

Figure 10.30(b) shows the generic structure of the MOS static RAM cell, consisting of two cross-coupled inverters and two access transistors. The load devices may be polysilicon resistors, depletion-type nMOS transistors, or pMOS transistors, depending on the type of the memory cell. The pass gates acting as data access switches are enhancement-type nMOS transistors.

The use of resistive-load inverters with undoped polysilicon resistors in the latch structure typically results in a significantly more compact cell size, compared with the other alternatives (Fig. 10.30(c)). This is true since the resistors can be stacked on top of the cell (using double-polysilicon technology), thereby reducing the cell size to four transistors, as opposed to the six transistor cell topologies. If multiple polysilicon

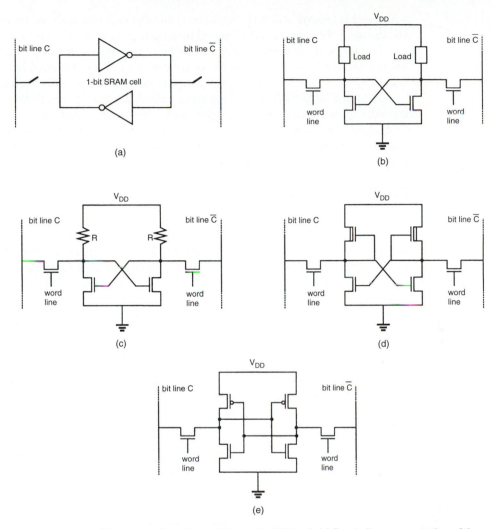

Figure 10.30 Various configurations of the static RAM cell. (a) Symbolic representation of the two-inverter latch circuit with access switches. (b) Generic circuit topology of the MOS static RAM cell. (c) Resistive-load SRAM cell. (d) Depletion-load nMOS SRAM cell. (e) Full CMOS SRAM cell.

layers are available, one layer can be used for the gates of the enhancement-type nMOS transistors, while another can be used for the load resistors and interconnects.

In order to attain acceptable noise margins and output pull-up times for the resistive-load inverter, the value of the load resistor has to be kept relatively low, as already examined in Section 5.2. On the other hand, a high-valued load resistor is required in order to reduce the amount of standby current being drawn by each memory cell. Thus, there is a trade-off between the high resistance required for low power and the requirement to provide wider noise margins and high speed. The power consumption issue will be addressed in more detail. The six-transistor depletion-load nMOS SRAM cell shown in Fig. 10.30(d) can be easily implemented with one polysilicon and one metal layer, and the cell size tends to be relatively small, especially with the use of buried metal-diffusion contacts. The static characteristics and the noise margins of this memory cell are typically better than those of the resistive-load cell. The static power consumption of the depletion-load SRAM cell, however, makes it an unsuitable candidate for high density SRAM arrays.

The full CMOS SRAM cells shown in Fig. 10.30(e) are currently most popular due to the lowest static power dissipation among the various circuit configurations and compatability with current logic process. In addition, the CMOS cell offers superior noise margins and switching speed as well. The comparative advantages and disadvantages of the static RAM cells will be investigated in depth later in this section.

Full CMOS SRAM Cell

A low-power SRAM cell may be designed simply by using cross-coupled CMOS inverters. In this case, the stand-by power consumption of the memory cell will be limited to relatively small leakage currents of both CMOS inverters. The possible drawback of using CMOS SRAM cells, on the other hand, is that the cell area tends to be slightly larger than that of the other alternatives shown in Fig. 10.30, in order to accommodate the n-well for the pMOS transistors and the polysilicon contacts.

The circuit structure of the full CMOS static RAM cell is shown in Fig. 10.31, along with the pMOS column pull-up transistors on the complementary bit lines. The memory cell consists of a simple CMOS latch (two inverters connected back-to-back), and two complementary access transistors (M3 and M4). The cell will preserve one of its two possible stable states, as long as the power supply is available. The access transistors are turned on whenever a word line (row) is activated for read or write operation, connecting the cell to the complementary bit-line columns.

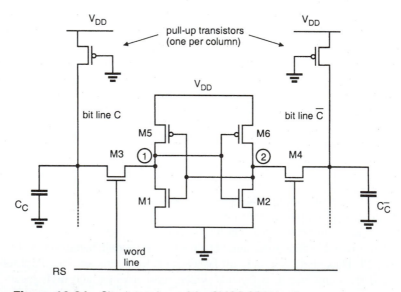

Figure 10.31 Circuit topology of the CMOS SRAM cell.

The most important advantage of this circuit topology is that the static power dissipation is very small; essentially, it is limited by the leakage current of the pMOS transistors. A CMOS memory cell thus draws current from the power supply only during a switching transition. The low stand-by power consumption has certainly been a driving force for the increasing prominence of high-density CMOS SRAMs.

Other advantages of CMOS SRAM cells include high noise immunity due to larger noise margins, and the ability to operate at lower power supply voltages than, for example, the resistive-load SRAM cells. The major disadvantages of CMOS memories historically were larger cell size, the added complexity of the CMOS process, and the tendency to exhibit "latch-up" phenomena. With the widespread use of multi-layer polysilicon and multi-layer metal processes, however, the area disadvantage of the CMOS SRAM cell has been reduced significantly in recent years. Considering the undisputable advantages of CMOS for low-power and low-voltage operation, the added process complexity and the required latch-up prevention measures do not present a substantial barrier against the implementation of CMOS cells in high density SRAM arrays. Figure 10.32 compares typical layouts of the four-transistor resistive-load SRAM cell and the six-transistor full CMOS SRAM cell.

Note that unlike the nMOS column pull-up devices used in resistive-load SRAM, the pMOS column pull-up transistors shown in Fig. 10.31 allow the column voltages to reach full V_{DD} level. To further reduce power consumption, these transistors can also be driven by a periodic precharge signal, which activates the pull-up devices to charge-up column capacitances.

CMOS SRAM Cell Design Strategy

To determine the (W/L) ratios of the transistors in a typical CMOS SRAM cell as shown in Fig. 10.31, a number of design criteria must be taken into consideration. The two basic requirements which dictate the (W/L) ratios are: (a) the data-read

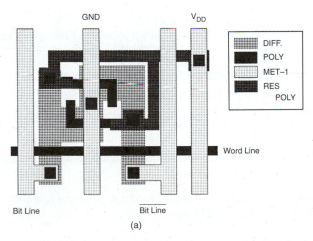

Figure 10.32   (a) Layout of the resistive-load SRAM cell.

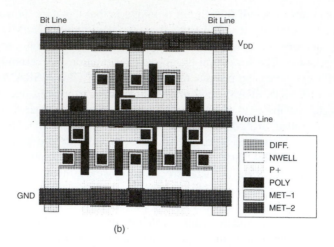

(b)

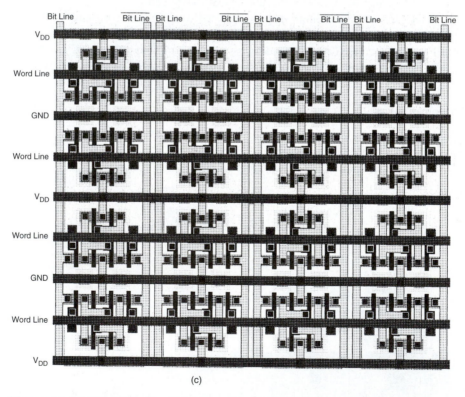

(c)

Figure 10.32 *(continued)*   (b) Layout of the CMOS SRAM cell. (c) Layout of a 4-bit ×
4-bit SRAM array, consisting of 16 CMOS SRAM cells.

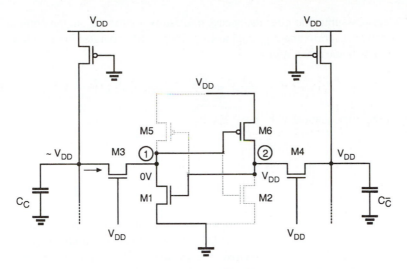

Figure 10.33 Voltage levels in the SRAM cell at the beginning of the "read" operation.

operation should not destroy the stored information in the SRAM cell, and (b) the cell should allow modification of the stored information during the data-write phase. Consider the data-read operation first, assuming that a logic "0" is stored in the cell. The voltage levels in the CMOS SRAM cell at the beginning of the "read" operation are depicted in Fig. 10.33. Here, the transistors M2 and M5 are turned off, while the transistors M1 and M6 operate in the linear mode. Thus, the internal node voltages are $V_1 = 0$ and $V_2 = V_{DD}$ *before* the cell access (or pass) transistors M3 and M4 are turned on. The active transistors at the beginning of the data-read operation are highlighted in Fig. 10.33.

After the pass transistors M3 and M4 are turned on by the row selection circuitry, the voltage level of column $\overline{C}$ will not show any significant variation since no current will flow through M4. On the other half of the cell, however, M3 and M1 will conduct a nonzero current and the voltage level of column C will begin to drop slightly. Note that the column capacitance C_C is typically very large; therefore, the amount of decrease in the column voltage is limited to a few hundred millivolts during the read phase. The data-read circuitry to be examined later in this chapter is responsible for detecting this small voltage drop and amplifying it as a stored "0." While M1 and M3 are slowly discharging the column capacitance, the node voltage V_1 will increase from its initial value of 0 V. Especially if the (W/L) ratio of the access transistor M3 is large compared to the (W/L) ratio of M1, the node voltage V_1 may exceed the threshold voltage of M2 during this process, forcing an unintended change of the stored state. The key design issue for the data-read operation is then to guarantee that the voltage V_1 does not exceed the threshold voltage of M2, so that the transistor M2 remains turned off during the read phase, i.e.,

$$V_{1,max} \leq V_{T,2} \tag{10.5}$$

We can assume that after the access transistors are turned on, the column voltage V_C remains approximately equal to V_{DD}. Hence, M3 operates in saturation while M1 operates in the linear region.

$$\frac{k_{n,3}}{2}(V_{DD} - V_1 - V_{T,n})^2 = \frac{k_{n,1}}{2}\left(2(V_{DD} - V_{T,n})V_1 - V_1^2\right) \qquad (10.6)$$

Combining this equation with (10.3) results in:

$$\frac{k_{n,3}}{k_{n,1}} = \frac{\left(\dfrac{W}{L}\right)_3}{\left(\dfrac{W}{L}\right)_1} < \frac{2(V_{DD} - 1.5V_{T,n})V_{T,n}}{(V_{DD} - 2V_{T,n})^2} \qquad (10.7)$$

The upper limit of the aspect ratio found above is actually more conservative, since a portion of the drain current of M3 will also be used to charge-up the parasitic node capacitance of node (1). To summarize, the transistor M2 will remain in cut-off during the read "0" operation if condition (10.5) is satisfied. A symmetrical condition also dictates the aspect ratios of M2 and M4.

Now consider the write "0" operation, assuming that a logic "1" is stored in the SRAM cell initially. Figure 10.34 shows the voltage levels in the CMOS SRAM cell at the beginning of the data-write operation. The transistors M1 and M6 are turned off, while the transistors M2 and M5 operate in the linear mode. Thus, the internal node voltages are $V_1 = V_{DD}$ and $V_2 = 0$ V *before* the cell access (or pass) transistors M3 and M4 are turned on.

The column voltage V_C is forced to logic "0" level by the data-write circuitry; thus, we may assume that V_C is approximately equal to 0 V. Once the pass transistors M3 and M4 are turned on by the row selection circuitry, we expect that the node voltage V_2 remains *below* the threshold voltage of M1, since M2 and M4 are designed according to condition (10.5). Consequently, the voltage level at node (2) would not be

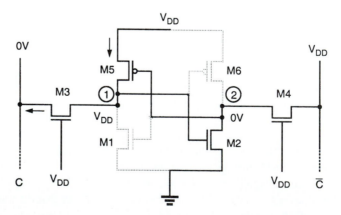

Figure 10.34   Voltage levels in the SRAM cell at the beginning of the "write" operation.

sufficient to turn on M1. To change the stored information, i.e., to force V_1 to 0 V and V_2 to V_{DD}, the node voltage V_1 *must be reduced below* the threshold voltage of M2, so that M2 turns off first (in fact, this condition may be relaxed in most cases; M2 will turn off when V_1 is reduced below the inversion threshold). When $V_1 = V_{T,n}$, the transistor M3 operates in the linear region while M5 operates in saturation.

$$\frac{k_{p,5}}{2}(0 - V_{DD} - V_{T,p})^2 = \frac{k_{n,3}}{2}\left(2(V_{DD} - V_{T,n})V_{T,n} - V_{T,n}^2\right) \qquad (10.8)$$

Rearranging this condition results in:

$$\frac{k_{p,5}}{k_{n,3}} < \frac{2(V_{DD} - 1.5V_{T,n})V_{T,n}}{(V_{DD} + V_{T,p})^2}$$

$$\frac{\left(\dfrac{W}{L}\right)_5}{\left(\dfrac{W}{L}\right)_3} < \frac{\mu_n}{\mu_p} \cdot \frac{2(V_{DD} - 1.5V_{T,n})V_{T,n}}{(V_{DD} + V_{T,p})^2} \qquad (10.9)$$

To summarize, the transistor M2 will be *forced* into cut-off mode during the write "0" operation if condition (10.7) is satisfied. This will guarantee that M1 subsequently turns on, changing the stored information. Note that a symmetrical condition also dictates the aspect ratios of M6 and M4.

Operation of SRAM

Figure 10.35 and Fig. 10.36 show the memory structure of a SRAM cell array and the timing diagrams for the read and write operations, respectively.

In the read operation shown in Fig. 10.36, a word line is selected by a row address. Typically the voltage of the word line is V_{DD} instead of a boosted voltage (V_{PP}) as in DRAM. In the read operation of DRAM, the voltage level on the bit line is changed by the charge sharing between the cell and bit line capacitances. Thus, the cell data is destructive and restoring operation should be followed after every read operation. However, the SRAM cell has a latch structure and the cell data is kept during the read operation without degradation. Consequently, a boosted voltage is not required. Another feature of the SRAM device is that a so-called address multiplexing scheme is not used; that is, a row and a column address are provided simultaneously by the external control devices. This is one of reasons why the access time of SRAM is faster than that of DRAM. The typical access time of a fast SRAM is several nanoseconds while that of DRAM is tens of nanoseconds. Before the read operation, the bit lines are usually precharged to V_{DD} or $V_{DD} - V_{TN}$ when an nMOS load transistor is used. Depending on the applications of SRAM devices (ultra low power for PDAs or high speed for caches), the load transistor is turned off or remains on during the read operation. When the wordline is enabled, one of the bit lines is discharged through an nMOS transistor connected to a "0" node of the cell. The current driving capability of the cell is very small (tens of μA) and the voltage change on the bit line is just tens of mV when the load transistor is turned on. The sense amplifier

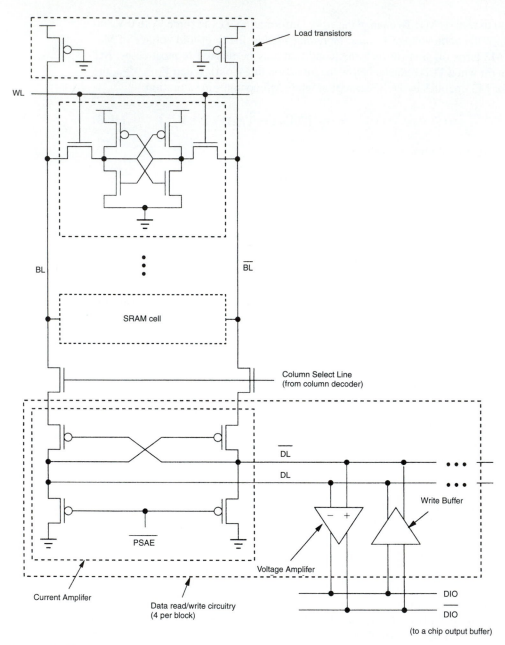

Figure 10.35 Memory structure of SRAM with read and write circuitry.

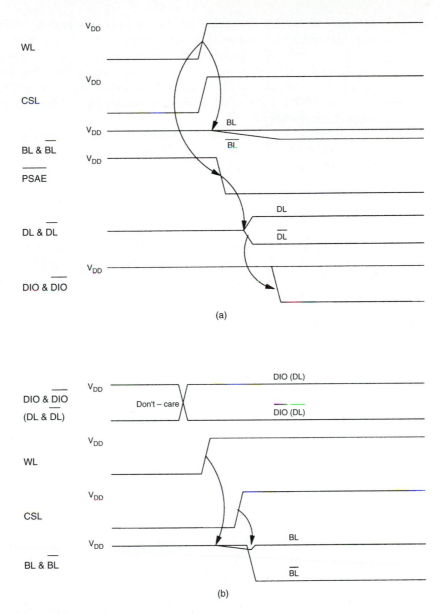

Figure 10.36 SRAM core operations. (a) Read timing diagram, (b) write timing diagram.

to detect the voltage difference on the bit lines is shared by multiple bit lines (e.g., 32) and the connection of the sense amplifier to a bit line pair is controlled by a column selection line. Typically a multi-stage amplifier is used to improve the read speed. A high gain voltage or current-mode amplifier, shown in Fig. 10.35, is used as the first-stage amplifier and the voltage-mode amplifier to generate CMOS level signals (e.g., *DIO* and $\overline{DIO}$) with a large current driving capability is used as the last stage of the amplifiers.

In the write operation, data to be written into a cell and provided by external devices and *DIO* and $\overline{DIO}$ and *DL* and $\overline{DL}$ lines are set according to the data. A word line is selected by a row address and one of the bit lines is discharged as in the read operation. When the column gates are opened, the write buffer starts to write data into the cell as shown in Fig. 10.36(b). Since the write buffer has much larger current driving capability than that of the cell, the write operation takes less time than the read operation.

Figure 10.37 shows waveforms of SRAM read operation for the cell data "0". An address with TTL level is converted into a CMOS level signal by an address buffer and decoded ($\overline{DA}$). Even though the operating voltage of the chip is 3.3 V, the amplitude of internal signals is scaled down to 1.2 V by the internal voltage regulator for

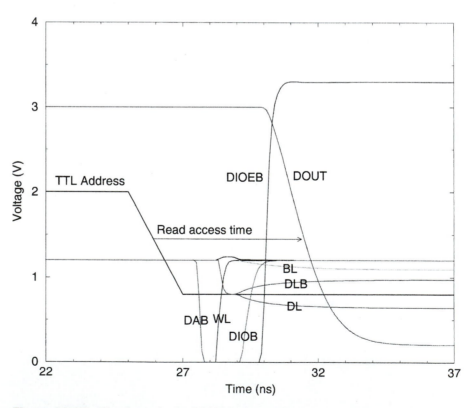

Figure 10.37   Waveforms for the SRAM read operation.

reduced power dissipation and for improved reliability. The word line (*WL*) is enabled and *BL* is discharged slightly. The signal difference on the bit lines are amplified by the first-stage amplifier (*DL* and $\overline{DL}$). A CMOS level signal ($\overline{DIO}$) is generated to drive the heavy-loaded line. The amplitude of the signal (e.g., 3.3 V) is changed to drive an output buffer effectively ($\overline{DIOE}$) and the read data are sent to the external devices (*DOUT*).

Leakage Currents in SRAM Cells

While leakage currents in DRAM cells affect the refresh interval of the DRAM chip, resulting in increased active and standby power consumption, leakage in SRAM cells typically contributes a major portion of the standby current of the chip. Since several millions of memory cells are integrated in the SRAM device, leakage current of each cell is accumulated to consume a fairly large amount of standby power as the feature size becomes scaled-down. This standby power is the key design parameter for low power chips used in hand-held devices such as PDAs. The leakage current of the SRAM cells consists of a junction current (I_j) at the data "1" node to the substrate, subthreshold leakage current (I_{nsub} and I_{psub}) flowing through turned-off nMOS and pMOS transistors and tunneling current ($I_{tunneling}$) cross the thin gate oxide as shown in Fig. 10.38. Unlike DRAM, measures such as increasing threshold voltage of transistors in the cell by either applying a negative substrate bias or using high threshold transistors should be considered carefully since there is a trade-off between the reduction of leakage current and the degradation of performance (read and write time).

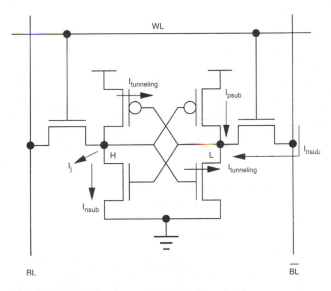

Figure 10.38   Leakage currents in SRAM cells.

SRAM Read/Write Circuits

Instead of the voltage sense amplifiers shown in Fig. 10.26, current-mode sense amplifiers are widely used in SRAM to improve signal sensing speed independent of the bit line load capacitance. A feature of the current-mode sensing scheme is the small voltage swing of the signal line (BL or $\overline{BL}$). A typical structure is shown in Fig. 10.39. In contrast to voltage sense amplifiers, the signal line is connected to the source of latch transistors (M1 and M2). When $\overline{PSAE}$ goes to low, the sense amplifier is activated. Since the levels of BL and $\overline{BL}$ and DL and $\overline{DL}$ are maintained about V_{DD} and

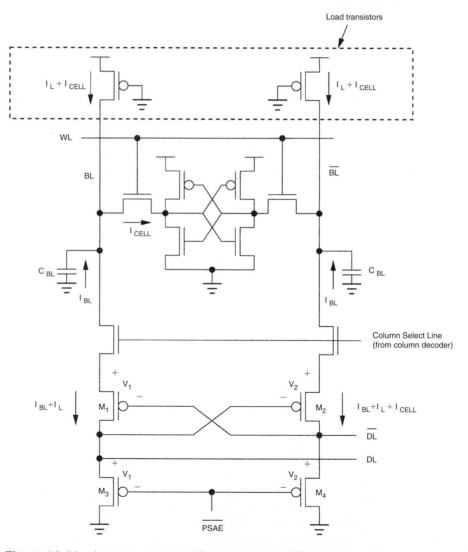

Figure 10.39 A current sense amplifier circuit used in SRAM.

V_{TP}, all transistors of the sense amplifiers (M1, M2, M3, and M4) are biased in saturation mode. Gate-source voltages of M1 and M3 (M2 and M4) are almost identical since the same current flows in each branch. When $\overline{PSAE}$ goes to low, the left and right bit lines have the same potential, $V_1 + V_2$. Since the bit line voltages are the same, the bit line load ($I_L + I_{CELL}$) and capacitive (I_{BL}) current is also equal. As the memory cell draws the cell current (I_{CELL}), more current by amount of (I_{CELL}) flows in the right leg to keep the bit line voltage to be equal. Therefore, the current difference appears on DL and $\overline{DL}$ lines. The sensing speed is almost independent of the bit line load capacitance since there is no capacitive discharging operations of BL and $\overline{BL}$. Thus, BL and $\overline{BL}$ need no precharge and equalizing operations which causes speed and cycle time penalty.

The open-loop gain of the circuit shown in Fig. 10.39 is expressed as

$$Gain_{open-loop} = \frac{gm(m3) * gm(m4)}{gm(m1) * gm(m2)} \qquad (10.10)$$

where $gm(m1)$, $gm(m2)$, $gm(m3)$ and $gm(m4)$ are the transconductance of M1, M2, M3, and M4, respectively. One drawback of current sense amplifiers is larger power consumption than voltage sense amplifiers.

These amplifiers are typically used in high speed SRAM and high density DRAM such as 256 Mb due to large load capacitance of data lines (BL_IO and BL_IOB in Fig. 10.15).

10.4 Nonvolatile Memory

A drawback of the MOS memory structures such as DRAM and SRAM is that the stored data are lost in the absence of the power supply. To overcome this problem, various kinds of nonvolatile and programmable (except the mask ROM) memory have been proposed. Recently, the *Flash* memory based on the floating-gate concept has become the most popular nonvolatile memory due to its small cell size and better functionality. Therefore, we describe in detail the basic structure and operation of the mask ROM and the Flash memory in this section.

The read-only memory array can also be seen as a simple combinational Boolean network which produces a specified output value for each input combination, i.e., for each address. Thus, storing binary information at a particular address location can be achieved by the presence or absence of a data path from the selected row (word line) to the selected column (bit line), which is equivalent to the presence or absence of a device at that particular location. In the following, we will examine two different implementations for MOS ROM arrays. Consider first the 4-bit × 4-bit memory array shown in Fig. 10.40. Here, each column consists of a pseudo-nMOS NOR gate driven by some of the row signals, i.e., the word lines.

As described in the previous section, only one word line is activated (selected) at a time by raising its voltage to V_{DD}, while all other rows are held at a low voltage level. If an active transistor exists at the cross point of a column and the selected row, the column voltage is pulled down to the logic low level by that transistor. If no active transistor exists at the cross point, the column voltage is pulled high by the

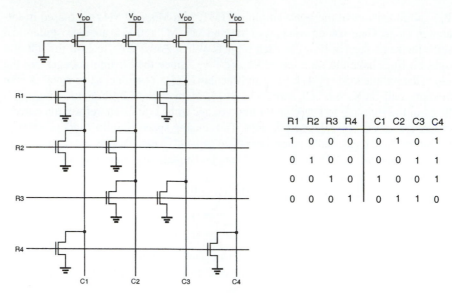

R1	R2	R3	R4	C1	C2	C3	C4
1	0	0	0	0	1	0	1
0	1	0	0	0	0	1	1
0	0	1	0	1	0	0	1
0	0	0	1	0	1	1	0

Figure 10.40   Example of a 4-bit × 4-bit NOR-based ROM array.

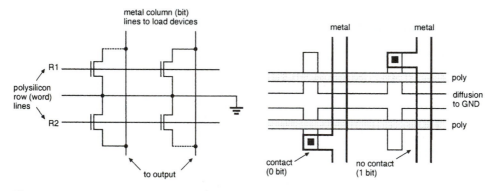

Figure 10.41   Layout example of a NOR ROM array.

pMOS load device. Thus, a logic "1"-bit is stored as the absence of an active transistor, while a logic "0"-bit is stored as the presence of an active transistor at the crosspoint. To reduce static power consumption, the pMOS load transistors in the ROM array shown in Fig. 10.40 can also be driven by a periodic precharge signal, resulting in a *dynamic* ROM.

In actual ROM layout, the array can be initially manufactured with nMOS transistors at every row-column intersection. The "1"-bits are then realized by omitting the drain or source connection, or the gate electrode of the corresponding nMOS transistors in the final metallization step. Figure 10.41 shows four nMOS transistors

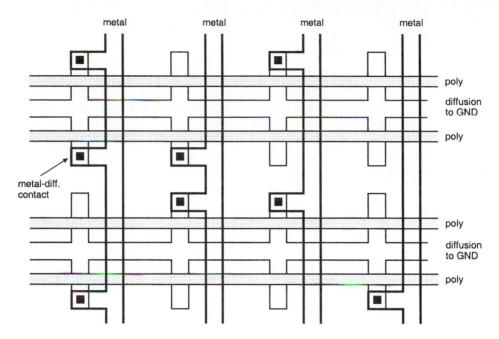

Figure 10.42 Layout of the 4-bit × 4-bit NOR ROM array example shown in Fig. 10.40.

in a NOR ROM array, forming the intersection of two metal bit lines and two polysilicon word lines. To save silicon area, the transistors in every two adjacent rows are arranged to share a common ground line, also routed in n-type diffusion. To store a "0"-bit at a particular address location, the drain diffusion of the corresponding transistor must be connected to the metal bit line via a metal-to-diffusion contact. Omission of this contact, on the other hand, results in a stored "1"-bit.

Figure 10.42 shows a larger portion of the ROM array, except for the pMOS load transistors connected to the metal columns. Here, the 4-bit × 4-bit ROM array shown in Fig. 10.40 is realized using the contact-mask programming methodology described above. Note that only 8 of the 16 nMOS transistors fabricated in this structure are actually connected to the bit lines via metal-to-diffusion contacts. In reality, the metal column lines are laid out directly on top of diffusion columns to reduce the horizontal dimension of the ROM array.

A different NOR ROM layout implementation is based on deactivation of the nMOS transistors by raising their threshold voltages through channel implants. Figure 10.43 shows the circuit diagram of a NOR ROM array in which every two rows of nMOS transistors share a common ground connection, and every drain diffusion contact to the metal bit line is shared by two adjacent transistors. In this case, all nMOS transistors are already connected to the column lines (bit lines); therefore, storing a "1"-bit at a particular location by omitting the corresponding drain contact is not possible. Instead, the nMOS transistor corresponding to the stored "1"-bit can

metal columns

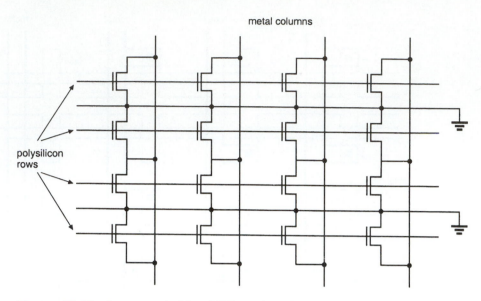

polysilicon
rows

Figure 10.43   Arrangement of the nMOS transistors in the implant-mask programmable NOR ROM array. Every metal-to-diffusion contact is shared by two adjacent devices.

be deactivated, i.e., permanently turned off, by raising its threshold voltage *above* the V_{OH} level through a selective channel implant during fabrication.

The alternative layout of the 4-bit × 4-bit bit ROM array example (Fig. 10.40), which is based on implant-mask programming, is shown in Fig. 10.44. Note that in this case, each threshold voltage implant signifies a stored "1"-bit, and all other (non-implanted) transistors correspond to stored "0"-bits. Since each diffusion-to-metal contact in this structure is shared by two adjacent transistors, the implant-mask ROM layout can yield a higher core density, i.e., a smaller silicon area per stored bit, compared to the contact-mask ROM layout.

Next, we will examine a significantly different ROM array design, which is also called a NAND ROM (Fig. 10.45). Here, each bit line consists of a depletion-load NAND gate, driven by some of the row signals, i.e., the word lines. In normal operation, all word lines are held at the logic-high voltage level except for the selected line, which is pulled *down* to logic-low level. If a transistor exists at the crosspoint of a column and the selected row, that transistor is turned off and the column voltage is pulled high by the load device. On the other hand, if no transistor exists (shorted) at that particular crosspoint, the column voltage is pulled low by the other nMOS transistors in the multi-input NAND structure. Thus, a logic "1"-bit is stored by the presence of a transistor that can be deactivated, while a logic "0"-bit is stored by a shorted or normally on transistor at the crosspoint.

As in the NOR ROM case, the NAND-based ROM array can be fabricated initially with a transistor connection present at every row-column intersection. A "0"-bit is then stored by lowering the threshold voltage of the corresponding nMOS transistor

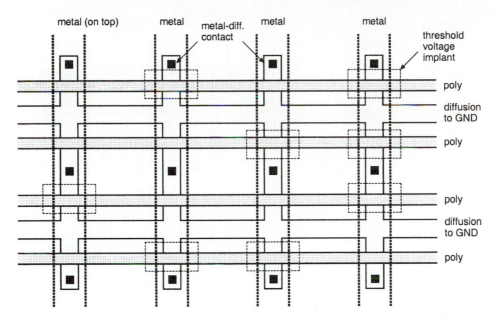

Figure 10.44 Layout of the 4-bit × 4-bit NOR ROM array example shown in Fig. 10.40. The threshold voltages of "1"-bit transistors are raised above V_{DD} through implant.

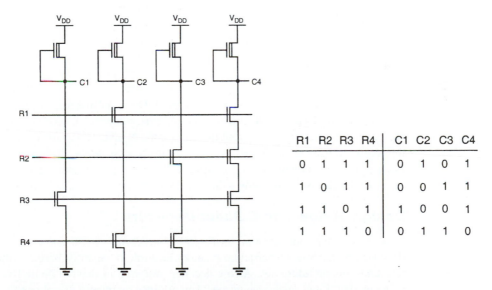

R1	R2	R3	R4	C1	C2	C3	C4
0	1	1	1	0	1	0	1
1	0	1	1	0	0	1	1
1	1	0	1	1	0	0	1
1	1	1	0	0	1	1	0

Figure 10.45 A 4-bit × 4-bit NAND-based ROM array.

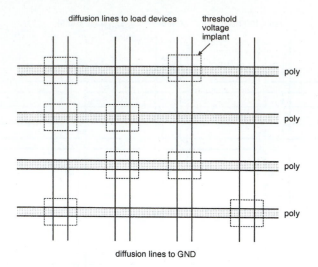

Figure 10.46   Implant-mask layout of the NAND ROM array example in Fig. 10.45. The threshold voltages of "0"-bit transistors are lowered below 0 V through implant.

at the crosspoint through a channel implant, so that the transistor remains on regardless of the gate voltage (i.e., the nMOS transistor at the intersection becomes a depletion-type device). The availability of this process step is also the reason why depletion-type nMOS load transistors are used instead of pMOS loads in the example shown before. Figure 10.46 shows a sample 4-bit × 4-bit layout of the implant-mask NAND ROM array. Here, vertical columns of n-type diffusion intersect at regular intervals with horizontal rows of polysilicon, which results in an nMOS transistor at each intersection point. The transistors with threshold voltage implants operate as normally-on depletion devices, thereby providing a continuous current path regardless of the gate voltage level. Since this structure has no contacts embedded in the array, it is much more compact than the NOR ROM array. However, the access time is usually slower than the NOR ROM, due to multiple series-connected nMOS transistors in each column. An alternative layout method for the NAND ROM array is not to place the nMOS transistors at "0"-bit locations, as in the case of the PLA (Programmable Logic Array) layout generation. In this case, the missing transistor is simply replaced by a metal line, instead of using a threshold voltage implant at that location.

Design of Row and Column Decoders

Now we will turn our attention to the circuit structures of row and column address decoders, which select a particular memory location in the array, based on the binary row and column addresses. A row decoder designed to drive a NOR ROM array must, by definition, select one of the 2^N word lines by raising its voltage to V_{OH}. As an example, consider the simple row address decoder shown in Fig. 10.47, which decodes a two-bit row address and selects one out of four word lines by raising its level.

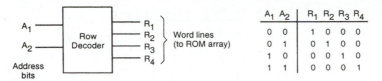

| A₁ A₂ | R₁ R₂ R₃ R₄ |

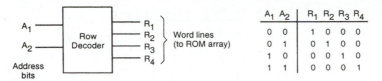

A_1 A_2	R_1 R_2 R_3 R_4
0 0	1 0 0 0
0 1	0 1 0 0
1 0	0 0 1 0
1 1	0 0 0 1

Figure 10.47 Row address decoder example for 2 address bits and 4 word lines.

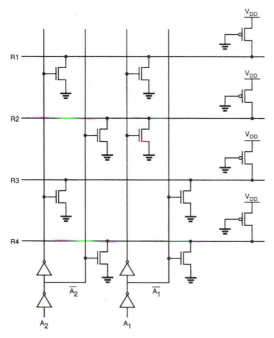

Figure 10.48 NOR-based row decoder circuit for 2 address bits and 4 word lines.

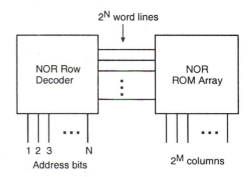

Figure 10.49 Implementation of the row decoder circuit and the ROM array as two adjacent NOR planes.

A most straightforward implementation of this decoder is another NOR array, consisting of 4 rows (outputs) and 4 columns (two address bits and their complements). Note that this NOR-based decoder array can be built just like the NOR ROM array, using the same selective programming approach (Fig. 10.48). The ROM array and its row decoder can thus be fabricated as two adjacent NOR arrays, as shown in Fig. 10.49.

A row decoder designed to drive a NAND ROM, on the other hand, must lower the voltage level of the selected row to logic "0" while keeping all other rows at a logic-high level. This function can be implemented by using an N-input NAND gate for each of the row outputs. The truth table of a simple address decoder for four rows and the double NAND-array implementation of the decoder and the ROM are shown

A_1 A_2	R_1 R_2 R_3 R_4
0 0	0 1 1 1
0 1	1 0 1 1
1 0	1 1 0 1
1 1	1 1 1 0

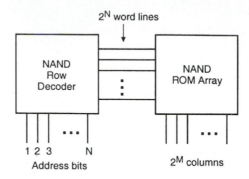

Figure 10.50   Truth table of a row decoder for the NAND ROM array, and implementation of the row decoder circuit and the ROM array as two adjacent NAND planes.

in Fig. 10.50. As in the NOR ROM case, the row address decoder of the NAND ROM array can thus be realized using the same layout strategy as the memory array itself.

The column decoder circuitry is designed to select one out of 2^M bit lines (columns) of the ROM array according to an M-bit column address, and to route the data content of the selected bit line to the data output. A straightforward but costly approach would be to connect an nMOS pass transistor to each bit-line (column) output, and to selectively drive one out of 2^M pass transistors by using a NOR-based column address decoder, as shown in Fig. 10.51. In this arrangement, only one nMOS pass transistor is turned on at a time, depending on the column address bits applied to the decoder inputs. The conducting pass transistor routes the selected column signal to the data output. Similarly, a number of columns can be chosen at a time, and the selected columns can be routed to a *parallel* data output port.

Note that the number of transistors required for this column decoder implementation is $2^M(M+1)$, i.e., 2^M pass transistors for each bit line and $M\,2^M$ transistors for the decoder circuit. This number can quickly become excessive for large M, i.e., for a large number of bit lines.

An alternative design of the column decoder circuit is to build a binary selection tree consisting of consecutive stages, as shown in Fig. 10.52. In this case, the pass transistor network is used to select one out of every two bit lines at each stage (level), whereas the column address bits drive the gates of the nMOS pass transistors. Notice that a NOR address decoder is not needed for this decoder tree structure, thereby

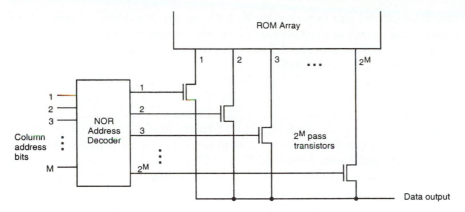

Figure 10.51 Bit-line (column) decoder arrangement using a NOR address decoder and nMOS pass transistors for every bit line.

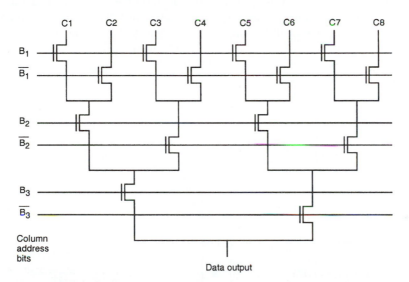

Figure 10.52 Column decoder circuit for eight bit lines, implemented as a binary tree decoder which is driven directly by the three column address bits.

reducing the number of transistors significantly although it requires M additional inverters ($2M$ transistors) for complementing column address bits. The example shown in Fig. 10.52 is a column decoder tree for eight bit lines, which requires three column address bits (and their complements) to select one of the eight columns.

One drawback of the decoder tree approach is that the number of series-connected nMOS pass transistors in the data path is equal to the number of column address bits, M. This situation can cause a long data access time, since the decoder delay time depends on the equivalent series resistance of the decoder branch that directs the column data to the output. To overcome this constraint, column address decoders can be built as a combination of the two structures presented here, i.e., consisting of relatively shallow, partial tree decoders and of additional selection circuits similar to that shown in Fig. 10.51.

EXAMPLE 10.1

In the following example, the design of a 32-kbit NOR ROM array will be considered, and the relevant design issues related to access time analysis will be examined.

A 32-kbit ROM array consists of $2^{15} = 32,768$ individual memory cells, which are arranged into a certain number of the rows and columns, as already explained in the beginning of this chapter. Note that the *sum* of row address bits and the column address bits must be equal to 15 in a 32-kbit array; the actual number of rows and columns in the memory array may be determined according to this and other constraints, as will be demonstrated in the following.

Assume that the ROM array under consideration has 7 row address bits and 8 column address bits, which results in a memory array of 128 rows and 256 columns. A section of the memory cell layout is shown in Fig. 10.53, in which the programming is done by an implant-mask to adjust the threshold voltages of those transistors which are to be left inactive (cf. Fig. 10.44). To provide a compact layout, the drain contact shown here is actually shared by two adjacent transistors, and the metal bit

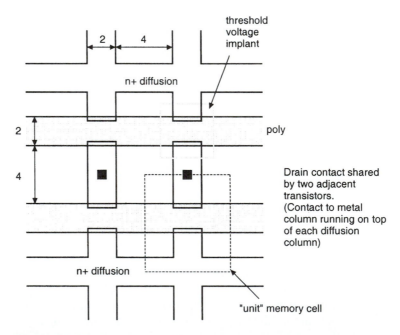

Figure 10.53 Simplified mask layout of the implant-mask programmable NOR ROM array considered in the example. All dimensions are in micrometers ($W = 2$ μm, $L = 1.5$ μm).

line (column) runs directly on top of the diffusion columns shown in this example. To simplify the mask view, the metal column line is not shown in Fig. 10.53.

Other relevant parameters of the structure are:

$$\mu_n C_{ox} = 20 \ \mu\text{A/V}^2$$
$$C_{ox} = 3.47 \ \mu\text{F/cm}^2$$
Poly sheet resistance $= 20 \ \Omega/\text{square}$

First, we calculate the row resistance and row capacitance per bit, i.e., per memory cell. It is assumed that the row capacitance of each memory cell is dictated primarily by the thin oxide capacitance of the nMOS transistor, and that the polysilicon capacitance outside the active region is negligible. The row resistance associated with each memory cell, on the other hand, is calculated by summing the number of polysilicon squares (in this case, three) over the unit cell.

$$C_{row} = C_{ox} \cdot W \cdot L = 10.4 \ \text{fF/bit}$$
$$R_{row} = (\text{\# of squares}) \times (\text{Poly sheet resistance}) = 60 \ \Omega/\text{bit}$$

We note that each polysilicon row (word line) in this memory array is actually a distributed RC transmission line. Figure 10.54 depicts a few components of this structure, which ultimately affects the row access time of the memory array in consideration. It can be seen that once the row is selected by the row address decoder, i.e., the row voltage is forced to a logic-high level at one end of the word line, the gate voltage of the last (256th) transistor in that row will rise *last* because of the RC line delay. The propagation delay time of the gate voltage of the 256th transistor in the row determines the row access time, t_{row} (Fig. 10.55).

By neglecting the signal propagation delay associated with the row address decoder circuit, and assuming that the row (word line) is driven by an ideal step voltage waveform, the row access time can be approximated by using the following empirical formula,

$$t_{row} \approx 0.38 \cdot R_T \cdot C_T = 15.53 \ \text{ns}$$

where

$$R_T = \sum_{all \ columns} R_i = 15.36 \ \text{k}\Omega$$

$$C_T = \sum_{all \ columns} C_i = 2.66 \ \text{pF}$$

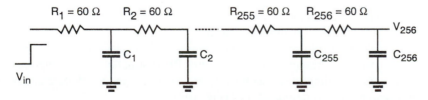

Figure 10.54 RC transmission line representation of the polysilicon word line.

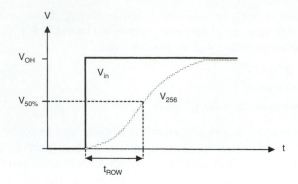

Figure 10.55 Definition of the row access time in a memory array with 256 columns.

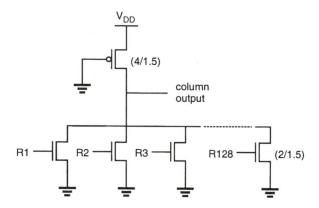

Figure 10.56 128-input NOR gate representation of a column (bit line) in the ROM array.

A more accurate RC delay value can be calculated by using the *Elmore time constant* for RC ladder circuits, as follows.

$$t_{row} = \sum_{k=1}^{256} R_{jk} C_k = 20.52 \text{ ns} \qquad \text{where} \quad R_{jk} = \sum_{j=1}^{k} R_j$$

The row access time t_{row} is the time delay associated with selecting and activating one of the 128 word lines in this ROM array.

To calculate the column access time, we need to consider one of the 128-input NOR gates, which represent the bit lines in this ROM structure. The pseudo-nMOS NOR gate corresponding to each column is designed by using a pMOS load transistor, in which the (W/L) ratio is $(4/1.5)$ (Fig. 10.56).

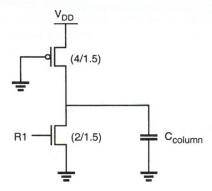

Figure 10.57 Equivalent inverter circuit representing the bit line (column). Note that only one word line (row) is activated at a time.

The combined column capacitance loading the output node of the 128-input NOR gate can be approximated by adding up the parasitic capacitances of each driver transistor.

$$C_{column} = 128 \times (C_{gd,driver} + C_{db,driver}) \approx 1.5 \text{ pF}$$

where

$$C_{gd,driver} + C_{db,driver} = 0.0118 \text{ pF/word line}$$

Since only one word line (row) is activated at a time by the row address decoder, the NOR gate representing the column can actually be reduced to the inverter shown in Fig. 10.57. To calculate the column access time, we must consider the worst-case signal propagation delay τ_{PHL} (for falling output voltage) for this inverter. By using the propagation delay formula (6.22) in Chapter 6, the worst-case column access time can thus be calculated as $t_{column} = 18$ ns. Note that the propagation delay for rising output signals τ_{PLH} is not considered here because the bit line (column) is precharged high *before* each row access operation.

The total access time for this ROM array consisting of 128 rows and 256 columns is found by adding the row and column access times, as $t_{access} = 38.5$ ns.

At this point, we may also consider a different arrangement for the ROM array, which consists of 256 rows and 128 columns, i.e., using 8 row address bits and 7 column address bits. Since the number of columns in this arrangement is half of the number of columns in the previous example, the total row resistance and row capacitance values will be approximately half of those obtained for the 256-column arrangement. Consequently, the row access time for this structure will be *one-fourth* of the row access time found for the 256-column ROM array, i.e., approximately 5 ns. The number of rows in the new arrangement, on the other hand, is 256, which results in a column capacitance approximately *twice* that in the previous case. Thus, the column access time of the 256-by-128 memory array will be twice as large, approximately 36 ns. In conclusion, we find that the total access time of the 128-by-256 array examined initially is shorter than that of the 256-by-128 array.

10.5 Flash Memory

The *Flash* memory cell consists of one transistor with a floating gate, whose threshold voltage can be changed (programmed) repeatedly by applying an electrical field to its gate. The memory cell (transistor) can have two threshold voltages (two states) corresponding to the existence of charges (electrons) at the floating gate. When electrons are accumulated at the floating gate, the threshold voltage of the memory cell becomes higher and the memory cell is considered to be in a "1" state as a convention. This is because the memory cell is not turned on with the read signal voltage (e.g., 5 V) applied to the control gate and the bit line precharge level (e.g., V_{DD}), is maintained. The threshold voltage of the memory cell can be lowered by removing electrons from the floating gate and the memory cell is regarded to be in a "0" state. In this case, the cell transistor is turned on with the applied voltage and the bit line is discharged to ground. Therefore, the cell data of *Flash* memory is programmed by either storing or ejecting electrons in the floating gate of a MOS transistor through channel hot-electron injection or Fowler-Nordheim tunneling mechanisms.

The schematic cross-section views of the *Flash* memory cell for two programming mechanisms are shown in Fig. 10.58(a) and (b), respectively. When the high voltage (e.g., 12 V) is applied to the control gate and across the drain to source (e.g., 6 V), electrons are *heated* by the high lateral electric field. Avalanche breakdown occurs at the near of the drain and electron-hole pairs are generated by the impact ionization. The high voltage on the control gates attracts and injects electrons into the floating gate through the oxide and the holes flow to the substrate as the substrate current. Instead of using hot-electrons, the floating gate is programmed or erased by the tunneling current of the oxide by the high field of >10 MV/cm. When the 0 V and a high voltage (e.g., 12 V) are applied to the control gate and the source, electrons at the floating gate are ejected to the source by the tunneling effect.

Figure 10.59 shows equivalent capacitive-coupling circuit of a Flash memory cell. When the voltages (V_{CG} and V_D) are applied to the control gate and the drain, the voltage at the floating gate (V_{FG}) by capacitive coupling is expressed as

$$V_{FG} = \frac{Q_{FG}}{C_{total}} + \frac{C_{FC}}{C_{total}} V_{CG} + \frac{C_{FD}}{C_{total}} V_D, \qquad (10.11)$$

$$C_{total} = C_{FC} + C_{FS} + C_{FB} + C_{FD} \qquad (10.12)$$

where Q_{FG} is the charge stored at the floating gate; C_{total} the total capacitance; C_{FC} the capacitance between the floating and control gates; C_{FS}, C_{FB}, and C_{FD} are the

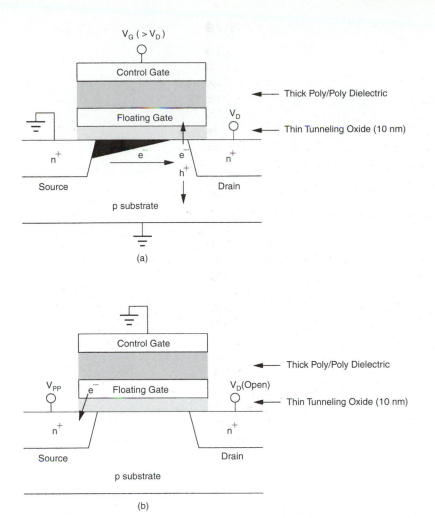

Figure 10.58 Data programming and erasing methods in the *Flash* memory Memory. (a) Hot electron injection mechanism, (b) Fowler-Nordheim tunneling mechanism.

capacitances between the floating gate and the source, bulk, and drain; and V_{CG} and V_D are the voltages at the control gate and the drain, respectively.

The minimum control gate voltage (V_{CG}) required to turn on the control gate transistor can be obtained by substituting $V_T(FG)$ into V_{FG} and arranging (10.11) as follows

$$V_T(CG) = \frac{C_{total}}{C_{FC}} V_T(FG) - \frac{Q_{FG}}{C_{FC}} - \frac{C_{FD}}{C_{FC}} V_D \qquad (10.13)$$

where $V_T(FG)$ is the threshold voltage to turn on the floating gate transistor.

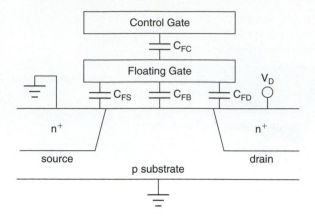

Figure 10.59   Equivalent capacitive-coupling circuit of a *Flash* memory cell.

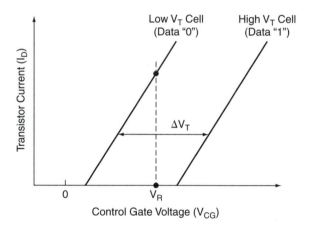

Figure 10.60   I-V characteristic curves of the Flash memory cells with low and high threshold voltages for the control gate voltage.

Also, the difference of threshold voltages between two memory data states ("0" and "1") can be expressed as

$$\Delta V_T(CG) = -\frac{\Delta Q_{FG}}{C_{FC}} \tag{10.14}$$

Figure 10.60 shows the I-V characteristic curves of the Flash memory cell for the different threshold voltages. A control gate voltage (V_R) is set to be large enough to turn on the low V_T transistor but insufficient to turn on the high V_T transistor for the cell read operation. Various types of the Flash memory cell structures and array architectures for the logical connection of cells using different programming mechanisms have been proposed. *NOR, NAND, AND, DINOR* (divided bit-line NOR), *HICR* (High Capacitive-coupling Ratio cell), 3-D, and multilevel cells are some examples of the Flash memory cells. In this section, the cell organization and basic operations of the most popular cell structures (*NOR* and *NAND*) are explained followed by the multilevel cell concept.

NOR Flash Memory Cell

Figure 10.61 and Table 10.4 show the *NOR* cell configuration and bias conditions for erase, programming and read operations, respectively. The NOR cell uses the F-N tunneling mechanism for the erase operation and the hot-electron injection mechanism for the programming operation.

Table 10.4 Bias conditions of the NOR cells for erase, programming, and read operations

Signal	Operation		
	Erase	**Programming**	**Read**
Bit line 1	Open	6 V	1 V
Bit line 2	Open	0 V	0 V
Source line	12 V	0 V	0 V
Word line 1	0 V	0 V	0 V
Word line 2	0 V	12 V	5 V
Word line 3	0 V	0 V	0 V

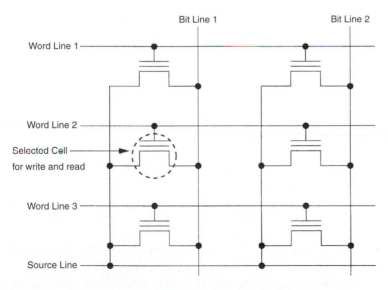

Figure 10.61 Bias conditions and configuration of the *NOR* cells.

In the erase operation, 0 V and a high voltage (e.g., 12 V) are simultaneously applied to all the control gates (word lines) and the sources of the cells, respectively. Any electrons existing in the floating gate are expelled to the source through the tunneling mechanism. Therefore, the data of all cells are erased and all memory cells are changed to be the low threshold voltage transistors.

The cell data can be programmed (written) by applying high voltages to the control gate and the drain of the selected cell (e.g., 12 V and 6 V for the control gate and drain of the circled cell, respectively). Hot-electrons generated at the near of the drain are injected into the floating gate and the memory cell becomes a high threshold voltage transistor due to the existence of electrons in the floating gate. Therefore, each cell can have either a "0" state without electrons due to the erase operation or a "1" state with electrons in the floating gate after the programming operation.

The read operation is performed by applying a moderate voltage (typically V_{DD}) to the control gate and a small voltage (e.g., 1 V) to the drain to avoid generating hot-electrons. When the cell data is "0" (the low threshold voltage) the cell transistor is turned on and current flows through the cell transistor. Otherwise, the cell transistor is not turned on due to its high threshold voltage with the applied control gate voltage and no current flows through the cell. By detecting current flow and amplifying the signal difference, cell data can be read out.

NAND Flash Memory Cell

By placing eight or sixteen cells in series, the memory cell area can be reduced by the elimination of contacts in cells. The cross-section view and equivalent circuit of the eight bit *NAND* cell structure are shown in Fig. 10.62 and Fig. 10.63, respectively. Bias conditions for erase, programming and read operations are shown in Table 10.5.

The *NAND* cell uses F-N tunneling mechanism for the erase operation. A high voltage (e.g., 20 V) is applied to source line, p-well 2 and n-substrate while 0 V to all the word lines to eject electrons from the floating gate to the p-well 2. All cells become low threshold voltage, strictly speaking, depletion, transistors in which current flows even when the gate voltage is 0 V unlike the *NOR* cell.

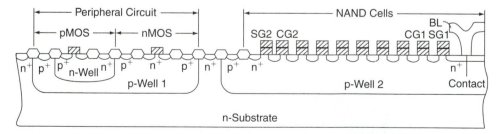

Figure 10.62   Cross-section view of the *NAND* cell structure.
(*Source:* F. Masuoka et al., "New ultra high density EEPROM and Flash EEPROM cell with NAND structure cell," *IEDM Dig. Tech.,* pp. 552–555, 1987.)

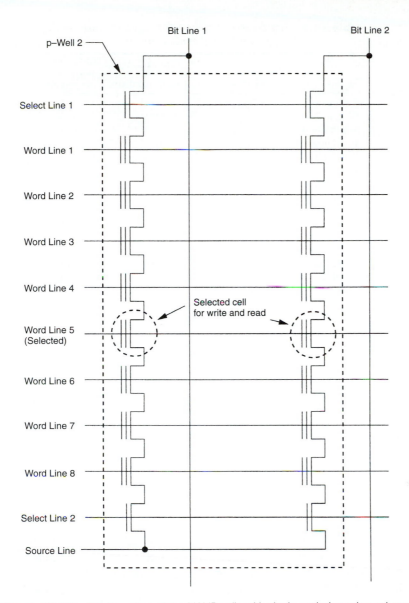

Figure 10.63 Configuration of the *NAND* cells with single-ended sensing scheme.

In the programming operation, a high voltage (e.g., 20 V) is applied only to the selected word line (word line 5) and a moderate voltage (e.g., 10 V) is applied to all the unselected word lines. V_{DD} (e.g., 5 V) is applied to the select line 1 to connect all the cells to the bit line while 0 V to the select line 2. P-well 2, the source line, and n-well are biased to 0 V. Since only the voltage applied to the selected word line (20 V) is large enough to turn on the floating gate transistor by capacitive coupling, electrons

Table 10.5   Bias conditions of the NAND cells for erase, programming, and read operations

Signal	Operation		
	Erase	**Programming**	**Read**
Bit line 1	Open	0 V	1 V
Bit line 2	Open	0 V	1 V
Select line 1	Open	5 V	5 V
Word line 1	0 V	10 V	5 V
Word line 2	0 V	10 V	5 V
Word line 3	0 V	10 V	5 V
Word line 4	0 V	10 V	5 V
Word line 5	0 V	20 V	0 V
Word line 6	0 V	10 V	5 V
Word line 7	0 V	10 V	5 V
Word line 8	0 V	10 V	5 V
Select line 2	Open	0 V	5 V
Source line	Open	0 V	0 V
p-well 2	20 V	0 V	0 V
n-sub	20 V	0 V	0 V

Table 10.6   Characteristic comparison between the NOR and NAND cells

	NOR	**NAND**
Erase method	Fowler-Nordheim tunneling	Fowler-Nordheim tunneling
Programming method	Hot electron injection	Fowler-Nordheim tunneling
Erase speed	Slow	Fast
Program speed	Fast	Slow
Read speed	Fast	Slow
Cell size	Large	Small
Scalability	Difficult	Easy
Application	Embedded system	Mass storage

are attracted to the floating gates connected to the selected word line from the channel (substrate). Therefore, the cell changes to the high threshold voltage transistor. The *NAND* cell uses the P-N tunneling mechanism to program the cell.

To read the cell data, 0 V is applied to the selected word line while 5 V is applied to the select lines and all the other unselected word lines. When the cell data is "1" (high threshold voltage), the cell transistor is turned off with the applied word line voltage. Thus, the precharge level of the bit line (e.g., 1 V) is maintained. With the cell with data "0" (negative threshold voltage), a current path is formed from the bit line to *Ground* since the cell transistor is normally turned on with the 0 V word line voltage. Therefore, the bit line is pulled down from the precharge level.

Table 10.6 shows the characteristic comparison between the *NOR* and *NAND* cells. The *NOR* cell structure shows faster programming and read speed but larger area than the *NAND* cell structure.

Multilevel Cell Concept

To increase memory density, reducing the cell size has been the primary technique. By storing multilevel (e.g., 4) data in the cell, the effective memory density can be also improved. All memory cells have only two discrete states in the cell, "0" or "1". However, the inherent feature of the flash memory to control the threshold voltage of the cell transistor makes it more feasible than other memory devices to store multiple states in the cell. If the programming operation can be done accurately enough, the cell can have four discrete charge states, resulting in a structure that achieves 2 bits per cell. Figure 10.64 shows the threshold voltage distribution of 2 bits/cell storage. The number of possible states is limited by several constraints such as available charge range, the accuracy of programming and read operations, and disturbance of a state over time.

Flash Memory Circuit

Since the high voltage (V_{PP}) is required to program the cell, the on-chip charge pump circuit is used to generate the programming voltage. Charge pump circuits are implemented with nMOS or pMOS transistors for the generation of a high positive or a negative voltage, respectively. Figure 10.65 shows a typical circuit implementation of the charge pump circuit with nMOS transistors to generate a high positive voltage. The charge pump circuit consists of a chain of the diode and capacitors to be charged or discharged successively each half-clock cycle.

When *Clock* goes to low, the capacitors connected to *Clock* (i.e., C_1, C_3, ..., $C_{(n-1)}$) are discharged and the corresponding nodes (i.e., V_1, V_3, ..., $V_{(n-1)}$) are pulled down. These nodes are charged to a voltage equal to the voltage of a preceeding node minus a diode voltage (nMOS threshold voltage). For example, V_1 is charged to $V_{in} - V_T(\text{MN1})$ and V_3 to $V_2 - V_T(\text{MN3})$, and so forth.

As *Clock* goes high ($\overline{Clock}$ goes low) the nodes coupled to *Clock* are boosted while the nodes connected to $\overline{Clock}$ (i.e., V_2, V_4, ..., V_n) are pulled down. Therefore, charges at the nodes connected to *Clock* are transferred to the nodes connected to $\overline{Clock}$.

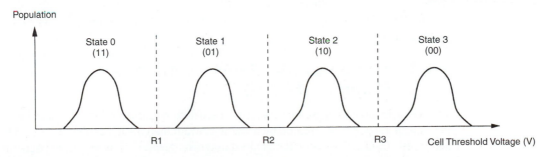

Figure 10.64 Multilevel-cell threshold voltage distribution.

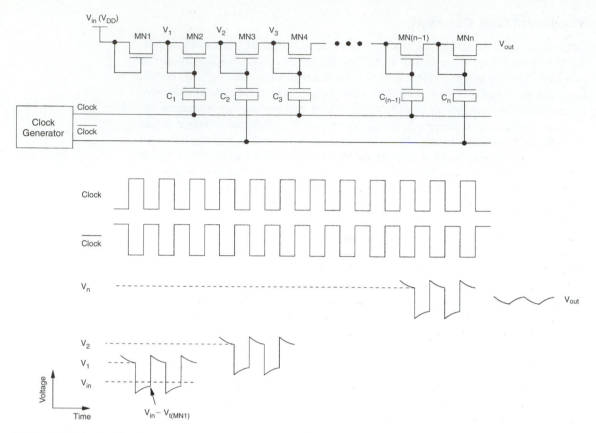

Figure 10.65   Positive charge pump circuit with nMOS transistors and voltage waveforms.

In the following half clock, charges at the nodes connected to $\overline{Clock}$ are transferred to the nodes connected to $Clock$ in the same manner. Thus, the voltage of V_{out} is expressed as

$$V_{out} = V_{in} + (\gamma V_{DD} - V_T(\text{MN1})) + \cdots + (\gamma V_{DD} - V_T(\text{MNn})) \qquad (10.15)$$

where γ is a boosting efficiency factor.

In the charge pump circuit, the current driving capability is independent of the number of stages.

10.6   Ferroelectric Random Access Memory (FRAM)

Ferroelectric memories use a ferroelectric capacitor where the dielectric material of the typical capacitor is replaced by a ferroelectric material such as $\text{Pb}(\text{Zr}_x\text{Ti}_{1-x})\text{O}_3$ and $\text{SbBi}_2\text{Ta}_2\text{O}_9$. Prior to FRAM, ferromagnetic memories using ferromagnetic cores with hysteresis characteristics were developed in 1950s. But those vanished due to large cell area and power consumption.

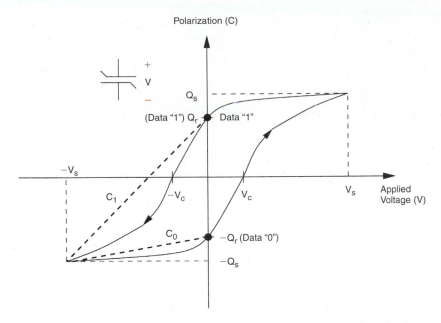

Figure 10.66 Hysteresis characteristic of a ferroelectric capacitor. Q_r and $-Q_r$: remnant charge; Q_s and $-Q_s$: saturation charge; V_c: coercive voltage; V_s: saturation voltage; C_0 and C_1: linear capacitance for data "0" and "1", respectively.

The hysteresis loop of a ferroelectric capacitor is shown in Fig. 10.66. The total charge of the ferroelectric capacitor varies as a function of the applied voltage and does not disappear in the absence of the electric field (Q_r for data "1" and $-Q_r$ for data "0") unlike that of the dielectric capacitor. This is because a spontaneous polarization occurs in the crystal structure of the ferroelectric material. The net change in the direction of the polarization happens when the applied electric field is larger than V_c. No further reorientation can occur after V_s and the polarization is saturated (Q_s).

The structure and operation of FRAM are similar to those of DRAM except *PL* (plate line). Figure 10.67(a) and (b) show the core structure and the timing diagram to read cell data "1" with the *step-sensing* scheme, respectively. Reference cells which always store only data "0" are placed per column to generate a reference signal used for read operation. Bit lines (*BL* and *BLB*) are precharged at V_{SS} since *PPRE* is high. A word line (*WL0*) is selected and activated to a boosted voltage (V_{PP}) for full data restoring as in DRAM. A reference word line connected to the other bit line (*RWL1*) is activated at the same time. Step signals are applied to normal and reference plate lines (*PL* and *RPL*) together. The total charge of the ferroelectric capacitor is changed from Q_r to $-Q_s$ and a positive charge appears on the bit line to maintain charge neutrality. When the word lines are enabled, capacitor dividers are formed by C_F and C_{BL} and C_{RF} and C_{BLB}. C_F, C_{RF}, C_{BL}, and C_{BLB} are the capacitances of the normal cell, reference cell, bit line, and bit line bar, respectively. Therefore, the resulting voltage

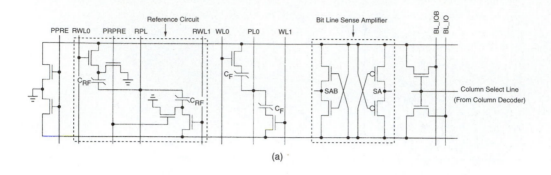

(a)

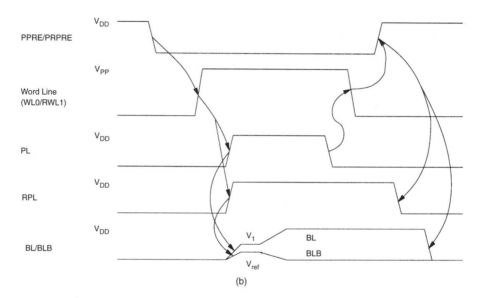

(b)

Figure 10.67 Memory structure of FRAM. (a) Cell array, (b) read timing diagram with a step-sensing scheme.

difference for data "1" is approximately expressed as

$$\Delta V_1 = \frac{C_1}{C_1 + C_{BL}} V_{DD} \qquad (10.16)$$

where C_1 is the linearly modeled capacitance of the ferroelectric capacitor shown in Fig. 10.66.

Similarly, the voltage difference for data "0" is expressed as

$$\Delta V_0 = \frac{C_0}{C_0 + C_{BL}} V_{DD} \qquad (10.17)$$

where C_0 is the linearly modeled capacitance of the ferroelectric capacitor shown in Fig. 10.66. The capacitance of the reference cell is set to make sure the voltage signal generated by the reference cell is between ΔV_0 and ΔV_1 and is larger than that of the normal cell. After voltage signals are generated on the bit line and bit line bar by the normal and the reference cells, the bit line sense amplifier detects and amplifies the signal difference on bit line pairs to V_{DD} and V_{SS} for full data restoring. The total charge changes to Q_s after the sensing operation and becomes Q_r after the applied filed vanishes.

Since the *step-sensing* scheme can cause some reliability issues, a *pulsed-sensing* scheme is also widely used with some read speed penalty, where a pulse is applied instead of keeping the voltage at the plate lines. FRAM suffers from two problems inherent to the ferroelectric material, so-called *fatigue* and *imprint*. The capacitance charge is gradually degraded with the repeated use of the capacitor (*fatigue*) and the ferroelectric capacitor tends to stay at one state preferably over the other when that state is maintained for a long time (*imprint*).

Exercise Problems

10.1 Consider the DRAM circuit shown in Fig. P10.1a. The threshold voltage of the two precharge transistors is 2 V. Calculate the steady-state voltages of

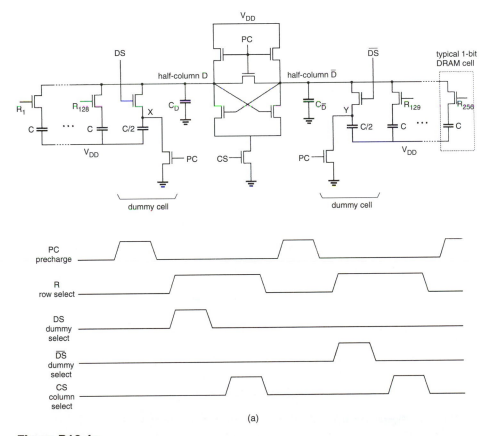

(a)

Figure P10.1a

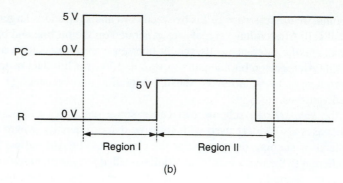

(b)

Figure P10.1b

V_D in Region I and Region II of Fig. P10.1b by assuming the following:

$$C = 50 \text{ fF}$$
$$C_D = 400 \text{ fF}$$
$$V(C) = V(C/2) = V_Y = 0 \text{ V in region I}$$

While PC is high, no other transistor connected to D or $\overline{D}$ is on.

10.2   A single-transistor DRAM cell is represented by the following circuit diagram (Fig. P10.2). The bit line can be precharged to $V_{DD}/2$ by using a clocked precharge circuit. Also the WRITE circuit is assumed here to bring the potential of the bit line to V_{DD} or 0 V during the WRITE operation with word line at V_{DD}. Using the parameters given:

$$V_{T0} = 1.0 \text{ V}$$
$$\gamma = 0.3 \text{ V}^{1/2}$$
$$|2\phi_F| = 0.6 \text{ V}$$

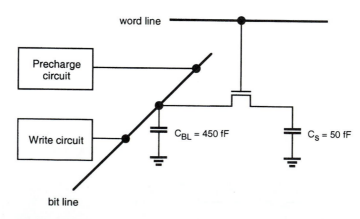

Figure P10.2

a. Find the maximum voltage across the storage capacitor C_S after WRITE-1 operation, i.e., when the bit line is driven to $V_{DD} = 5$ V.

b. Assuming zero leakage current in the circuit, find the voltage at the bit line during READ-1 operation after the bit line is first precharged to $V_{DD}/2$.

10.3 A dynamic CMOS Read Only Memory (ROM) has been designed with a core array consisting of 64 rows with a pitch of 12 μm and 64 columns with a pitch of 10 μm, as shown in Fig. P10.3. Each column is precharged to 5 V by a pMOS transistor during the interval of zero clock phase and then is pulled down after the clock signal switches to 5 V by one or more nMOS transistors (i.e., NOR implementation) with high gate inputs on the appropriate rows. All of the nMOS transistors have channel widths $W = 4$ μm and source/drain lengths $Y = 5$ μm. As a designer, you are to determine the propagation delay time from a particular input (on row 64) going high to a particular bit-line output (on column 64) going low τ_{PHL} between 50% points. Assume that the input signal to row 64 becomes valid-high only after the precharge operation is finished, as shown in the timing diagram. Also, assume that row 64 is running over 30 nMOS transistors and column 64 has 20 nMOS transistors connected to it. For delay calculation, assume that only one nMOS transistor is pulling down. Also assume that pMOS is strong enough to fully charge the precharging node during the precharge phase of the clock signal and to neglect its

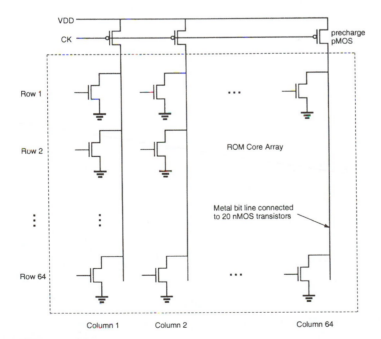

Figure P10.3

drain parasitic capacitance. Device parameters are given:

$$C_{jsw} = 250 \text{ pF/m}$$

$$C_{j0} = 80 \ \mu\text{F/m}^2$$

$$C_{ox} = 350 \ \mu\text{F/m}^2$$

$$L_D = 0.5 \ \mu\text{m}$$

$$K_{eq} = 1.0 \text{ for worst-case capacitance}$$

$$C_{metal} = 2.0 \text{ pF/cm and } R_{metal} = 0.03 \ \Omega/\text{sq.}$$

$$C_{poly} = 2.2 \text{ pF/cm and } R_{poly} = 25 \ \Omega/\text{sq.}$$

$$\text{Polysilicon line width} = 2 \ \mu\text{m}$$

$$\text{Metal line width} = 2 \ \mu\text{m}$$

$$k'_n = 20 \ \mu\text{A/V}^2$$

$$k'_p = 10 \ \mu\text{A/V}^2$$

$$V_{T,n} = -V_{T,p} = 1.0 \text{ V}$$

10.4   Consider the CMOS SRAM cell shown in Fig. P10.4. Transistors M1 and M2 have (W/L) values of 4/4. Transistors M3 and M4 have (W/L) values of 2/4. M5 and M6 are to be sized such that the state of the cell can be changed for $V_C \leq 0.5$ V. Assuming that M5 and M6 are the same size, calculate the required (W/L). Use the following parameters:

$$V_{T0,n} = 0.7 \text{ V}$$

$$V_{T0,p} = -0.7 \text{ V}$$

$$k'_n = 20 \ \mu\text{A/V}^2$$

$$k'_p = 10 \ \mu\text{A/V}^2$$

$$\gamma = 0.4 \text{ V}^{1/2}$$

$$|2\phi_F| = 0.6 \text{ V}$$

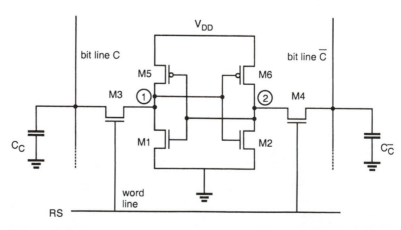

Figure P10.4

10.5 Draw circuit diagrams of the row decoder and the column decoder for an EPROM with 4 rows and 2 columns. Use nMOS technology. Develop formulas for the row and column delays in the EPROM. Define any terms in your formulas which are not obvious.

10.6 Consider an 8k × 8k SRAM, which has 64k (= 65536) memory cells and 8 output lines. In the particular SRAM under discussion, 7 address bits go to the row decoder and 6 address bits go to the column decoder. Bit lines are precharged to $V_{DD} = 5$ V before each read operation. A read operation is complete when the bit line has discharged by 0.5 V. A memory cell can provide 1.0 mA of pull-down current to discharge the bit line.

   a. Word line resistance is 390 Ω per memory cell. What formula was used to calculate this resistance?

   b. Word line capacitance is 22 fF per memory cell. What formula was used to calculate this capacitance?

   c. Bit line capacitance is 6 fF per memory cell. What formula was used to calculate this capacitance?

   d. Calculate the access time (row delay + column delay) for this SRAM.

   e. Describe the operation and design of the word line decoder and the bit line decoder.

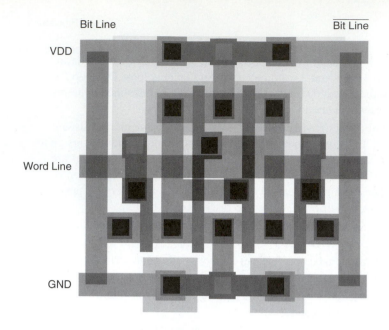

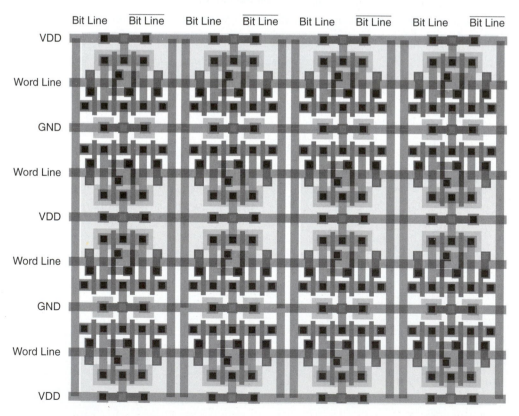

Plate 10 (Top) Layout of a CMOS SRAM cell.
(Bottom) Layout of a 4 × 4 SRAM array, consisting of 16 cells.

480

CHAPTER 11

Low-Power CMOS Logic Circuits

11.1 Introduction

The increasing prominence of portable systems and the need to limit power consumption (and hence, heat dissipation) in very high density ULSI chips have led to rapid and innovative developments in low-power design during the recent years. The driving forces behind these developments are portable device applications requiring low power dissipation and high throughput, such as notebook computers, portable communication devices, and personal digital assistants (PDAs). In most of these cases, the requirements for low power consumption must be met along with equally demanding goals of high chip density and high throughput. Hence, low-power design of digital integrated circuits has emerged as a very active and rapidly developing field.

The limited battery lifetime typically imposes very strict demands on the overall power consumption of the portable system. Although new rechargeable battery types such as Nickel–Metal Hydride (NiMH) are being developed with higher energy capacity than that of the conventional Nickel-Cadmium (NiCd) batteries, revolutionary increase of the energy capacity is not expected in the near future. The energy density (amount of energy stored per unit weight) offered by the new battery technologies (e.g., NiMH) is about 30 Wh/lb, which is still low in view of the expanding applications of portable systems. Therefore, reducing the power dissipation of integrated circuits through design improvement is a major challenge in portable systems design.

The need for low-power design is also becoming a major issue in high-performance digital systems, such as microprocessors, digital signal processors (DSPs), and other applications. The common traits of high-performance chips are the high integration density and the high clock frequency. The power dissipation of the chip, and thus, the temperature, increases with the increasing clock frequency. Since the dissipated heat must be removed effectively to keep the chip temperature at an acceptable level, the cost of packaging, cooling, and heat removal becomes a significant factor in these circuits. Several high-performance microprocessor chips designed in the early 1990s (e.g., Intel Pentium, DEC Alpha, and PowerPC) operated at clock frequencies in the range of 100 to 300 MHz, and their typical power consumption was between 20 and 50 W. Modern microprocessors are running at clock frequencies above 1 GHz with 100 W power dissipation.

ULSI reliability is yet another concern which points to the need for low-power design. There is a close correlation between the peak power dissipation of digital circuits and reliability problems such as electromigration and hot-carrier induced device degradation. Also, the thermal stress caused by heat dissipation on chip is a major reliability concern. Consequently, the reduction of power consumption is also crucial for reliability enhancement.

The methodologies which are used to achieve low power consumption in digital systems span a wide range, from device/process level to algorithm level. Device characteristics (e.g., threshold voltage), device geometries, and interconnect properties are significant factors in lowering the power consumption. Circuit-level measures such as the proper choice of circuit design styles, reduction of the voltage swing, and clocking strategies can be used to reduce power dissipation at the transistor level. Architecture-level measures include smart power management of various system blocks, utilization of pipelining and parallelism, and design of bus structures. Finally, the power consumed by the system can be reduced by a proper selection of the data processing algorithms, specifically to minimize the number of switching events for a given task.

In this chapter, we will primarily concentrate on the circuit- or transistor-level design measures which can be applied to reduce the power dissipation of digital integrated circuits. Various sources of power consumption will be discussed in detail, and design strategies will be introduced to reduce the power dissipation. We will discuss system-level considerations such as pipelining and hardware replication (parallel processing) and we will explore the influence of such measures upon power consumption. The concept of adiabatic logic will also be examined since it emerges as an effective means for reducing the power consumption.

11.2 Overview of Power Consumption

In the following, we will examine various sources (components) of time-averaged power consumption in CMOS circuits. The average power consumption in conventional CMOS digital circuits can be expressed as the sum of three main components, namely, (i) the dynamic (switching) power consumption, (ii) the short-circuit power consumption, and (iii) the leakage power consumption. If the system or chip includes circuits other than conventional CMOS gates that have continuous current paths between the power supply and the ground, a fourth (static) power component should also be considered. We will limit our discussion to the conventional static and dynamic CMOS logic circuits.

Switching Power Dissipation

This component represents the power dissipated during a switching event, i.e., when the output node voltage of a CMOS logic gate makes a logic transition. In digital

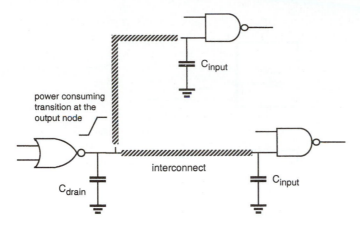

Figure 11.1 A NOR gate driving two NAND gates through interconnection lines.

CMOS circuits, switching power is dissipated when energy is drawn from the power supply to charge up the output node capacitance. During this charge-up phase, the output node voltage typically makes a full transition from 0 to V_{DD} and one-half of the energy drawn from the power supply is dissipated as heat in the conducting pMOS transistors. Note that no energy is drawn from the power supply during the charge-down phase, yet the energy stored in the output capacitance during charge-up is dissipated as heat in the conducting nMOS transistors, as the output voltage drops from V_{DD} to 0. To illustrate the dynamic power dissipation during switching, consider the circuit example given in Fig. 11.1. Here, a two-input NOR gate drives two NAND gates, through interconnection lines. The total capacitive load at the output of the NOR gate consists of (i) the output node capacitance of the gate itself, (ii) the total interconnect capacitance, and (iii) the input capacitances of the driven gates.

The output node capacitance of the gate consists mainly of the junction parasitic capacitances, which are due to the drain diffusion regions of the MOS transistors in the circuit. The physical properties and the calculation of these capacitances were already examined in detail in Chapter 3. The amount of diffusion capacitance is approximately a linear function of the junction area. Consequently, the size of the total drain diffusion area dictates the amount of parasitic capacitance at the output node. The interconnect lines between the gates contribute to the second component of the total capacitance. The estimation of parasitic interconnect capacitance was discussed thoroughly in Chapter 6. Note that especially in sub-micron technologies, the interconnect capacitance can become the dominant component, compared to the transistor-related capacitances. Finally, the input capacitances are mainly due to gate oxide capacitances of the transistors connected to the input terminal. Again, the amount of the gate oxide capacitance is determined primarily by the gate area of each transistor.

Any CMOS logic gate making an output voltage transition can thus be represented by its nMOS network, pMOS network, and the total load capacitance connected to its output node, as seen in Fig. 11.2. As already shown in Chapter 6, the

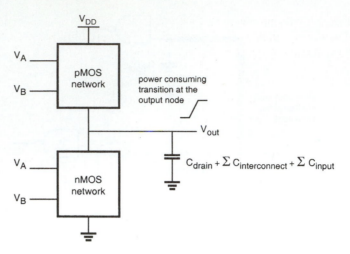

Figure 11.2 Generic representation of a CMOS logic gate for switching power calculation.

average power dissipation of the CMOS logic gate, driven by a periodic input voltage waveform with ideally zero rise- and fall-times, can be calculated from the energy required to charge up the output node to V_{DD} and charge down the total output load capacitance to ground level.

$$P_{avg} = \frac{1}{T} \left[\int_0^{T/2} V_{out} \left(-C_{load} \frac{dV_{out}}{dt} \right) dt + \int_{T/2}^{T} (V_{DD} - V_{out}) \left(C_{load} \frac{dV_{out}}{dt} \right) dt \right]$$
(11.1)

Evaluating this integral yields the well-known expression for the average dynamic (switching) power consumption in CMOS logic circuits.

$$P_{avg} = \frac{1}{T} C_{load} V_{DD}^2$$
(11.2)

or

$$P_{avg} = C_{load} \cdot V_{DD}^2 \cdot f_{CLK}$$
(11.3)

Note that the average switching power dissipation of a CMOS gate is essentially independent of all transistor characteristics and transistor sizes as long as a full voltage swing is achieved. Hence, given an input pattern, the switching delay times have no relevance to the amount of power consumption during the switching events as long as the output voltage swing is between 0 and V_{DD}.

The analysis of switching power dissipation presented above is based on the assumption that the output node of a CMOS gate undergoes one power-consuming transition (0-to-V_{DD} transition) in each clock cycle. This assumption, however, is not always correct; the node transition rate can be slower than the clock rate, depending on the circuit topology, logic style, and the input signal statistics. To better represent this behavior, we will introduce α_T (node transition factor), which is the effective number of power-consuming voltage transitions experienced per clock cycle. Then, the average switching power consumption becomes

$$P_{avg} = \alpha_T \cdot C_{load} \cdot V_{DD}^2 \cdot f_{CLK} \tag{11.4}$$

The estimation of switching activity and various measures to reduce its rate will be discussed in detail in Section 11.4.

The switching power expressions (11.3) and (11.4) are derived by taking into account the charge-up and charge-down process of the output node load capacitance C_{load}. In complex CMOS logic gates, however, most of the internal circuit nodes also make full or partial voltage transitions during switching. Since there is a parasitic capacitance associated with each internal node, these internal transitions contribute to the overall power dissipation of the circuit. In fact, an internal node may undergo several transitions while the output node voltage of the circuit remains unchanged, as illustrated in Fig. 11.3. Thus, an underestimation of the overall switching power consumption may result if only output node voltage transitions are considered.

In general terms, internal node voltage transitions can be partial, i.e., the node voltage swing may be only V_i which is smaller than the full voltage swing of V_{DD}. Thus, the generalized expression for the average switching power dissipation can be written as

$$P_{avg} = \left(\sum_{i=1}^{\text{\# of nodes}} \alpha_{Ti} \cdot C_i \cdot V_i \right) \cdot V_{DD} \cdot f_{CLK} \tag{11.5}$$

where C_i represents the parasitic capacitance associated with each node in the circuit (including the output node) and α_{Ti} represents the corresponding node transition factor associated with that node. Hence, the terms in the parentheses in (11.5) represent the total amount of *charge* which is drawn from the power supply during each

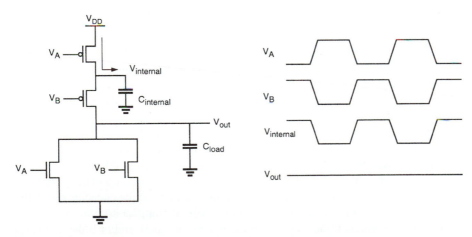

Figure 11.3 Switching of the internal node in a two-input NOR gate results in dynamic power dissipation even if the output node voltage remains unchanged.

switching event. While (11.5) is a more exact representation of the switching power dissipation in a CMOS logic gate, its evaluation may be relatively complicated. Therefore, we will mainly rely on (11.4) to express the switching power dissipation in CMOS logic circuits.

Observations on Switching Power Reduction

Expressions (11.4) and (11.5) derived above for the average switching power dissipation of CMOS logic gates suggest that we have several different means for reducing the power consumption. These measures include (i) reduction of the power supply voltage V_{DD}, (ii) reduction of the voltage swing in all nodes, (iii) reduction of the switching probability (transition factor), and (iv) reduction of the load capacitance. Note that the switching power dissipation is also a linear function of the clock frequency, yet simply reducing the frequency would significantly diminish the overall system performance. Thus, the reduction of clock frequency would be a viable option only in cases where the overall throughput of the system can be maintained by other means.

The reduction of power supply voltage is one of the most widely practiced measures for low-power design. While such reduction is usually very effective, several important issues must be addressed so that the system performance is not sacrificed. In particular, we need to consider that reducing the power supply voltage leads to an increase of delay. Also, the input and output signal levels of a low-voltage circuit or module should be made compatible with the peripheral circuitry, in order to maintain correct signal transmission.

The reduction of switching activity requires a detailed analysis of signal transition probabilities, and implementation of various circuit-level and system-level measures such as logic optimization, use of gated clock signals, and prevention of glitches. Finally, the load capacitance can be reduced by using certain circuit design styles and by proper transistor sizing. These and other methods for the reduction of switching power dissipation will be examined in detail in the following Sections.

Short-Circuit Power Dissipation

The switching power dissipation examined above is purely due to the energy required to charge up the parasitic load capacitances in the circuit, and the switching power is independent of the rise and fall times of the input signals. Yet, if a CMOS inverter (or a logic gate) is driven with input voltage waveforms with finite rise and fall times, both the nMOS and the pMOS transistors in the circuit may conduct *simultaneously* for a short amount of time during switching, forming a direct current path between the power supply and the ground, as shown in Fig. 11.4.

The current component which passes through both the nMOS and the pMOS devices during switching does not contribute to the charging of the capacitances in the circuit, and hence, it is called the *short-circuit current* component. This component is especially prevalent if the output load capacitance is small, and/or if the input signal rise and fall times are large, as seen in Fig. 11.5. Here, the input/output voltage waveforms and the components of the current drawn from the power supply are illustrated for a symmetrical CMOS inverter with small capacitive load. The nMOS

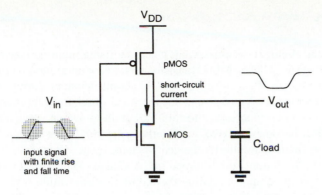

Figure 11.4 Both nMOS and pMOS transistors may conduct (simultaneously) a short-circuit current during switching.

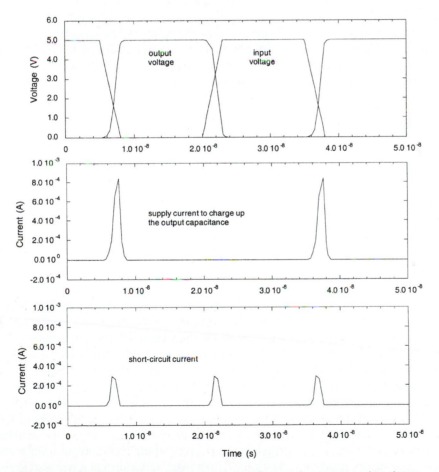

Figure 11.5 Input-output voltage waveforms, the supply current used to charge up the load capacitance, and the short-circuit current in a CMOS inverter with small capacitive load. The total current drawn from the power supply is the sum of both current components.

transistor in the circuit starts conducting when the rising input voltage exceeds the threshold voltage $V_{T,n}$. The pMOS transistor remains on until the input reaches the voltage level $(V_{DD} - |V_{T,p}|)$. Thus, there is a time window during which both transistors are turned on. As the output capacitance is discharged through the nMOS transistor, the output voltage starts to fall. The drain-to-source voltage drop of the pMOS transistor becomes nonzero, which allows the pMOS transistor to conduct as well. The short circuit current is terminated when the input voltage transition is completed and the pMOS transistor is turned off. A similar event is responsible for the short-circuit current component during the falling input transition, when the output voltage starts rising while both transistors are on.

Note that the magnitude of the short-circuit current component will be approximately the same during both the rising-input transition and the falling-input transition, assuming that the inverter is symmetrical and the input rise and fall times are identical. The pMOS transistor also conducts the current which is needed to charge up the small output load capacitance, but only during the falling-input transition (the output capacitance is discharged through the nMOS device during the rising-input transition). This current component, which is responsible for the switching power dissipation of the circuit (current component to charge up the load capacitance), is also shown in Fig. 11.5. The average of both of these current components determines the total amount of power drawn from the supply.

For a simple analysis consider a symmetric CMOS inverter with $k_n = k_p = k$ and $V_{T,n} = |V_{T,p}| = V_T$, and with a very small capacitive load. If the inverter is driven with an input voltage waveform with equal rise and fall times ($\tau_{rise} = \tau_{fall} = \tau$), it can be derived that the time-averaged short circuit current drawn from the power supply is

$$I_{avg}(short\text{-}circuit) = \frac{1}{12} \cdot \frac{k \cdot \tau \cdot f_{CLK}}{V_{DD}}(V_{DD} - 2V_T)^3 \qquad (11.6)$$

Hence, the short-circuit power dissipation becomes

$$P_{avg}(short\text{-}circuit) = \frac{1}{12} \cdot k \cdot \tau \cdot f_{CLK} \cdot (V_{DD} - 2V_T)^3 \qquad (11.7)$$

Note that the short-circuit power dissipation is linearly proportional to the input signal rise and fall times, and also to the transconductance of the transistors. Hence, reducing the input transition times will decrease the short-circuit current component.

Now consider the same CMOS inverter with a larger output load capacitance and smaller input transition times. During the rising input transition, the output voltage will effectively remain at V_{DD} until the input voltage completes its swing and the output will start to drop only after the input has reached its final value. Although both nMOS and pMOS transistors are on simultaneously during the transition, the pMOS transistor cannot conduct a significant amount of current since the voltage drop between its source and drain terminals is almost zero. Similarly, the output voltage will remain at approximately 0 V during the falling input transition and it will start to rise only after the input voltage completes its swing. Again, both transistors will be on simultaneously during the input voltage transition, yet the nMOS transistor will not be able to conduct a significant amount of current since its drain-to-source voltage is

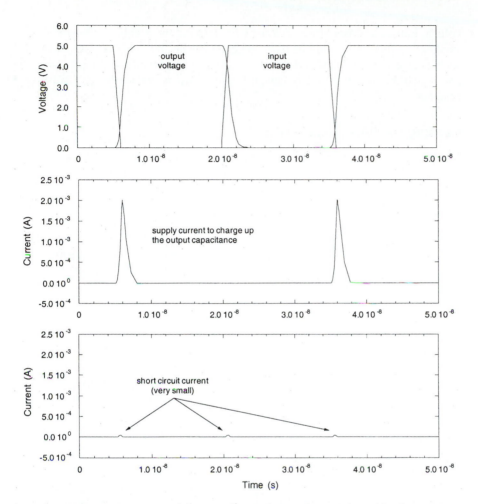

Figure 11.6 Input-output voltage waveforms, the supply current used to charge up the load capacitance, and the short-circuit current in a CMOS inverter with larger capacitive load and smaller input transition times. The total current drawn from the power supply is approximately equal to the charge-up current.

approximately zero. This situation is illustrated in Fig. 11.6, which shows the simulated input and output voltage waveforms of the inverter as well as the short-circuit and dynamic current components drawn from the power supply. Notice that the peak value of the supply current to charge up the output load capacitance is larger in this case. The reason for this is that the pMOS transistor remains in saturation during the entire input transition, as opposed to the previous case shown in Fig. 11.5 where the transistor leaves the saturation region before the input transition is completed.

The discussion concerning the magnitude of the short-circuit current may suggest that the short-circuit power dissipation can be reduced by making the output voltage transition times larger and/or by making the input voltage transition times smaller. Yet this goal should be balanced carefully against other performance goals such as propagation delay, and the reduction of the short-circuit current should be considered as one of the many design requirements that must be satisfied by the designer.

Leakage Power Dissipation

The nMOS and pMOS transistors used in a CMOS logic gate generally have nonzero reverse leakage and subthreshold currents. In a CMOS VLSI chip containing a very large number of transistors, these currents can contribute to the overall power dissipation even when the transistors are not undergoing any switching event. The magnitude of the leakage currents is determined mainly by the processing parameters.

Of the two main leakage current components in a MOSFET, the reverse diode leakage occurs when the pn-junction between the drain and the bulk of the transistor is reverse-biased. The reverse-biased drain junction then conducts a reverse saturation current which is drawn from the power supply. Consider a CMOS inverter with a high input voltage, where the nMOS transistor is turned on and the output node voltage is discharged to zero. Although the pMOS transistor is turned off, there will be a reverse potential difference of V_{DD} between its drain and the n-well, causing a diode leakage through the drain junction. The n-well region of the pMOS transistor is also reverse-biased with V_{DD}, with respect to the p-type substrate. Therefore, another significant leakage current component exists because of the n-well junction (Fig. 11.7).

A similar situation can be observed when the input voltage is zero, and the output voltage is charged up to V_{DD} through the pMOS transistor. Then, the reverse potential difference between the nMOS drain region and the p-type substrate causes a reverse leakage current which is also drawn from the power supply (through the pMOS transistor).

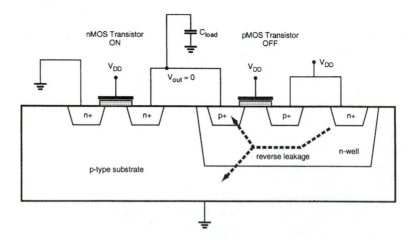

Figure 11.7 Reverse leakage current paths in a CMOS inverter with high input voltage.

The reverse leakage current of a pn-junction is expressed by

$$I_{reverse} = A \cdot J_S \left(e^{\frac{q V_{bias}}{kT}} - 1 \right) \qquad (11.8)$$

where V_{bias} is the reverse bias voltage across the junction, J_S is the reverse saturation current density and the A is the junction area. The typical reverse saturation current density is 1–5 pA/μm^2, and it increases quite significantly with temperature. Note that the reverse leakage occurs even during the stand-by operation when no switching takes place. Hence, the power dissipation due to this mechanism can be significant in a large chip containing several million transistors.

Another component of leakage currents in CMOS circuits is the subthreshold current, which is due to carrier diffusion between the source and the drain regions of the transistor in weak inversion. The behavior of an MOS transistor in the subthreshold operating region is similar to a bipolar device, and the subthreshold current exhibits an exponential dependence on the gate voltage. The amount of the subthreshold current may become significant when the gate-to-source voltage is smaller than, but very close to, the threshold voltage of the device. In this case, the power dissipation due to subthreshold leakage can become comparable in magnitude to the switching power dissipation of the circuit. The subthreshold leakage current is illustrated in Fig. 11.8.

Note that the subthreshold leakage current can occur even when there is no switching activity in the circuit and that this component must be carefully considered for estimating the total power dissipation in the stand-by operation mode. The subthreshold current expression given in Section 3.5 is repeated below, in order to illustrate the exponential dependence of the current on terminal voltages.

$$I_D(subthreshold) \cong \frac{q D_n W x_c n_0}{L_B} \cdot e^{\frac{q \phi_r}{kT}} \cdot e^{\frac{q}{kT}(A \cdot V_{GS} + B \cdot V_{DS})} \qquad (11.9)$$

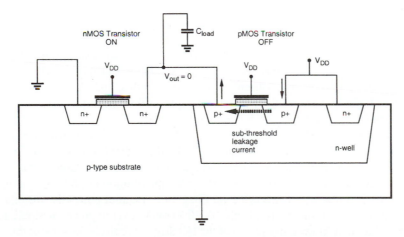

Figure 11.8 Subthreshold leakage current path in a CMOS inverter with high input voltage.

One relatively simple measure to limit the subthreshold current component is to avoid very low threshold voltages, albeit with some speed penalty, so that the V_{GS} of the nMOS transistor remains safely below $V_{T,n}$ when the input is logic zero, and the $|V_{GS}|$ of the pMOS transistor remains safely below $|V_{T,p}|$ when the input is logic one.

In addition to the three major sources of power consumption in CMOS digital integrated circuits discussed in this Section, some chips may also contain circuits which consume static power. One example is the pseudo-nMOS logic circuits which utilize a pMOS transistor as the pull-up device. This circuit type was already examined in Section 7.4, and it was shown that the circuit exhibits a nonzero static power dissipation due to the load current. The presence of such circuits should be taken into account when estimating the overall power dissipation.

To summarize our discussion, the total power dissipation in CMOS digital circuits can be expressed as the sum of four components,

$$P_{total} = \alpha_T \cdot C_{load} \cdot V_{DD}^2 \cdot f_{CLK} + V_{DD}(I_{short\text{-}circuit} + I_{leakage} + I_{static}) \qquad (11.10)$$

where $I_{short\text{-}circuit}$ denotes the average short-circuit current, $I_{leakage}$ denotes the reverse leakage and subthreshold leakage currents, and I_{static} denotes the DC current component drawn from the power supply. The switching power dissipation, which is the first term in (11.10), is the dominating component in most CMOS logic gates.

Examples of Actual Power Dissipation

In this section, we have examined various components of power dissipation in CMOS digital integrated circuits and we have identified the physical origins of power dissipation. In the following, we will apply this analysis to reduce the power dissipation in CMOS logic gates. It is also very important to understand the power consumption profile of a large-scale system or chip, i.e., how much power is dissipated in different parts of the chip. Understanding the dominant components of power consumption in a CMOS VLSI chip can help us to develop a more realistic perspective, and guide our efforts to reduce the power dissipation of large-scale systems.

In the following example, we will examine the power consumption of actual chips, and how the total consumption is divided among four major parts: logic circuits, clock generation and distribution, interconnections, and off-chip driving (I/O circuits). Table 11.1 shows the percentage of power dissipated in each one of these four parts, for three different CMOS digital VLSI chips.

Notice that all three chips presented here have very different characteristics, in terms of the minimum feature size, operating frequency and supply voltage. Consequently, the amount of total power dissipation also shows a very large variation. Yet we can recognize a number of similarities in the distribution of power consumption among the four major parts in each chip. First, the amount of power dissipated in the logic gates is only a fraction (typically 10–30%) of the overall power dissipation. On the other hand, the power dissipation of the I/O driver circuits is 30–40% of the overall consumption, which is a significant portion of the power budget. Also, note that a very large amount of power can be dissipated in the clock network, which includes

Table 11.1 Power dissipation statistics for various CMOS digital VLSI chips

Chip	Intel 80386	DEC Alpha 21064	Cell based ASIC
Minimum feature size	1.5 μm	0.75 μm	0.5 μm
Number of gates	36,808	263,666	10,000
Clock frequency f_{CLK}	16 MHz	200 MHz	110 MHz
Supply voltage	5 V	3.3 V	3 V
Total power dissipation	1.41 W	32 W	0.8 W
Logic gates	32%	14%	9%
Clock distribution	9%	32%	30%
Interconnect	28%	14%	15%
I/O drivers	26%	37%	43%

the clock drivers, the distribution wires, and the input capacitances of all latches driven by the global clock signal.

These observations are very valuable in identifying the parts which are critical for the overall power consumption in a chip. For example, the I/O circuits are typically designed to drive large off-chip capacitive loads, and the signal levels are almost always dictated by the external devices. Therefore, innovative low-power methods for driving large off-chip loads always have a significant impact on the overall power consumption. Similarly, a power-saving clock distribution strategy can be a very valuable asset.

11.3   Low-Power Design Through Voltage Scaling

Equation (11.3) shows that the average switching power dissipation is proportional to the square of the power supply voltage; hence, reduction of V_{DD} will significantly reduce the power consumption. In the following, we will examine the effects of reducing the power supply voltage V_{DD} upon switching power consumption and dynamic performance of the gate.

Influence of Voltage Scaling on Power and Delay

Although the reduction of power supply voltage significantly reduces the dynamic power dissipation, the inevitable design trade-off is the increase of delay. This can be seen easily by examining the following propagation delay expressions for the CMOS inverter circuit, which were derived in Chapter 6.

$$\tau_{PHL} = \frac{C_{load}}{k_n(V_{DD} - V_{T,n})}\left[\frac{2V_{T,n}}{V_{DD} - V_{T,n}} + ln\left(\frac{4(V_{DD} - V_{T,n})}{V_{DD}} - 1\right)\right] \quad (11.11a)$$

$$\tau_{PLH} = \frac{C_{load}}{k_p(V_{DD} - |V_{T,p}|)}\left[\frac{2|V_{T,p}|}{V_{DD} - |V_{T,p}|} + ln\left(\frac{4(V_{DD} - |V_{T,p}|)}{V_{DD}} - 1\right)\right]$$

$$(11.11b)$$

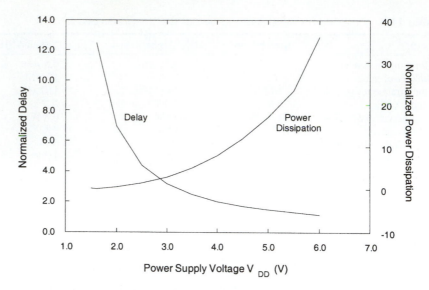

Figure 11.9   Normalized propagation delay and average switching power dissipation of a CMOS inverter, as a function of the power supply voltage V_{DD}.

If the power supply voltage is scaled down while all other parameters are kept constant, the propagation delay time would increase. Figure 11.9 shows the normalized variation of the delay as a function of V_{DD}, where the threshold voltages of the nMOS and the pMOS transistor are $V_{T,n} = 0.8$ V and $V_{T,p} = -0.8$ V, respectively. The normalized variation of the average switching power dissipation as a function of the power supply voltage is also shown on the same plot.

Notice that the dependence of circuit speed on the power supply voltage may also influence the relationship between the dynamic power dissipation and the supply voltage. Equation (11.3) suggests a quadratic improvement (reduction) of power consumption as the power supply voltage is reduced. However, this interpretation assumes that the switching frequency (i.e., the number of switching events per unit time) remains constant. If the circuit is always operated at the maximum frequency allowed by its propagation delay, the number of switching events per unit time (i.e., the operating frequency) will drop as the propagation delay becomes larger with the reduction of the power supply voltage. The net result is that the dependence of switching power dissipation on the power supply voltage becomes stronger than a simple quadratic relationship, shown in Fig. 11.9.

It is important to note that the voltage scaling discussed in this Section is distinctly different from *constant-field scaling* (cf. Chapter 3), where the power supply voltage as well as the critical device dimensions (channel length, gate oxide thickness) and doping densities are scaled by the same factor. Here, we examine the effects of reducing the power supply voltage for a given technology; hence, key device parameters and the load capacitances are assumed to be constant.

The propagation delay expressions (11.11) show that the negative effect of reducing the power supply voltage upon delay can be compensated for, if the threshold

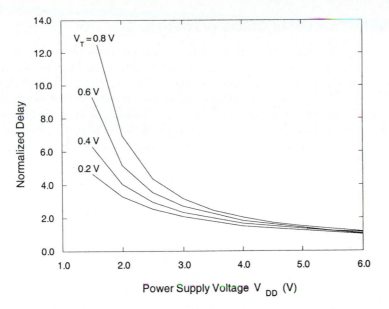

Figure 11.10 Variation of the normalized propagation delay of a CMOS inverter, as a function of the power supply voltage V_{DD} and the threshold voltage V_T.

voltage of the transistors (V_T) is scaled down accordingly. However, this approach is limited because the threshold voltage may not be scaled to the same extent as the supply voltage. When scaled linearly, reduced threshold voltages allow the circuit to produce the same speed-performance at a lower V_{DD}. Figure 11.10 shows the variation of the propagation delay of a CMOS inverter as a function of the power supply voltage, and for different threshold voltage values.

We can see, for example, that reducing the threshold voltage from 0.8 V to 0.2 V can improve the delay at $V_{DD} = 2$ V by a factor of 2. The positive influence of threshold voltage reduction upon propagation delay is especially pronounced at low power supply voltages, for $V_{DD} < 2$ V. It should be noted, however, that using low-V_T transistors raises significant concerns about noise margins and subthreshold conduction. Smaller threshold voltages lead to smaller noise margins for the CMOS logic gates. The subthreshold conduction current also sets a severe limitation against reducing the threshold voltage. For threshold voltages smaller than 0.2 V, leakage due to subthreshold conduction in stand-by, i.e., when the gate is not switching, may become a very significant component of the overall power consumption. In addition, propagation delay becomes more sensitive to process-related fluctuations of the threshold voltage.

Next, we will examine two circuit design techniques which can be used to overcome the difficulties (such as leakage and high stand-by power dissipation) associated with the low-V_T circuits. These techniques are called Variable-Threshold CMOS (VTCMOS) and Multiple-Threshold CMOS (MTCMOS).

Variable-Threshold CMOS (VTCMOS) Circuits

We have seen that using a low supply voltage (V_{DD}) and a low threshold voltage (V_T) in CMOS logic circuits is an efficient method for reducing the overall power dissipation, while maintaining high-speed performance. Yet designing a CMOS logic gate entirely with low-V_T transistors will inevitably lead to increased subthreshold leakage, and consequently, to higher stand-by power dissipation when the output is not switching. One possible way to overcome this problem is to *adjust* the threshold voltages of the transistors in order to avoid leakage in the stand-by mode, by changing the substrate bias.

The threshold voltage V_T of an MOS transistor is a function of its source-to-substrate voltage V_{SB}, as we have already discussed in Chapter 3. In conventional CMOS logic circuits, the substrate terminals of all nMOS transistors are connected to ground potential, while the substrate terminals of all pMOS transistors are connected to V_{DD}. This ensures that the source and drain diffusion regions always remain reverse-biased with respect to the substrate, and that the threshold voltages of the transistors are not significantly influenced by the body (backgate-bias) effect. In VTCMOS circuit technique, on the other hand, the transistors are designed inherently with a low threshold voltage, and the substrate bias voltages of nMOS and pMOS transistors are generated by a variable substrate bias control circuit, as shown in Fig. 11.11.

When the inverter circuit in Fig. 11.11 is operating in its *active* mode, the substrate bias voltage of the nMOS transistor is $V_{Bn} = 0$ and the the substrate bias voltage of the pMOS transistor is $V_{Bp} = V_{DD}$. Thus, the inverter transistors do not experience any backgate-bias effect. The circuit operates with low V_{DD} *and* low V_T, benefiting from both low power dissipation (due to low V_{DD}) and high switching speed (due to low V_T).

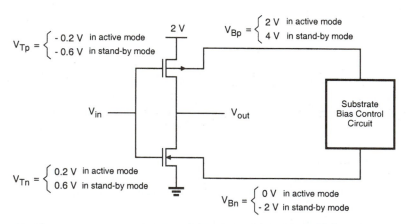

Figure 11.11 A variable-threshold CMOS (VTCMOS) inverter circuit. The threshold voltages of nMOS and pMOS transistors are increased by adjusting the substrate bias voltage, in order to reduce subthreshold leakage currents in the stand-by mode.

When the inverter circuit is in the *stand-by* mode, however, the substrate bias control circuit generates a lower substrate bias voltage for the nMOS transistor and a higher substrate bias voltage for the pMOS transistor. As a result, the magnitudes of the threshold voltages V_{Tn} and V_{Tp} both increase in the stand-by mode, due to the backgate-bias effect. Since the subthreshold leakage current drops exponentially with increasing threshold voltage, the leakage power dissipation in the stand-by mode can be significantly reduced with this technique.

The VTCMOS technique can also be used to automatically control the threshold voltages of the transistors in order to reduce leakage currents, and to compensate for process-related fluctuations of the threshold voltages. This approach is also called the Self-Adjusting Threshold-Voltage Scheme (SATS).

The variable-threshold CMOS circuit design techniques are very effective for reducing the subthreshold leakage currents and for controlling threshold voltage values in low V_{DD}–low V_T applications. However, this technique usually requires twin-well or triple-well CMOS technology in order to apply different substrate bias voltages to different parts of the chip. Also, separate power pins may be required if the substrate bias voltage levels are not generated on-chip. The additional area occupied by the substrate bias control circuitry is usually negligible compared to the overall chip area.

Figure 11.12 shows the block-diagram of a typical low-power chip with low internal supply voltage V_{DDL} and with threshold voltage control. Note that the input/output circuits of the chip usually operate with a higher external supply voltage, in order to increase the noise margins and to enable communication with the peripheral devices. An on-chip DC-DC voltage converter generates the low internal supply voltage V_{DDL}, which is used by the internal circuitry. Two signal swing converters (level converters) are used to reduce the voltage swing of the incoming input signals and to increase the voltage swing of the outgoing output signals, respectively. The internal low-voltage circuitry can be designed using VTCMOS techniques, where the threshold voltage control unit adjusts the substrate bias in order to suppress leakage currents.

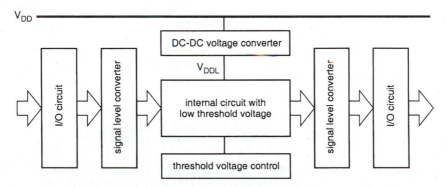

Figure 11.12 Block diagram of a typical low-power chip. The internal supply voltage is generated on-chip, by a DC-DC converter circuit.

Multiple-Threshold CMOS (MTCMOS) Circuits

Another technique which can be applied for reducing leakage currents in low-voltage circuits in the stand-by mode is based on using two different types of transistors (both nMOS and pMOS) with two different threshold voltages in the circuit. Here, low-V_T transistors are typically used to design the logic gates where switching speed is essential, whereas high-V_T transistors are used to effectively isolate the logic gates in stand-by and to prevent leakage dissipation. The generic circuit structure of the MTCMOS logic gate is shown in Fig. 11.13.

In the active mode, the high-V_T transistors are turned on and the logic gates consisting of low-V_T transistors can operate with low switching power dissipation and small propagation delay. When the circuit is driven into stand-by mode, on the other hand, the high-V_T transistors are turned off and the conduction paths for any subthreshold leakage currents that may originate from the internal low-V_T circuitry are effectively cut off. Figure 11.14 shows a simple D-latch circuit designed with the MTCMOS technique. Note that the critical signal propagation path from the input to the output consists exclusively of low-V_T transistors, while a cross-coupled inverter pair consisting of high-V_T transistors is used for preserving the data in the stand-by mode.

The MTCMOS technique is conceptually easier to apply and to use compared to the VTCMOS technique, which usually requires a sophisticated substrate bias control mechanism. It does not require a twin-well or triple-well CMOS process; the only significant process-related overhead of MTCMOS circuits is the fabrication of MOS transistors with different threshold voltages on the same chip. One of the disadvantages of the MTCMOS circuit technique is the presence of series-connected stand-by transistors, which increase the overall circuit area and also add extra parasitic capacitance and delay.

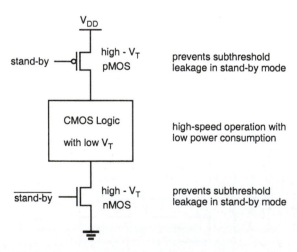

Figure 11.13 Generic structure of a multiple-threshold CMOS (MTCMOS) logic gate.

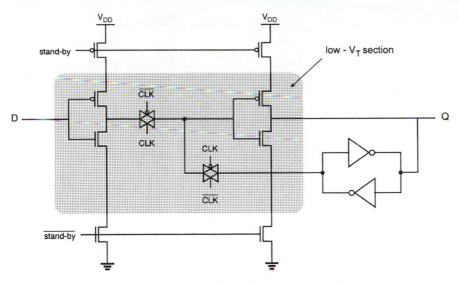

Figure 11.14 Low-power/low-voltage D-latch circuit designed with MTCMOS technique.

While the VTCMOS and MTCMOS circuit techniques can be very effective in designing low-power/low-voltage logic gates, they may not be used as a universal solution to low-power CMOS logic design. In certain types of applications where variable threshold voltages and multiple threshold voltages are infeasible due to technological limitations, system-level architectural measures such as pipelining and hardware replication techniques offer feasible alternatives for maintaining the system performance (throughput) despite voltage scaling. In the following, we will examine some examples of these architectural design techniques for reducing power dissipation.

Pipelining Approach

First, consider the single functional block shown in Fig. 11.15 which implements a logic function **F(INPUT)** of the input vector, **INPUT.** Both the input and the output vectors are sampled through register arrays, driven by a clock signal **CLK.** Assume that the critical path in this logic block (at a power supply voltage of V_{DD}) allows a maximum sampling frequency of f_{CLK}; in other words, the maximum input-to-output propagation delay $\tau_{P,max}$ of this logic block is equal to or less than $T_{CLK} = 1/f_{CLK}$. Figure 11.15 also shows a simplified timing diagram of the circuit. A new input vector is latched into the input register array at each clock cycle, and the output data become valid with a latency of one cycle. Let C_{total} be the total capacitance switched every clock cycle. Here, C_{total} consists of (i) the capacitance switched in the input register array, (ii) the capacitance switched to implement the logic function, and (iii) the capacitance switched in the output register array. Then,

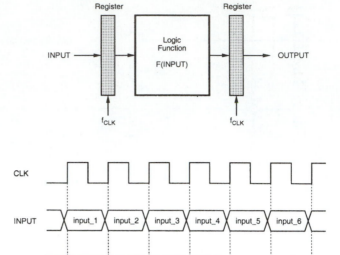

Figure 11.15   Single-stage implementation of a logic function and its simplified timing diagram.

the dynamic power consumption of this structure can be found as

$$P_{reference} = C_{total} \cdot V_{DD}^2 \cdot f_{CLK} \tag{11.12}$$

Now consider an N-stage pipelined structure for implementing the same logic function, as shown in Fig. 11.16. The logic function **F(INPUT)** has been partitioned into N successive stages, and a total of $(N - 1)$ register arrays have been introduced, in addition to the original input and output registers, to create the pipeline. All registers are clocked at the original sample rate, f_{CLK}. If all stages of the partitioned function have approximately equal delays of

$$\tau_P(pipeline\_stage) = \frac{\tau_{P,max}(input\text{-}to\text{-}output)}{N} = T_{CLK} \tag{11.13}$$

Then the logic blocks between two successive registers can operate N-times slower while *maintaining the same functional throughput* as before. This implies that the power supply voltage can be reduced to a value of $V_{DD,new}$, to effectively slow down the circuit by a factor of N. The supply voltage needed for this reduction can be found by solving (11.11).

The dynamic power consumption of the N-stage pipelined structure with a lower supply voltage and with the same functional throughput as the single-stage structure can be approximated by

$$P_{pipeline} = [C_{total} + (N - 1)C_{reg}] \cdot V_{DD,new}^2 \cdot f_{CLK} \tag{11.14}$$

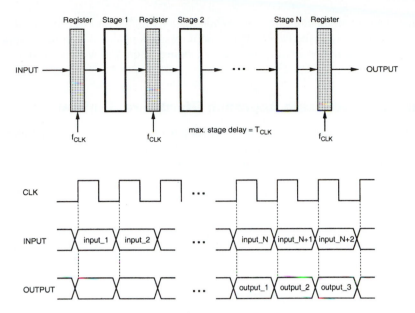

Figure 11.16 N-stage pipeline structure realizing the same logic function as in Fig. 11.15. The maximum pipeline stage delay is equal to the clock period, and the latency is N clock cycles.

where C_{reg} represents the capacitance switched by each pipeline register. Then, the power reduction factor achieved in a N-stage pipeline structure is

$$\frac{P_{pipeline}}{P_{reference}} = \frac{[C_{total} + (N-1)C_{reg}] \cdot V_{DD,new}^2 \cdot f_{CLK}}{C_{total} \cdot V_{DD}^2 \cdot f_{CLK}}$$

$$= \left[1 + \frac{C_{reg}}{C_{total}}(N-1)\right] \frac{V_{DD,new}^2}{V_{DD}^2} \qquad (11.15)$$

As an example, consider replacing a single-stage logic block ($V_{DD} = 5$ V, $f_{CLK} = 20$ MHz) with a four-stage pipeline structure, running at the same clock frequency. This means that the propagation delay of each pipeline stage can be increased by a factor of 4 without sacrificing the data throughput. Assuming that the magnitude of the threshold voltage of all transistors is 0.8 V, the target speed reduction can be achieved by reducing the power supply voltage from 5 V to approximately 2 V (see Fig. 11.10). With $(C_{reg}/C_{total}) = 0.1$, the overall power reduction factor is found from (11.15) as 0.2. This means that replacing the original single-stage logic block with a four-stage pipeline running at the same clock frequency and reducing the power supply voltage from 5 V to 2 V will provide a switching power saving of about 80%, while maintaining the same throughput as before.

The architectural modification described here has a relatively small area overhead. A total of (N − 1) register arrays have to be added to convert the original single-stage structure into a pipeline. While trading off area for lower power, this approach also increases the latency from one to N clock cycles. Yet in many applications such as signal processing and data encoding, latency is not a very significant concern.

Parallel Processing Approach (Hardware Replication)

Another method for trading off area for lower power dissipation is to use parallelism, or hardware replication. This approach could be useful especially when the logic function to be implemented is not suitable for pipelining. Consider N identical processing elements, each implementing the logic function **F(INPUT)** in parallel, as shown in Fig. 11.17. Assume that the consecutive input vectors arrive at the same rate as in the single-stage case examined earlier. The input vectors are routed to all the registers of the N processing blocks. Gated clock signals, each with a clock period of (NT_{CLK}), are used to load each register every N clock cycles. This means that

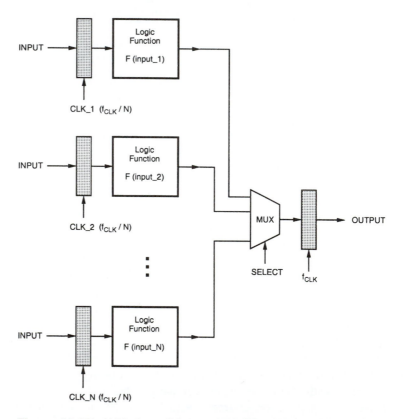

Figure 11.17   N-block parallel structure realizing the same logic function as in Fig. 11.15. Notice that the input registers are clocked at a lower frequency of (f_{CLK}/N).

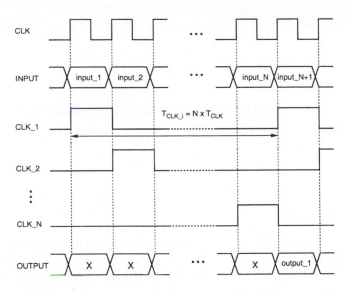

Figure 11.18 Simplified timing diagram of the N-block parallel structure shown in Fig. 11.17.

the clock signals to each input register are skewed by T_{CLK}, such that each one of the N consecutive input vectors is loaded into a different input register. Since each input register is clocked at a lower frequency of (f_{CLK}/N), the time allowed to compute the function for each input vector is increased by a factor of N. This implies that the power supply voltage can be reduced until the critical path delay equals the new clock period of (NT_{CLK}). The outputs of the N processing blocks are multiplexed and sent to an output register which operates at a clock frequency of f_{CLK}, ensuring the same data throughput rate as before. The timing diagram of this parallel arrangement is given in Fig. 11.18.

Since the time allowed to compute the function for each input vector is increased by a factor of N, the power supply voltage can be reduced to a value of $V_{DD,new}$, to effectively slow down the circuit. The new supply voltage can be found, as in the pipelined case, by solving (11.11). The total dynamic power dissipation of the parallel structure (neglecting the dissipation of the multiplexor) is found as the sum of the power dissipated by the input registers and the logic blocks operating at a clock frequency of (f_{CLK}/N), and the output register operating at a clock frequency of f_{CLK}.

$$P_{parallel} = N \cdot C_{total} \cdot V_{DD,new}^2 \cdot \frac{f_{CLK}}{N} + C_{reg} \cdot V_{DD,new}^2 \cdot f_{CLK}$$

$$= \left(1 + \frac{C_{reg}}{C_{total}}\right) \cdot C_{total} \cdot V_{DD,new}^2 \cdot f_{CLK} \qquad (11.16)$$

Note that there is also an additional overhead which consists of the input routing capacitance, the output routing capacitance and the capacitance of the output multiplexor structure, all of which are increasing functions of N. If this overhead is neglected, the amount of power reduction achievable in a N-block parallel implementation is

$$\frac{P_{parallel}}{P_{reference}} = \frac{V_{DD,new}^2}{V_{DD}^2} \cdot \left(1 + \frac{C_{reg}}{C_{total}}\right) \tag{11.17}$$

The lower bound of switching power reduction realizable with architecture-driven voltage scaling is found, assuming zero threshold voltage, as

$$\frac{P_{parallel}}{P_{reference}} \geq \frac{1}{N^2} \tag{11.18}$$

Two obvious consequences of this approach are the increased area and the increased latency. A total of N identical processing blocks must be used to slow down the operation (clocking) speed by a factor of N. In fact, the silicon area will grow even faster than the number of processors because of signal routing and the overhead circuitry. The timing diagram in Fig. 11.18 shows that the parallel implementation has a latency of N clock cycles, as in the N-stage pipelined implementation. Considering its smaller area overhead, however, the pipelined approach offers a more efficient alternative for reducing the power dissipation while maintaining the throughput.

11.4   Estimation and Optimization of Switching Activity

In the previous section, we have discussed methods for minimizing dynamic power consumption in CMOS digital integrated circuits by supply voltage scaling. Another approach to low-power design is to reduce the switching activity and the amount of the switched capacitance to the minimum level required to perform a given task. The measures to accomplish this goal can range from optimization of algorithms to logic design, and finally to physical mask design. In the following, we will examine the concept of switching activity, and introduce some of the approaches used to reduce it. In Section 11.5, we will also examine the various measures used to minimize the amount of capacitance which must be switched to perform a given task.

The Concept of Switching Activity

It was already discussed in Section 11.2 that the dynamic power consumption of a CMOS logic gate depends, among other parameters, on the node transition factor α_T, which is the effective number of power-consuming voltage transitions experienced by the output capacitance per clock cycle. This parameter, also called the switching activity factor, depends on the Boolean function performed by the gate, the logic family, and the input signal statistics.

We can easily investigate the output transition probabilities for different types of logic gates. First, we introduce two signal probabilities, P_0 and P_1. P_0 corresponds to the probability of having a logic "0" at the output, and $P_1 = (1 - P_0)$ corresponds

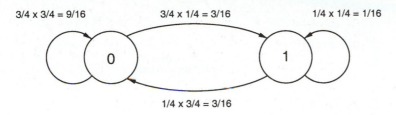

3/4 x 3/4 = 9/16   3/4 x 1/4 = 3/16   1/4 x 1/4 = 1/16

1/4 x 3/4 = 3/16

Figure 11.19 State transition diagram and state transition probabilities of a NOR2 gate.

to the probability of having a logic "1" at the output. Therefore, the probability that a power-consuming (0-to-1) transition occurs at the output node is the product of these two output signal probabilities. Consider, for example, a static CMOS NOR2 gate. If the two inputs are independent and uniformly distributed, the four possible input combinations (00, 01, 10, 11) are equally likely to occur. Thus, we can find from the truth table of the NOR2 gate that $P_0 = 3/4$, and $P_1 = 1/4$. The probability that a power-consuming transition occurs at the output node is therefore

$$P_{0 \to 1} = P_0 \cdot P_1 = \frac{3}{4} \cdot \frac{1}{4} = \frac{3}{16} \tag{11.19}$$

The transition probabilities can be shown on a state transition diagram which consists of the two possible output states and the possible transitions among them (Fig. 11.19). In the general case of a CMOS logic gate with n input variables, the probability of a power-consuming output transition can be expressed as a function of n_0, which is the number of zeros in the output column of the truth table.

$$P_{0 \to 1} = P_0 \cdot P_1 = \left(\frac{n_0}{2^n}\right) \cdot \left(\frac{2^n - n_0}{2^n}\right) \tag{11.20}$$

The output transition probability is shown as a function of the number of inputs in Fig. 11.20, for different types of logic gates for equal input probabilities. For a NAND or NOR gate, the truth table contains only one "0" or "1", respectively, regardless of the number of inputs. Therefore, the output transition probability drops as the number of inputs is increased. In a XOR gate, on the other hand, the truth table always contains an equal number of logic "0" and logic "1" values. The output transition probability therefore remains constant at 0.25.

In multi-level logic circuits, the distribution of input signal probabilities is typically not uniform, i.e., one cannot expect to have equal probabilities for the occurrence of a logic "0" and a logic "1". Then, the output transition probability becomes a function of the input probability distributions. As an example, consider the NOR2 gate examined above. Let $P_{1,A}$ represent the probability of having a logic "1" at the input A, and $P_{1,B}$ represent the probability of having a logic "1" at the input B. The probability of obtaining a logic "1" at the output node is

$$P_1 = (1 - P_{1,A}) \cdot (1 - P_{1,B}) \tag{11.21}$$

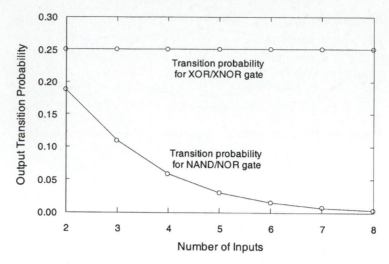

Figure 11.20   Output transition probabilities of different logic gates, as a function of the number of inputs. Note that the transition probability of the XOR gate is independent of the number or inputs.

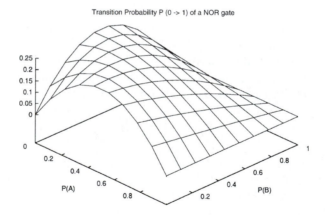

Figure 11.21   Output transition probability of NOR2 gate as a function of two input probabilities.

Using this expression, the probability of a power-consuming output transition is found as a function of $P_{1,A}$ and $P_{1,B}$.

$$P_{0 \to 1} = P_0 \cdot P_1 = (1 - P_1) \cdot P_1$$
$$= (1 - (1 - P_{1,A}) \cdot (1 - P_{1,B})) \cdot ((1 - P_{1,A}) \cdot (1 - P_{1,B})) \qquad (11.22)$$

Figure 11.21 shows the distribution of the output transition probability in NOR2 gate, as a function of two input probabilities. It can be seen that the evaluation of switching activity becomes a complicated problem in large circuits, especially when sequential elements, reconvergent nodes (where the output of a logic gate is used as the input variable of two or more logic gates, propagated along several independent paths and eventually merged at the input of another logic gate), and feedback loops are involved. The designer must therefore rely on computer-aided design (CAD) tools for correct estimation of switching activity in a given network.

In dynamic CMOS logic circuits, the output node is precharged during every clock cycle. If the output node was discharged (i.e., if the output value was equal to "0") in the previous cycle, the pMOS precharge transistor will draw a current from the power supply during the precharge phase. This means that the dynamic CMOS logic gate will consume power every time the output value equals "0", regardless of the preceding or following values. Therefore, the power consumption of dynamic logic gates is determined by the signal-value probability of the output node and not by the transition probability. From the discussion above, we can see that signal-value probabilities are always larger than transition probabilities; hence, the power consumption of dynamic CMOS logic gates is typically larger than static CMOS gates under the same conditions.

Reduction of Switching Activity

Switching activity in CMOS digital integrated circuits can be reduced by algorithmic optimization, by architecture optimization, by proper choice of logic topology, or by circuit-level optimization. In the following, we will briefly discuss some of the measures that can be applied to optimize the switching probabilities, and hence the dynamic power consumption.

Algorithmic optimization depends heavily on the application and on the characteristics of the data such as dynamic range, correlation, and statistics of data transmission. Some of the techniques can be applied only for specific applications such as Digital Signal Processing (DSP) and cannot be used for general purpose processing. For example, a proper vector quantization (VQ) algorithm can be chosen for minimum switching activity. Also, the number of memory accesses, the number of multiplications, and the number of additions can be reduced by about a factor of 30 if the *differential tree search* algorithm is used instead of the *full search* algorithm.

The representation of data can have a significant impact on switching activity at the system level. In applications where data bits change sequentially and are highly correlated (such as the address bits to access instructions), the use of Gray coding leads to a reduced number of transitions compared to simple binary coding. Another example is the use of *sign-magnitude* representation instead of the conventional *two's complement* representation for signed data. A change in sign will cause transitions of the higher-order bits in the two's complement representation, whereas only the sign bit will change in *sign-magnitude* representation. Therefore, the switching activity can be reduced by using the *sign-magnitude* representation in applications where the data sign changes are frequent.

Glitch Reduction

An important architecture-level measure to reduce switching activity is based on delay balancing and the reduction of glitches. In multi-level logic circuits, the propagation delay from one logic block to the next can cause spurious signal transitions, or glitches, as a result of *critical races* or *dynamic hazards*. In general, if all input signals of a gate change simultaneously, no glitching occurs. But a dynamic hazard or glitch can occur if input signals change at different times. Thus, a node can exhibit multiple transitions in a single clock cycle before settling to the correct logic level (Fig. 11.22). In some cases, the signal glitches are only partial, i.e., the node voltage does not make a full transition between the ground and V_{DD} levels, yet even partial glitches can have a significant contribution to dynamic power dissipation.

Glitches occur primarily due to a mismatch or imbalance in the path lengths in the logic network. Such a mismatch in path lengths results in a mismatch of signal timing with respect to the primary inputs. As an example, consider the simple parity network shown in Fig. 11.23. If all XOR gates have the same delay and four input signals arrive at the same time, the network in Fig. 11.23(a) will suffer from glitching due to the wide disparity between the arrival times of the input signals. In the network shown in Fig. 11.23(b), on the other hand, all input arrival times are uniformly identical because the delay paths are balanced. Such redesign can significantly reduce the glitches, and consequently, the dynamic power dissipation in complex multi-level networks. Also notice that the tree structure shown in Fig. 11.23(b)

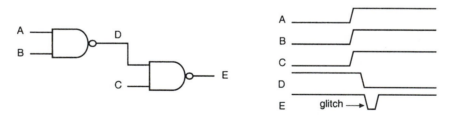

Figure 11.22 Signal glitching in multi-level static CMOS circuits.

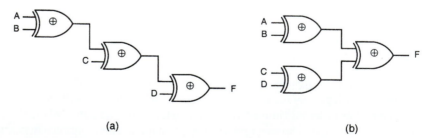

(a)   (b)

Figure 11.23 (a) Implementation of a four-input parity (XOR) function using a chain structure. (b) Implementation of the same function using a tree structure which will reduce glitching transitions.

results in smaller overall propagation delay. Finally, it should be noted that glitching is not a significant issue in multi-level dynamic CMOS logic circuits, since each node undergoes at most one transition per clock cycle.

Gated Clock Signals

Another effective design technique for reducing the switching activity in CMOS logic circuits is the use of conditional, or *gated* clock signals. We have already seen in Section 11.2 that the switching power dissipation in the clock distribution network can be very significant. If certain logic blocks in a system are not immediately used during the current clock cycle, temporarily diabling the clock signals of these blocks will obviously save switching power that is otherwise wasted. Note that the design of an effective clock gating strategy requires careful analysis of the signal flow and of the interrelations among the various operations performed by the circuit. Figure 11.24 shows the block diagram of an N-bit number comparator circuit which is designed using the gated clock technique.

The circuit compares the magnitudes of two unsigned N-bit binary numbers (A and B) and produces an output to indicate which one is larger. In the conventional approach, all input bits are first latched into two N-bit registers, and subsequently applied to the comparator circuit. In this case, two N-bit register arrays dissipate power in every clock cycle. Yet if the most significant bits, A[N − 1] and B[N − 1], of the

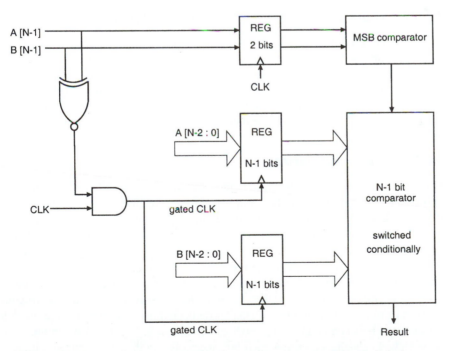

Figure 11.24 Block diagram of an N-bit number comparator with gated clock scheme.

two binary numbers are different from each other, then the decision can be made by comparing the most significant bits (MSBs) only. The circuit shown in Fig. 11.24 exploits this simple strategy for saving switching power which would otherwise be wasted to latch and to process the lower-order bits. The two MSBs are latched in a two-bit register which is driven by the original system clock. At the same time, these two bits are applied to an XNOR gate and the output of the XNOR is used to generate the gated clock signal with an AND gate. If the two MSBs are different (i.e., "01" or "10"), the XNOR produces a logic "0" at the output, disabling the clock signal of the lower-order registers. In this case, a separate MSB comparator circuit is used to decide which one of the two numbers is larger. If the two MSBs are identical (i.e., "00" or "11"), the gated clock signal is applied to the lower-order registers and the decision is made by the $(N-1)$-bit comparator circuit.

The amount of power dissipated in the lower-order registers and the $(N-1)$-bit comparator circuit can be quite significant, especially if the bit-length (N) is large. Assuming that the incoming binary numbers are randomly distributed, we can see that the gated clock strategy effectively reduces the overall switching power dissipation of the system by approximately 50%, since a large portion of the system is disabled for half of all input combinations. This example shows that the gated clock strategy can be very effective for reducing switching power dissipation in certain logic circuits.

11.5   Reduction of Switched Capacitance

It was already established in the previous sections that the amount of switched capacitance plays a significant role in the dynamic power dissipation of the circuit. Hence, reduction of this parasitic capacitance is a major goal for low-power design of digital integrated circuits. In this Section, we will consider various techniques at the system level, circuit level and physical design (mask) level which can be used to reduce the amount of switched capacitance.

System-Level Measures

At the system level, one approach to reduce the switched capacitance is to limit the use of shared resources. A simple example is the use of a global bus structure for data transmission between a large number of operational modules (Fig. 11.25). If a single shared bus is connected to all modules as in Fig. 11.25(a), this structure results in a large bus capacitance due to (i) the large number of drivers and receivers sharing the same transmission medium, and (ii) the parasitic capacitance of the long bus line. Obviously, driving the large bus capacitance will require a significant amount of power consumption during each bus access. Alternatively, the global bus structure can be partitioned into a number of smaller dedicated local buses to handle the data transmission between neighboring modules, as shown in Fig. 11.25(b). In this case, the switched capacitance during each bus access is significantly reduced, although multiple buses may increase the overall routing area on the chip.

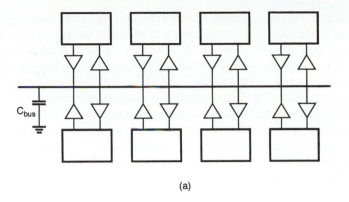

(a)

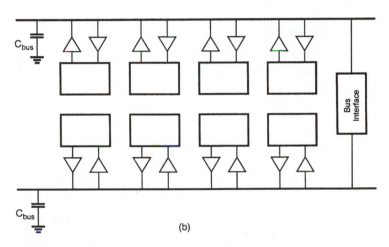

(b)

Figure 11.25 (a) Using a single global bus structure for connecting a large number of modules on chip results in large bus capacitance and large dynamic power dissipation. (b) Using smaller local buses reduces the amount of switched capacitance, at the expense of additional chip area.

Circuit-Level Measures

The type of logic style used to implement a digital circuit also affects the output load capacitance of the circuit. The capacitance is a function of the number of transistors that are required to implement a given function. For example, one approach to reduce the load capacitance is to use transfer gates (pass-transistor logic) instead of conventional CMOS logic gates to implement logic functions. Pass-gate logic design is attractive since fewer transistors are required for certain functions such as XOR and XNOR. Therefore, this design style has emerged as a promising alternative to conventional CMOS, for low-power design. Still, a number of important issues must be considered for pass-gate logic.

The threshold-voltage drop through nMOS transistors while transmitting a logic "1" makes swing restoration necessary in order to avoid static currents in subsequent inverter stages or logic gates (cf. Chapter 9). In order to provide acceptable output driving capabilities, inverters are usually attached to pass-gate outputs, which increases the overall area, time delay and the switching power dissipation of the logic gate. Because pass-transistor structures typically require complementary control signals, dual-rail logic is used to provide all signals in complementary form. As a consequence, two complementary nMOS pass-transistor networks are necessary in addition to swing restoration and output buffering circuitry, effectively diminishing the inherent advantages of pass-transistor logic over conventional CMOS logic. Thus, the use of pass-transistor logic gates to achieve low power dissipation must be carefully considered, and the choice of logic design style must ultimately be based on a detailed comparison of all design aspects such as silicon area, overall delay as well as switching power dissipation.

Mask-Level Measures

The amount of parasitic capacitance that is switched (i.e., charged up or charged down) during operation can be also reduced at the physical design level, or mask level. The parasitic gate and diffusion capacitances of MOS transistors in the circuit typically constitute a significant amount of the total capacitance in a combinational logic circuit. Hence, a simple mask-level measure to reduce power dissipation is keeping the transistors (especially the drain and source regions) at minimum dimensions whenever possible and feasible, thereby minimizing the parasitic capacitances. Designing a logic gate with minimum-size transistors certainly affects the dynamic performance of the circuit, and this trade-off between dynamic performance and power dissipation should be carefully considered in critical circuits. Especially in circuits driving a large *extrinsic* capacitive loads, e.g., large fan-out or routing capacitances, the transistors must be designed with larger dimensions. Yet in many other cases where the load capacitance of a gate is mainly *intrinsic,* the transistor sizes can be kept at a minimum. Note that most standard cell libraries are designed with larger transistors in order to accommodate a wide range of capacitive loads and performance requirements. Consequently, a standard-cell based design may have considerable overhead in terms of switched capacitance in each cell.

11.6   Adiabatic Logic Circuits

In conventional level-restoring CMOS logic circuits with rail-to-rail output voltage swing, each switching event causes an energy transfer from the power supply to the output node, or from the output node to the ground. During a 0-to-V_{DD} transition of the output, the total output charge $Q = C_{load}V_{DD}$ is drawn from the power supply at a constant voltage. Thus, an energy of $E_{supply} = C_{load}V_{DD}^2$ is drawn from the power supply during this transition. Charging the output node capacitance to the voltage level V_{DD} means that at the end of the transition, the amount of stored energy in the output node is $E_{stored} = C_{load}V_{DD}^2/2$. Thus, half of the injected energy from the power supply is dissipated in the pMOS network while only one half is delivered to the output node. During a subsequent V_{DD}-to-0 transition of the output node, no charge is drawn from the power supply and the energy stored in the load capacitance is dissipated in the nMOS network.

To reduce the dissipation, the circuit designer can minimize the switching events, decrease the node capacitance, reduce the voltage swing, or apply a combination of these methods. Yet in all these cases, the energy drawn from the power supply is *used* only once before being dissipated. To increase the energy efficiency of logic circuits, other measures can be introduced for *recycling* the energy drawn from the power supply. A novel class of logic circuits called *adiabatic logic* offers the possibility of further reducing the energy dissipated during switching events, and the possibility of recycling, or reusing, some of the energy drawn from the power supply. To accomplish this goal, the circuit topology and the operation principles have to be modified, sometimes drastically. The amount of energy recycling achievable using adiabatic techniques is also determined by the fabrication technology, switching speed, and voltage swing.

The term "adiabatic" is typically used to describe thermodynamic processes that have no energy exchange with the environment, and therefore, no energy loss in the form of dissipated heat. In our case, the electric charge transfer between the nodes of a circuit will be viewed as the process, and various techniques will be explored to minimize the energy loss, or heat dissipation, during charge transfer events. It should be noted that fully adiabatic operation of a circuit is an ideal condition which may only be approached asymptotically as the switching process is slowed down. In practical cases, energy dissipation associated with a charge transfer event is usually composed of an adiabatic component and a non-adiabatic component. Therefore, reducing all energy loss to zero may not be possible, regardless of the switching speed.

Adiabatic Switching

Consider the simple circuit shown in Fig. 11.26 where a load capacitance is charged by a constant current source. This circuit is similar to the equivalent circuit used to model the charge-up event in conventional CMOS circuits, with the exception that in conventional digital CMOS circuits, the output capacitance is charged by a *constant voltage* source and not by a constant current source. Here, R represents the on-resistance of the pMOS network. Also note that a constant charging current corresponds to a linear

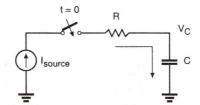

Figure 11.26 Constant-current source charging a load capacitance C, through a resistance R.

voltage ramp. Assuming that the capacitance voltage V_C is zero initially, the variation of the voltage as a function of time can be found as

$$V_C(t) = \frac{1}{C} \cdot I_{source} \cdot t \qquad (11.23)$$

Hence, the charging current can be expressed as a simple function of V_C and time t.

$$I_{source} = C \frac{V_C(t)}{t} \qquad (11.24)$$

The amount of energy dissipated in the resistor R from $t = 0$ to $t = T$ can be found as

$$E_{diss} = R \int_0^T I_{source}^2 \, dt = R \cdot I_{source}^2 \cdot T \qquad (11.25)$$

Combining (11.24) and (11.25), we can also express the dissipated energy during this charge-up transition as follows.

$$E_{diss} = \frac{RC}{T} C V_C^2(T) \qquad (11.26)$$

Now, a number of simple observations can be made based on (11.26). First, the dissipated energy is smaller than for the conventional case if the charging time T is larger than $2RC$. In fact, the dissipated energy can be made arbitrarily small by increasing the charging time, since E_{diss} is inversely proportional to T. Also, we observe that the dissipated energy is proportional to the resistance R, as opposed to the conventional case where the dissipation depends on the capacitance and the voltage swing. Reducing the on-resistance of the pMOS network will reduce the energy dissipation.

We have seen that the constant-current charging process efficiently transfers energy from the power supply to the load capacitance. A portion of the energy thus stored in the capacitance can also be reclaimed by reversing the current source direction, allowing the charge to be transferred from the capacitance back into the supply. This possibility is unique to the adiabatic operation, since in conventional CMOS circuits the energy is dissipated after being used only once. The constant-current power supply must certainly be capable of retrieving the energy back from the circuit. Adiabatic logic circuits thus require non-standard power supplies with time-varying voltage, also called pulsed-power supplies. The additional hardware overhead associated with these specific power supply circuits is one of the design trade-offs that must be considered when using the adiabatic logic.

Adiabatic Logic Gates

In the following, we will examine simple circuit configurations which can be used for adiabatic switching. Note that most of the research on adiabatic logic circuits is rather recent. Therefore, the circuits presented here should be considered as simple examples. Other circuit topologies are also possible, but the overall theme of energy recycling should be enforced, regardless of specific circuit configuration.

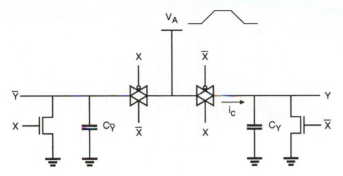

Figure 11.27 Adiabatic amplifier circuit which transfers the complementary input signals to its complementary outputs through CMOS transmission gates.

First, consider the adiabatic amplifier circuit shown in Fig. 11.27, which can be used to drive capacitive loads. It consists of two CMOS transmission gates and two nMOS clamp transistors. Both the input (X) and the output (Y) are dual-rail encoded, which means that the inverses of both signals are also available, to control the CMOS transmission gates.

When the input signal X is set to a valid value, one of the two transmission gates becomes transparent. Next, the amplifier is *energized* by applying a slow voltage ramp V_A, rising from zero to V_{DD}. The load capacitance at one of the two complementary outputs is adiabatically charged to V_{DD} through the transmission gate, while the other output node remains clamped to ground potential. When the charging process is completed, the output signal pair is valid and can be used as an input to other similar circuits. Next, the circuit is de-energized by ramping the voltage V_A back to zero. Thus, the energy that was stored in the output load capacitance is retrieved by the power supply. Note that the input signal pair must be valid and stable throughout this sequence.

The simple circuit principle of the adiabatic amplifier can be extended to allow the implementation of arbitrary logic functions. Figure 11.28 shows a general circuit topology for conventional CMOS logic gates and adiabatic counterparts. To convert a conventional CMOS logic gate into an adiabatic gate, the pull-up and pull-down networks must be replaced with complementary transmission-gate (T-gate) networks. The T-gate network implementing the pull-up function is used to drive the true output of the adiabatic gate, while the T-gate network implementing the pull-down function drives the complementary output node. Note that all inputs should also be available in complementary form. Both networks in the adiabatic logic circuit are used to charge-up as well as charge-down the output capacitances, which ensures that the energy stored at the output node can be retrieved by the power supply, at the end of each cycle. To allow adiabatic operation, the DC voltage source of the original circuit must be replaced by a pulsed-power supply with ramped voltage output. Note that the circuit modifications which are necessary to

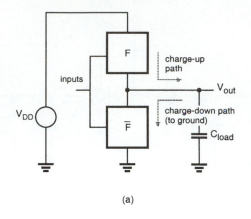

(a)

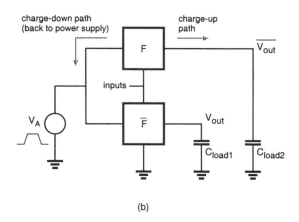

(b)

Figure 11.28 (a) The general circuit topology of a conventional CMOS logic gate. (b) The topology of an adiabatic logic gate implementing the same function. Note the difference in charge-up and charge-down paths for the output capacitance.

convert a conventional CMOS logic circuit into an adiabatic logic circuit increase the device count by a factor of two or even more. In general, the reduction of energy dissipation comes at the cost of slower switching speed, a trade-off in all adiabatic methods.

Figure 11.29 shows the circuit diagram of an adiabatic two-input AND/NAND gate as an example, which consists of two complementary T-gate networks. Note that the network consisting of two CMOS T-gates connected in series is used for realizing the AND function of the two input variables, while the network consisting of two CMOS T-gates connected in parallel is used for realizing the complementary NAND function.

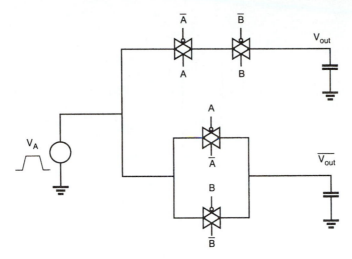

Figure 11.29 Circuit diagram of an adiabatic CMOS AND/NAND gate.

Stepwise Charging Circuits

We have seen earlier that the dissipation during a charge-up event can be minimized, and in the ideal case it can be reduced to zero by using a constant-current power supply. This requires that the power supply be able to generate linear voltage ramps. Practical supplies can be constructed by using resonant inductor circuits to approximate the constant output current and the linear voltage ramps with sinusoidal signals. But the use of inductors presents several difficulties at the circuit level, especially in view of chip-level integration and overall efficiency.

An alternative to using pure voltage ramps is to use stepwise supply voltage waveforms, where the output voltage of the power supply is increased and decreased in small increments during charging and discharging. Since the energy dissipation depends on the average voltage drop traversed by the charge that flows onto the load capacitance, using smaller voltage steps, or increments, should reduce the dissipation considerably.

Figure 11.30 shows a CMOS inverter driven by a stepwise supply voltage waveform. Assume that the output voltage is equal to zero initially. With the input voltage set to logic low level, the power supply voltage V_A is increased from 0 to V_{DD}, in n equal voltage steps (Fig. 11.31). Since the pMOS transistor is conducting during this transition, the output load capacitance will be charged up in a stepwise manner. The on-resistance of the pMOS transistor can be represented by the linear resistor R. Thus, the output load capacitance is being charged up through a resistor, in small voltage increments. For the ith time increment, the amount of capacitor current can be expressed as

$$ i_C = C \frac{d V_{out}}{dt} = \frac{V_A^{(i+1)} - V_{out}}{R} \qquad (11.27) $$

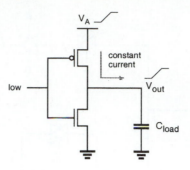

Figure 11.30 A CMOS inverter
circuit with a stepwise-increasing
supply voltage.

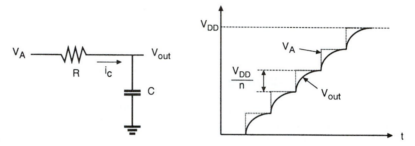

Figure 11.31 Equivalent circuit, and the input and output voltage waveforms
of the CMOS inverter circuit shown in Fig. 11.30 (stepwise charge-up case).

Solving this differential equation with the initial condition $V_{out}(t_i) = V_A^{(i)}$ yields

$$V_{out}(t) = V_A^{(i+1)} - \frac{V_{DD}}{n}e^{-t/RC} \qquad (11.28)$$

Here, n is the number of steps of the supply voltage waveform. The amount of energy
dissipated during one voltage step increment can now be found as

$$E_{step} = \int_0^\infty i_C^2 R \, dt = \frac{1}{n^2}C\frac{V_{DD}^2}{2} \qquad (11.29)$$

Since n steps are used to charge up the capacitance to V_{DD}, the total dissipation is

$$E_{total} = n \cdot E_{step} = \frac{1}{n}C\frac{V_{DD}^2}{2} \qquad (11.30)$$

According to this simplified analysis, charging the output capacitance with n voltage
steps or increments reduces the energy dissipation per cycle by a factor of n. Therefore,
the total power dissipation is also reduced by a factor of n using stepwise charging.
This result implies that if the voltage steps can be made very small and the number of

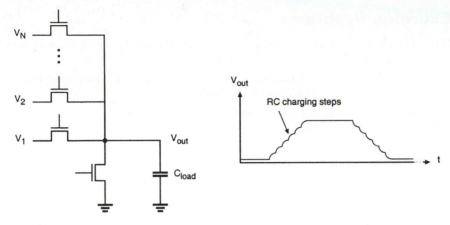

Figure 11.32 Stepwise driver circuit for capacitive loads. The load capacitance is successively connected to constant voltage sources V_i through an array of switch devices.

voltage steps n approaches infinity (i.e., if the supply voltage is a slow linear ramp), the energy dissipation will approach zero.

Another example for simple stepwise charging circuits is the stepwise driver for capacitive loads, implemented with nMOS devices as shown in Fig. 11.32. Here, a bank of n constant voltage supplies with evenly distributed voltage levels is used. The load capacitance is charged up by connecting the constant voltage sources V_1 through V_N to the load successively, using an array of switch devices. To discharge the load capacitance, the constant voltage sources are connected to the load in the reverse sequence.

The switch devices are shown as nMOS transistors in Fig. 11.32, yet some of them may be replaced by pMOS transistors to prevent the undesirable threshold-voltage drop problem and the substrate-bias effects at higher voltage levels. One of the most significant drawbacks of this circuit configuration is the need for multiple supply voltages. A power supply system capable of efficiently generating n different voltage levels would be complex and expensive. Also, the routing of n different supply voltages to each circuit in a large system would create a significant overhead. In addition, the concept is not easily extensible to general logic gates. Therefore, stepwise charging driver circuits can be best utilized for driving a few critical nodes in the circuit that are responsible for a large portion of the overall power dissipation, such as output pads and large buses.

So far, we have seen that adiabatic logic circuits can offer significant reduction of energy dissipation, but usually at the expense of switching times. Therefore, such adiabatic logic circuits can be best utilized in cases where delay is not critical. Moreover, the realization of unconventional power supplies needed in adiabatic circuit configurations typically results in an overhead in terms of both overall energy dissipation and silicon area. These issues should be carefully considered when adiabatic logic is used as a method for low-power design.

Exercise Problems

11.1   For a typical CMOS inverter circuit (you can use the device parameters used in Problem 6.6):

   a.   Plot the voltage waveform across the pMOS transistor during the low-to-high output transition, for a step input.

   b.   Plot the instantaneous power dissipation during this switching event and calculate the *average* power dissipation.

11.2   For the CMOS inverter considered in Problem 11.1, replace its constant power supply voltage by a ramp voltage source, with 3 ns transition time from 0 to 3 V. Repeat the steps (a) and (b) above for this circuit.

11.3   Design an adiabatic 2-input AND/NAND gate (circuit topology given in Fig. 11.29). Simulate its power dissipation using power supply ramps with (a) 3 ns and (b) 30 ns transitions from 0 to 3 V.

11.4   Design an adiabatic full adder circuit. Compare its transistor count with (a) the conventional CMOS full adder circuit and (b) the pass-transistor full adder circuit.

CHAPTER 12

BiCMOS Logic Circuits

12.1 Introduction

The signal propagation delay due to large interconnect capacitances is a major factor which limits the performance of CMOS digital integrated circuits. The system speed is ultimately limited by the current-driving capability of CMOS gates that drive large capacitive loads, such as the word lines in memory arrays, or data bus lines between large logic blocks. The problem of driving large off-chip as well as on-chip loads is traditionally solved by using specific CMOS buffer circuits with enhanced current-driving capabilities, which will be examined in further detail in Chapter 13. Most of these buffer configurations require a significant amount of silicon area for improvement in the signal propagation delay.

Consider a large capacitive load being driven by a CMOS inverter circuit, which in turn is driven by another CMOS inverter, or any CMOS logic gate in general, as shown in Fig. 12.1. Assume that the propagation delay of the output buffer is represented by t_{out}, and the propagation delay of the entire two-stage structure is represented by t_{total}. Simple dynamic analysis suggests that, to drive a large capacitive load with a specified propagation delay, we have to design the output buffer with sufficiently large transistor channel widths, W_n and W_p. However, simply increasing the transistor dimensions will not necessarily help to reduce the propagation delay time.

As we make the channel widths of the nMOS and the pMOS transistors larger, the proportionally increasing drain parasitics will eventually annihilate the speed improvement to be gained by larger transistor sizing. Thus, the output stage propagation delay may not be reducible beyond a certain limit due to the increase in drain parasitics, as illustrated in Fig. 12.2. In fact, the overall delay of the two-stage structure shown in Fig. 12.1 will even *increase* with larger W_n and W_p, because of the large gate capacitance being imposed on the first stage.

The only feasible solution using conventional CMOS structures is to build a tapered or scaled buffer chain, consisting of several stages which gradually increase in size from the input stage toward the output. The detailed design issues associated with scaled buffer chains will be investigated in Chapter 13. Yet scaled or tapered

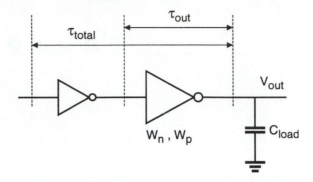

Figure 12.1 A two-stage CMOS buffer structure used for driving a large capacitive load.

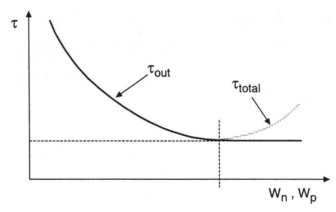

Figure 12.2 Propagation delay of the CMOS buffer vs. transistor size in the output stage.

buffer structures typically require a large silicon area to implement, which increases the overall cost, especially if many on-chip and off-chip loads must be considered.

In comparison, bipolar junction transistors (BJTs) have more current-driving capability, and hence, can overcome such speed bottlenecks using less silicon area. However, the power dissipation of bipolar logic gates is typically one or two orders of magnitude larger than that of comparable CMOS gates. Therefore, such all-bipolar high-speed VLSI circuits are difficult to realize and require very elaborate heat-sink arrangements.

An alternative solution to the problem of driving large capacitive loads can be provided by merging CMOS and bipolar devices (BiCMOS) on chip. Taking advantage of the low static power consumption of CMOS and the high current-driving capability of the bipolar transistor during transients, the BiCMOS configuration can combine the "best of both worlds" (Fig. 12.3). In view of the limited driving capabilities of MOS transistors in general, the BiCMOS combination has significant

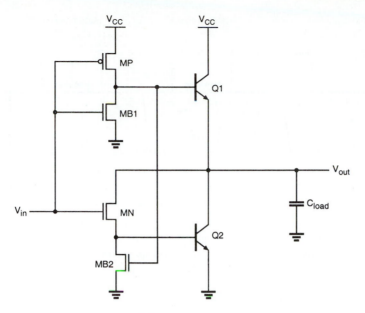

Figure 12.3 A typical BiCMOS inverter circuit, with four MOSFETs and two BJTs.

advantages to offer, such as improved switching speed and less sensitivity with respect to the load capacitance. In general, BiCMOS logic circuits are not bipolar-intensive; i.e., most logic operations are performed by conventional CMOS subcircuits, while the bipolar transistors are used only when high on-chip or off-chip drive capability is required.

The most significant drawback of the BiCMOS circuits lies in the increased fabrication process complexity. The fabrication of the bipolar transistors involves more process steps beyond the CMOS process. But many of the conventional CMOS process steps can be utilized to concurrently fabricate bipolar transistor structures along with the MOS transistors, as shown in the simplified cross-section in Fig. 12.4. For example, the process used to create the n-well (n-tub) on a p-type substrate can also be used to create the n-type collector region of the npn bipolar transistor. The source and drain diffusion steps in the CMOS process can be used to form the emitter region and the base contact region of the bipolar transistor. The silicided polysilicon gate of the MOS transistor forms the emitter contact. In fact, the only bipolar fabrication step which cannot be adopted from the CMOS process is the creation of the base region. The BiCMOS fabrication process typically requires only 3–4 masks in addition to the well-established CMOS process.

In the following sections, we will present the basic building blocks of BiCMOS logic circuits, discuss the static and dynamic behavior of BiCMOS circuits, and examine some applications. To provide the necessary analytical basis for the upcoming discussions, we start by investigating the structure and the operation of the bipolar junction transistor (BJT).

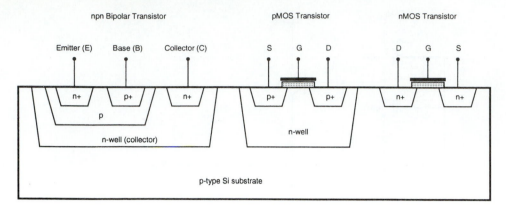

npn Bipolar Transistor

pMOS Transistor

nMOS Transistor

Figure 12.4 Simplified cross-section showing an npn bipolar transistor, an n-channel MOS transistor, and a p-channel MOS transistor fabricated on the same p-type silicon substrate. Notice that many standard CMOS process steps can be used to create bipolar and MOS transistors side-by-side on the chip.

12.2 Bipolar Junction Transistor (BJT): Structure and Operation

The simplified cross-section of an npn bipolar junction transistor (BJT) is shown in Fig. 12.5. The structure consists of several regions and layers of doped silicon, which essentially form the three terminals of the device: emitter (E), base (B), and collector (C). The npn-type bipolar transistor shown in Fig. 12.5 is fabricated on a p-type Si substrate. In full-bipolar ICs, a lightly doped n-type epitaxial layer created on top of the p-type substrate serves as the collector (note: in BiCMOS ICs, the n-well is used as the collector region). The collector contact region is doped with a higher concentration of n-type impurities in order to reduce the contact resistance. The base region is created by forming a p-type diffusion in the n-type epitaxial layer, and the heavily doped n-type emitter is formed within this p-type base diffusion region.

The rectangular slice highlighted in Fig. 12.5 represents the basic functional BJT device, which essentially consists of two back-to-back connected pn junctions. This basic structure is shown separately in Fig. 12.6 with applied bias voltages, V_{BE} and V_{CB}. Notice that with the voltage source polarities shown here, the base-emitter junction is forward-biased, whereas the base-collector junction is reverse-biased. We will call this particular operating mode the *forward active mode*. The circuit symbol of the npn transistor is also shown in Fig. 12.6.

BJT Operation: A Qualitative View

Under the bias conditions shown in Fig. 12.6, the positive base-emitter voltage V_{BE} causes the base-emitter pn junction to become forward-biased and to conduct a current which corresponds to the emitter terminal current I_E. The emitter current consists of two components: electrons injected from the emitter into the base region, and

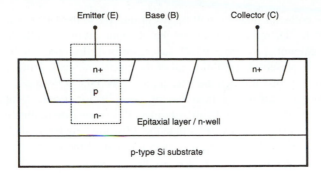

Figure 12.5 Cross-section of an npn-type bipolar junction transistor (BJT).

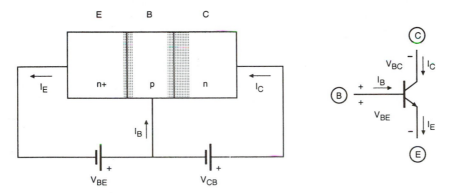

Figure 12.6 The simplified view of the bipolar junction transistor (BJT) biased in the forward active mode with V_{BE} and V_{CB}, and the corresponding circuit symbol. Note that the depletion regions at both pn junctions are also highlighted.

holes injected from the base into the emitter. Since the doping concentration of the emitter region is much higher than that of the base region, the magnitude of the injected electron current component is larger than the injected hole current component.

The injected electrons which enter the very thin base region are the *minority carriers* in the p-type base. Their concentration is highest at the emitter side of the base and lowest (zero) at the collector side of the base region. The reason for the zero concentration at the collector side is that the positive collector voltage V_{CB} causes the electrons at the collector-base junction to be swept across the depletion region, and into the collector. Thus, electrons injected from the emitter into the base region will *diffuse* toward the collector, since the local electron concentration profile *decreases* in that direction. The diffusion current made up of electrons in the base is directly proportional to the minority carrier concentration difference between the emitter side and the collector side of the base. Electrons that diffuse through the thin base region and reach the collector junction will be swept into the collector, making up the collector current I_C. Therefore, we can see that the amount of the collector current depends on the amount of electrons which successfully diffuse through the base region.

Ideally, it is preferred that almost all of the electrons injected into the base at the emitter junction diffuse through the base region, reach the collector junction and are swept into the collector region, thus making the collector current I_C approximately equal to the emitter current I_E. In order to meet this requirement, the thickness W of the base region must be much smaller than the diffusion length L_D of electrons in the base, i.e., $W \ll L_D$. Also, the doping concentration of the emitter region must be much larger than that of the base region, so that the emitter terminal current is dominated by the electron injection current component. If these conditions are satisfied, the terminal behavior of the BJT will be a good approximation of a controlled current source. The emitter current and therefore the collector current flow can be controlled by modifying the bias at the base-emitter junction.

Note that some of the electrons that are diffusing through the base region will recombine with holes, i.e., with the majority carriers in the base. Also, a small number of holes will be injected from the base into the emitter through the forward-biased base-emitter junction. The injected holes constitute the base current I_B. Although the magnitude of the base current is very small in comparison to the collector current, it nevertheless plays an important role in the current-controlled operation. It can be shown that the collector current is in fact proportional to the base current, with a very large proportionality factor, also called the *current gain*. Thus, by simply modifying the amount of the small base current, one can control the collector current of a bipolar transistor operating in the forward active mode.

This brief discussion is not intended to provide a complete picture of the operation of BJTs, but rather to illustrate qualitatively the pertinent characteristics of current-controlled operation in bipolar devices. Our objective is to provide a sufficient basis for understanding the analytical models to be presented next.

BJT Current-Voltage Models

The operational characteristics of the npn bipolar transistor under various conditions will be analyzed in the following by using the well-known Ebers-Moll model. We will not attempt to derive the model equations here; they are simply presented as a complete set of equations describing the current-voltage behavior of the bipolar transistor under certain bias conditions.

The collector current I_C and the emitter current I_E are expressed as

$$I_C = \alpha_F I_F - I_R \tag{12.1}$$

$$I_E = I_F - \alpha_R I_R \tag{12.2}$$

where α_F and α_R are two dimensionless coefficients, and the two currents I_F and I_R are given by the following diode current equations.

$$I_F = I_{ES}\left(e^{\frac{qV_{BE}}{kT}} - 1 \right) \tag{12.3}$$

$$I_R = I_{CS}\left(e^{\frac{qV_{BC}}{kT}} - 1 \right) \tag{12.4}$$

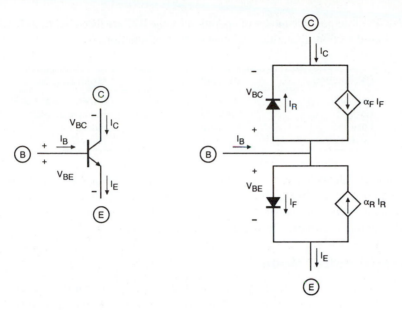

Figure 12.7   The Ebers-Moll equivalent circuit diagram of the npn BJT.

Here, I_{ES} and I_{CS} represent the reverse saturation currents of the base-emitter and the base-collector junctions, respectively. The equivalent circuit diagram corresponding to the Ebers-Moll model equations is shown in Fig. 12.7.

Note that the BJT terminal current directions used in this chapter do *not* match the current directions used in SPICE BJT models, where all terminal currents are directed into the device. By combining (12.1) through (12.4), we obtain the following form of the Ebers-Moll current-voltage model equations for the bipolar transistor.

$$I_C = \alpha_F I_{ES}\left(e^{\frac{V_{BE}}{V_T}} - 1\right) - I_{CS}\left(e^{\frac{V_{BC}}{V_T}} - 1\right) \qquad (12.5)$$

$$I_E = I_{ES}\left(e^{\frac{V_{BE}}{V_T}} - 1\right) - \alpha_R I_{CS}\left(e^{\frac{V_{BC}}{V_T}} - 1\right) \qquad (12.6)$$

where V_T represents the thermal voltage,

$$V_T = \frac{kT}{q} \qquad (12.7)$$

It is also obvious that the sum of the three terminal currents must always be equal to zero, therefore,

$$I_E = I_C + I_B \qquad (12.8)$$

Now we can examine the external current-voltage behavior of the npn bipolar junction transistor under different terminal bias conditions, also called *operating modes*. The four possible modes of operation for the BJT are listed below as functions of the bias directions (polarities) applied to the two junctions.

BE junction	BC junction	Operating mode
Forward	Reverse	Forward active
Reverse	Forward	Reverse active
Forward	Forward	Saturation
Reverse	Reverse	Cut-off

In analog circuit applications where linear current and/or voltage amplification is the primary concern, the bipolar transistor is usually operated only in the forward active mode. In digital circuit applications where the BJT is used primarily as a switching device, on the other hand, all four operating modes may be involved.

Forward Active Mode

In the forward active mode, the base-emitter junction is forward-biased, with $V_{BE} > 0$, and the base-collector junction is reverse-biased, with $V_{BC} < 0$. Under these conditions, the current across the reverse-biased base-collector junction is limited to the reverse saturation current I_{CS}, which is relatively small in magnitude (on the order of approximately 10^{-15} A). The Ebers-Moll equivalent circuit diagram for the forward active mode is shown in Fig. 12.8.

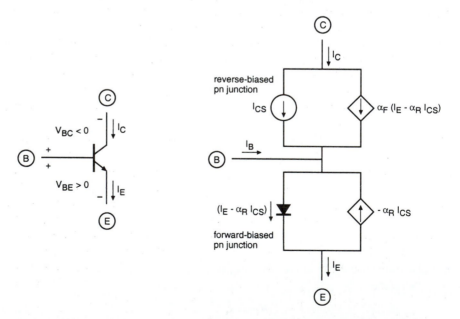

Figure 12.8 The Ebers-Moll equivalent circuit diagram of the npn BJT operating in the forward active mode.

As seen from Fig. 12.8, the two diode current components I_F and I_R used in the original model can now be expressed as

$$I_F = I_E - \alpha_R I_{CS}$$
$$I_R = -I_{CS} \tag{12.9}$$

The collector current is then found as follows.

$$I_C = \alpha_F(I_E - \alpha_R I_{CS}) + I_{CS} \tag{12.10}$$

$$I_C = \alpha_F I_E + I_{CO} \quad \text{where} \quad I_{CO} = I_{CS}(1 - \alpha_F \alpha_R) \tag{12.11}$$

Using (12.8), the collector current can also be written in terms of the base current.

$$I_C = \alpha_F(I_C + I_B) + I_{CO} \tag{12.12}$$

$$I_C = \frac{\alpha_F}{1 - \alpha_F} I_B + \frac{I_{CO}}{1 - \alpha_F} \tag{12.13}$$

At this point, we will define a new current coefficient, β_F. Since the magnitude of α_F is usually less than but very close to unity, the magnitude of β_F is on the order of 100 to 1000 for a typical npn BJT.

$$\beta_F = \frac{\alpha_F}{1 - \alpha_F} \tag{12.14}$$

Finally, the collector current of the npn bipolar transistor operating in the forward active mode is found as

$$I_C = \beta_F I_B + (1 + \beta_F) I_{CO} \approx \beta_F I_B \tag{12.15}$$

This current equation shows that the collector current in the forward active mode is proportional to the much smaller base current, with a large proportionality (current gain) factor of β_F.

Reverse Active Mode

In the reverse active mode, the base-emitter junction is reverse-biased, with $V_{BE} < 0$, and the base-collector junction is forward-biased, with $V_{BC} > 0$. Note that this bias condition can be viewed as the complementary mode to the forward active mode examined above. Under these conditions, the current across the reverse-biased base-emitter junction is limited to the small reverse saturation current I_{ES}. The Ebers-Moll equivalent circuit diagram of the BJT operating in reverse active mode is shown in Fig. 12.9.

It is seen from Fig. 12.9 that the two diode current components I_F and I_R can now be expressed as follows:

$$I_F = -I_{ES}$$
$$I_R = -I_C + \alpha_F I_{ES} \tag{12.16}$$

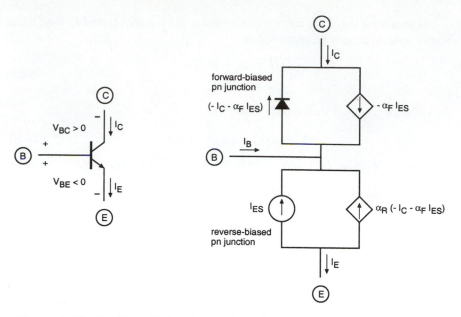

Figure 12.9 The Ebers-Moll equivalent circuit diagram of the npn BJT operating in reverse active mode.

The emitter current is then found from (12.2):

$$I_E = -\alpha_R(-I_C - \alpha_F I_{ES}) - I_{ES} \tag{12.17}$$

$$I_E = \alpha_R I_C - I_{EO} \qquad \text{where} \quad I_{EO} = I_{ES}(1 - \alpha_R \alpha_F) \tag{12.18}$$

Using (12.8), the emitter current I_E can also be written in terms of the base current I_B as follows:

$$I_E = \alpha_R(I_E - I_B) - I_{EO} \tag{12.19}$$

$$I_E = \frac{-\alpha_R}{1 - \alpha_R} I_B - \frac{I_{EO}}{1 - \alpha_R} \tag{12.20}$$

At this point, let us define a new current coefficient, β_R, with

$$\beta_R = \frac{\alpha_R}{1 - \alpha_R} \tag{12.21}$$

Typical values for α_R are between 0.4 and 0.9, and for β_R between 0.6 and 10. The emitter current of the npn bipolar transistor operating in the reverse active mode is found as

$$I_E = -\beta_R I_B - (1 + \beta_R)I_{EO} \approx -\beta_R I_B \tag{12.22}$$

This equation shows us that the emitter current in the reverse active mode is proportional to the much smaller base current, with a proportionality factor of β_R.

It can be observed that there is a close analogy between the current equations derived for the forward active mode and those for the reverse active mode. From a practical point of view, the emitter and collector terminals of the bipolar transistor appear to be swapped, and the current flow directions for the emitter and the collector currents are reversed accordingly. Note, however, that unlike the MOS transistor, the physical structure of the BJT is not symmetrical. Since the doping concentrations in the collector and the emitter regions are quite different, the reverse active mode characteristics of the BJT will not be exact replicas of the forward active mode characteristics. In fact, the reverse active mode current gain is significantly smaller than the forward active mode current gain.

Cut-off Mode

In this case, both the base-emitter and the base-collector junctions are reverse-biased, i.e., $V_{BE} < 0$ and $V_{BC} < 0$. Consequently, the pn junction currents I_F and I_R are both reduced to the corresponding reverse saturation current values.

$$I_F = -I_{ES}$$
$$I_R = -I_{CS}$$
(12.23)

The Ebers-Moll equivalent circuit diagram of the BJT operating in the cut-off mode is shown in Fig. 12.10. The emitter and the collector currents of the bipolar transistor in cut-off mode are expressed as follows:

$$I_E = -I_{ES} + \alpha_R I_{CS}$$
$$I_C = -\alpha_F I_{ES} + I_{CS}$$
(12.24)

Using the following equality

$$\alpha_F I_{ES} = \alpha_R I_{CS}$$
(12.25)

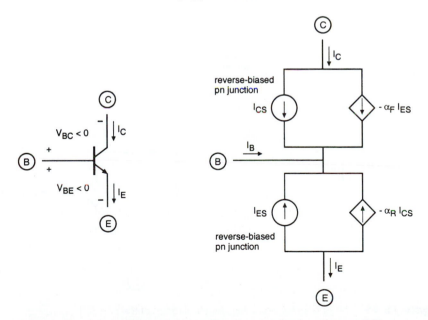

Figure 12.10 The Ebers-Moll equivalent circuit diagram of the npn BJT operating in cut-off mode.

the terminal currents can be found as

$$I_E = -(1 - \alpha_F)I_{ES}$$
$$I_C = (1 - \alpha_R)I_{CS}$$

$$(12.26)$$

Both the emitter current and the collector current of the bipolar transistor in cut-off mode are even smaller than the junction reverse saturation currents, I_{CS} and I_{ES}. Although the non-ideal effects of leakage and thermal generation in the depletion regions are neglected in this derivation, the actual values of the emitter and collector currents of the BJT in cut-off mode are nevertheless very small. It can thus be assumed that there is an open circuit between each of the transistor terminals.

Saturation Mode

In the saturation mode, both the base-emitter junction and the base-collector junction are forward-biased, with $V_{BE} > 0$ and $V_{BC} > 0$. The Ebers-Moll equivalent circuit diagram of the BJT operating in the saturation mode is shown in Fig. 12.11.

Since both of the junctions are conducting in the forward mode, the currents I_F and I_R can be expressed as

$$I_F = I_{ES}e^{\frac{V_{BE,sat}}{V_T}}$$
$$I_R = I_{CS}e^{\frac{V_{BC,sat}}{V_T}}$$

$$(12.27)$$

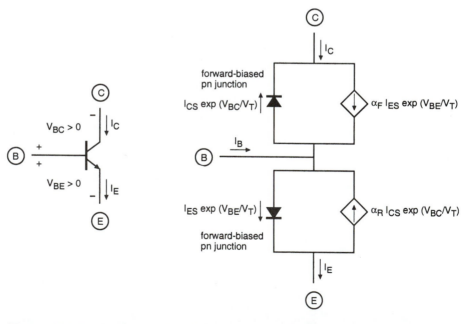

Figure 12.11   The Ebers-Moll equivalent circuit diagram of the npn BJT operating in saturation mode.

In this case, the emitter and the collector currents are found as follows.

$$I_{E,sat} = I_F - \alpha_R I_R$$
$$I_{C,sat} = \alpha_F I_F - I_R \tag{12.28}$$

$$I_{E,sat} = I_{ES}e^{\frac{V_{BE,sat}}{V_T}} - \alpha_R I_{CS}e^{\frac{V_{BC,sat}}{V_T}}$$
$$I_{C,sat} = \alpha_F I_{ES}e^{\frac{V_{BE,sat}}{V_T}} - I_{CS}e^{\frac{V_{BC,sat}}{V_T}} \tag{12.29}$$

For a simple analysis of the bipolar transistor operating in saturation, we can assume the voltages across both forward-biased junctions to be constant. Since the collector-emitter voltage V_{CE} of the bipolar transistor appears as the output voltage in most digital circuit configurations, we also have to determine the exact value of V_{CE} in the saturation mode. Writing the simple voltage loop equation around the three terminals of the bipolar transistor, we obtain

$$V_{CE,sat} = V_{BE,sat} - V_{BC,sat} \tag{12.30}$$

Combining the current equations (12.28) and (12.29) yields

$$I_{E,sat} - \alpha_R I_{C,sat} = I_{ES}e^{\frac{V_{BE,sat}}{V_T}}(1 - \alpha_F \alpha_R) \tag{12.31}$$

which allows us to express the base-emitter voltage in the saturation mode in terms of the collector and emitter currents, as follows.

$$V_{BE,sat} = V_T \ln\left(\frac{I_{E,sat} - \alpha_R I_{C,sat}}{I_{ES}(1 - \alpha_F \alpha_R)}\right)$$
$$= V_T \ln\left(\frac{I_{B,sat} + I_{C,sat}(1 - \alpha_R)}{I_{EO}}\right) \tag{12.32}$$

Similarly, the following current equation can be used to express the base-collector voltage in the saturation mode in terms of the collector and emitter currents.

$$I_{C,sat} - \alpha_F I_{E,sat} = -I_{CS}e^{\frac{V_{BC,sat}}{V_T}}(1 - \alpha_F \alpha_R) \tag{12.33}$$

$$V_{BC,sat} = V_T \ln\left(\frac{I_{C,sat} - \alpha_F I_{E,sat}}{-I_{CS}(1 - \alpha_F \alpha_R)}\right)$$
$$= V_T \ln\left(\frac{\alpha_F I_{B,sat} - I_{C,sat}(1 - \alpha_F)}{I_{CO}}\right) \tag{12.34}$$

Finally, the collector-emitter voltage of the BJT in saturation can be found using (12.30), (12.32), and (12.34).

$$V_{CE,sat} = V_T \ln\left(\frac{\frac{1}{\alpha_R} + \left(\frac{I_{C,sat}}{I_{B,sat}}\right)\frac{1 - \alpha_R}{\alpha_R}}{1 - \left(\frac{I_{C,sat}}{I_{B,sat}}\right)\frac{1 - \alpha_F}{\alpha_F}}\right) \tag{12.35}$$

Here, the ratio of the saturation collector current $I_{C,sat}$ and the saturation base current $I_{B,sat}$ is called the *saturation* β, or *forced* β.

$$\left(\frac{I_{C,sat}}{I_{B,sat}} \right) \equiv \beta_{sat} < \beta_F \tag{12.36}$$

Note that the value of β_{sat} is not a constant for a transistor operating in saturation, and that it changes with the particular operating point.

With $\alpha_F = 0.99, \alpha_R = 0.66$, and $(I_{C,sat}/I_{B,sat}) = 10$, the value of the collector-emitter voltage in saturation can be found by using (12.35) as $V_{CE,sat} = 55$ mV. We must emphasize that the $V_{CE,sat}$ value calculated with (12.35) is the *intrinsic saturation voltage,* which does not take into account the additional voltage drops across the emitter and collector bulk resistances. The actual value of $V_{CE,sat}$ measured between the collector and emitter terminals is therefore larger, typically around 200 mV.

BJT Inverter Circuit: Static Characteristics

Now that we have introduced the basic current-voltage model for the bipolar junction transistor, we can embark on analyzing the simplest bipolar digital circuit, the resistive-load inverter. The circuit diagram of the resistive-load BJT inverter circuit is shown in Fig. 12.12. The circuit consists of one npn bipolar transistor, which has a resistor R_C connected between its collector terminal and the power supply voltage source V_{CC}, and another resistor R_B between its base terminal and the input node.

Assuming that the inverter is not *loaded* by another similar BJT inverter, i.e., assuming that the output current I_{out} is equal to zero, we can calculate the output voltage V_{out} as a function of the collector current I_C.

$$V_{out} = V_{CC} - R_C I_C \tag{12.37}$$

In the following, we will examine the voltage transfer characteristic of the BJT inverter circuit, and highlight its main features.

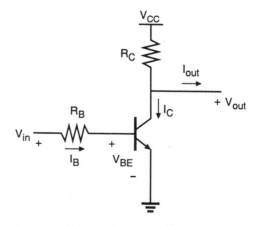

Figure 12.12   The resistive-load BJT inverter circuit.

If the input voltage V_{in} is smaller than the base-emitter junction turn-on voltage ($V_{BE,turn\text{-}on}$), the transistor is in the cut-off mode. Hence, the collector current will be approximately equal to zero, and the logic-high output voltage V_{OH} will be equal to the power supply voltage V_{CC}, according to (12.37). Note that the logic-high output voltage level will change if there is a nonzero output current flowing through the output node, i.e., if the inverter is loaded by another stage in cascade.

Once the applied input voltage is increased above $V_{BE,turn\text{-}on}$, the transistor enters the forward active mode and starts conducting a positive collector current. The output voltage, therefore, starts to decrease after this point. Thus, for a first-order approximation, we can assume that the voltage V_{IL} is defined as the input voltage at which the transistor enters the forward active operation mode.

As the input voltage is further increased, the output voltage continues to decrease, until the transistor enters the saturation region. At this point, the collector-emitter voltage of the transistor, which is also equal to the output voltage of the inverter, assumes a relatively small and constant value, corresponding to the logic-low output voltage level of V_{OL}. The typical static voltage transfer characteristic of the BJT inverter circuit is illustrated in Fig. 12.13, showing the significant voltage points and the corresponding operating modes of the transistor.

The values of the three important voltage points on the VTC are summarized below, based on the simple discussion outlined in the preceding paragraphs.

$$V_{OH} = V_{CC} \ (no \ output \ load)$$
$$V_{OL} = V_{CE,sat} \qquad\qquad (12.38)$$
$$V_{IL} = V_{BE,turn\text{-}on}$$

To calculate the value of the fourth voltage point (V_{IH}) on the VTC, we have to determine the input voltage value at which the transistor enters the saturation mode.

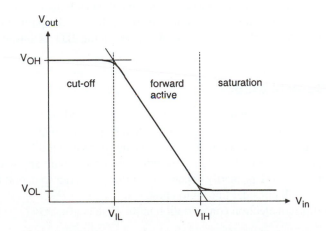

Figure 12.13 Typical input-output voltage transfer characteristic of the resistive-load BJT inverter circuit.

Notice that the input voltage V_{in} can be expressed as

$$V_{in} = V_{BE} + R_B I_B \tag{12.39}$$

The collector current of the transistor at the edge of saturation mode is found as follows.

$$I_C \ (edge \ of \ saturation) = \frac{V_{CC} - V_{CE,sat}}{R_C} \tag{12.40}$$

Note that, since at this point the transistor is also operating at the edge of the forward active mode, by definition, the collector current is still proportional to the base current.

$$I_C \ (edge \ of \ saturation) = \beta_F I_B \ (edge \ of \ saturation) \tag{12.41}$$

Using (12.39), the voltage V_{IH} can be expressed as follows.

$$V_{IH} = V_{BE,sat} + R_B I_B \ (edge \ of \ saturation) \tag{12.42}$$

$$V_{IH} = V_{BE,sat} + \frac{R_B}{R_C} \frac{1}{\beta_F} (V_{CC} - V_{CE,sat}) \tag{12.43}$$

The simple VTC analysis above has been carried out by assuming that the inverter under consideration has no output load; therefore, $I_{out} = 0$. If we assume more realistic conditions where one inverter is driving one or more inverters in cascade, however, the logic-high output voltage level will decrease to a lower value, since subsequent logic stages draw a nonzero output current from the inverter.

12.3 Dynamic Behavior of BJTs

To examine the dynamic behavior of bipolar junction transistors under transient terminal voltage conditions, we have to use a simple, universal model that describes the current-voltage relationships of the transistor regardless of the particular operating mode. For this reason, the charge-control model of the BJT is introduced in the following.

Charge-Control Model

The minority carrier concentrations in the emitter, base, and collector regions of an npn-type BJT operating in forward active mode are shown in Fig. 12.14. Here, p_{eo}, n_{bo}, and p_{co} represent the minority carrier concentrations at equilibrium, and $p_e(x)$, $n_b(x)$, and $p_c(x)$ represent the actual concentration profiles with applied bias. Since the operation of the BJT primarily depends on minority carrier transport through the base region, we will consider this region first.

The equilibrium electron concentration in the base is given as

$$n_{bo} = \frac{n_i^2}{N_A(base)} \tag{12.44}$$

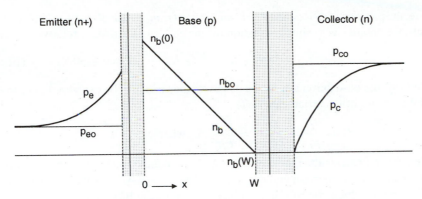

Figure 12.14 Minority carrier concentrations in the emitter, base, and collector of an npn-type BJT operating in forward active mode.

With applied bias, the *excess minority carrier concentration* in the base region is described by

$$n'_b = n_b(x) - n_{bo} \tag{12.45}$$

Note that the boundary condition for the excess electron concentration at the base-emitter junction is

$$n'_b(x = 0) = n_{bo}\left(e^{\frac{V_{BE}}{V_T}} - 1\right) \tag{12.46}$$

In forward active operation mode, the collector current I_C is related to the excess minority carrier concentration as follows.

$$I_C = qAD_b\frac{dn'_b(x)}{dx} \approx qAD_b\frac{n'_b(x = 0)}{W} \tag{12.47}$$

$$I_C = qAD_b\frac{n_{bo}}{W}\left(e^{\frac{V_{BE}}{V_T}} - 1\right) \tag{12.48}$$

where A represents the cross-sectional area of the junction, and D_b is the base diffusion constant. Let us now define the excess minority carrier charge in the base region, Q_F.

$$Q_F = \frac{qAWn'_b(x = 0)}{2} = \frac{qAWn_{bo}\left(e^{\frac{V_{BE}}{V_T}} - 1\right)}{2} \tag{12.49}$$

Comparing (12.49) with (12.47), we see that the excess minority carrier charge can also be written in terms of the collector current.

$$Q_F = \tau_F I_C \quad \text{where} \quad \tau_F = \frac{W^2}{2D_b} \tag{12.50}$$

Here, the parameter τ_F is called the *mean forward transit time* of electrons in the base. Thus, we obtain a very simple equation for the collector current, as follows.

$$I_C = \frac{Q_F}{\tau_F} \tag{12.51}$$

Similarly, the base current in forward active mode can also be expressed in terms of the excess minority carrier charge Q_F.

$$I_B = \frac{I_C}{\beta_F} = \frac{Q_F}{\tau_F \beta_F} = \frac{Q_F}{\tau_{BF}} \qquad \text{where} \quad \tau_{BF} = \tau_F \beta_F \tag{12.52}$$

Here, τ_{BF} represents the time constant which combines the effects of base recombination and carrier injection into the emitter. Thus, Q_F may be viewed as an independent variable which controls the terminal currents of the BJT.

For transient analysis, we have to express the instantaneous current values in terms of the excess minority carrier charge. Note that any time-dependent variation of the base minority carrier charge must be attributed to the instantaneous base current, which actually supplies the carriers responsible for charge variation. For this reason, an additional derivative term is used in the base current expression. The emitter current is simply obtained from i_C and i_B, by using Kirchhoff's current law.

$$i_C(t) = \frac{Q_F(t)}{\tau_F}$$

$$i_B(t) = \frac{Q_F(t)}{\tau_{BF}} + \frac{dQ_F(t)}{dt} \tag{12.53}$$

$$i_E(t) = \frac{Q_F(t)}{\tau_F} + \frac{Q_F(t)}{\tau_{BF}} + \frac{dQ_F(t)}{dt}$$

The analysis until this point considered only the forward active mode operation. Using simple analogy, we can define the excess minority carrier charge in the base for reverse active mode as

$$Q_R = \frac{qAWn_b'(x = W)}{2} = \frac{qAWn_{bo}\left(e^{\frac{V_{BC}}{V_T}} - 1\right)}{2} \tag{12.54}$$

Also, we define the reverse transit time for minority carriers in the base as follows.

$$\tau_{BR} = \beta_R \tau_R \qquad \text{where} \quad \beta_R = \frac{\alpha_R}{1 - \alpha_R} \tag{12.55}$$

Now, we can devise a set of current equations which describe the terminal currents of the BJT operating in the reverse active mode.

$$i_{E,reverse}(t) = -\frac{Q_R(t)}{\tau_R}$$

$$i_{B,reverse}(t) = \frac{Q_R(t)}{\tau_{BR}} + \frac{dQ_R(t)}{dt} \tag{12.56}$$

$$i_{C,reverse}(t) = -\frac{Q_R(t)}{\tau_R} - \frac{Q_R(t)}{\tau_{BR}} - \frac{dQ_R(t)}{dt}$$

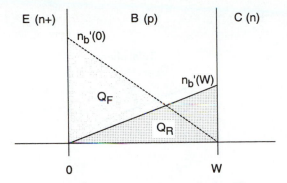

Figure 12.15 Minority carrier charge distribution in the base region for forward active and for reverse active modes.

Figure 12.15 shows the minority carrier charge distribution in the base region for forward active and for reverse active modes. In general, the forward active mode base charge Q_F is much larger than the reverse active mode charge Q_R. Also note that the reverse-mode excess minority carrier charge reaches its maximum value at the base-collector junction, as expected.

The charge-controlled current equations derived for the forward active mode and for the reverse active mode can now be combined into one unified set of model equations that is applicable for all possible cases. Note that the time-dependent variation of the depletion region charges (Q_{jC} and Q_{jE}) at both junctions are also included in the equations below, to complete the picture.

$$i_C(t) = \underbrace{\frac{Q_F(t)}{\tau_F}}_{forward} - \underbrace{Q_R(t)\left(\frac{1}{\tau_R} + \frac{1}{\tau_{BR}}\right) - \frac{dQ_R(t)}{dt}}_{reverse} - \underbrace{\frac{dQ_{jC}(t)}{dt}}_{depletion\ charge}$$

$$i_B(t) = \underbrace{\frac{Q_F(t)}{\tau_{BF}} + \frac{dQ_F(t)}{dt}}_{forward} + \underbrace{\frac{Q_R(t)}{\tau_{BR}} + \frac{dQ_R(t)}{dt}}_{reverse} + \underbrace{\frac{dQ_{jC}(t)}{dt} + \frac{dQ_{jE}(t)}{dt}}_{depletion\ charge} \qquad (12.57)$$

$$i_E(t) = \underbrace{Q_F(t)\left(\frac{1}{\tau_F} + \frac{1}{\tau_{BF}}\right) + \frac{dQ_F(t)}{dt}}_{forward} - \underbrace{\frac{Q_R(t)}{\tau_R}}_{reverse} + \underbrace{\frac{dQ_{jE}(t)}{dt}}_{depletion\ charge}$$

If the transistor is operating in the saturation mode, the total excess minority carrier charge in the base region is the superposition of the forward and reverse mode base charges Q_F and Q_R, since both junctions are forward-biased and inject carriers into the base region. Because the voltages across the two junctions are nearly constant in saturation, the depletion charge variation terms are usually negligible.

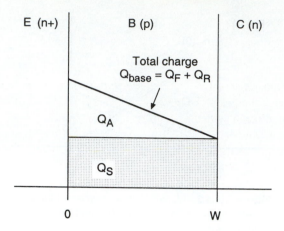

Figure 12.16 Total excess minority carrier charge in the base region of a BJT operating in saturation.

Thus, the collector current and the base current of a BJT operating in saturation mode can be expressed as

$$i_C = \frac{Q_F}{\tau_F} - Q_R \left(\frac{1}{\tau_R} + \frac{1}{\tau_{BR}} \right) - \frac{dQ_R}{dt}$$

$$i_B = \frac{Q_F}{\tau_{BF}} + \frac{Q_R}{\tau_{BR}} + \frac{d}{dt}(Q_F + Q_R)$$

(12.58)

The total base region excess minority charge Q_{base} is equal to the sum of the forward mode and the reverse mode excess minority charges. This sum can also be expressed as the sum of two charges, Q_A and Q_S, as follows (Fig. 12.16).

$$Q_{base} = Q_F + Q_R = Q_A + Q_S$$

(12.59)

where the new charge components are defined as

$$Q_A = \tau_F I_{C,edge\,of\,saturation}$$

(12.60)

and

$$Q_S = \tau_S I_{B,overdrive}$$

(12.61)

Note that Q_A represents the amount of base charge which is required to bring the transistor to the edge of saturation, according to (12.60). The charge component Q_S represents the *overdrive* base charge that drives the BJT into saturation, with

$$I_{B,overdrive} = I_{B,saturation} - I_{B,edge\,of\,saturation}$$

(12.62)

Now, the base current in the saturation mode can be expressed using only one time constant, i.e., the saturation time constant τ_S

$$i_B(t) = \frac{I_{C,edge\,of\,saturation}}{\beta_F} + \frac{Q_S}{\tau_S} + \frac{dQ_S}{dt}$$

(12.63)

where τ_S is expressed by

$$\tau_S = \frac{\alpha_F(\tau_F + \alpha_R \tau_R)}{1 - \alpha_F \alpha_R} \qquad (12.64)$$

The saturation time constant τ_S, which characterizes the recombination process for the overdrive base charge, is typically much larger than the forward transit time τ_F. This causes a significant switching delay for the bipolar transistor if it is already operating in saturation mode, i.e., if there is a nonzero overdrive base charge present in the base region. The charge-controlled current equations derived in this section can now be applied to transient analysis of bipolar logic circuits.

BJT Inverter Delay Times

Consider the resistive-load BJT inverter circuit with no output load shown in Fig. 12.17, which is driven with an ideal pulse voltage waveform applied to the input terminal. The output voltage waveform is expected to follow the low-to-high and the high-to-low transitions of the input voltage, with certain time delay. Typical input and output voltage waveforms associated with the inverter are plotted in Fig. 12.18, and the relevant delays are identified on the time axis.

When the input voltage switches from low to high, the transistor is initially in cut-off. After the switching of the input voltage, the transistor remains in cut-off mode, and the output voltage remains at its logic-high level for a time period of τ_1, during which the base-emitter and the base-collector junction capacitances are charged up by the base current, so that the BJT can subsequently operate in forward active mode.

Once the transistor enters forward active operation mode, a positive collector current starts to flow through the collector resistor, and the output voltage falls accordingly. The time delay τ_2 is primarily due to junction capacitance effects, and also due to the time required for minority carrier charge buildup in the base region. By the

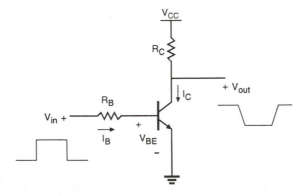

Figure 12.17 BJT inverter with ideal pulse voltage waveform applied to the input.

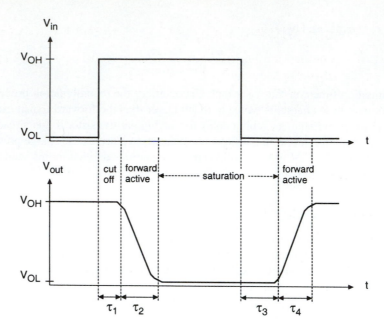

Figure 12.18 Typical input and output voltage waveforms of the BJT inverter during transient operation.

time the excess minority carrier charge in the base becomes sufficiently large, the BJT goes into saturation, and the output voltage stabilizes at the logic-low value of $V_{OL} = V_{CE,sat}$.

The BJT will remain in saturation as long as the input voltage is equal to V_{OH}. When the input voltage falls from high to low, the transistor cannot leave the saturation mode immediately. A certain saturation recovery time delay (τ_3) is required for the removal of the overdrive base charge, through the base current. Only after the saturation charge (Q_S) is removed from the base region can the transistor enter the forward active mode again. Once in forward active mode, the collector current of the BJT starts to decrease, causing the output voltage to rise from logic-low to logic-high voltage level. After a time delay of τ_4, the output voltage stabilizes again at $V_{out} = V_{OH}$.

Note that the delay times examined here are primarily due to finite carrier transit times and internal junction capacitances of the BJT. These delay times can therefore be called *intrinsic delays* of the transistor. If the BJT inverter is driving a capacitive load, on the other hand, the time required to charge and discharge the external load capacitance must also be taken into consideration. Assuming that the output load capacitance is connected to the collector node, it will appear in parallel with the collector junction capacitance in (12.57), thereby increasing all delay times except the saturation recovery delay time τ_3.

12.4 Basic BiCMOS Circuits: Static Behavior

The obvious advantages of using npn-type bipolar transistors jointly with MOS devices in various digital circuit structures were briefly outlined in Section 12.1. Having examined the operation of the bipolar transistor in detail in the preceding sections, we can now consider static and dynamic operating characteristics of the basic BiCMOS logic circuits.

Consider the simple BiCMOS inverter circuit shown in Fig. 12.19, which consists of two MOS transistors and two npn-type bipolar transistors which drive a large output capacitance C_{load}. The operation concept of this circuit can be very briefly summarized as follows.

The complementary pMOS and nMOS transistors MP and MN supply base currents to the bipolar transistors and thus act as "trigger" devices for the bipolar output stage. The bipolar transistor Q1 can effectively pull up the output voltage in the presence of a large output capacitance, whereas Q2 pulls down the output voltage, similar to the well-known *totem pole* configuration. Depending on the logic level of the input voltage, either MN or MP can be turned on in steady state, therefore assuring a fully complementary push-pull operation mode for the two bipolar transistors. In this very simplistic configuration, two resistors are used to remove the base charge of the bipolar transistors when they are in cut-off mode.

In general, the superiority of the BiCMOS gate lies in the high current drive capability of the bipolar output transistors, the zero static power dissipation, and the high input impedance provided by the MOSFET configuration. To reduce the

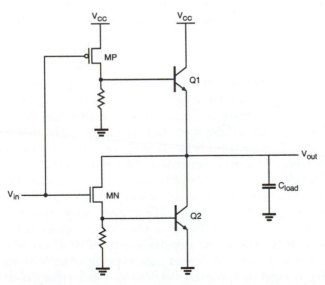

Figure 12.19 Simple BiCMOS inverter circuit with resistive base pull-down.

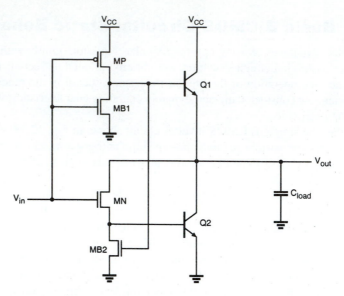

Figure 12.20 Conventional BiCMOS inverter circuit with active base pull-down.

turn-off times of the bipolar transistors during switching, two minimum-size nMOS transistors (MB1 and MB2) are usually added to provide the necessary base discharge path, instead of the two resistors. The resulting six-transistor inverter circuit, shown in Fig. 12.20, is the most widely used conventional BiCMOS inverter configuration.

The typical input-output voltage transfer characteristic of a BiCMOS inverter circuit is shown in Fig. 12.21. When the input voltage is very close to zero, the nMOS transistors MN and MB1 are off, whereas the pMOS transistor MP and thus the nMOS transistor MB2 are on. Since the base voltage of Q2 is equal to zero, the bipolar output pull-down transistor is in cut-off mode. Consequently, both of the bipolar transistors Q1 and Q2 are not able to conduct any current at this point. Notice that the pMOS transistor MP is operating in the linear region, but its drain-to-source voltage is equal to zero because no base current is flowing through Q1.

When the input voltage is increased beyond $(V_{T,n} + V_{BE,2})$, the nMOS transistor MN starts to conduct a nonzero base current for Q2. Thus, the bipolar transistors Q1 and Q2 enter the forward active region. The base-emitter junction voltages of both transistors rise very sharply and cause the observed *step-down* in the DC voltage transfer characteristic. The second step-down in the VTC occurs when the pMOS transistor is turned off, and both bipolar transistors are abruptly driven into cut-off mode. For higher input voltage levels, the drain-to-source voltage of MN is equal to zero and no base current can be supplied to Q2.

The logic threshold voltage of the BiCMOS inverter circuit can be adjusted by modifying the aspect ratio between the pMOS and nMOS transistors, MN and MP.

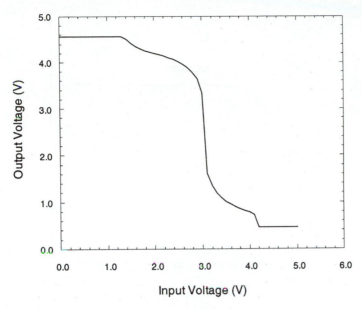

Figure 12.21   Typical voltage transfer characteristic of the BiCMOS inverter circuit.

12.5   Switching Delay in BiCMOS Logic Circuits

As we already mentioned in the preceding sections, the large current-driving capability of the BiCMOS inverter is one of its most significant advantages over the conventional CMOS buffer circuits for driving large capacitive loads. Figure 12.22 shows the simulated transient output voltage waveforms of a conventional CMOS buffer circuit and a BiCMOS buffer circuit, both of which occupy the same amount of silicon area, and drive the same capacitive load of 5 pF. It can be seen that the BiCMOS buffer can pull up the output voltage in about one-fourth of the time required for the CMOS inverter to pull up the output.

It should be emphasized that the speed advantage of BiCMOS circuits becomes more pronounced for larger capacitive loads. In fact, for driving small load capacitances, the conventional CMOS inverter still provides a better and more viable option than the BiCMOS buffer (Fig. 12.23). The reason for this is that the parasitic junction capacitances associated with the bipolar transistors tend to cancel out any speed improvement to be gained by using a BiCMOS buffer when driving a small capacitive load. For larger capacitive loads, the influence of internal device parasitics becomes negligible, and current-driving advantage of BiCMOS starts to dominate.

Since the typical BiCMOS logic gate does not dissipate any significant amount of static power during steady-state operation, it also enjoys a substantial advantage in terms of power consumption over the conventional bipolar logic circuit families, e.g., TTL or ECL. Figure 12.24 shows that the BiCMOS logic family has about the

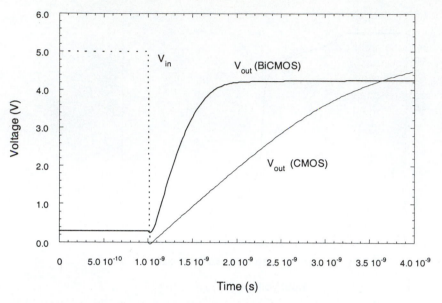

Figure 12.22   Simulated output voltage waveforms of a conventional CMOS inverter and a BiCMOS inverter, both driving a capacitive load of 5 pF. Note that both circuits also occupy approximately the same amount of silicon area.

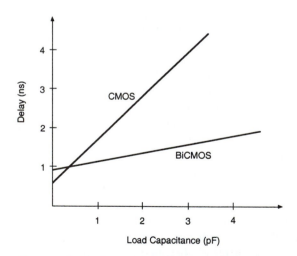

Figure 12.23   Delay versus load capacitance for conventional CMOS and BiCMOS buffers.

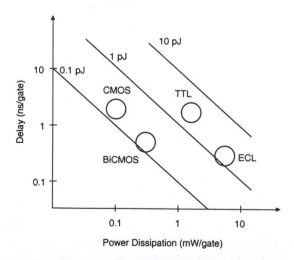

Figure 12.24   Power-delay products of BiCMOS, CMOS, and conventional bipolar logic families, using 2-μm technology.

same power-delay product as the conventional CMOS, but the gate delay is much smaller. Compared with fully bipolar alternatives, the power dissipation of the BiCMOS is at least one order of magnitude smaller for the same time delay.

In the following, we will examine in detail the transient behavior of the BiCMOS inverter circuit. Consider first the output pull-up transient response, which starts with the input voltage abruptly falling from V_{OH} to V_{OL} at $t = 0$. The initial condition of the output node voltage is assumed to be $V_{out} = V_{OH}$. The inverter circuit during this switching event is depicted in Fig. 12.25, where the active (conducting) devices are highlighted.

As the input voltage drops, the pMOS transistor MP is turned on and starts operating in the saturation region. The nMOS transistors MN and MB1 are turned off; thus, the lower "pull-down" part of the inverter circuit can be ignored except for the corresponding parasitic capacitances of the nMOS transistors and the bipolar transistor Q2. The base pull-down transistor MB2 is turned on, which effectively drains the excess base minority carrier charge of Q2 and assures that Q2 remains in cut-off mode. At the same time, MP is supplying the base current of Q1, which starts to charge up with its emitter current. Here, C'_{load} is the combined output capacitance consisting of the external load and the parasitic capacitances of Q1. The pull-up circuit, which is responsible for charging up the output load capacitance during the pull-up transient event, can be represented by the equivalent circuit shown in Fig. 12.26.

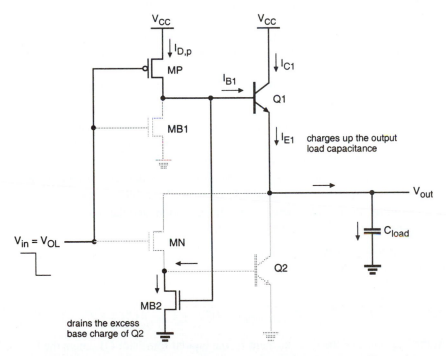

Figure 12.25 BiCMOS inverter during transient output pull-up event. The active devices in the circuit are highlighted (darker).

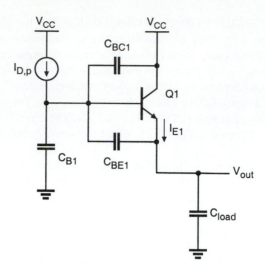

Figure 12.26   Equivalent circuit used for the pull-up delay analysis.

It is obvious that both the pMOS transistor MP and the bipolar transistor Q1 will switch operating modes more than once during the pull-up transient event. To simplify the delay analysis, the output pull-up time period can be partitioned into three time segments depending on the operating modes of the transistors. Figure 12.27 shows the typical output voltage waveform during the pull-up event, and the three time segments (τ_1, τ_2, and τ_3).

The three time segments are defined according to the operating modes of the transistors, as follows:

- During τ_1, the base-emitter voltage of Q1 increases from its initial value to $V_{BE,turn\text{-}on}$. The bipolar transistor is in cut-off during this time segment, and the pMOS transistor operates in the saturation region.
- During τ_2, the bipolar transistor starts conducting a nonzero current in the forward active mode, and the pMOS transistor MP operates in saturation.
- During τ_3, the pMOS operates in the linear region, and the bipolar transistor continues to operate in the forward active mode.

The differential equation describing the base-emitter voltage variation of the bipolar transistor during τ_1 is as follows:

$$\frac{dV_{BE}}{dt} = I_{D,p} \cdot \frac{C'_{load}}{C'_{load} \cdot C_{BE1} + (C_{B1} + C_{BC1})(C'_{load} + C_{BE1})} \quad (12.65)$$

At the beginning of the time segment τ_2, the bipolar transistor Q1 enters the forward active mode while the pMOS is operating in saturation. The combined output capacitance begins to charge up due to the emitter current. The differential equations

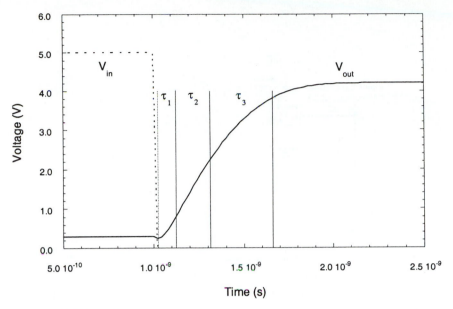

Figure 12.27 Output voltage waveform during the pull-up transient event, and the partitioning of the pull-up delay time into three segments.

describing the time dependence of the output and the base voltages are given in the following.

$$\frac{dV_{out}}{dt} = \frac{i_C + i_B}{C'_{load}} \tag{12.66}$$

$$\frac{dV_B}{dt} = \frac{I_{D,p} - i_B}{C_{B1} + C_{BC1}} \tag{12.67}$$

Note that high-level injection phenomena were found to be dominant in bipolar transistors during the time segments τ_2 and τ_3. The dependence of the current gain factor upon the collector current under high-level injection conditions is described by the Gummel-Poon model as

$$\beta_F = \frac{\beta_{FO}}{1 + \dfrac{I_C}{I_k}} \tag{12.68}$$

where I_k is called the *knee current*. Hence, the relationship between the forward transit time and the collector current can be described by

$$\tau_F = \tau_{FO}\left(1 + \frac{I_C}{I_k}\right) \tag{12.69}$$

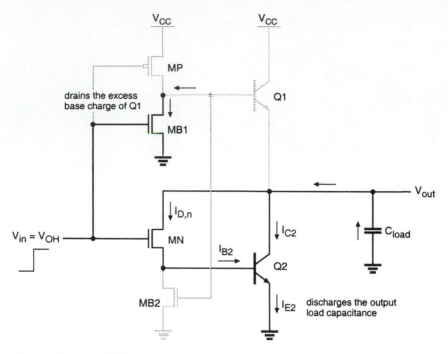

Figure 12.28 BiCMOS inverter during a transient output pull-down event. The active devices in the circuit are highlighted (darker).

The time-dependent variation of the output voltage during τ_2 can thus be found accurately by combining (12.66) and (12.67) with the modified charge-control model equations given in (12.57). Finally, the time segment τ_3 is calculated by replacing the saturation current with the linear region current expression for the pMOS transistor in (12.67).

Now consider the output pull-down transient response, which starts with the input voltage abruptly rising from V_{OL} to V_{OH} at $t = 0$. The initial condition of the output node voltage is assumed to be $V_{out} = V_{OL}$. The inverter circuit during this switching event is depicted in Fig. 12.28, where the active (conducting) devices are highlighted.

As the input voltage rises, the pMOS transistor MP is turned off and the nMOS transistors MN and MB1 are turned on. The bipolar pull-up transistor Q1 immediately ceases to conduct because its base current drops to zero, and MB1 starts to remove the excess minority carrier base charge of Q1. The nMOS transistor MN operates initially in the saturation region and supplies the base current of the bipolar pull-down transistor Q2. Note that the base pull-down transistor MB2 will continue to conduct during the initial phase of this transition, because MB1 cannot pull down the base node of Q1 immediately after the input pulse arrives.

The pull-down circuit which is responsible for discharging the output capacitance during the pull-down transient event can be represented by the equivalent

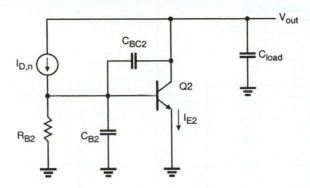

Figure 12.29 Equivalent circuit used for the pull-down delay analysis.

circuit shown in Fig. 12.29. The calculation of the pull-down delay time is analogous to the calculation of the pull-up delay, which is outlined on the previous pages.

12.6 BiCMOS Applications

The basic BiCMOS inverter structure, which is studied extensively in Sections 12.4 and 12.5, can be used as the starting point to construct complex logic gates, simply by replacing the pMOS and nMOS base driver transistors MP and MN of the basic inverter circuit with complementary pMOS and nMOS logic blocks.

Figure 12.30 shows the circuit diagram of a BiCMOS NOR2 gate. Here, the base of the bipolar pull-up transistor Q1 is being driven by two series-connected pMOS transistors. Therefore, the pull-up device can be turned on only if both of the inputs are logic-low.

The base of the bipolar pull-down transistor Q2 is driven by two parallel-connected nMOS transistors. Therefore, the pull-down device can be turned on if either one or both of the inputs are logic-high. Also, the base charge of the pull-up device is removed by two minimum-size nMOS transistors connected in parallel between the base node and the ground. Notice that only one nMOS transistor, MB2, is being used for removing the base charge of Q2, when both inputs are logic-low.

Figure 12.31 shows the circuit diagram of a BiCMOS NAND2 gate. In this case, the base of the bipolar pull-up transistor Q1 is being driven by two parallel-connected pMOS transistors. Hence, the pull-up device is turned on when either one or both of the inputs are logic-low. The bipolar pull-down transistor Q2, on the other hand, is driven by two series-connected nMOS transistors between the output node and the base. Therefore, the pull-down device can be turned on only if both of the inputs are logic-high. For the removal of the base charges of Q1 during turn-off, two series-connected nMOS transistors are used, whereas only one nMOS transistor is utilized for removing the base charge of Q2.

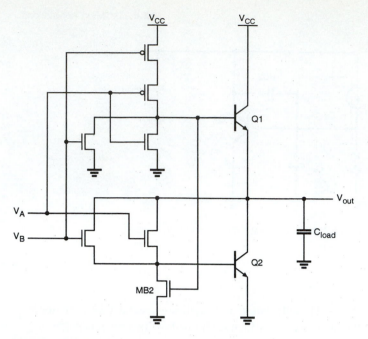

Figure 12.30 Circuit diagram of the BiCMOS NOR2 gate.

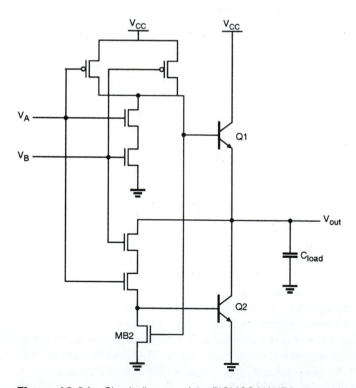

Figure 12.31 Circuit diagram of the BiCMOS NAND2 gate.

552

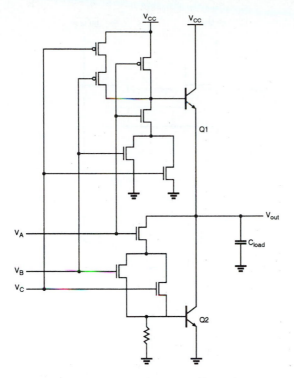

Figure 12.32 Circuit diagram of a BiCMOS complex logic gate.

The complex logic gate concept developed in Chapter 7 for nMOS and CMOS circuits can also be implemented in BiCMOS circuits with minimal hardware overhead, as illustrated in Fig. 12.32. Note that in this example, the base pull-down of the bipolar transistor Q2 is provided by a simple resistor, instead of an active base pull-down device used in the previous examples.

Finally, consider the following BiCMOS implementation in a high-performance bit-line read circuitry for memory arrays. It was shown in Chapter 10 that high-speed column sense-amplifier circuits are crucial for reducing the data access times in large memory arrays. Since the basic operations required from the column sense amplifier are to detect a small voltage difference between the two column (bit-line) voltages *and* to drive the data output during the read cycle, the well-known bipolar differential-pair (common emitter amplifier) circuit appears to be a good alternative.

The circuit diagram of a typical column pair in a memory array with the attached BiCMOS sense amplifier is shown in Fig. 12.33. During the data read cycle, the bit-line differential voltage appears between the respective base terminals of the bipolar differential pair. Collector current flows in the bipolar device that has the higher base current. Each bipolar device in the differential pair arrangement has its collector terminal connected to one of the local data lines.

Note that although each column pair in the memory array has its own differential-pair amplifier, current can flow only in the sense amplifier of the selected column. The

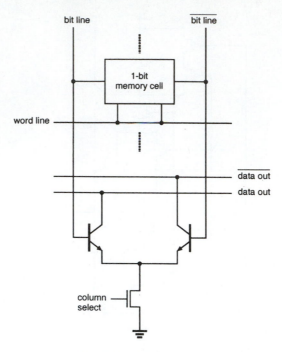

Figure 12.33 BiCMOS column sense amplifier for high-speed memory array.

current source of the bipolar differential pair is provided by an nMOS device, which is controlled by the column select signal. Also, the column pull-up operation is usually provided by bipolar transistors in this BiCMOS memory array.

The application areas for BiCMOS circuit techniques have grown steadily. However, the industry has also experienced that careful process technology development is required to achieve high-performance bipolar transistors. In this section, we provided some circuit examples in order to illustrate the circuit design possibilities and the performance advantages of the BiCMOS. Interested readers are referred to the growing literature on BiCMOS applications for a more comprehensive treatment of this subject.

Exercise Problems

12.1 A BJT transistor has $\alpha_F = 0.99$ and $\alpha_R = 0.2$. Calculate the collector junction reverse saturation current when its emitter junction reverse saturation current is 10^{-14} A. Determine the area of the collector junction compared to the area of the emitter junction.

12.2 A circuit for generating a reference voltage V_R is shown in Fig. P12.2. Find the value of V_R by assuming that the voltage drop across each diode and base-emitter junction in Q1 is 0.7 V.

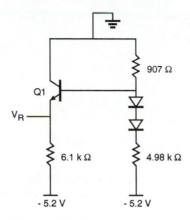

Figure P12.2

12.3 Derive the expression for τ_S in equation (12.64) in the text.

12.4 For the BJT inverter circuit in Figure 12.12 with the following parameters:

$$V_{CC} = 5 \text{ V}$$
$$R_B = 10 \text{ k}\Omega$$
$$R_C = 1 \text{ k}\Omega$$
$$\beta_F = 100$$
$$V_{BE(on)} = 0.7 \text{ V}$$
$$V_{BE(sat)} = 0.8 \text{ V}$$
$$V_{CE(sat)} = 0.1 \text{ V}$$

calculate V_{IL}, V_{IH}, and noise margins NM_L, NM_H.

12.5 Derive equation (12.65) for the equivalent circuit in Fig. 12.26.

12.6 Consider a logic gate for the Boolean function $Z = \overline{ABCDE + FGH}$.

  a.   Design a BiCMOS circuit to implement the Boolean function Z.

  b.   Design a domino CMOS circuit to implement the Boolean function Z.

  c.   Compare the pros and cons of the BiCMOS over the domino CMOS implementation.

12.7 For the bipolar transistor shown in Fig. P12.7, use the following parameters and neglect any junction capacitances.

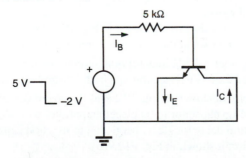

Figure P12.7

$$I_{ES} = 10^{-16} \text{ A}$$
$$I_{CS} = 2 \times 10^{-16} \text{ A}$$
$$\alpha_F = 0.98$$
$$\alpha_R = 0.49$$
$$\tau_F = 0.2 \text{ ns}$$
$$\tau_R = 20 \text{ ns}$$

a. Use the Ebers-Moll equations to find I_B, I_E, and I_C before and (long) after the transition.

b. Use the simplified charge-control equations to sketch V_B as a function of time. Include values for all important voltages and time intervals on the graph.

12.8 Compare the basic BiCMOS inverter and the basic BiNMOS inverter. Specifically, address voltage swing and propagation delay.

12.9 Use the following parameters for your calculations, assuming the emitter junction is an abrupt junction and the collector junction is a gradual junction:

$$V_{BE(on)} = 0.7 \text{ V}$$
$$V_{BE(sat)} = 0.8 \text{ V}$$
$$V_{CE(sat)} = 0.1 \text{ V}$$
$$\tau_F = 0.2 \text{ ns}$$
$$\tau_{BF} = 15 \text{ ns}$$
$$\tau_S = 20 \text{ ns}$$
$$C_{je0} = 0.5 \text{ pF} \qquad \text{(emitter junction capacitance at zero bias)}$$
$$C_{jc0} = 0.25 \text{ pF} \qquad \text{(collector junction capacitance at zero bias)}$$
$$\phi_e = 0.9 \text{ V}$$
$$\phi_c = 0.7 \text{ V}$$

a. For the bipolar inverter shown in Fig. P12.9(a), calculate the DC transfer characteristics (i.e., V_{OH}, V_{OL}, V_{IH}, V_{IL}, NM_H, NM_L). Repeat for the circuit shown in Fig. P12.9(b).

b. For the circuit shown in Fig. P12.9(a), use hand calculations to solve for τ_{PLH} and τ_{PHL}.
   Repeat for the circuit shown in Fig. P12.9(b). Describe how the pull-down resistor at the base affects propagation delay.

c. Compare your calculations for both circuits with SPICE results.
   Note: Run SPICE to simulate both DC and transient responses.

d. For the circuit shown in Fig. P12.9(a), let $R_C = 0$. Calculate the size of the speed-up capacitor to be placed in parallel with R_B to minimize the propagation delay (neglect charge due to junction capacitance). Repeat for the circuit shown in Fig. P12.9(b).

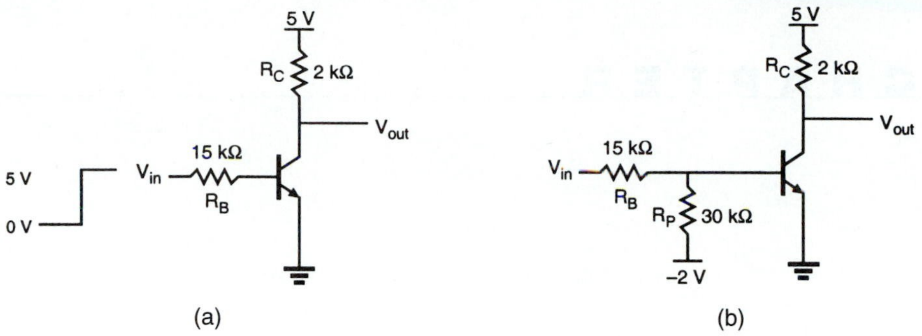

Figure P12.9

12.10   Calculate the τ_{PLH} and τ_{PHL} values for the BiCMOS gate shown in
Figure 12.20. Assume that charging of the MOSFET's parasitic
capacitances can be neglected. Use the following parameters for MOSFETs
and BJTs. Ignore the bias dependencies of the capacitances. Explain any
simplifying assumptions that you make.

$$V_{CC} = 5 \text{ V}$$
$$V_{BE(on)} = 0.7 \text{ V}$$
$$V_{T0} = 0.8 \text{ V}$$
$$k'_n = 200 \ \mu\text{A/V}^2$$
$$k'_p = 100 \ \mu\text{A/V}^2$$
$$\beta_F = 100$$
$$C_{je} = 20 \text{ fF}$$
$$C_{jc} = 22 \text{ fF}$$
$$C_L = 0.5 \text{ pF}$$
$$(W/L)_p = 6$$
$$(W/L)_n = 3$$
$$r_c = 75 \ \Omega$$

13 CHAPTER

Chip Input and Output (I/O) Circuits

13.1   Introduction

The input/output (I/O) circuits, clock generation, and distribution circuits are essential to VLSI chip design. The design quality of these circuits is a critical factor that determines the reliability, signal integrity, and interchip communication speed of the chip in a systems environment. If the package is considered a protection layer of the silicon chip, then the I/O frame containing input and output circuits and clock circuits can be considered a second protection layer. Any external hazards such as electrostatic discharge (ESD) and noises should be filtered out before propagating to the internal circuits for their protection. Also, some chips have to communicate with Transistor-Transistor Logic (TTL) or Emitter-Coupled Logic (ECL) bipolar chips, and in such cases, the input or output circuit must provide proper level shifting so that the transmitted signal contents can be correctly received or sent by the CMOS chip.

Most VLSI chips receive clock signals from a common clock source and then in turn must generate internal clock signals. Although the ideal location of such a clock module is the center of the chip, in most cases the clock module is placed in the I/O frame due to wire-bonding constraints. However, some recent chips which use bare flip-chip bonding onto printed circuit boards (PCBs) or multichip modules (MCMs) have their clock circuits placed in the center for on-chip distribution of clock signals to various locations with less skew. In this chapter, we discuss the design of electrostatic discharge damage-protection circuits, input circuits, on-chip clock generation and distribution, output circuits, on-chip noise due to parasitic inductance in bonding wires of output pads, super buffer circuit design, and the latch-up phenomenon in the I/O frame due to parasitic bipolar transistors in CMOS chips and its prevention method.

13.2 ESD Protection

Electrostatic discharge is one of the most prevalent causes for chip failures in both chip manufacturing and field operation. ESD can occur when the charges stored in machines or the human body are discharged to the chip on contact or by static induction. Figure 13.1 shows different models for ESD testing, namely the human body model (HBM), the machine model (MM), and the charged device model (CDM).

A human walking across synthetic carpet in 80% relative humidity can potentially induce 1.5 kV of static voltage stress. In the HBM (MIL-STD-883C, Method 3015, 1988) shown in Fig. 13.1(a), a touch of a charged person's finger is simulated by discharging a 100-pF capacitor through a 1.5-kΩ resistor. It is important that some protection network be designed into the I/O circuits of the chip so that the ESD effect can be filtered out before its propagation to the internal logic circuit. Effective protection networks can withstand as high as 8-kV HBM ESD stress.

In addition to human handling, contact with other machines can also cause ESD stress. Since body resistance is absent, the stress can be more severe with higher current levels. The schematic diagram of the machine model is shown in Fig. 13.1(b).

The third model is the charged device model shown in Fig. 13.1(c). It is intended to model the discharge of the packaged integrated circuits. The charge can be accumulated either during the chip assembly process or in the shipping tubes. The CDM ESD testers electrically charge the device under test (DUT) and then discharge it to ground, thus probing the high short-duration current pulse to DUT.

Simplified lumped-circuit element models of both HBM and MM ESD testers are shown in Fig. 13.2 along with corresponding parameter values.

The protection network (PN) usually consists of a *diffused* resistor-diode structure as shown in Fig. 13.3 along with its equivalent circuit model. The input

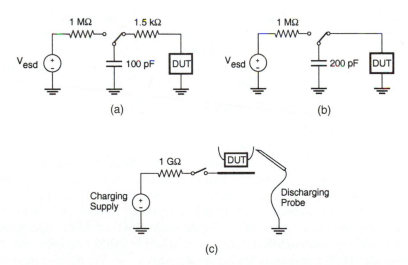

Figure 13.1 (a) Human body model, (b) machine model, and (c) charged device model, for ESD testing.

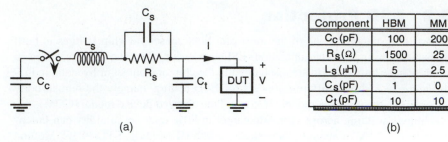

Component	HBM	MM
C_C (pF)	100	200
R_S (Ω)	1500	25
L_S (µH)	5	2.5
C_S (pF)	1	0
C_t (pF)	10	10

(a)   (b)

Figure 13.2   (a) Simplified lumped-element model of HBM-ESD and MM-ESD testers. (b) Model parameter values.

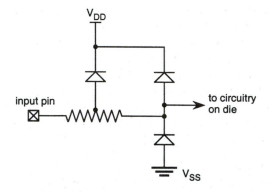

Figure 13.3   ESD protection network example.

resistance is normally between 1 and 3 kΩ. This resistance in conjunction with the capacitances in diffusion, diodes, and gate capacitance in input transistors integrates and clamps the voltage to a safe level. The RC time constant, however, should be small enough not to increase the circuit delay significantly.

In essence, the diodes clamp the signal level within a certain voltage range, in order to minimize the impact of ESD.

$$-0.7 \text{ V} < V_A < V_{DD} + 0.7 \text{ V} \tag{13.1}$$

This practice is adopted to meet the industry standard (JEDEC Standard No. 7) which is intended to avoid user-related damage to the chips. In order not to permanently damage the diode structure, the current through the diode should be limited to less than several tens of milliamperes. Past attempts to use polysilicon series resistors failed due to dielectric breakdown under high electric fields. The use of additional thick-oxide nMOS transistors as shown in Fig. 13.4 has proven to be very effective and yielded protection in excess of 3 kV in the HBM-ESD test. In this circuit, M1 is a thick-oxide punch-through device, M2 is a thick-oxide nMOS transistor, and M3 is a thin-oxide nMOS transistor operating in saturation mode. For positive input transients, M1 and M2 have threshold values of 20 to 30 V.

Figure 13.5 shows typical ESD failure modes caused by ESD-induced heat dissipation in an nMOS transistor along with a scanning electron microscopy (SEM)

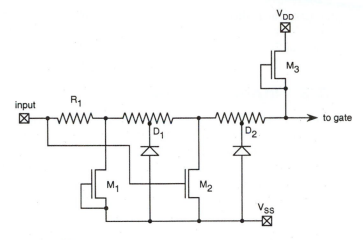

Figure 13.4 Protection network with thick-oxide transistor.

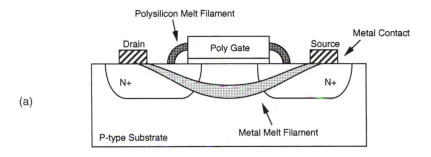

(a)

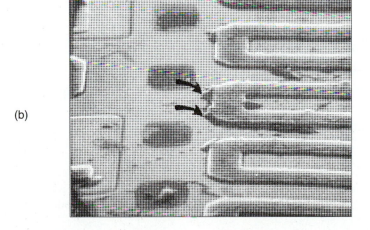

(b)

Figure 13.5 (a) Typical ESD failure modes. (b) SEM photograph of a failed nMOS transistor (from Diaz et al., 1994).

photograph of a failed nMOS transistor. Similar protection circuits can be used for the output circuit, although large driver transistors have intrinsic protection capability through diffusion and substrate or tub structures.

13.3   Input Circuits

A simple input circuit consisting of a transmission gate activated by an enable (E) signal and its complement is shown in Fig. 13.6. The incoming signal A is fed into the transmission gate through the protection network (PN) from the bonding pad of the chip. The enable signal is generated on-chip and controls the gating of the input signal as

- $X = A$, when $E = 0$
- $X =$ high-impedance state, otherwise

Any unused chip input terminals should be tied to V_{DD} or V_{SS} using pull-up or pull-down resistors externally. Some input pad circuit modules have a built-in internal pull-up or pull-down resistor or active load (normally-on transistor) with a resistance of 200 kΩ to 1 MΩ.

Figure 13.7 shows an inverting input circuit consisting of the protection network and a CMOS inverter. Typical values for V_{IL} and V_{IH} are $0.3V_{DD}$ and $0.7V_{DD}$, respectively for about 30% noise margins.

This basic input circuit can be designed to receive TTL signals for CMOS logic circuits by adjusting the ratio of the channel widths in pMOS and nMOS transistors in the inverter. Figure 13.8 shows the principle of level shifting from TTL to CMOS

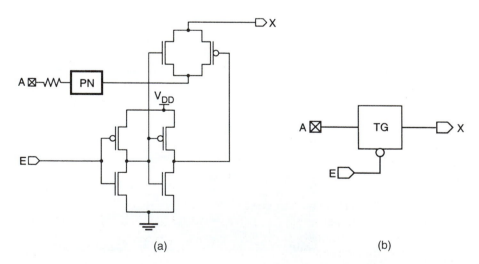

(a) (b)

Figure 13.6   (a) Input series transmission gate circuit and (b) its symbolic representation.

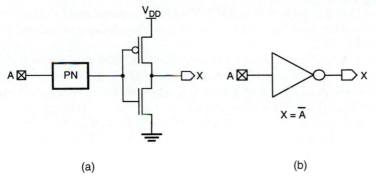

Figure 13.7 Inverting input circuit with (a) protection network, and (b) symbolic view.

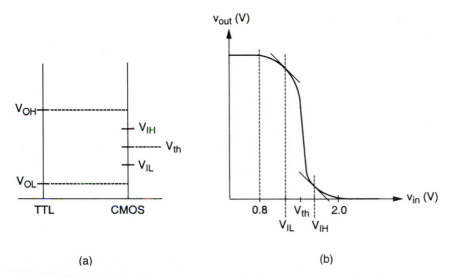

Figure 13.8 (a) TTL to CMOS level shifting and (b) the corresponding voltage transfer characteristic curve.

logic. In TTL, the worst-case output signal levels are

■ $V_{OL} = 0.8$ V
■ $V_{OH} = 2.0$ V

Therefore, input voltages less than or equal to 0.8 V should be interpreted low and input voltages greater than or equal to 2.0 V should be interpreted high.

After the input protection circuit, the incoming signals have to be level-shifted to a desirable level, depending on their voltage levels. For instance, if the incoming signal is from a TTL driver, then its low voltage can be as high as 0.8 V and its high output voltage can be as low as 2.0 V. Therefore a careful level shifting has to be done to translate such logic levels to corresponding MOS gate voltage levels as shown in Fig. 13.8.

The level shifting between a TTL driver and a CMOS gate can be achieved by properly designing the ratio between pMOS and nMOS transistors of the receiving CMOS inverter gate. A practical method is to adjust the transistor ratio in the inverter gate such that the saturation voltage at which both transistors operate in the saturation region is set at the midpoint between 0.8 V and 2.0 V. By using first-level models of MOS transistors, it can be shown that the saturation voltage of the inverter gate can be expressed by

$$V_{th} = \frac{V_{DD} + V_{Tp} + rV_{Tn}}{1 + r} \tag{13.2}$$

$$r = \sqrt{\frac{\mu_n C_{ox} W_n/L_n}{\mu_p C_{ox} W_p/L_p}} \tag{13.3}$$

From these two equations, we find that

$$\frac{W_n/L_n}{W_p/L_p} = \frac{\mu_p}{\mu_n} \left[\frac{V_{DD} + V_{Tp} - V_{sat}}{V_{sat} - V_{Tn}} \right]^2 \tag{13.4}$$

For example, if $\mu_n = 3\mu_p$ and $V_{Tn} = -V_{Tp} = 1.0$ V and $V_{DD} = 5$ V, then in order to achieve

$$V_{sat} = \frac{0.8 + 2.0}{2} = 1.4 \text{ V}$$

the nMOS-to-pMOS ratio must be

$$\frac{W_n/L_n}{W_p/L_p} = \frac{1}{3} \left[\frac{5 - 1 - 1.4}{1.4 - 1} \right]^2 = \frac{169}{12}$$

From the above calculation, we determine that $r = 6.5$ and

$$V_{IL} = \frac{2V_{out} - V_{DD} + r^2 V_{Tn} + V_{Tp}}{r^2 + 1} = \frac{2V_{out} + 36.25}{43.25}$$

where V_{out} satisfies the following current equation:

$$\frac{r^2}{2}(V_{IL} - V_{Tn})^2 = (V_{DD} - V_{IL} + V_{Tp})(V_{DD} - V_{out}) - \frac{1}{2}(V_{DD} - V_{out})^2$$

or

$$21.125(V_{IL} - 1)^2 = (4 - V_{IL})(5 - V_{out}) - \frac{1}{2}(5 - V_{out})^2.$$

Combining these two equations, we obtain

$$21.125\left[\frac{2V_{out}-7}{43.25}\right]^2 = \left[\frac{136.75-2V_{out}}{43.25}\right](5-V_{out}) - \frac{1}{2}(5-V_{out})^2$$

$$V_{out}=4.97\text{ V}$$

and, hence

$$V_{IL}=\frac{2\times4.97+36.25}{43.25}=1.07\text{ V}$$

Likewise,

$$V_{IH}=\frac{r^2(2V_{out}+V_{Tn})+V_{DD}+V_{Tp}}{r^2+1}=\frac{84.5V_{out}+47.25}{43.25}$$

where V_{out} satisfies the following current equation:

$$\frac{1}{2}(V_{DD}-V_{IH}+V_{Tp})^2 = r^2\left[(V_{IH}-V_{Tn})V_{out}-\frac{1}{2}V_{out}^2\right]$$

or

$$\frac{1}{2}(4-V_{IH})^2 = 6.5^2\left[(V_{IH}-1)V_{out}-\frac{1}{2}V_{out}^2\right]$$

Combining these two equations, we obtain

$$\frac{1}{2}\left(4-\frac{84.5V_{out}+47.25}{43.25}\right)^2 = 42.25\left[\left(\frac{84.5V_{out}+4}{43.25}\right)V_{out}-\frac{1}{2}V_{out}^2\right]$$

Solving for V_{out} and V_{IH} yields:

$$V_{out}=0.206\text{ V}\quad\text{and}\quad V_{IH}=1.47\text{ V}$$

This design appears to meet the design objective of a level-shifting CMOS inverter, providing logic 1 output level for TTL input voltages of up to 0.8 V (less than $V_{IL}=1.07$ V) and logic 0 output level for TTL input voltages not less than 2.0 V. The output voltage of 0.206 V at $V_{in}=1.47$ V is much less than the n-channel threshold voltages of the next stage. However, to assure that the circuit would function properly under all circumstances, careful circuit simulation should be performed by considering the variations in process conditions, device temperature, and power supply voltage level. Note that due to process variations, some chips can have strong pMOS (PH)-weak nMOS (NL), or weak pMOS (PL)-strong nMOS (NH) combinations for which the level-shift circuit performance would be somewhat different. This variation is illustrated in Fig. 13.9.

Figure 13.10 shows another non-inverting TTL level-shifting circuit. In this circuit, the level shifting is accomplished in the first stage, which is followed by the second-stage inverter.

Figure 13.11 shows an input pad circuit with a Schmitt trigger circuit and a 70-kΩ pull-down resistor. This circuit provides a negative-going logic threshold voltage of 1 V and a positive-going logic threshold voltage of 4 V, for a 5-V power supply.

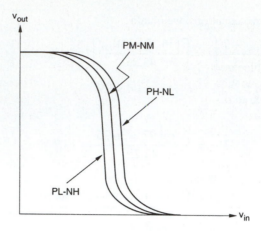

Figure 13.9   Variation of the level-shifter VTC due to process variations. Statistical analysis and design methods to overcome such variations will be discussed in Chapter 14.

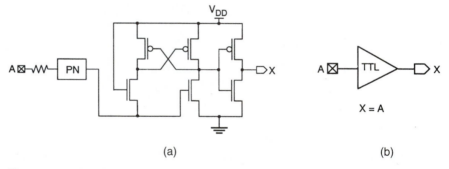

(a)   (b)

Figure 13.10   (a) Non-inverting TTL level-shifting circuit and (b) its symbolic view.

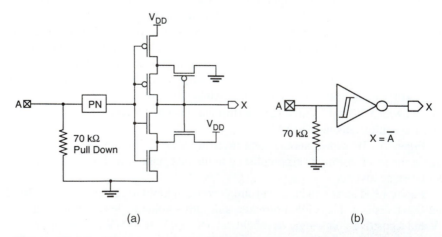

(a)   (b)

Figure 13.11   (a) Input pad circuit with Schmitt trigger and (b) its symbolic view.

13.4 Output Circuits and *L*(*di*/*dt*) Noise

The output circuits of VLSI chips are designed to be tristable as shown in Fig. 13.12. The circuit implementation 13.12(b) requires more transistors (12 transistors) than the circuit implementation 13.12(c), in which only four transistors are required if polarity is ignored. In terms of silicon area, however, the implementation in 13.12(b) may require less than the circuit in 13.12(c) since the last-stage transistors have to be sized large to provide sufficient current sinking and sourcing capability and also to reduce delay times. Unfortunately, such a requirement demands a high rate of change in the current *di*/*dt* and can cause significant on-chip noise problems due to the *L*(*di*/*dt*) drop across the bonding wire connecting the output pad to the package.

For illustration, consider a case in which the capacitor load is initially charged to $V_{DD} = 5$ V and the clock signal is set to turn on the nMOS transistor to sink the current into the ground. Figure 13.13 shows the current waveform during the switching period. The solid line represents a realistic current waveform whereas the dotted triangular line represents a simple approximation of the current waveform. By approximation as used in [1]

$$I_{max} \frac{t_s}{2} = C_{load} V_{DD} \qquad (13.5)$$

Also,

$$\left[\frac{di}{dt} \right]_{max} \geq \frac{I_{max}}{t_s/2} = \frac{2I_{max}}{t_s} \qquad (13.6)$$

Therefore, the following inequality holds.

$$\left[\frac{di}{dt} \right]_{max} \geq \frac{4C_{load} V_{DD}}{t_s^2} \qquad (13.7)$$

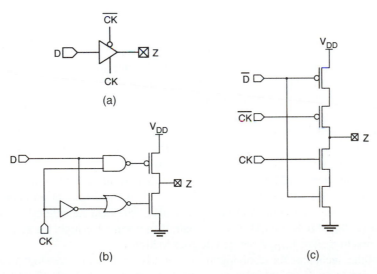

Figure 13.12 (a) Symbolic view of a tristable output circuit. (b), (c) Two different circuit implementations.

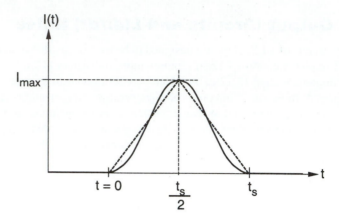

Figure 13.13   Typical output circuit current waveform during switching.

For example, if $C_{load} = 100$ pF and $t_s = 5$ ns, then

$$\left[\frac{di}{dt}\right]_{max} \geq \frac{4 \times 100 \times 10^{-12} \times 5}{(5 \times 10^{-9})^2} = 80 \, \frac{\text{mA}}{\text{ns}}$$

and for a bonding wire with $L = 2$ nH, the $L(di/dt)$ drop can be as high as

$$L\left[\frac{di}{dt}\right]_{max} \geq 160 \, \text{mV}$$

It should be noted that this voltage drop would be quadrupled if t_s were reduced by a factor of two. This shows a serious trade-off problem between the delay time and the noise. It has been observed in a 1.2-μm CMOS process chip that a current surge can be as high as 1100 mA/ns at power and ground terminals.

In high-end microprocessor chips with 32 bits or higher number of data bus lines, the noise problem can be significantly escalated if all output drivers are driven simultaneously. In such cases, it is desirable to stagger the switching times with built-in delays in the clock distribution network, which amounts to reducing the noise at the expense of chip speed.

An interesting circuit technique for reducing di/dt is shown in Fig. 13.14. This circuit requires an additional strobe signal and hence, complicates the timing design, but reduces the magnitude of di/dt significantly.

The role of two nMOS transistors controlled by the strobe signal (ST) is to precharge the gate potentials of the last-stage driver transistors at an approximate midpoint between the initial and final potentials of the load capacitor. For instance, if $r = 1$ for the pMOS and nMOS driver pair, then when ST is high, the gate voltages can be precharged to $V_{DD}/2$ before CK goes to high.

Another technique for resolving the output driver problem is to adopt a basic driver circuit that sends out only changes in the data pattern, as shown in Fig. 13.15. With a delay element, the circuit produces pulses at nodes B and C only when the polarity

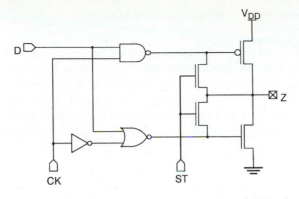

Figure 13.14 Circuit structure for reducing (di/dt) noise.

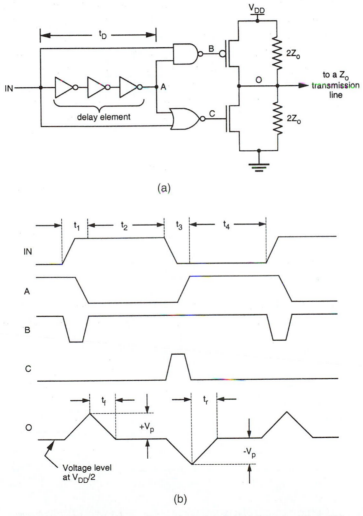

(a)

(b)

Figure 13.15 (a) Basic driver circuit that transmits only differential signals. (b) Timing diagram showing the voltage waveforms associated with the driver circuit.

of the input signal changes. As a result the driver transmits only differential signals rather than full digital waveforms. As shown in Fig. 13.15(b), the reference output voltage level is maintained at $V_{DD}/2$ during the quiescent periods, which are equivalent to tristate periods. The output driver uses a phase splitter to generate differential pairs. The corresponding receiver circuit has to sense, latch, and level-shift the differential data. The circuit shown in Fig. 13.16 performs these functions.

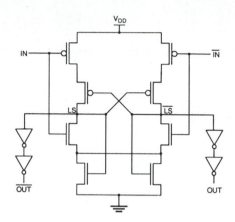

Figure 13.16   Receiver circuit designed to sense, latch, and level-shift differential data.

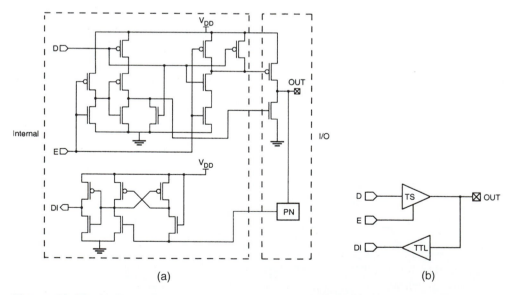

(a)   (b)

Figure 13.17   (a) Circuit diagram of a bidirectional buffer circuit with TTL input capability. (b) Block diagram of the bidirectional buffer.

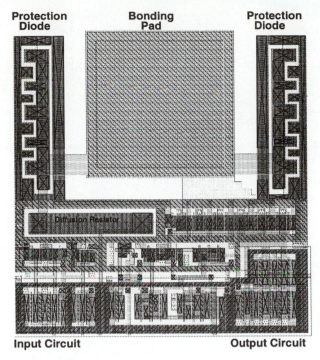

Figure 13.18 Layout of a bidirectional I/O pad circuit (courtesy of MOSIS).

A pair of input and output circuits can be combined into a single bidirectional I/O pad circuit as shown in Fig. 13.17. The layout of a sample bidirectional I/O pad is presented in Fig. 13.18, showing the bonding pad, protection diodes, diffusion resistor, and input and output circuits.

13.5 On-Chip Clock Generation and Distribution

Clock signals are the heartbeats of digital systems. Hence, the stability of clock signals is highly important. Ideally, clock signals should have minimum rise and fall times, specified duty cycles, and zero skew. In reality, clock signals have nonzero skews and noticeable rise and fall times; duty cycles can also vary. In fact, as much as 10% of a machine cycle time is expended to allow realistic clock skews in large computer systems. The problem is no less serious in VLSI chip design. A simple technique for on-chip generation of a primary clock signal would be to use a ring oscillator as shown in Fig. 13.19. Such a clock circuit has been used in low-end microprocessor chips.

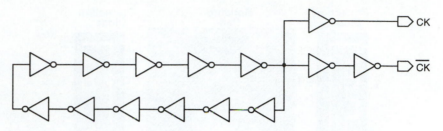

Figure 13.19   Simple on-chip clock generation circuit using a ring oscillator.

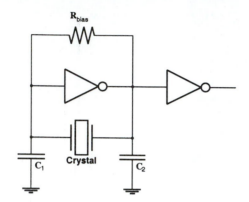

Figure 13.20   Circuit diagram of a Pierce crystal oscillator circuit.

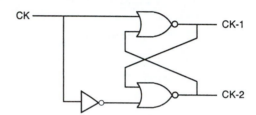

Figure 13.21   A simple circuit that generates a pair of non-overlapping clock signals from CK.

However, the generated clock signal can be quite process-dependent and unstable. As a result, separate clock chips which use crystal oscillators have been used for high-performance VLSI chip families. Figure 13.20 shows the circuit schematic of a Pierce crystal oscillator with good frequency stability. This circuit is a near series-resonant circuit in which the crystal sees a low load impedance across its terminals. Series resonance exists in the crystal but its internal series resistance largely the determines the oscillation frequency. In its equivalent circuit model, the crystal can be represented as a series RLC circuit; thus, the higher the series resistance, the lower the oscillation frequency. The external load at the terminals of the crystal also has a considerable effect on the frequency and the frequency stability. The inverter across the crystal provides the necessary voltage differential, and the external inverter provides the amplification to drive clock loads. Note that the oscillator circuit presented here is by no means a typical example of the state-of-the-art; design of high-frequency, high-quality clock oscillators is a formidable task, which is beyond the scope of this section.

Usually a VLSI chip receives one or more primary clock signals from an external clock chip and, in turn, generates necessary derivatives for its internal use. It is often necessary to use two non-overlapping clock signals. The logical product of such two clock signals should be zero at all times. Figure 13.21 shows a simple circuit

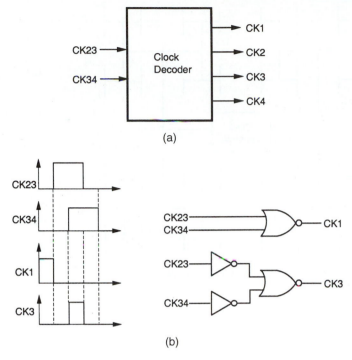

(a)

(b)

Figure 13.22 Clock decoder circuit: (a) symbolic representation and (b) sample waveforms and gate-level implementation.

that generates CK-1 and CK-2 from the original clock signal CK. Figure 13.22 shows a clock decoder circuit that takes in the primary clock signals and generates four phase signals.

Since clock signals are required almost uniformly over the chip area, it is desirable that all clock signals are distributed with a uniform delay. An ideal distribution network would be the H-tree structure shown in Fig. 13.23. In such a structure, the distances from the center to all branch points are the same and hence, the signal delays would be the same. However, this structure is difficult to implement in practice due to routing constraints and different fan-out requirements. A more practical approach for clock-signal distribution is to route main clock signals to macroblocks and use local clock decoders to carefully balance the delays under different loading conditions.

The reduction of clock skews, which are caused by the differences in clock arrival times and changes in clock waveforms due to variations in load conditions, is a major concern in high-speed VLSI design. In addition to uniform clock distribution (H-tree) networks and local skew balancing, a number of new computer-aided design techniques have been developed to automatically generate the layout of an optimum clock distribution network with zero skew. Figure 13.24 shows a zero-skew clock routing network that was constructed based on estimated routing parasitics.

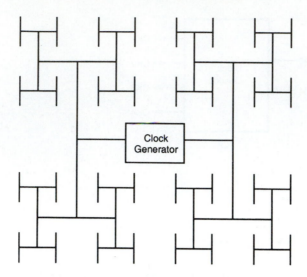

Figure 13.23 General layout of an H-tree clock distribution network.

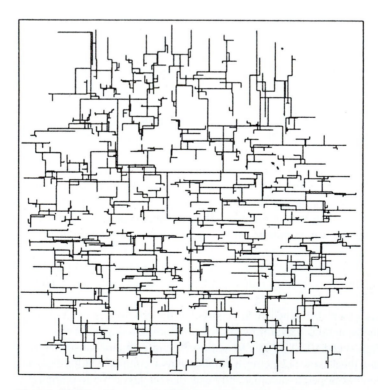

Figure 13.24 An example of the zero-skew clock routing network, generated by a computer-aided design tool.

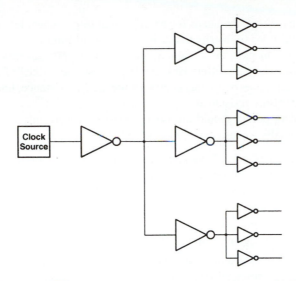

Figure 13.25 Three-level buffered clock distribution network.

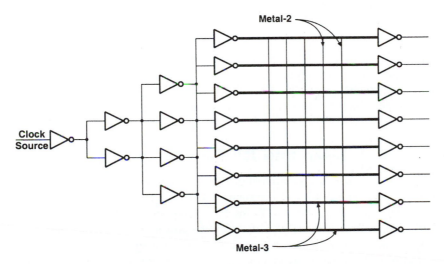

Figure 13.26 General structure of the clock distribution network used in DEC Alpha microprocessor chips.

Regardless of the exact geometry of the clock distribution network, the clock signals must be buffered in multiple stages as shown in Fig. 13.25 to handle the high fan-out loads. It is also essential that every buffer stage drives the same number of fan-out gates so that the clock delays are always balanced. In the configuration shown in Fig. 13.26 (used in the DEC Alpha chip designs), the interconnect wires are cross-connected with vertical metal straps in a mesh pattern, in order to keep the clock signals in phase across the entire chip.

So far we have seen the needs for having equal interconnect lengths and extensive buffering in order to distribute clock signals with minimal skews and healthy signal waveforms. In practice, designers must spend significant time and effort to tune the transistor sizes in buffers (inverters) and also the widths of interconnects. Widening the interconnection wires decreases the series resistance, but at the cost of increasing the parasitic resistance.

The following points should always be considered carefully in digital system design, but especially for successful high-speed VLSI design:

- Ideal duty cycle of a clock signal is 50%, and the signal can travel farther in a chain of inverting buffers with ideal duty cycle. The duty cycle of a clock signal can be improved, i.e., made closer to 50%, by using feedback based on the voltage average.

- To prevent reflection in the interconnection network, the rise time and the fall time of the clock signal should not be reduced excessively.

- The load capacitance should be reduced as much as possible, by reducing the fan-out, the interconnection lengths, and the gate capacitances.

- The characterictic impedance of the clock distribution line should be reduced by using properly increased (w/h)-ratios (the ratio of the line width to vertical separation distance of the line from the substrate).

- Inductive loads can be used to partially cancel the effects of parasitic capacitance of a clock receiver (matching network).

- Adequate separation should be maintained between high-speed clock lines in order to prevent cross-talk. Also, placing a power or ground rail between two high-speed lines can be an effective measure.

13.6 Latch-Up and Its Prevention

Latch-up is defined as the generation of a low-impedance path in CMOS chips between the power supply rail and the ground rail due to interaction of parasitic pnp and npn bipolar transistors. These BJTs form a silicon-controlled rectifier (SCR) with positive feedback and virtually short circuit the power rail to ground, thus causing excessive current flows and even permanent device damage. Although the use of an epitaxial layer and other process improvements have lessened the severity of latch-up problems, the reliability concerns persist on latch-up, especially in I/O circuits, since its packing density also increases with decreasing feature sizes and spacings. The latch-up susceptibility is inversely proportional to the product of the substrate doping level and the square of the spacing. In other words, if the spacing is reduced by half and the substrate doping is increased by two times, then the latch-up susceptibility would increase by a factor of two. Figure 13.27 shows a cross-sectional view of a CMOS inverter circuit with identification of parasitic npn and pnp bipolar transistors.

In the equivalent circuit, Q1 is a vertical double-emitter pnp transistor whose base is formed by the n-well with its base-to-collector current gain (β_1) as high as

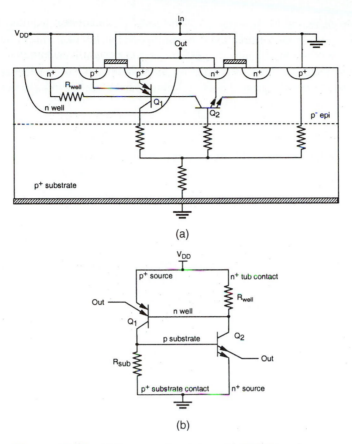

Figure 13.27 (a) Cross-sectional view of a CMOS inverter with parasitic bipolar transistors. (b) Circuit model for SCR formed of parasitic BJTs.

several hundred. Q2 is a lateral double-emitter npn transistor with its base formed by the p-type substrate. The base-to-collector current gain (β_2) of this lateral transistor may range from a few tenths to tens. R_{well} represents the parasitic resistance in the n-well structure with its value ranging from 1 kΩ to 20 kΩ. The substrate resistance R_{sub} strongly depends on the substrate structure, whether it is a simple p$^-$ or p$^-$ epitaxial layer grown on top of the p$^+$ substrate which acts as a ground plane. In the former case R_{sub} can be as high as several hundred ohms, whereas in the latter case the resistance can be as low as a few ohms.

To examine the latch-up event, first assume that the parasitic resistances R_{well} and R_{sub} are sufficiently large so that they can be neglected (open circuit). Unless the SCR is triggered by an external disturbance, the collector currents of both transistors consist of the reverse leakage currents of the collector-base junctions and therefore, their current gains are very low. If the collector current of one of the transistors is temporarily increased by an external disturbance, however, the resulting feedback loop causes this current perturbation to be multiplied by $(\beta_1 \cdot \beta_2)$. This event is called the *triggering* of the SCR. Once triggered, each transistor drives the other transistor with positive feedback, eventually creating and sustaining a low-impedance path between the power and the ground rails, resulting in latch-up. It can be seen that if the condition

$$\beta_1 \cdot \beta_2 \geq 1$$

is satisfied, both transistors will continue to conduct a high (saturation) current, even after the triggering perturbation is no longer available. This latch-up condition can also be written in terms of the collector-emitter current gains as follows.

$$\frac{\alpha_1}{1 - \alpha_1} \cdot \frac{\alpha_2}{1 - \alpha_2} \geq 1 \quad \Rightarrow \quad \alpha_1 + \alpha_2 \geq 1 \qquad (13.8)$$

Figure 13.28 shows the *I-V* characteristics of a typical SCR. At the onset of latch-up, the voltage drop across the SCR becomes

$$V_H = V_{BE1,sat} + V_{CE2,sat}$$
$$= V_{BE2,sat} + V_{CE1,sat}$$

where V_H is called the *holding voltage*. The low impedance state is sustained as long as the current through SCR is greater than I_H, the *holding current* value which is determined by the device structure. Also note that the slope of the *I-V* curve is determined by the total parasitic resistance in the current path, R_T.

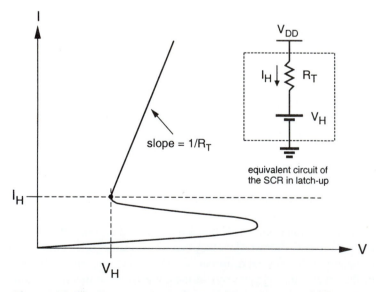

Figure 13.28   Current-voltage characteristics of a typical SCR.

Some of the causes for latch-up are:

- Slewing of V_{DD} during initial start-up can cause enough displacement currents due to the well junction capacitance in the substrate and the well. If the slew rate is large enough, latch-up can be induced. But, the SCR can have a *dynamic recovery* before latch-up when the slew rate is not very high.

- Large currents in the parasitic SCR in CMOS chips can occur when the input or output signal swings either far beyond the V_{DD} level or far below the V_{SS} (ground) level, thus injecting a triggering current. Such disturbance can happen due to impedance mismatches in transmission lines of high-speed circuits.

- ESD stress can also cause latch-up by the injection of minority carriers from the clamping device in the protection circuit into either the substrate or the well.

- Sudden transients in power or ground buses due to simultaneous switching of many drivers may turn on a BJT in SCR.

- Leakage currents in well junctions can cause large enough lateral currents.

- Radiation due to X-rays, cosmic rays, or alpha particles may generate enough electron-hole pairs in both the substrate and well regions and thus trigger the SCR.

In the following, we derive an expression for the SCR holding current I_H in terms of current gains in parasitic transistors, Q1 and Q2. For simple illustration, the circuit in Fig. 13.27(b) is redrawn in Fig. 13.29, with important circuit parameters. It can be observed that

$$I = I_{E1} + I_{RW} \tag{13.9}$$

$$I = I_{E2} + I_{RS} \tag{13.10}$$

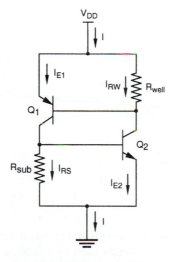

Figure 13.29 Equivalent circuit model for the SCR.

From the collector-emitter current gain (α) relations of Q1 and Q2,

$$I_{C1} = \alpha_1 I_{E1} = \alpha_1^0 I \tag{13.11}$$

$$I_{C2} = \alpha_2 I_{E2} = \alpha_2^0 I \tag{13.12}$$

where α_1^0 and α_2^0 denote the equivalent collector-to-emitter current gains absorbing the effects of parasitic resistances into transistors. Thus, when α_1^0 and α_2^0 are used, the resistors in Fig. 13.29 are effectively open-circuited. The SCR current, I, can be expressed by

$$I = I_{C1} + I_{C2} + (I_{CBO1} + I_{CBO2}) \tag{13.13}$$

where I_{CBO1} and I_{CBO2} represent the collector-base junction leakage currents, which can be combined into a single term I_{CBO}. Combining (13.9) through (13.13), we obtain the following relationship.

$$I = \frac{I_{CBO} - (I_{RS}\alpha_1 + I_{RW}\alpha_2)}{1 - (\alpha_1 + \alpha_2)} \tag{13.14}$$

The holding current, I_H, is defined as the current with zero I_{CBO}, i.e.,

$$I_H = \frac{I_{RS}\alpha_1 + I_{RW}\alpha_2}{\alpha_1 + \alpha_2 - 1} \tag{13.15}$$

Now it is clear that when the sum of the collector-to-emitter gains of Q1 and Q2 is close to unity, the value of the holding current will be large. Thus it is important to keep the gains of parasitic BJTs low. The parasitic resistances R_{sub} and R_{well} also play an important role in latch-up since their currents actually *reduce* the base currents of the parasitic transistors and thus, *weaken* the feedback loop which leads to latch-up. Therefore, reducing these resistances may prevent latch-up. Consider the SCR current at the onset of latch-up

$$I \geq I_H = (V_{DD} - V_H)/R_T \tag{13.16}$$

where both transistors are at the saturation boundary and hence, the holding voltage is $V_H = 2V_{BE}$, with $V_{BE1} = V_{BE2} = V_{BE}$. Here, the SCR is modeled by a DC voltage source of magnitude V_H in series with a resistor R_T. Combining the SCR current expression (13.15) with (13.16), and using $I_{RW} = V_{BE}/R_{well}$ and $I_{RS} = V_{BE}/R_{sub}$ in (13.15), we obtain

$$\alpha_1 + \alpha_2 \geq 1 + \left(\frac{\dfrac{R_T}{R_{well}}\alpha_1 + \dfrac{R_T}{R_{sub}}\alpha_2}{\left(\dfrac{V_{DD}}{V_{BE}} - 2\right)} \right) \tag{13.17}$$

as the condition for the occurrence of latch-up in the presence of parasitic resistances. Compare this equation with the simple latch-up condition given in (13.8). The extra term on the right-hand side of (13.17) dictates that the sum of the two current gains must be *larger* than unity by that amount, in order to satisfy the latch-up condition and

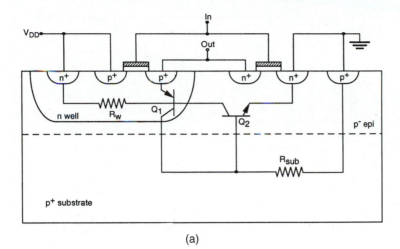

(a)

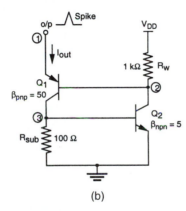

(b)

Figure 13.30 CMOS inverter circuit used in the latch-up simulation example.

to trigger the SCR. Therefore, to avoid latch-up, this extra term should be made as large as possible; i.e., the resistances R_{sub} and R_{well} should be reduced as much as possible.

The following simulation example (Fig. 13.31) illustrates latch-up in the CMOS inverter structure shown in Fig. 13.30, which is triggered by a pulse at the output node of the circuit.

Guidelines for Avoiding Latch-Up

■ Reduce the gains of BJTs by lowering the minority carrier lifetime through gold doping of the substrate (but without causing excessive leakage currents) or reducing the minority carrier injection efficiency of BJT emitters by using Schottky source/drain contacts.

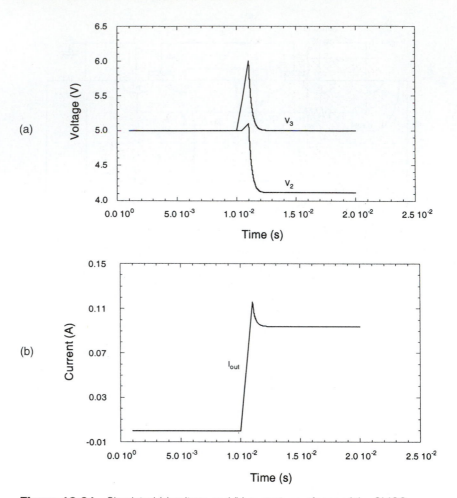

Figure 13.31   Simulated (a) voltage and (b) current waveforms of the CMOS inverter circuit during latch-up triggered through the output node.

- Use p^+ guardband rings connected to ground around nMOS transistors and n^+ guard rings connected to V_{DD} around pMOS transistors to reduce R_w and R_{sub} and to capture injected minority carriers before they reach the base of the parasitic BJTs.

- Place substrate and well contacts as close as possible to the source connections of MOS transistors to reduce the values of R_w and R_{sub}.

- Use minimum area p-wells (in case of twin-tub technology or n-type substrate) so that the p-well photocurrent can be minimized during transient pulses.

- Source diffusion regions of pMOS transistors should be placed so that they lie along equipotential lines when currents flow between V_{DD} and p-wells. In some n-well I/O circuits, wells are eliminated by using only nMOS transistors.

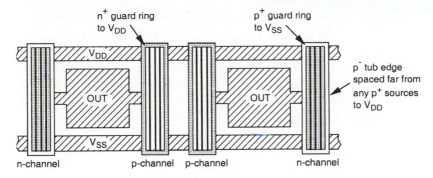

Figure 13.32 I/O cell layout with latch-up guidelines.

■ Avoid the forward biasing of source/drain junctions so as not to inject high currents; the use of a lightly doped epitaxial layer on top of a heavily doped substrate has the effect of shunting lateral currents from the vertical transistor through the low-resistance substrate.

■ Lay out n- and p-channel transistors such that all nMOS transistors are placed close to V_{SS} and pMOS transistors are placed close to V_{DD} rails. Also maintain sufficient spacings between pMOS and nMOS transistors.

For prevention of latch-up, chip manufacturers typically specify limiting operating conditions. As an example, Mitel's octal interface device, MD74SC540AC, requires 1.9 V over V_{DD} and 200 mA, and 1.0 V below V_{SS} and 90 mA to trigger output latch-up. An I/O cell layout which is designed using some of the latch-up guidelines listed here is shown in Fig. 13.32, where the same types of transistors are placed side-by-side.

Exercise Problems

13.1 For low-power design, multiple power supply voltages may be used on a chip by using on-chip voltage converters. A chip may take in a 5-V power supply and then in turn generate and use 3.3-V power rails besides 5-V power rails. Design a level shifter which can interface 3.3-V logic with a 5-V logic circuit. Use $|V_{T0}| = 1.0$ V, $\mu_n/\mu_p = 3$ in your calculation.

13.2 The distribution of clock signals without clock skew is usually desirable in order to lessen the design complexity. However, in some cases, clock skews can be utilized to resolve very tight timing budget problems. Find an example wherein clock skews can be utilized.

13.3 Design a clock decoder circuit which generates four clock phases from two primary clock signals.

13.4 Since the fan-out count of a typical clock signal is very high, it is important to size the interconnection wire dimensions properly. The parasitic interconnection resistance and capacitance are discussed in Chapter 6. The parasitic resistance in metallic wire is assumed to be 0.03 Ω/square.

  a. For $t = 0.4\ \mu$m, $h = 1\ \mu$m, l (length) $= 1000\ \mu$m, and w (width) $= 2\ \mu$m, calculate the interconnection delay for a fan-out capacitance load of 5 pF by using the Elmore delay formula. For consideration of distributed parasitic effect, the total length can be divided into 10 segments of 100-μm length.

  b. Verify the answer in part (a) using SPICE simulation.

13.5 The bonding pads in I/O circuits are implemented in the topmost metal layer with a dimension of 75 μm $\times$ 75 μm. If the separation of the topmost metal layer with SiO_2 from the common substrate layer (ground plane) is 1 μm,

  a. What is the parasitic capacitance of the bonding pad?

  b. What is the total parasitic capacitance of the bonding pad node if it is connected to a CMOS inverter gate ($W_p = 10\ \mu$m, $W_n = 5\ \mu$m, $L_M = 1\ \mu$m) and also to the output of a tristable buffer ($W_p = 1000\ \mu$m, $W_n = 500\ \mu$m, $L_M = 1\ \mu$m).

The other dimension of the drain regions is 3 μm and the parasitic capacitance in the drain is $C_{j0} = 0.3$ fF/μm^2, $C_{jsw} = 0.5$ fF/μm.

13.6 Verify the correctness of the TTL to CMOS level shifter by SPICE simulation.

13.7 The transistors in the final stage of the chip output buffer are usually chosen to be very large in order to provide sufficient current-driving capability. Discuss the layout strategy for implementing such huge transistors in the bonding pad area.

13.8 The switching noise in the power and ground rails of the chip output driver circuit can be very large to the extent that it may upset the logic level of nearby internal circuits due to coupled noise. Discuss whether this problem can be avoided with the use of separate power and ground rails for I/O circuits (noisy) and internal circuits (quiet).

13.9 The bonding wire inductance is 2 nH, load capacitance is 100 pF, and the 50% switching delay time is 5 ns.

  a. Estimate the maximum $L(di/dt)$ noise.

  b. Explain how this noise would change at a lower operating temperature and higher power supply voltage.

  c. Calculate the total noise voltage peaks when 32 such output pads are switching simultaneously and when 32 output pads are switching with 3.2 ns skew from the first bit to the last bit.

  d. Verify your results using proper models and SPICE simulation.

13.10 Discuss how the sensitivity of the level-shifting I/O circuit to process variations, especially the variation in the channel length, can be reduced by the mask design of particular (W/L) values. Would you choose the minimum L allowed?

13.11 Discuss the pros and cons of having pull-up and pull-down resistors connected to I/O pads in view of impedance matching in high-speed circuits.

CHAPTER 14

Design for Manufacturability

14.1 Introduction

Digital circuits should be designed such that fabricated circuits meet the specifications for their performance, such as speed and power dissipation, under all operating conditions. However, random fluctuations in fabrication processes cause an undesirable spread in the circuit performance. Also, random variations in the circuit operating conditions, such as the power supply voltage V_{DD} and the operating temperature, result in variations in the circuit performance. Excessive deviations of performance can lead to a significant yield loss and thus can increase the unit cost of the product. It is, therefore, important that the effect of these inevitable statistical variations in the processing and environmental conditions be considered early on during the design of the circuit. The circuit performance should be made least sensitive to these variations and should have enough margins that a large fraction of the manufactured circuits pass the acceptability criteria. This is the essential motivation for *design for manufacturability* (DFM).

The term design for manufacturability (also called *statistical design* in computer-aided design literature) encompasses many methodologies and techniques. In this chapter, we briefly discuss some important issues in design for manufacturability which have an impact on digital circuit design techniques. In particular, these issues are parametric yield estimation, parametric yield maximization, worst-case analysis, and variability minimization. We will discuss formulation of the problems and some simple solution strategies for each of these issues.

The relationship between processing and device parameters and their effect on circuit and system performances are shown in Fig. 14.1.

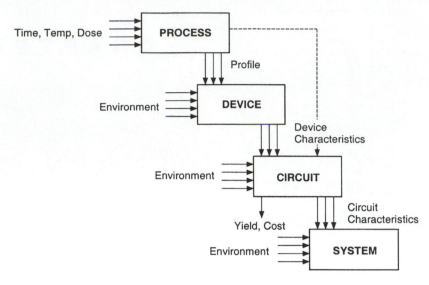

Figure 14.1 Relationship between process and device parameters and circuit and system performance.

14.2   Process Variations

Figure 14.2 shows a plot of SPICE simulation results for the output waveform of a 4-bit adder circuit for nominal $V_{DD} = 5$ V at room temperature. It can be seen that the output varies significantly due to variations in device parameters caused by process fluctuations. Although a common set of masks is used to fabricate integrated circuits, some chips will have short delay times while others will have long delay times. In other words, circuit performances can vary significantly due to uncontrollable variations in the fabrication lines. Thus, an important design task is to minimize the impact of process variations on the circuit performances.

As discussed in Chapter 2, the fabrication procedure of CMOS integrated circuits is highly complex. Most sub-micron MOS technologies use more than 10 masks for over 100 steps of chemical processes to deposit oxide layers and photoresist materials, to transfer mask patterns to wafers with optical lithography, followed by chemical etchings. Even with computer-controlled high-precision fabrication steps, some deviations in mask alignment, doping or implantation of targeted amounts of impurities, chemical etching of polysilicon gate lengths of MOS transistors, and thickness control of the thin-gate oxide layer are inevitable.

The performances of digital circuits critically depend on the I-V characteristics of MOS transistors and the parasitics in interconnection lines, both of which can vary due to process variations. A particular transistor on a chip can have a much higher or lower drain current than its counterpart on a different chip on the same wafer. This variation can be even larger for chips on different wafers. A similar statement can be made for the parasitic resistances and capacitances in interconnection structures. In

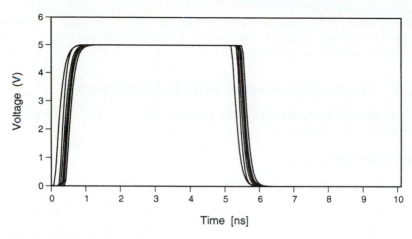

Figure 14.2 Process-related variation in output waveform of a 4-bit adder.

this chapter, we will concentrate on the variation of transistor characteristics because it has a larger impact, although the variation of interconnection parasitics is becoming non-negligible.

We recall that the drain current of an MOS transistor can be described by

$$I_d = \mu C_{ox} \frac{W}{L} f(V_{DS}, V_{GS}, V_T) \qquad (14.1)$$

where μ is the mobility of electrons in nMOS (holes in pMOS),

$C_{ox} = \varepsilon_{ox}/t_{ox}$ is the gate oxide capacitance per unit area,

W/L is the ratio of the channel width to channel length, and

V_T is the threshold voltage of the transistor.

The mobility of majority carriers in the surface channel depends on the doping level in the substrate or tub. The threshold voltage depends on the flatband voltage of the MOS system and the substrate doping. Also, C_{ox} is inversely proportional to the gate oxide layer thickness t_{ox}. Thus, random fluctuations in $\mu, t_{ox}, W/L$, and V_T will cause corresponding variations in the drain current even under the same biasing conditions. These random variations in the drain current may be translated into random variations in circuit performances such as delay times, power consumption, and logic threshold voltage. Among the quantities in (14.1), the only designable parameter under the jurisdiction of the designer is the nominal value of the aspect ratio, W/L. Thus, in order to make the circuit performance less sensitive to process variations, the most obvious design choice is to determine the optimal values of W and L for various MOS transistors in the circuit. Also, a more careful layout with proper orientation of transistors can be done to make the processed circuit less sensitive to process variations (this is even more important for analog circuits). In digital circuits, the value of L is usually chosen to be the minimum value allowed, except for sensitive circuits such as the level shifter described in Chapter 13 and memory cells for which the channel leakage is a great concern. In analog circuits, the L value may be made an order of magnitude larger than the minimum value to minimize the sensitivity of circuit performance to process variations.

14.3 Basic Concepts and Definitions

In this section, we introduce some basic concepts and define terms commonly used in design for manufacturability.

Circuit Parameters

Due to random variations in manufacturing processes and operating conditions, it is expected that the actual values of circuit parameters will be different from their nominal or target values. For instance, the actual channel width W of a MOS transistor can be decomposed into a statistically varying component ΔW and a nominal component W^o, i.e., $W = W^o + \Delta W$. In general, any circuit parameter can be considered to have a nominal component and an uncontrollable statistically varying component as shown in Table 14.1.

In this table, the geometrical parameters have a nominal component which can be set to particular values by the circuit designer. Such a nominal component is said to be *designable* or *controllable,* e.g., W^o and L^o. The statistically varying component of the geometrical parameters is called the *noise* component, and it represents the uncontrollable fluctuation of a circuit parameter about its designable component, e.g., ΔW and ΔL. For the device model parameters and operating conditions, the nominal component is not under the control of the designer and is set by nominal processing and operating conditions. For these parameters, the nominal and random components are together called the noise component, e.g., V_T and V_{DD}. In general, we can express any circuit parameter x_i as

$$x_i = d_i + s_i \tag{14.2}$$

where d_i is the designable component and s_i is the random noise component. For circuit parameters which do not have a designable component, d_i is set to zero. Similarly, for circuit parameters which are completely controllable, s_i is set to zero.

Table 14.1 Circuit parameters as sum of the nominal and random components

	Actual	**=**	**Nominal**	**+**	**Random**
Geometrical Parameters					
MOS channel width	W	$=$	W^o	$+$	ΔW
MOS channel length	L	$=$	L^o	$+$	ΔL
Device Model Parameters					
Threshold voltage	V_T	$=$	V_T^o	$+$	ΔV_T
Gate oxide thickness	t_{ox}	$=$	t_{ox}^o	$+$	Δt_{ox}
Mobility	μ	$=$	μ^o	$+$	$\Delta \mu$
Operating Conditions					
Power supply voltage	V_{DD}	$=$	V_{DD}^o	$+$	ΔV_{DD}
Temperature	T	$=$	T^o	$+$	ΔT

All of the designable components are commonly grouped to form a set of parameters called *designable parameters*. These parameters are denoted by the vector **d**. Similarly, all of the noise components are grouped to form the set of *noise parameters*. These are denoted by the random vector **s**. Vectorially, (14.2) can be written as

$$\mathbf{x} = \mathbf{d} + \mathbf{s} \tag{14.3}$$

Other commonly used terminology classifies the noise parameters based on whether their variation is related to fluctuations in manufacturing processes or operating conditions. The former are referred to as *internal noise parameters* and the latter as *external noise parameters*. For instance, V_T is considered to be an internal noise parameter, while V_{DD} is an external noise parameter.

Noise Parameter Distributions

A noise parameter is treated as a random variable. Since each circuit parameter consists of a designable component and a noise component, it can also be treated as a random variable. Any random variable is characterized by a *probability density function* (and by a mean and a standard deviation which depends on the density function). The vector of noise parameters **s** can be considered to be a random vector and is characterized by a *joint probability density function* (*jpdf*). We will denote the *jpdf* of the noise parameters by $f(\mathbf{s})$. The vector of circuit parameters **x** is also a random vector and its *jpdf* is denoted by $f(\mathbf{x})$ or $f(\mathbf{d} + \mathbf{s})$. The second notation emphasizes that the variability of circuit parameters comes from the noise components, but the *jpdf* may depend on the designable parameters.

The statistical distributions of the internal noise parameters can be obtained via test structure measurements and parameter extraction. However, to simplify the analysis, a commonly used assumption is that the internal noise parameters have a Gaussian distribution, while the external noise parameters are uniformly distributed random variables (Fig. 14.3). The internal noise parameters are *statistically*

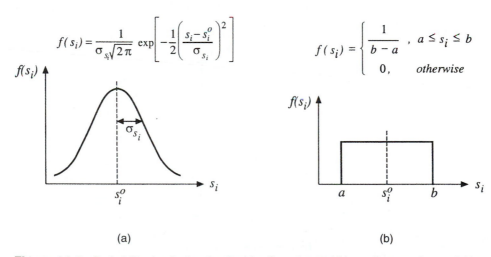

Figure 14.3 Probability density function for (a) a Gaussian and (b) a uniform random variable.

correlated due to the sequential nature of the processing steps. The external noise parameters are, however, *statistically independent* random variables. Hence, the internal noise parameters are treated as a correlated Gaussian random vector and the external noise parameters as an independent uniform random vector. When noise vector **s** is multivariate Gaussian, then its distribution can be completely characterized by its mean vector μ and the covariance matrix **Q**, i.e., $\mathbf{s} \sim \text{MVG}(\mu, \mathbf{Q})$.

Circuit Performance Measures

A circuit performance measure is used to monitor the functionality of the circuit. For example, the propagation delay of an inverter or the signal skew between various branches of a clock distribution tree can be considered to be performance measures.

Consider the simple CMOS inverter circuit shown in Fig. 14.4. Let us make the following assumptions about the circuit parameters to illustrate some basic concepts:

(i)   The widths and lengths of the MOS transistors are fixed and are not subject to statistical variations; i.e., W_1, L_1, W_2, and L_2 are designable parameters.

(ii) The internal noise parameters are the threshold voltages of MN and MP, $V_{T,n}$ and $V_{T,p}$, and the common gate oxide thickness t_{ox}. We assume for simplicity that these random variables are uncorrelated (independent) and Gaussian. The threshold voltage $V_{T,n}$ has a mean of 0.8 V and a standard deviation of 0.17 V, $V_{T,p}$ has a mean of -0.9 V and a standard deviation of 0.17 V, and t_{ox} has a mean of 27.5 nm and a standard deviation of 4.17 nm.

(iii) The external noise parameters are the power supply voltage V_{DD}, which is uniformly distributed in the range [4.8 V, 5.2 V], and the operating temperature T, which is uniformly distributed in the range [30°C, 90°C]. Moreover, V_{DD} and T are considered to be independent random variables.

Suppose that we are interested in the propagation delay of the inverter as the performance measure. The propagation delay of a CMOS inverter τ_P is given by (6.4), while the expressions for the high-to-low and low-to-high propagation delays are

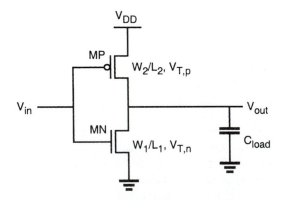

Figure 14.4   A CMOS inverter circuit.

given in (6.22b) and (6.23b) and are repeated here for convenience.

$$\tau_{PHL} = \frac{C_{load}}{k_n(V_{DD} - V_{T,n})} \left[\frac{2V_{T,n}}{V_{DD} - V_{T,n}} + \ln\left(\frac{4(V_{DD} - V_{T,n})}{V_{DD}} - 1 \right) \right] \qquad (14.4a)$$

$$\tau_{PLH} = \frac{C_{load}}{k_p(V_{DD} - |V_{T,p}|)} \left[\frac{2|V_{T,p}|}{V_{DD} - |V_{T,p}|} + \ln\left(\frac{4(V_{DD} - |V_{T,p}|)}{V_{DD}} - 1 \right) \right] \quad (14.4b)$$

It can be seen from the above equations that the two propagation delays are dependent on the widths and lengths of the MOS transistors, as well as on their threshold voltages and gate oxide thicknesses. The dependence on the power supply voltage is also evident. Although the dependence on the operating temperature is not shown explicitly, the temperature influences the propagation delay by affecting device parameters such as the mobility and threshold voltage of the MOS transistors. One can generalize from the above observation that a performance measure r is a function of the designable, internal noise and external noise parameters of the circuit, i.e.,

$$r = r(\mathbf{d} + \mathbf{s}) = r(\mathbf{x}) \qquad (14.5)$$

In some cases (as in the propagation delays above), it may be possible to express a performance measure by a closed-form analytical equation in terms of the circuit parameters of interest. In many cases, however, especially for large circuits, it is not possible to express the performances as explicit functions of the circuit parameters. In such cases, the value of a circuit performance for given values of the circuit parameters can be obtained by circuit simulation.

When the internal and external noise parameters are assumed to be fixed at their mean values, the corresponding circuit performance value is called the *nominal value*. Since it depends on the designable parameters alone, the nominal value of a performance r is denoted by

$$r^o(\mathbf{d}) = r(\mathbf{d} + \mathbf{s}^o) \qquad (14.6)$$

where $\mathbf{s}^o$ is the mean vector of the noise parameters. For the inverter example, the nominal value of the propagation delay τ_P is computed to be $\tau_P^o = 0.186$ ns for a load capacitance of $C_{load} = 0.5$ pF.

Due to inevitable statistical variations, there is a spread in the performance values about the nominal value. To demonstrate the variability in circuit performances, we fix the designable parameters (MOS channel lengths and widths) and vary the internal and external noise parameters. First, we vary $V_{T,n}$, $V_{T,p}$, and t_{ox} according to their statistical distribution while keeping the external noise parameters fixed. For each of the 1000 samples obtained in this manner, the τ_P value is computed from SPICE simulation results. Figure 14.5 compares the transient waveforms for the best (smallest delay) and worst (largest delay) cases among these samples with the nominal case. The histogram of the propagation delay distribution is plotted in Fig. 14.6. This histogram shows that the τ_P values are distributed mostly in the range [0.1 ns, 0.25 ns]. Next, we vary the external noise parameters V_{DD} and T, while keeping the

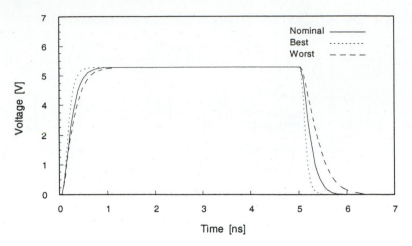

Figure 14.5   Comparison of waveforms for the nominal, best, and worst cases.

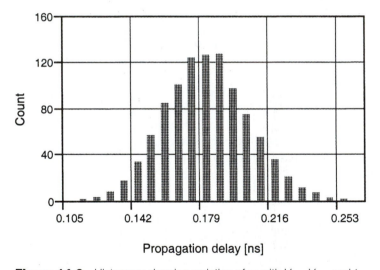

Figure 14.6   Histogram showing variation of τ_P with $V_{T,n}$, $V_{T,p}$, and t_{ox}.

internal noise parameters fixed at their nominal values. The histogram for this case is plotted in Figure 14.7 and shows that the τ_P values are distributed in a much smaller range of [0.17 ns, 0.19 ns].

The circuit performance measure, which is a function of the random circuit parameters, is also a random variable. Therefore, a performance measure will have a mean value and a standard deviation. The histograms in Figs. 14.6 and 14.7 show the approximate form of the distribution of τ_P. The relationship between a performance measure and the circuit parameters is often not known explicitly. Thus, the

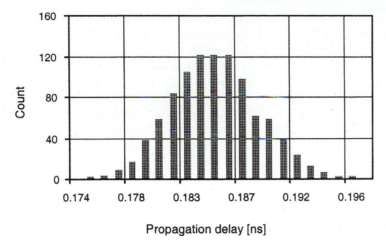

Figure 14.7 Histogram showing variation of τ_P with V_{DD} and T. Note that the *range* of the distribution of delay values is much smaller than in Fig. 14.6.

probability distribution of a performance is also not known explicitly, and its mean and standard deviation have to be estimated. For the inverter example and for the case of internal noise parameter variations alone (i.e., the case corresponding to the histogram of Fig. 14.6), the mean and standard deviation of τ_P are estimated to be 0.184 ns and 0.023 ns, respectively. Note that the nominal value of a performance is not the same as its mean. However, like the nominal value, the mean and standard deviation of a performance measure are functions of the designable parameters alone, since the effects of the noise parameters have been "averaged out."

Parametric Yield and Performance Variability

Another important concept related to circuit performance is the *yield*. Minimization of yield loss is the central notion of design for manufacturability. There are a large number of factors which lead to yield losses: material defects, lithography misalignments, processing variations, and insufficient design margins. *Catastrophic faults* are shorts and opens which cause malfunctioning of the circuit, whereas *parametric faults* are shifts in device and circuit performances. A circuit under parametric faults may be logically functional but may fail to meet some performance specifications. Each circuit performance is associated with a specification which determines its acceptability. A circuit is said to be *acceptable* if each of its performances satisfies the corresponding specifications. The yield (also called *parametric yield* to distinguish it from the functional yield) is defined as the fraction of the total number of manufactured circuits that is acceptable, i.e.,

$$\text{parametric yield} = \frac{\text{total number of acceptable circuits}}{\text{total number of manufactured circuits}} \tag{14.7}$$

The yield of a circuit directly determines the profitability of the product; hence, the maximization of yield is the primary motivation of design for manufacturability. In the preceding inverter example, suppose that we are considering variations in the internal noise parameters alone and that the acceptability criterion is that τ_P should be less than 0.19 ns. From the histogram of Fig. 14.5, it can be seen that a large fraction of the sampled circuits would fail the acceptability criterion, which would lead to a poor yield of 61.2%. Note that a poor parametric yield may result even when the nominal value of the performance meets the specifications.

The *variability* of a circuit performance is a measure of the spread in the performance values due to uncontrollable statistical variations in the circuit parameters. The minimization of variability is another important task in design for manufacturability, since it results in uniform products which are more desirable. Several measures of variability are used in practice: the standard deviation or variance, the ratio of the standard deviation and the mean, the range of the performances, etc.

To summarize the various points and observations made in this section:

- Circuit parameters are composed of designable and noise components.
- Noise parameters represent the statistical variations due to manufacturing and environmental fluctuations.
- Designable parameters are deterministic, while noise parameters are random.
- Circuit performances are functions of designable and noise parameters.
- Circuit performances are characterized by a probability distribution, which usually has to be estimated.
- A circuit with acceptable nominal performances may have low parametric yield.

14.4 Design of Experiments and Performance Modeling

Suppose that there are n circuit parameters of interest denoted by $\mathbf{x} = (x_1, x_2, \ldots, x_n)$. These circuit parameters can be designable parameters or noise parameters. As noted earlier, a circuit performance r is a function $r(\mathbf{x})$ of these parameters. Usually, this function is not known explicitly, and for particular values of $\mathbf{x}$, r has to be evaluated using a circuit simulator such as SPICE. Circuit simulations are computationally expensive, especially if the circuit size is large and transient simulations are required. An attractive alternative would be to construct a compact model of the circuit performance in terms of the parameters $\mathbf{x}$ and then use the performance model instead of the circuit simulator to evaluate the performance. The utility of such an approach depends on two criteria. First, the model should be computationally efficient to construct and evaluate so that substantial computational savings can be achieved. Second, the model should be accurate. Clearly, these two features are conflicting requirements, and some optimal trade-offs should be made in order to develop a good model.

The next question is how to build such a model. Since the model is to act as a surrogate to the circuit simulator, it should be built from values of r obtained by

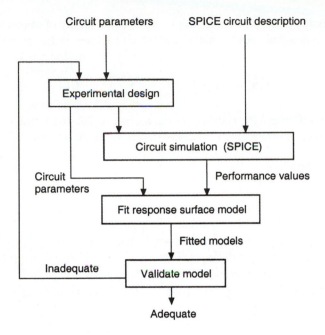

Figure 14.8 Performance modeling procedure.

circuit simulations. Figure 14.8 shows the procedure for developing such a model. The model building procedure consists of four steps. First, *m training points* are selected in the **x**-space. The ith training point is denoted by $\mathbf{x}_i = (x_{1i}, x_{2i}, \ldots, x_{ni})$. In the second step, the circuit is simulated at the m training points, and the values of the performance measure are obtained from the simulation results as $r(\mathbf{x}_1), \ldots, r(\mathbf{x}_m)$. In the third step, a pre-assigned function of r in terms of **x** is "fitted" to the data. In the final step, the model is validated for accuracy. If the model is not sufficiently accurate, then the modeling procedure is repeated with a larger number of training points or with different models.

The model is called the *response surface model* (RSM) of the performance. The computational cost of modeling depends on m, the number of training points, and on the procedure of fitting the model to the simulation data. The accuracy of the model is calibrated by computing error measures which quantify the "goodness of fit." The accuracy of the derived model would be influenced greatly by the manner in which the training points are selected from the **x**-space. *Experimental design* techniques are systematic means of selecting the training points in the "best" possible manner, in the sense that the smallest number of training points will be required and the most accurate model can be obtained from them.

Design of experiments (DOE) is an established branch of statistics, and has been used successfully in many manufacturing fields since the 1920s. In this chapter, we provide a brief glimpse into this discipline by discussing some commonly used DOE techniques in the context of design for manufacturability of integrated circuits. To illustrate certain points, it is convenient to assume that the RSM of the performance measure r is a quadratic polynomial in the circuit parameters x_i, $i = 1, 2, \ldots, n$. In particular,

$$r'(\mathbf{x}) = \alpha_0 + \sum_{i=1}^{n} \alpha_i x_i + \sum_{i=1}^{n} \sum_{j=i}^{n} \alpha_{ij} x_i x_j \qquad (14.8)$$

is the RSM used, where the coefficients α_0, α_i, and α_{ij} are the fitting parameters in the model. Note, however, that the discussion is valid for any other RSM as well.

Factorial Design

In this experimental design technique, each of the parameters $x_1, x_2, \ldots, x_n$ is quantized into two levels or settings (the smallest and largest values in its range). Without any loss of generality, we can assume that these values are -1 and $+1$ for each parameter after normalization. A full factorial design consists of all possible combinations of values for the n parameters. Therefore, a full factorial design for n parameters has 2^n training points or experimental runs. The design matrix for the case of $n = 3$ is shown in Table 14.2 and the design is depicted in Fig. 14.9. The perfomance

Table 14.2 Full factorial design for $n = 3$

	Parameters			**Interactions**				
Run	x_1	x_2	x_3	$x_1 \times x_2$	$x_1 \times x_3$	$x_2 \times x_3$	$x_1 \times x_2 \times x_3$	r
1	-1	-1	-1	$+1$	$+1$	$+1$	-1	r_1
2	-1	-1	$+1$	$+1$	-1	-1	$+1$	r_2
3	-1	$+1$	-1	-1	$+1$	-1	$+1$	r_3
4	-1	$+1$	$+1$	-1	-1	$+1$	-1	r_4
5	$+1$	-1	-1	-1	-1	$+1$	$+1$	r_5
6	$+1$	-1	$+1$	-1	$+1$	-1	-1	r_6
7	$+1$	$+1$	-1	$+1$	-1	-1	-1	r_7
8	$+1$	$+1$	$+1$	$+1$	$+1$	$+1$	$+1$	r_8

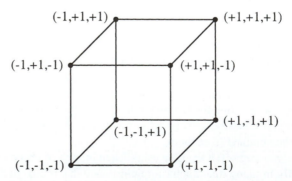

Figure 14.9 Pictorial representation of full factorial design for $n = 3$.

data for the kth run are denoted by r_k. The full factorial design provides much information on the relationship between the parameters x_i and the performance r. For example, one can evaluate the *main* or *individual effect* of a parameter x_i, which quantifies how much the parameter affects the performance. The main effect is the difference between the average performance value when the parameter is at the high level ($+1$) and the average performance value when the parameter is at the low level (-1). The main effect of x_i is the coefficient of the x_i term in the RSM of (14.8). One can also determine the *interaction effects* of two or more parameters which quantify how those factors jointly affect the performance. Two factor interactions can be computed as the difference between the average performance value when both parameters are at the same level and the average performance value when the parameters are at different levels. The two-factor interaction effect of parameters x_i and x_j is the coefficient of the $x_i x_j$ term in (14.8). Higher-order multifactor interactions can be computed in a recursive manner.

Thus, the full factorial design allows us to estimate all the first-order and cross-factor second-order coefficients in the RSM of (14.8). However, it does not allow us to estimate the coefficients of the pure quadratic terms x_i^2. Moreover, the number of experimental runs increases exponentially with the number of parameters. In most modeling situations, the information about the high-order multifactor interaction effects is often unnecessary and unimportant. It is possible to reduce the size of a full factorial design without compromising the accuracy of the main effects and the low-order interaction effects that are of primary interest. This is accomplished by considering only a fraction of the original full factorial design by systematically eliminating some of the runs. Such designs are called *fractional factorial designs*. If k is the degree of the fraction ($k = 1$ is a half fraction, $k = 2$ is a quarter fraction, etc.), then such designs have 2^{n-k} experimental runs and are called 2^{n-k} designs. One of the half-fractions of the full factorial design of Table 14.2 is shown in Table 14.3. We observe that some of the columns of Table 14.3 are identical. It is not possible to distinguish between effects that correspond to identical columns. Such effects are said to be *confounded* or *aliased* with each other. In the fractional design of Table 14.3, we see that the main effect of x_3 is confounded with the interaction effect of x_1 and x_2. Moreover, the column $x_1 \times x_2 \times x_3$ is identical to a column of 1s. This implies that the three-factor interaction effect is confounded with the grand average of the performances. Confounding, however, is not really a problem since in most applications, high-order multifactor interactions are negligible and may be ignored. In the quadratic RSM in (14.8), only main effects and two-factor interaction effects are

Table 14.3 Half-fraction of full factorial design for $n = 3$

Run	Parameters			Interactions				r
	x_1	x_2	x_3	$x_1 \times x_2$	$x_1 \times x_3$	$x_2 \times x_3$	$x_1 \times x_2 \times x_3$	
1	-1	-1	$+1$	$+1$	-1	-1	$+1$	r_1
2	-1	$+1$	-1	-1	-1	-1	$+1$	r_2
3	$+1$	-1	-1	-1	$+1$	$+1$	$+1$	r_3
4	$+1$	$+1$	$+1$	$+1$	$+1$	$+1$	$+1$	r_4

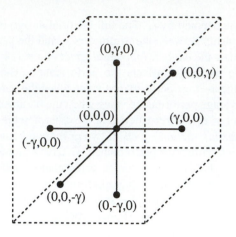

Figure 14.10 Central composite
design for $n = 3$.

important, and it can be assumed that all higher-order interactions are absent. One
of the most important attributes of fractional factorial designs is that these designs
are orthogonal, which allows the model coefficients to be estimated with minimum
errors.

Central Composite Designs

As mentioned above, one of the problems of factorial designs in regard to the RSM
of (14.8) is that the coefficients of the pure quadratic terms cannot be estimated. This
can be done with a *central composite design,* which is a combination of a factorial
(full or fractional) and a "star" design. Figure 14.10 shows the central composite design for the case of $n = 3$. The factorial design is the "cube" design shown in dotted
lines, while the star design is shown in solid lines. Each parameter in this design can
take on five levels, $0, \pm 1, \pm \gamma$, where $0 < \gamma < 1$, and the star section of the design
consists of $(2n + 1)$ runs, which includes

- one *center* point, where all parameters are set to 0, and
- $2n$ *axial* points, with the axial-pair values set to $-\gamma$ and $+\gamma$ while all other
  parameters are set to 0.

The parameter γ is usually selected by the designer. The main advantage of the central composite design is that all the coefficients of (14.8) can be estimated using a
reasonable number of simulations.

Taguchi's Orthogonal Arrays

Another popular experimental design technique is Taguchi's method using *orthogonal arrays* (OAs). An orthogonal array is a fractional factorial design matrix which
allows a balanced and fair comparison of levels of any parameter or interaction of

Table 14.4 Taguchi's L18 orthogonal array

	Parameters							
Run	1	2	3	4	5	6	7	8
1	1	1	1	1	1	1	1	1
2	1	1	2	2	2	2	2	2
3	1	1	3	3	3	3	3	3
4	1	2	1	1	2	2	3	3
5	1	2	2	2	3	3	1	1
6	1	2	3	3	1	1	2	2
7	1	3	1	2	1	3	2	3
8	1	3	2	3	2	1	3	1
9	1	3	3	1	3	2	1	2
10	2	1	1	3	3	2	2	1
11	2	1	2	1	1	3	3	2
12	2	1	3	2	2	1	1	3
13	2	2	1	2	3	1	3	2
14	2	2	2	3	1	2	1	3
15	2	2	3	1	2	3	2	1
16	2	3	1	3	2	3	1	2
17	2	3	2	1	3	1	2	3
18	2	3	3	2	1	2	3	1

parameters. These orthogonal arrays are readily available in a table format and they are of two types. The first category of OAs corresponds to parameters which have been quantized to two levels, and the second corresponds to OAs where the parameters are quantized to three levels each. These arrays are often available as tables in textbooks which discuss Taguchi techniques. As an example, we show the L18 OA in Table 14.4. The number in the designation of the array indicates the number of trials or experimental runs. The L18 OA belongs to the second category of designs; i.e., each parameter has three levels. In the Taguchi technique, the experimental design matrix for the designable or controllable parameters is called the *inner array,* while that for the noise parameters is called the *outer array.* The L18 design of Table 14.4 can be applied to the inner array as well as to the outer array.

Latin Hypercube Sampling

The factorial and central composite experimental designs described above set the parameters to certain levels or quantized values within their ranges. As a result, most of the parameter space remains unsampled. A more "space filling" sampling strategy would be more desirable. The most obvious method of obtaining more complete coverage of the parameter space would be to use the simplest sampling strategy, *random sampling,* to obtain the training points in the parameter space. Technically, a random sample is drawn from a probability density function. We have already noted that the internal and external noise parameters are random variables and have associated density functions. The designable parameters are, however, deterministic variables. For the purpose of sampling, we assume for a designable parameter that any value within its range is equally likely. In other words, we assume that the designable parameters are independent and uniformly distributed random variables for sampling. Once the random samples are generated, the circuit is simulated and the values of the performance measure are computed. Random samples are theoretically convenient and easy to generate, and many inferences can be drawn about the probabilistic distribution of the performance. This sort of random sampling is also known as *Monte Carlo sampling*. The problem with random sampling is that a large number of samples is required to estimate quantities with sufficiently small errors. Moreover, a random sample may not be very space-filling either. This can be understood by considering the bell-shaped Gaussian density function of Fig. 14.3(a). In random sampling, values near the peak of the bell-shaped curve would be more likely to be selected in the sample since such values have a greater probability of occurrence. Stated differently, values away from the central region of the distribution would not be well represented in the sample.

Latin hypercube sampling (LHS) is a sampling strategy that alleviates this problem. It ensures that each parameter x_i has all portions of its distribution represented by sample values. If S is the sample size, then the range of each x_i, $i = 1, 2, \ldots, n$, is divided into S non-overlapping intervals of equal marginal probability $1/S$. Each such interval is sampled once to obtain S values for each of the parameters. The S values for one parameter are then randomly paired with the S values of another, and so on. The process is illustrated in Fig. 14.11, where x_1 is a uniform random variable, x_2 is a Gaussian random variable, and $S = 5$. Note that the (marginal) probability is the area under the individual probability density curves. Therefore, intervals with equal probability are those with equal areas under the probability density curve. For the uniform random variable x_1, such intervals are of equal length. For x_2, the central intervals have smaller lengths (since the density there is higher) than the intervals

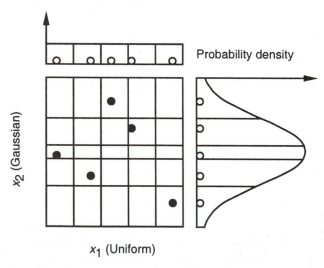

Figure 14.11 Latin hypercube sampling for a uniform and a Gaussian parameter.

away from the center (where the density is lower). Figure 14.11 also shows that $S = 5$ values are chosen from each region for both x_1 and x_2 (shown as open circles). These values are then randomly paired so as to obtain the sample points (shown as filled dots). This example also illustrates some of the advantages of Latin hypercube sampling. First, LHS provides more uniform coverage of the input parameter space than other experimental design techniques. Second, a sample of any size can be easily generated and all types of probability densities are handled.

Model Fitting

Once the experimental design is done to select the training points in the **x**-space, the circuit is simulated at the training points and the values of the performance measure are extracted from the simulation results. Let S denote the number of training points. Now, we have S points in the **x**-space and corresponding values of r. We will describe the technique of fitting the quadratic RSM of (14.8) with the understanding that the same techniques can be used to fit any RSM. The aim in model fitting is to determine the coefficients in the model such that the model fits the data as accurately as possible, i.e., the fitting error is minimized. The number of coefficients in the quadratic RSM of (14.8) is $C = (n + 1)(n + 2)/2$, where n is the number of model parameters. There are interpolation methods which can be used to fit the quadratic RSM with a smaller number of data points than coefficients, i.e., with $S < C$. But we will restrict ourselves to the case of $S \geq C$. If $S = C$, there are as many equations as unknowns, and simple simultaneous equation solving will provide the values of the coefficients. In such a case, there is no attempt to minimize any fitting error. It is, therefore, a better practice to collect more data points than the number of coefficients so that the optimal set of model coefficients can be determined. This method is called *least-squares fitting* (in a numerical analysis context) or *linear regression* (in a statistical context). The error measure used is called the sum of squared errors and is given as

$$\varepsilon = \sum_{k=1}^{S} (r(\mathbf{x}_k) - r'(\mathbf{x}_k))^2 \qquad (14.9)$$

where $r(\mathbf{x}_k)$ is the simulated performance value at the kth data point, and $r'(\mathbf{x}_k)$ is the model-predicted value. Note that $r'(\mathbf{x}_k)$, and therefore ε, depends on the model coefficients. The aim of least-squares fitting is to obtain the values of the coefficients such that the error is minimized. Formally stated, least-squares fitting is the following optimization problem:

$$\underset{\alpha_i}{\text{minimize}} \left[\varepsilon = \sum_{k=1}^{S} (r(\mathbf{x}_k) - r'(\mathbf{x}_k))^2 \right] \qquad (14.10)$$

The error ε is used to determine the adequacy of the model. If the model is considered to be inaccurate, the modeling procedure has to be repeated with a larger number of training points or a different design strategy, or a different model of the performance may have to be used. There are several measures of accuracy of a model, but these are available in many statistics textbooks and are not discussed here.

14.5   Parametric Yield Estimation

The parametric yield is used to characterize the manufacturability of a circuit. As defined in (14.7), parametric yield is the fraction of manufactured circuits that meet all acceptability criteria. Let $\mathbf{r} = (r_1, r_2, \ldots, r_p)$ denote the p circuit performance measures of interest. Each performance has acceptability specifications which can be expressed as

$$a_k \leq r_k \leq b_k, \quad k = 1, 2, \ldots, p \qquad (14.11)$$

where a_k denotes the lower limit and b_k the upper limit of acceptability for the kth performance. The specifications on the circuit performances define an *acceptability region* in the p-dimensional *performance space* denoted by A_r and expressed as

$$A_r = \{\mathbf{r} \,|\, a_k \leq r_k \leq b_k, \quad k = 1, 2, \ldots, p\} \qquad (14.12)$$

For example, consider an adder circuit where the performances are the power dissipation P_d and the propagation delay τ_P. Suppose the specifications are as follows:

$$\begin{aligned} P_d &\leq 0.5 \text{ mW} \\ \tau_P &\leq 0.16 \text{ ns} \end{aligned} \qquad (14.13)$$

Figure 14.12 illustrates the acceptability region in the performance space for this case. The parametric yield is defined as the fraction of manufactured circuits that are acceptable, and since the circuit performances are random variables, the parametric yield can be expressed as

$$Y = \Pr(\mathbf{r} \in A_r) \qquad (14.14)$$

Since the probability density functions of the circuit performances are not known explicitly, the probability above is difficult to evaluate. Therefore, an alternative approach is used to estimate the yield.

In addition to the performance specifications, the circuit parameters may also be restricted to a subset of the parameter space. These restrictions may arise from

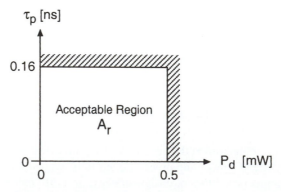

Figure 14.12   Region of acceptability in performance space for adder example.

physical considerations such as nonnegative resistance values. Let the allowed circuit parameter space be represented by $\mathbf{X}$. Then, we can define a region in the circuit parameter space called the *acceptable region* or *feasible region* A_x as follows:

$$A_x = \{\mathbf{x} \mid a_k \le r_k(\mathbf{x}) \le b_k, \quad k = 1, 2, \ldots, p \text{ and } \mathbf{x} \in \mathbf{X}\} \tag{14.15}$$

Note the distinction between A_r and A_x: A_r denotes the acceptable region in the performance space, while A_x denotes the acceptable region in the circuit parameter space. Clearly, A_x is a subset of $\mathbf{X}$, i.e., $\mathbf{X} \supset A_x$. The mapping $\mathbf{x} \to \mathbf{r}$ from the circuit parameter space to the performance space is determined by the functions $r_k(\mathbf{x})$, $k = 1, 2, \ldots, p$. These functions are known implicitly (i.e., they have to be evaluated by circuit simulation) or can be approximated by explicit response surface models as explained in the previous section. The inverse mapping, $\mathbf{r} \to \mathbf{x}$, from the performance space to the circuit parameter space is generally not known. Therefore, the boundaries that define the acceptable region in the parameter space cannot be determined easily.

Figure 14.13 shows an example of a hypothetical acceptable region in the parameter space for $p = 2$ and $n = 2$. Consider the designable parameter vector (also referred to as the *design point*) marked $P = (d_1, d_2)$ on the figure. The circle around this point represents the actual values (x_1, x_2) of the circuit parameters that may be realized as a result of statistical variations. It is easy to see that the circuit parameters of acceptable circuits for this design point lie in the intersection of the circle and the acceptable regions A_x. In other words, the parametric yield Y of the circuit for this

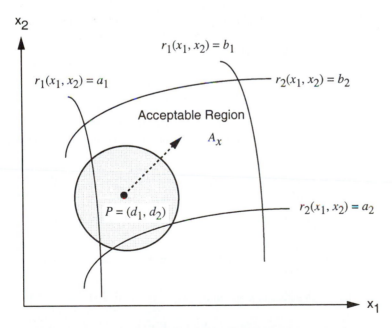

Figure 14.13 Acceptable region in the circuit parameter space for $p = 2$ and $n = 2$.

design point is the area (in a probabilistic sense) of the intersection. Note that for a given process technology, Y can be represented as a function of the designable parameters alone, since the noise components are averaged out when the area is computed. This can be understood from Fig. 14.13 by realizing that as the design point (which is the center of the circle that represents parameter variations) is moved toward the center of the acceptable region, the parametric yield increases.

Given that the acceptable region A_x consists of circuit parameter values for which the circuit performances meet specifications, the parametric yield Y at a design point $\mathbf{d}$ can be defined as the probability that the actual circuit parameter $\mathbf{x}$ belongs to A_x. Mathematically,

$$Y(\mathbf{d}) = \Pr(\mathbf{x} \in A_x) = \Pr(\mathbf{d} + \mathbf{s} \in A_x) = \int_{A_x} f(\mathbf{d} + \mathbf{s}) \, d\mathbf{s} \qquad (14.16)$$

We define an indicator function $I(r_1, r_2, \ldots, r_p)$ as follows:

$$I(r_1, r_2, \ldots, r_p) = \begin{cases} 1, & \text{if } a_k \leq r_k \leq b_k \text{ for all } k \\ 0, & \text{otherwise} \end{cases} \qquad (14.17)$$

Then the yield expression can be rewritten as

$$Y(\mathbf{d}) = \int I(r_1(\mathbf{d} + \mathbf{s}), r_2(\mathbf{d} + \mathbf{s}), \ldots, r_p(\mathbf{d} + \mathbf{s})) f(\mathbf{d} + \mathbf{s}) \, d\mathbf{s} \qquad (14.18)$$

Next, we discuss two simple methods for parametric yield estimation.

Direct Monte Carlo Method

Monte Carlo or simple random sampling is the most widely used method for parametric yield estimation. The various steps are outlined below.

Step 1: Generate a (large) number of samples of the noise parameters $\mathbf{s}_i$, $i = 1, 2, \ldots, N_{MC}$, from the joint probability distribution of the noise parameters $f(\mathbf{s})$.

Step 2: Obtain by circuit simulation $r_1(\mathbf{d} + \mathbf{s}_i), r_2(\mathbf{d} + \mathbf{s}_i), \ldots, r_p(\mathbf{d} + \mathbf{s}_i)$, $i = 1, 2, \ldots, N_{MC}$.

Step 3: Calculate the estimated yield as the fraction of samples that are acceptable.

$$Y'(\mathbf{d}) = \frac{1}{N_{MC}} \sum_{i=1}^{N_{MC}} I(r_1(\mathbf{d} + \mathbf{s}_i), r_2(\mathbf{d} + \mathbf{s}_i), \ldots, r_p(\mathbf{d} + \mathbf{s}_i)) \qquad (14.19)$$

The error of the yield estimate is given by

$$\sigma(Y') = \left[\frac{Y'(1 - Y')}{N_{MC} - 1} \right]^{1/2} \qquad (14.20)$$

Thus, the accuracy of the yield estimate is inversely proportional to the square root of the sample size. The Monte Carlo method has the following advantages:

- It provides the error of the yield estimate.
- The sample size N_{MC} is independent of the number of circuit parameters.
- It makes no restrictive assumptions regarding the nature of the joint probability density function $f(\mathbf{s})$.
- It makes no restrictive assumptions on the nature of the relationship between the circuit parameters and the circuit performances.

Although various strategies to reduce the sample size, such as variance reduction techniques, importance sampling, control variates, and stratified sampling, have been considered, Monte Carlo sampling is still expensive in terms of computational cost due to the large number of samples and the high cost of circuit simulations.

Performance Model Method

In this method, the response surface models of the circuit performance measures are used to evaluate the performance values for a given set of circuit parameter values. This avoids expensive circuit simulations and makes yield analysis feasible. Since the designable parameter vector $\mathbf{d}$ is fixed during yield estimation, the response surface models of the circuit performances are expressed in terms of the noise parameters alone. These response surface models are denoted by $r_k'(\mathbf{s})$, $k = 1, 2, \ldots, p$. For example, the quadratic RSMs of (14.8) can be used with x_i replaced by s_i. The various steps of the yield estimation procedure are enumerated below.

Step 1: Design an experiment and simulate the circuit at the training points.

Step 2: Fit the RSMs of the performances $r_1'(\mathbf{s})$, $r_2'(\mathbf{s})$, $\ldots$, $r_p'(\mathbf{s})$, and improve model accuracy if required.

Step 3: The parametric yield Y is obtained by Monte Carlo sampling using the RSMs. Generate a (large) sample of the noise parameters, i.e., $\mathbf{s}_i$, $i = 1, 2, \ldots, N_{MC}$, from the joint density function of $\mathbf{s}$.

Step 4: Compute $r_k'(\mathbf{s}_i)$, $k = 1, 2, \ldots, p$ and $i = 1, 2, \ldots, N_{MC}$.

Step 5: The estimated parametric yield is given by (14.19) with r_k replaced by r_k'.

This method preserves the advantages of the Monte Carlo method of yield estimation and at the same time considerably reduces the computational cost.

A Simple Example of Parametric Yield Estimation

Let us consider the simple inverter example of Fig. 14.4. To illustrate the yield estimation procedure, we make the following assumptions:

(i) The operating conditions are not subject to statistical variations; i.e., there are no external noise parameters.

(ii)   The channel lengths of MN and MP are also not subject to statistical variations. Moreover, we assume that both channel lengths are fixed at 0.8 μm. The channel width of an MOS transistor is the sum of a designable (nominal) component and a noise component. The designable components are W_1^o and W_2^o for transistors MN and MP, respectively. The corresponding noise components are ΔW_1 and ΔW_2.

(iii) Two additional noise parameters arise from random variations in the threshold voltages of MN and MP ($V_{T,n}$ and $V_{T,p}$). We assume that all four noise parameters are independent Gaussian random variables. Their means and standard deviations (s.d.) are as follows:

$$\Delta W_1: \quad \text{mean} = 0 \ \mu\text{m}, \qquad \text{s.d.} = 0.03 \ \mu\text{m}$$
$$\Delta W_2: \quad \text{mean} = 0 \ \mu\text{m}, \qquad \text{s.d.} = 0.06 \ \mu\text{m}$$
$$V_{T,n}: \quad \text{mean} = 0.8 \ \text{V}, \qquad \text{s.d.} = 0.067 \ \text{V}$$
$$V_{T,p}: \quad \text{mean} = -0.9 \ \text{V}, \quad \text{s.d.} = 0.067 \ \text{V}$$

(iv)   There are two performance measures of interest. The first is the propagation delay τ_P defined earlier. The second is the area of the circuit. Since it is not possible to evaluate the area accurately without layout data, we assume that the area is given by the sum of the products of the widths and lengths of all MOS transistors in the circuit. Furthermore, since the transistor length is assumed to be fixed, the area measure can be simplified to be the sum of all transistor widths. Therefore, the area measure A_m is given by

$$A_m = W_1 + W_2 = W_1^o + \Delta W_1 + W_2^o + \Delta W_2 \qquad (14.21)$$

(v)   The specifications on the two performances are: $\tau_P \leq 0.172$ ns and $A_m \leq 35 \ \mu$m.

With these assumptions, the parametric yield is computed using the two methods outlined above at the design point $\mathbf{d}_{init} = (W_1^o = 10, W_2^o = 20)$, where all of the widths are shown in units of μm. The value of N_{MC} used in both methods is 1000. In the direct Monte Carlo method, N_{MC} circuit simulations are used to estimate the parametric yield at $\mathbf{d}$ to be 79.5% (the error of the estimate, according to (14.20), is 1.28%). In the performance modeling method, 10 circuit simulations are used to construct the following linear RSM for the propagation delay τ_P in terms of the noise parameters:

$$\tau_P' = 0.169 + 0.0069V_{T,n} - 0.0071V_{T,p} - 0.0007\Delta W_1 - 0.0008\Delta W_2 \qquad (14.22)$$

Note that an RSM for the area measure is not required since (14.21) already expresses it explicitly in terms of the circuit parameters. Furthermore, as discussed previously, the RSM for the propagation delay is given in terms of the noise parameters alone since the design point is fixed. Based on this RSM, the parametric yield is estimated to be 79.6% (with an error of 1.27%). Ten circuit simulations were required in the performance modeling approach as opposed to 1000 simulations used in the direct Monte Carlo method. This example shows that the performance modeling approach can be quite accurate and provides a substantial saving in the number of circuit simulations.

14.6   Parametric Yield Maximization

As noted in the previous section, the parametric yield is a function of the designable parameter values. Therefore, the designable parameters may be adjusted to maximize the yield. This is the basic idea of parametric yield maximization. Various yield maximization approaches have been proposed over the years. These approaches can be classified into two categories: *Monte Carlo-based methods* and *geometrical methods*.

Monte Carlo-Based Methods

In these approaches, the yield integral of (14.16) or (14.18) is computed by the Monte Carlo method as before (or improved variants of the Monte Carlo methods), and the yield is then numerically maximized. Several choices of optimization techniques are available in the maximization of yield. Some techniques do not require the derivatives of the yield with respect to the designable parameters. These are, however, slower than those that require derivative information. Since the yield is a multi-dimensional integral involving statistical distributions, analytical formulae for the derivatives are usually not available. There are many sophisticated methods for approximating these derivatives, but the simplest one uses finite difference methods. In the Monte Carlo-based approaches, the acceptable region is not explicitly characterized. Rather, the decision as to whether a set of circuit parameter values belongs to the acceptable region is made by evaluating the performances and verifying whether they meet the specifications. The evaluation of circuit performances can be done both by actual circuit simulation as well as by constructing analytical response surface models for the performances.

Geometrical Methods

These methods build an approximation to the acceptable region A_x and this approximation is used in the yield maximization. There are two techniques that can be used to approximate A_x. The first technique progressively builds a geometric approximation, such as a simplex, to A_x. This technique is called *simplicial approximation*. The simplicial approximation method has one important drawback: the cost of constructing the approximation to A_x and maximizing the yield grows exponentially as the number of circuit parameters increases. This problem is often called the "curse of dimensionality." The second technique uses analytical models of the circuit performances, such as the response surface models introduced earlier. Then the boundaries of A_x are given by the constraint equations, $r'_k = a_k$ and $r'_k = b_k$, where r'_k denotes the performance model for the kth performance and a_k and b_k are the lower and upper acceptability limits, respectively. Many methods use analytically approximated acceptability regions as they are relatively inexpensive for a large number of circuit parameters. Once the acceptability region approximation has been constructed, yield maximization involves a technique called *design centering,* which attempts to move the design point towards the center of the acceptability region. Design centering is an attractive method and many approaches to design centering have been proposed.

A Simple Yield Maximization Method

Since a detailed description of the yield maximization approaches is beyond the scope of this book, a simple method for parametric yield maximization is outlined below. This method belongs to the class of Monte Carlo-based methods described before.

Step 1: Assume models for all circuit performances in terms of both the designable and the noise parameters. These are denoted by $r'_k(\mathbf{x})$, or alternatively by $r'_k(\mathbf{d} + \mathbf{s})$, for $k = 1, 2, \ldots, p$.

Step 2: Design an experiment and simulate the circuit at the training points.

Step 3: Fit the models for the performances and validate the models.

Step 4: The parametric yield of a circuit at a design point $\mathbf{d}$ is obtained by drawing a Monte Carlo sample of the noise parameters and using (14.19) to estimate the yield.

Step 5: Maximize the estimated yield $Y'(\mathbf{d})$ with respect to $\mathbf{d}$ using some optimization algorithm. The same Monte Carlo sample of noise parameters is used at each new design point encountered during the optimization. Let the final design point be $\mathbf{d}^*$. Obtain a confirmatory yield estimate at $\mathbf{d}^*$.

In the above procedure, the experimental design of Step 2 and the model of Step 3 are given in terms of both the designable and the noise parameters. However, the Monte Carlo sample in the yield estimation of Step 4 is given in terms of the noise parameters alone.

A Simple Example of Parametric Yield Maximization

Let us consider the familiar inverter example. We make the same assumptions that were used to illustrate the yield estimation procedures in Section 14.5. The designable parameters are the nominal channel widths of MN and MP, W_1^o and W_2^o, respectively. Earlier, we had estimated the parametric yield at the point $\mathbf{d}_{init} = (W_1^o = 10, W_2^o = 20)$ to be $Y'(\mathbf{d}_{init}) = 79.6\%$. The yield maximization procedure described above is used. We construct an RSM for τ_P in terms of both of the designable parameters W_1^o and W_2^o, and the noise parameters $V_{T,n}$, $V_{T,p}$, ΔW_1, and ΔW_2. At each design point, the yield is estimated from a Monte Carlo sample of the four noise parameters. The final design point obtained from the optimization is $\mathbf{d}_{final} = (W_1^o = 11, W_2^o = 22)$. At $\mathbf{d}_{final}$, the parametric yield is estimated to be 100%. Figure 14.14 shows a comparison of the scatter of τ_P and A_m values before and after yield maximization. This plot shows that at $\mathbf{d}_{init}$, the τ_P specification was violated

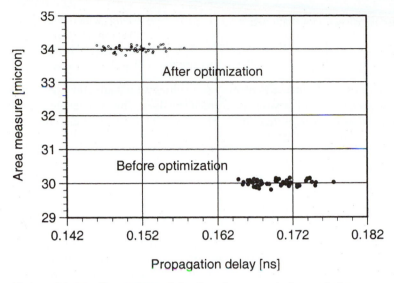

Figure 14.14 Comparison of circuit performances before and after optimization. During the course of yield maximization, delay and area are traded off so that at d_{final}, all points satisfy both the delay and the area specifications.

at many of the points while the area measure satisfied its specifications at all the points.

14.7 Worst-Case Analysis

Worst-case analysis is the most commonly used technique in industry for considering manufacturing process tolerances in the design of digital integrated circuits. These approaches are relatively inexpensive compared to the yield maximization approaches in terms of computational cost and designer effort, and they also provide high parametric yields. At any design point, uncontrollable fluctuations in the circuit parameters cause circuit performances to deviate from their nominal designed values. The goal of worst-case analysis is to determine the worst values that the performances may have under these statistical fluctuations. In addition to finding the worst-case values of the circuit performances, this analysis also finds the corresponding worst-case values of the noise parameters. The worst-case noise parameter vector is used in circuit simulation to verify whether circuit performances are acceptable under these conditions. Similar to worst-case analysis, one can also perform *best-case analysis*. In fact, industrial designs are often simulated under best, worst, and nominal noise parameter conditions, which provides designers with quick estimates of the range of variation of circuit performances.

For some circuit performances, e.g., delay, the larger their values are, the worse they become. For other performances, e.g., power dissipation, the smaller the values are the better they become. For each performance r, we can therefore define a *worst-case direction* as follows:

$$w = \begin{cases} +1, & \text{if larger values are worse than smaller values} \\ -1, & \text{if smaller values are worse than larger values} \end{cases} \tag{14.23}$$

The worst-case performance is defined as a value that bounds some desired percentage of the population of possible performance values. This percentage or probability is called the *worst-case probability* ρ. Therefore, the worst-case circuit performance value r^{wc} can be defined as the value for which

$$\rho = \begin{cases} \Pr(r \geq r^{wc}), & \text{if } w = +1 \\ \Pr(r \leq r^{wc}), & \text{if } w = -1 \end{cases} \tag{14.24}$$

The definition is illustrated in Fig. 14.15 where the probability density function of the performance r is denoted by $f(r)$. Here, the worst-case performance value is the value for which a fraction ρ of the distribution is worse than that value. Note that for any random variable, there exists a *probability distribution function* (to be distinguished from the probability density function) which expresses the probability that the random variable is less than some specified value. For the performance r, we can define a probability distribution function $F()$ as

$$F(a) = \Pr(r \leq a) \tag{14.25}$$

The worst-case performance value r^{wc} can now be defined in terms of the probability distribution function

$$\rho = \begin{cases} 1 - F(r^{wc}), & \text{if } w = +1 \\ F(r^{wc}), & \text{if } w = -1 \end{cases} \tag{14.26}$$

One may raise a question about the relationship between worst-case analysis and the parametric yield of a circuit. Traditional worst-case analysis does not solve the yield maximization problem. Instead, it provides the designer with a single-ended test for

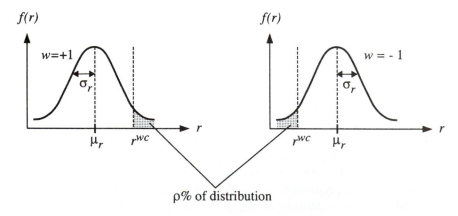

Figure 14.15   Definition of worst-case performance value r^{wc}.

the variability in a performance measure. If a circuit is designed such that all the worst-case performance values satisfy their specifications, i.e., if

$$a_k \le r_k^{wc} \le b_k, \quad \text{for } k = 1, 2, \ldots, p \tag{14.27}$$

the parametric yield of the circuit is at least $(1 - \rho)\%$. However, if all of the specifications are not met under worst-case conditions, no definitive statement can be made regarding the parametric yield of the circuit.

We will describe two techniques for worst-case analysis. The first, often called the "corners" technique, is a simplistic and commonly used procedure. In many cases, however, this technique is too conservative (overly pessimistic) and produces design bottlenecks. The second technique is introduced for a more accurate and realistic analysis. The reader should keep in mind that the designable parameters are fixed during worst-case analysis and therefore do not appear in the discussion below. Further, since the procedures discussed apply uniformly to all performance measures, only one performance measure r is used in the discussion.

The Corners Technique

In this technique, each noise parameter is *independently* set to its extreme value. For a noise parameter s_i with mean s_i^o and standard deviation σ_i, a typical extreme value is

$$s_i = s_i^o \pm 3\sigma_i, \quad i = 1, 2, \ldots, n_s \tag{14.28}$$

where n_s is the number of noise parameters. The direction of deviation of a noise parameter from its mean (i.e., whether $+$ or $-$ is used in the above formula) depends on whether increasing or decreasing the noise parameter s_i makes the performance worse. This is determined by *sensitivity analysis*, i.e., by computing the derivative of the performance with respect to a noise parameter and by considering the sign of that derivative. For each of the n_s noise parameters, the sensitivity is computed at its nominal (or mean) value. Since this technique sets each noise parameter to its extreme or corner value, it is called the "corners" or the one-at-a-time technique.

Once the worst-case values of the noise parameters have been obtained, the worst-case performance value r^{wc} is computed by simulating the circuit with those values (at a fixed design point). This worst-case analysis technique is very conservative, since all of the noise parameters are independently set at their extreme values. The probability that such a combination of noise parameter values will actually occur in a fabricated circuit is extremely small. Therefore, the predicted worst-case value of the circuit performance is overly pessimistic. If the circuit can be designed such that all of the specifications are met even under these pessimistic conditions, then the parametric yield is guaranteed to be very high. In many cases, however, such a design cannot be obtained, which causes severe bottlenecks.

A More Realistic Worst-Case Analysis Technique

In this technique, the worst-case performance value is computed first. How does one compute it? We have two possible cases depending on whether the performance is Gaussian or not.

Gaussian Performance Measure   If the performance r is Gaussian, its density function is completely characterized by its mean μ_r and its standard deviation σ_r. Moreover, given a worst-case probability ρ, one can refer to readily available tables for the distribution function of a standard Gaussian random variable to determine the value of r^{wc}. A standard Gaussian random variable has mean 0 and standard deviation 1. The random variable r can be converted into a standard Gaussian random variable q via the transformation $q = (r - \mu_r)/\sigma_r$. $\Phi()$ is the standard notation for the distribution function of a standard Gaussian random variable, and $\Phi^{-1}()$ is its inverse. In this case, it can be shown that the solution of (14.26) is

$$r^{wc} = \mu_r + \Phi^{-1}(1 - \rho)w\sigma_r \tag{14.29}$$

Non-Gaussian Performance Measure   If the performance measure r is not Gaussian, (14.26) cannot be used to compute r^{wc}. In this case, the distribution function $F()$ can be estimated using Monte Carlo techniques. Since a large number of samples are required in the Monte Carlo technique, an RSM for the performance measure is usually used. Given the value of ρ and the estimated $F()$, (14.26) is numerically solved for r^{wc} using the well-known Newton-Raphson procedure.

After r^{wc} has been obtained, one has to compute the corresponding set of noise parameter values that would result in the worst-case performance. Recall that the designable parameters are fixed and that the circuit performance is a function of the noise parameters alone, i.e., $\tilde{r}(\mathbf{s})$. Therefore, the worst-case noise parameter vector $\mathbf{s}^{wc}$ is the solution to the following equation.

$$\tilde{r}(\mathbf{s}^{wc}) = r^{wc} \tag{14.30}$$

Note that equation (14.30) above has n_s unknowns and there are infinitely many solutions (or a surface of solutions) for the worst-case noise vector. Any combination of noise parameter values on that surface produces the worst-case performance; thus, an additional condition is required to uniquely identify $\mathbf{s}^{wc}$. The most intuitively appealing choice for the worst-case vector would be the solution of (14.30) which is most probable. This corresponds to the most likely combination of noise parameter values that would be seen in a population of manufactured circuits which would produce the worst-case performance value. Recall that the noise parameter vector is characterized by the joint probability density function ($jpdf$) $f(\mathbf{s})$. Therefore, one can formulate the following constrained maximization problem for determining the worst-case noise vector:

$$\begin{aligned} \text{maximize} \quad & f(\mathbf{s}) \\ \text{subject to} \quad & \tilde{r}(\mathbf{s}) = r^{wc} \end{aligned} \tag{14.31}$$

In the above problem, the maximization of the density is equivalent to the maximization of probability, and the constraint ensures that the worst-case performance value is produced.

If the $jpdf$ of the noise parameters is Gaussian, the optimization has a simple implication. In this case, the solution $\mathbf{s}^{wc}$ of (14.31) is the point on the surface (14.30) which is (in a probabilistic sense) closest to the mean vector $\mathbf{s}^o$. This is illustrated in

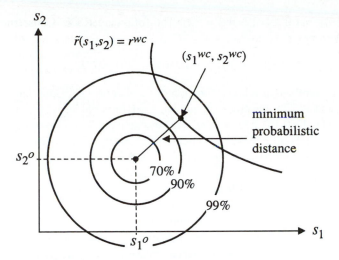

Figure 14.16 Illustration of worst-case analysis for two Gaussian noise parameters.

Fig. 14.16 for the case of $n_s = 2$. If we assume that the noise parameters s_1 and s_2 are independent, then the contours of equal probability density $f(s_1, s_2)$ are circles centered at the mean point (s_1^o, s_2^o). The solution of (14.31) is akin to moving along the surface $r(s_1, s_2) = r^{wc}$, and looking for the point that is closest to the mean point.

If, in addition to the fact that the noise parameters are Gaussian, the performance r is linear with respect to the noise parameters, (14.31) can be solved analytically to provide a closed-form solution for the worst-case noise vector. In other cases, the constrained maximization problem has to be solved numerically using an optimization method.

A Simple Worst-Case Analysis Example

To illustrate the worst-case analysis procedure, we will again consider the example of the inverter shown in Fig. 14.4. We have four noise parameters $V_{T,n}$, $V_{T,p}$, ΔW_1, and ΔW_2, whose statistical distributions have been stated in Section 14.5. The propagation delay τ_P is a smaller-the-better performance measure and so its worst-case direction is $w = +1$. For this circuit, the propagation delay τ_P is relatively insensitive to variations in ΔW_1 and ΔW_2. This is also evident from the RSM of (14.22), where the coefficients of the linear ΔW_1 and ΔW_2 terms are small. Hence, for this example, we drop these two noise parameters. This is an example of a more general technique called *parameter screening*, which is utilized in many design-for-manufacturability methods. Screening helps to reduce the number of parameters being considered and thereby the computational cost.

The fixed design point used in the worst-case analysis is the final design point $\mathbf{d}_{final}$ obtained in the parametric yield maximization example of the previous section: $\mathbf{d}_{final} = (W_1^o = 12, W_2^o = 22)$. The following RSM is constructed for the propagation delay τ_P:

$$\tau_P' = 0.151 + 0.0056V_{T,n} - 0.0073V_{T,p} \tag{14.32}$$

When τ_P is expressed as a linear function of the independent Gaussian random variables $V_{T,n}$ and $V_{T,p}$, it becomes a Gaussian random variable. Its mean and variance are estimated to be 0.151 ns and 0.0028 ns, respectively. For a worst-case probability ρ of 0.13%, we can compute the worst-case value using (14.29) to be $\tau_P^{wc} = 0.159$ ns. The corresponding worst-case values of $V_{T,n}$ and $V_{T,p}$ are found by solving (14.31) to be $V_{T,n}^{wc} = 0.91$ V and $V_{T,p}^{wc} = -1.06$ V. Note that these values are different from those that would be obtained from the corners method.

The results from the worst-case analysis using the linear response surface of (14.32) must be verified. For verification, the worst-case value of the propagation delay is obtained by solving (14.26) numerically using the procedure described earlier for a non-Gaussian performance measure. This procedure yields $\tau_P^{wc} = 0.159$ ns. To find the corresponding set of worst-case noise parameter values, we first draw a large Monte Carlo sample of the noise parameters and simulate the inverter circuit to obtain the propagation delay value for each sample. Next, we isolate the sample point for which the performance is within a small tolerance of the worst-case value (we use 0.0005 ns) and for which the probability density is the largest. Then, another Monte Carlo sample is drawn in a small region around this point and the process is repeated until sufficient accuracy in the estimate is achieved. A plot of the values of the propagation delay and the probability density is shown for the last set of sample points in Fig. 14.17. The point with the largest value of the probability density is chosen (rightmost point on Fig. 14.17) as the worst-case noise parameter vector: $V_{T,n}^{wc} = 0.89$ V and $V_{T,p}^{wc} = -1.07$ V. The worst-case analysis using the response surface method is therefore verified.

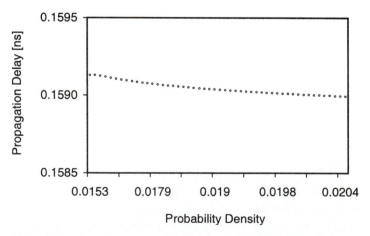

Figure 14.17   Monte Carlo confirmation of worst-case analysis.

14.8 Performance Variability Minimization

Variabilities in circuit performances are caused by processing and environmental fluctuations. Minimization of performance variability is another important topic in design for manufacturability. Circuits with low variabilities will result in products which are more uniform and have higher quality. As mentioned earlier, there are several quantities which may be used to measure the variability in a performance. We will use the variance of a performance r, denoted by σ^2, to quantify its variability. Two points which were made in Section 14.3 are reiterated here. First, the mean and variance of a performance measure are functions of the designable parameters alone since the effects of the noise parameters have been averaged out. This implies that the variability of a performance measure can be minimized by properly choosing the values of the designable parameters. This is the essence of variability minimization. Second, the mean and variance of a performance have to be estimated since the probability density of the performance is not known. To estimate these two quantities at a design point, a Monte Carlo sample of the noise parameters is drawn: $\mathbf{s}_i, i = 1, 2, \ldots, N_{MC}$. Then the mean μ is estimated by the sample mean given by

$$\mu(\mathbf{d}) = \frac{1}{N_{MC}} \sum_{i=1}^{N_{MC}} r(\mathbf{d} + \mathbf{s}_i) \tag{14.33}$$

and the variance σ^2 is estimated by the sample variance given by

$$\sigma^2(\mathbf{d}) = \frac{1}{N_{MC} - 1} \sum_{i=1}^{N_{MC}} r(\mathbf{d} + \mathbf{s}_i)^2 - \frac{N_{MC}}{N_{MC} - 1} \mu^2 \tag{14.34}$$

The specifications on the variability of the circuit performances can be expressed as

$$\sigma_k^2 \le c_k, \quad \text{for } k = 1, 2, \ldots, p \tag{14.35}$$

where σ_k^2 denotes the variance of performance r_k and p is the number of circuit performances. In addition to the variability specifications, there may also be specifications on the nominal values of the circuit performances:

$$a_k \le r_k^o \le b_k, \quad \text{for } k = 1, 2, \ldots, p \tag{14.36}$$

Thus, the variability minimization problem is to determine the optimal set of designable parameter values such that the variability and nominal performance specifications are met. In the parametric yield maximization methods, we had only one objective function to optimize, namely the yield. In the variability minimization problem, however, there may be multiple objective functions since for each performance there are two objectives, and for a particular circuit, multiple performances may be of interest. Such optimization problems are called *multi-criteria optimization problems*. There is a wide body of literature on methods to solve such problems; in this section, we provide a simple (although not the most efficient) solution strategy.

For each performance measure r_k, we define a *variability penalty* A_k using the variance σ_k^2 as follows:

$$A_k(\mathbf{d}) = 100 \left(\frac{\sigma_k^2(\mathbf{d})}{c_k} \right) \tag{14.37}$$

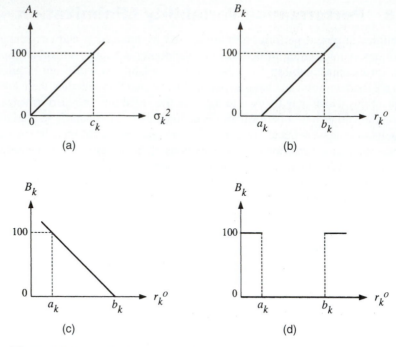

Figure 14.18 Definitions of (a) variability penalty, (b) performance penalty for smaller-the-better case, (c) performance penalty for larger-the-better case, and (d) performance penalty when all acceptable values are equally good.

This linear relationship is shown in Fig. 14.18(a). The best value of σ_k^2 is zero, and the corresponding penalty is zero; the penalty is higher for larger values of σ_k^2. Next, we define a *performance penalty* B_k which depends on the nominal value of the performance, r_k^o, according to one of the relationships shown in Fig. 14.18(b)–(d). Figure 14.18(b) shows a definition for B_k that can be used if smaller values of performance are better than larger values, and Fig. 14.18(c) shows a definition for the opposite case. If all values within the specification are equally desirable, then the definition of Fig. 14.18(c) may be used. Note that each of the above definitions for the penalty is linear; various other nonlinear relationships may also be used, based on the relative desirability of the values. The variability and performance penalties for each performance measure are defined as above.

Next, we define a *circuit penalty* at a design point, $Z(\mathbf{d})$, as

$$Z(\mathbf{d}) = \max_{1 \leq k \leq p} \{A_k(\mathbf{d}), B_k(\mathbf{d})\} \qquad (14.38)$$

The circuit penalty expresses the overall quality of the design with respect to the variability and nominal performance specifications. Now, we can formulate the variability minimization problem as the minimization of the circuit penalty. Formally, the variability minimization problem is to

$$\text{minimize} \quad Z(\mathbf{d}) \qquad (14.39)$$

Thus, we have converted the multi-criteria optimization into an optimization involving a single objective function. In so doing, however, our objective function is no longer differentiable. Therefore, the choice of optimization methods to perform the minimization of (14.39) must be restricted to those that do not require the objective functions to be differentiable.

An Example of Variability Minimization

Let us consider the CMOS clock driver circuit shown in Fig. 14.19. The top branch has three inverters and the lower branch has two inverters to illustrate the problem of signal skew that is present in many clock trees.

For our purpose, we shall define the skew to be the difference in the times at which CLK and its complement cross $0.5 V_{DD}$. The definition is illustrated in Fig. 14.20, where we show two skews, one corresponding to the rising edge of CLK

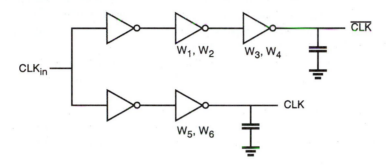

Figure 14.19 Clock distribution circuit.

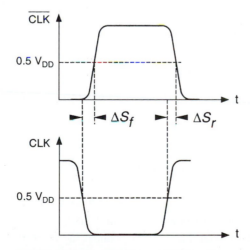

Figure 14.20 Definition of rising and falling clock skews.

(ΔS_r) and the other corresponding to the falling edge of CLK (ΔS_f). The signal skew ΔS is defined as the larger of the two skews and is the performance measure of interest in this circuit.

The variability minimization problem is to ensure that

(i)   the range of ΔS is less than 0.05 ns, and
(ii) the nominal value of ΔS is less than 0.5 ns.

The range is defined as the difference between the maximum and minimum values of ΔS.

The designable parameters in the circuit are the nominal channel widths of the two transistors in the second and third inverters of the top branch and the last inverter of the lower branch. The designable parameters are marked on Fig. 14.19. The internal noise parameters are the noise components of the channel widths and lengths of all of the nMOS and pMOS transistors in the circuit, and the common gate-oxide thickness of all the transistors in the circuit. The external noise parameters are the power supply voltage V_{DD} and the operating temperature. To perform this variability minimization, we construct a response surface model for ΔS in terms of the designable, internal, and external noise parameters. We estimate the range and the nominal value of ΔS using the RSM. The initial design point is $\mathbf{d}_{init} = (W_1 = 3, W_2 = 6, W_3 = 3, W_4 = 6, W_5 = 3, W_6 = 6)$, where all the widths are shown in μm. The range and nominal value of ΔS at $\mathbf{d}_{init}$ are estimated to be 0.311 ns and 1.398 ns, respectively. At the end of variability minimization, we obtain the final design point as $\mathbf{d}_{final} = (1.7, 11.1, 16.3, 26.9, 2.5, 38.6)$. The range and nominal value of ΔS at $\mathbf{d}_{final}$ are estimated to be 0.04 ns and 0.384 ns, respectively. If the acceptability criteria for the circuit are taken to be the specifications on the range and nominal value of ΔS, then the parametric yield at $\mathbf{d}_{init}$ is 0%, which implies that the nominal performance at the initial design point is not acceptable. At $\mathbf{d}_{final}$, the parametric yield is computed to be 100%. We have achieved a 72.5% reduction in the nominal value of ΔS, with a simultaneous reduction of 87% in the range of ΔS.

Exercise Problems

14.1   Consider a simple RC circuit model for a point-to-point interconnect in CMOS chips. The 50% delay time for a step input pulse is

$$\tau_{50\%} = 0.38RC$$

Both R and C values are subject to random fluctuations due to process variations.

a.   Express the sensitivities of $\tau_{50\%}$ with respect to R and C.

b.   Express the percentage change in the delay in terms of percentage changes in both R and C values under the assumption that their values can vary independently.

c.   Determine the largest delay increase possible due to $\pm 10\%$ fluctuations in both R and C values.

14.2 An empirically determined relationship between the bulk mobility of majority carriers and the doping level is

$$\mu = \mu_{min} + \frac{\mu_{max} - \mu_{min}}{1 + \left(\dfrac{N}{N_0}\right)^{\alpha}} \ [\text{cm}^2/\text{V} \cdot \text{s}]$$

where $\mu_{max} = 1360 \ \text{cm}^2/\text{V} \cdot \text{s}$, $\mu_{min} = 92 \ \text{cm}^2/\text{V} \cdot \text{s}$, $N_0 = 1.3 \times 10^{17}/\text{cm}^3$, $\alpha = 0.91$ for electrons, and the corresponding numbers for holes are $\mu_{max} = 495 \ \text{cm}^2/\text{V} \cdot \text{s}$, $\mu_{min} = 48 \ \text{cm}^2/\text{V} \cdot \text{s}$, $N_0 = 6.3 \times 10^{16}/\text{cm}^3$ and $\alpha = 0.76$.

a. Express the sensitivity of the mobility μ with respect to the doping level N.

b. Determine the percentage change in μ for 10% change in doping level from the nominal value for electrons and holes, respectively.

14.3 The dependence of the mobility on temperature is determined empirically as

$$\mu(T) = \frac{\mu(300 \ \text{K})}{\left(\dfrac{T}{300}\right)^{1.5}}$$

a. Derive an expression for $\Delta\mu/\mu$ in terms of $\Delta T/T$.

b. Calculate the percentage change in μ due to a 5% increase in temperature from 300 K for $\mu(300 \ \text{K}) = 980 \ \text{cm}^2/\text{V} \cdot \text{s}$.

14.4 Let us consider the device transconductance parameter

$$k = \mu C_{ox} \frac{W}{L}$$

It is known that all four variables μ, C_{ox}, W, and L are random variables due to process variations.

a. Derive an expression for the percentage change in k due to the percentage changes in μ, C_{ox}, W, and L.

b. If you assume that all four parameters change independently (corners technique), what would be the maximum percentage change expected due to 5% changes in all four parameters for nominal values of $(\mu^o, C_{ox}^o, W^o, L^o) = (980 \ \text{cm}^2/\text{V} \cdot \text{s}, 35 \times 10^{-8} \ \text{F/cm}^2, 5 \ \mu\text{m}, 1.0 \ \mu\text{m})$?

14.5 The logic threshold voltage V_{th} of a CMOS inverter from (5.87) is

$$V_{th} = \frac{V_{T0,n} + \sqrt{\dfrac{1}{k_R}}(V_{DD} + V_{T0,p})}{1 + \sqrt{\dfrac{1}{k_R}}}$$

where

$$k_R = \frac{k_n}{k_p} = \frac{\mu_n C_{ox} \dfrac{W_n}{L_n}}{\mu_p C_{ox} \dfrac{W_p}{L_p}}$$

a. Assuming that $\mu_n = 2.5\mu_p$, $L_n = L_p$ for simplicity, derive an expression for the percentage change in V_{th} in terms of percentage changes in $V_{T0,n}$, $V_{T0,p}$, and W_p/W_n.

b. For nominal values of $V_{T0,n} = 0.8$ V, $V_{T0,p} = -0.8$ V, and $W_p = 2.5W_n$, determine the maximum (worst-case) deviation of V_{th} for a ± 0.1 V change in both $V_{T0,n}$ and $V_{T0,p}$, and a $\pm 15\%$ change in W_p/W_n.

14.6   A CMOS chip with three signal input terminals, one output terminal, one power supply terminal, and one ground terminal is presented to you. The input signal voltages can vary from 0 to 5 V, the power supply voltage can vary from 4.5 V to 5.5 V, and the device operating temperature can range from 0 to 85°C.

We want to design an experiment so that the DC response of the chip can be described by a quadratic response surface model for three inputs, power supply voltage, and temperature.

a. Design a full factorial experiment.

b. Design a Latin hypercube experiment.

c. Design Taguchi's orthogonal array.

14.7   Design a CMOS full adder circuit and its layout using $W/L = 5/2$ for nMOS transistors and $W/L = 10/2$ for pMOS transistors. For a capacitive load of 1 pF, determine the worst-case delay of your particular design when the range of power supply variations is 4.5 to 5.5 V and the operating temperature ranges from 25°C to 85°C. Also assume that the magnitude ranges of the threshold voltages are from 0.8 to 1.2 V for both transistor types. All other parameters are assumed to be at their nominal values in this problem.

14.8   Determine the worst-case power dissipation for the design in Problem 14.7.

14.9   Suppose that the noise parameters are jointly Gaussian. The joint probability density function of a Gaussian random vector $\mathbf{s}$ with mean $\mathbf{s}^o$ and variance/covariance matrix $\mathbf{Q}$ is given by

$$f(\mathbf{s}) = \frac{1}{(\sqrt{2\pi})^{n_s}(\sigma_1\sigma_2\cdots\sigma_{n_s})} \exp\left[-\frac{1}{2}(\mathbf{s}-\mathbf{s}^o)^T\mathbf{Q}^{-1}(\mathbf{s}-\mathbf{s}^o)\right]$$

Consider the RC circuit model of Problem 14.1 with the performance measure being the 50% delay time $\tau_{50\%}$. Suppose R and C are independent Gaussian random variables. R has a mean of $R^o = 10$ kΩ and a standard deviation of $\sigma_R = 1$ kΩ. C has a mean of $C^o = 10$ pF and a standard deviation of $\sigma_C = 1$ pF. The worst-case direction for $\tau_{50\%}$ is +1.

a. Determine the mean and standard deviation of $\tau_{50\%}$.

b. $\tau_{50\%}$ is not a Gaussian random variable. To verify this, draw a Monte Carlo sample of R and C, use the formula given in Problem 14.1 to compute $\tau_{50\%}$, and plot the density of $\tau_{50\%}$. Compare the density to that of a Gaussian random variable having the mean and standard deviation of $\tau_{50\%}$ that you computed in (a).

c. The worst-case value of $\tau_{50\%}$ can be computed using the procedure for a non-Gaussian performance measure given in the discussion of worst-case analysis. Suppose the value is denoted by $\tau_{50\%}^{wc}$. To find the worst-case values of R and C, one must solve (14.31). Using the formula for the (*jpdf*) of a Gaussian random vector, reformulate the maximization of (14.31) into a minimization. The objective function for this minimization is called the "probabilistic distance." For this example, the minimization problem can be solved analytically using pencil and paper. Obtain the condition for minimization.

d. Use the fact that $R^o = C^o$ and $\sigma_R = \sigma_C$ to obtain R^{wc} and C^{wc} in terms of $\tau_{50\%}^{wc}$.

14.10 a. List the pros and cons of using the pessimistic worst-case analysis results in the design of integrated circuits in view of parametric yield, design effort, and project schedule.

b. Would the use of the simple-minded, pessimistic worst-case analysis technique such as the corners technique always shorten the development time?

c. What can be the most serious problem that can be faced by the pessimistic analysis in a big project which involves many different organizations?

14.11 Worst-case models even for the same circuit are different for different performances. For instance, the worst-case MOS model for delay time will be different from the worst-case MOS model for power dissipation. Yet, most design practices have been carried out using *slow, medium, fast* transistor models for both nMOS and pMOS transistors. Discuss how one would simulate the clock skew in a CMOS clock distribution circuit under such circumstances. What would be the correct way of performing circuit simulation of clock skews, if the transistor models can be custom-ordered by designers?

14.12 Discuss the strategy for developing manufacturable specifications for chip processing in terms of required control on threshold voltages, variations in channel lengths and widths, gate oxide thickness, and substrate and tub doping profiles.

14.13 The so-called "throw-over-the-wall" approach overstrains the effort needed by the counterpart organization. Discuss how one can avoid such overstraining requirements with shared responsibility in order to make the whole technology development process more cost-effective.

14.14 Taguchi's orthogonal array approach appeals to designers due to its simplicity; the design of the experiment is readily available in table form. Apply the L18 in MOS circuit design by using an example of your choice to demonstrate the principle of statistical design.

14.15 From (14.31) and (14.32) derive the expression for the worst-case noise vector for two independent Gaussian random variables, each with a mean of 0 and standard deviation of 1.

15 CHAPTER

Design for Testability

15.1 Introduction

The task of determining whether fabricated chips are fully functional is highly complex and can be very time-consuming. However, when faulty chips pass an improperly designed test, they can cause system failures and enormous difficulty in system debugging. It is known that the debugging cost increases by about tenfold from chip level to board level, and also from board level to system level. Thus, it is of great importance to detect faults as early as possible. As the number of transistors integrated into a single chip increases, the task of chip testing to ensure correct functionality becomes increasingly more difficult. However, in a production environment, many chips must be tested within a short time for timely delivery to customers. To overcome such difficult issues, *design for testability* has become ever more critical. In this chapter, we discuss types of faults, the corresponding fault models, design of testable circuits, and self-testing circuits. The testability will be defined in terms of observability and controllability, which are also commonly used in control and system theory. Introductory material in this chapter is largely based on a tutorial by Patel. For in-depth treatment of this topic, the reader is encouraged to consult Abramovici et al.

15.2 Fault Types and Models

Chip testing, in the conventional sense, is usually multi-purpose and attempts to detect faults in fabrication, design, and failures due to stressful operating conditions, namely the reliability problems. Input test vectors are devised and applied to the device under test (DUT) or circuit under test (CUT) as its stimuli. Then the measured outputs are compared with the expected correct responses to determine whether DUT is good (go) or bad (no-go). The main difficulty in testing is caused by the fact that only input and output pins of DUT are accessible, although at the test bench in the development laboratory, the internal nodes of unpackaged chips can be probed at the topmost metal level before passivation is done. As the operating clock frequency of the chips increases the at-speed test has also become a difficult problem. The difficulty stems from the signal integrity (transient ringing) problem in sending test signals from the tester to DUT and in detecting response signals from DUT due to impedance mismatch and transmission line problems in the tester interconnects. The impedance mismatch problem has been partially addressed in chip I/O design or by using a table lookup technique to correct delay measurement errors. In addition to the tester problem, the generation of correct test vectors to detect all modeled faults and design errors in complex chips, either manually or through an automatic test pattern generator (ATPG), has become a difficult task. In this chapter, we will limit our discussion to faults caused by physical defects.

Examples of *physical defects* include:

- Defects in silicon substrate
- Photolithographic defects
- Mask contamination and scratches
- Process variations and abnormalities
- Oxide defects

The physical defects can cause electrical faults and logical faults. The *electrical faults* include:

- Shorts (bridging faults)
- Opens
- Transistor stuck-on, stuck-open
- Resistive shorts and opens
- Excessive change in threshold voltage
- Excessive steady-state currents

The electrical faults in turn can be translated into logical faults. The *logical faults* include:

- Logical stuck-at-0 or stuck-at-1
- Slower transition (delay fault)
- AND-bridging, OR-bridging

The relationships between physical defects, electrical faults, and logical faults can be explained using a simple NOR2 gate as shown in Fig. 15.1. A metallic blob (physical defect) between the common drain terminal in the n-diffusion region and the ground bus line shown in Fig. 15.1(a) can be modeled as a *resistive short* between the output node Z and the ground as shown in Fig. 15.1(b), and also by a *stuck-at-0* (s-a-0) fault of output Z when the resistance is low or a *pull-up delay fault* when the resistance is high, as shown in Fig. 15.1(c).

Figure 15.2 shows other types of faults in a CMOS circuit consisting of NOR2, NAND2, and inverter gates. In this circuit, the input line B can be stuck-at-1 (s-a-1), since some part of the input line is shorted to the power rail. The pMOS transistor of the first stage NOR2 gate is stuck-on due to a process problem that causes a short between its source and drain terminals. The top nMOS transistor in the NAND2 gate,

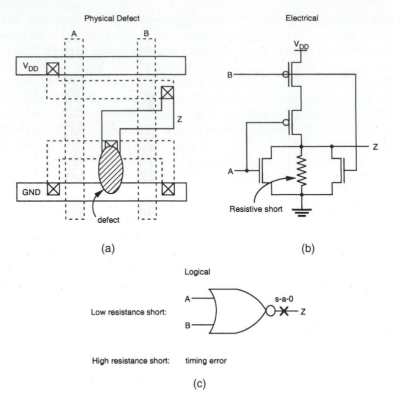

Figure 15.1   (a) Physical defect in NOR2 fabrication, (b) its electrical
fault model, and (c) its logical fault models.

on the other hand, is stuck-open due to either an incomplete contact (open) of the
source or drain node or due to a large separation of drain or source diffusion from the
gate, which causes permanent turn-off of the transistor regardless of the input C
value. The stuck-on and stuck-open faults are elaborated on in Fig. 15.3. The bridging fault between the output line of the inverter and the input line C can be due to a
fabrication defect which causes a short between any two parts of the two lines.
Although in the circuit diagram, these two lines are seemingly far apart, in the actual
layout, some parts of these two lines can be close to each other. In such a layout,
these two lines can be shorted due to underetching in the line patterning process.

   The single stuck-at fault models are used frequently, although the DUT can have
defects that do not map to a single stuck-at fault. Some of the reasons are:

■   Complexity of test generation is greatly reduced.

■   Single stuck-at fault is independent of technology, design style.

■   Single stuck-at tests cover a large percentage of multiple stuck-at faults.

■   Single stuck-at tests cover a large percentage of *unmodeled* physical defects.

In fact, it has been shown that in a two-level circuit with no redundancy, any complete
test set for all single stuck-at faults can cover all stuck-at faults. Multiple stuck-at fault

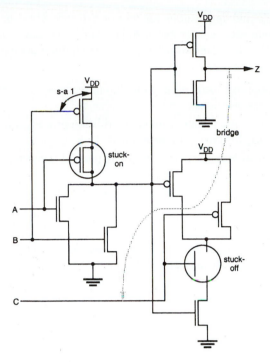

Figure 15.2 Some process-related defects in a CMOS circuit consisting of NOR2, NAND2, and inverter gates.

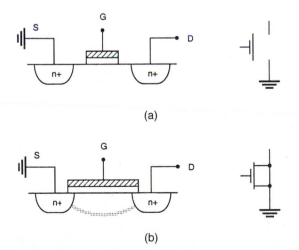

Figure 15.3 MOS transistor with (a) stuck-open (off) fault and (b) stuck-on (short) fault.

models find applications for fuse or anti-fuse based programmable designs such as programmable gate arrays, field programmable gate arrays (FPGAs), and RAMs.

The *delay fault* which causes timing failures at target speed can be due to several factors. To name a few,

- improper estimation of on-chip interconnect delays and other timing considerations,
- excessive variations in the fabrication process which cause significant variations in circuit delays and clock skews,
- opens in metal lines connecting parallel transistors which make the effective transistor size much smaller,
- aging effects such as hot-carrier induced delay increase.

The task of detecting delay faults is even more subtle than detecting functional faults in steady state. The functional test is usually done at speeds lower than the target speed due to the limitations of the testers. Special clocking is used to apply delay tests on a slow tester. The fault models mentioned above are used in fault simulation aimed at

- Test generation,
- Construction of fault dictionaries, or
- Circuit analysis in the presence of faults.

Each fault dictionary stores the expected output response of every faulty circuit to a particular test vector corresponding to a particular simulated fault.

15.3   Controllability and Observability

The *controllability* of a circuit is a measure of the ease (or difficulty) with which the controller (test engineer) can establish a specific signal value at each node by setting values at the circuit input terminals. The *observability* is a measure of the ease (or difficulty) with which one can determine the signal value at any logic node in the circuit by controlling its primary input and observing the primary output. Here the term *primary* refers to the I/O boundary of the circuit under test. The degree of controllability and observability and, thus, the degree of testability of a circuit, can be measured with respect to whether test vectors are generated deterministically or randomly. For example, if a logic node can be set to either logic 1 or 0 only through a very long sequence of random test vectors, the node is said to have a very *low random controllability* since the probability of generating such a vector in random test generation is very low. There exist time constraints in practice, and in such cases the circuit may not be considered testable. There are deterministic procedures for test generation for combinational circuits, such as the D-algorithm which uses a recursive search procedure advancing one gate at a time and backtracking, if necessary, until all the faults are detected. The D-algorithm requires a large amount of computer time. To overcome such shortcomings, many improved algorithms such as Path-Oriented DEcision Making (PODEM) and FAN-out-oriented test generation (FAN)

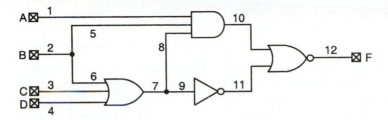

Figure 15.4 A simple circuit consisting of four gates with four primary inputs and one primary output.

have been introduced. Sequential circuit test generation is several orders of magnitude more difficult than these algorithms. To ease the task of ATPG, design-for-test (DFT) techniques are routinely employed.

Let us now consider the simple circuit in Fig. 15.4 consisting of four simple logic gates. To detect any defect on line 8, the primary inputs A and B must be set to logic 1. However, such a setting forces line 7 to logic 1. Thus, any stuck-at-1 (s-a-1) fault on line 7 *cannot be tested* at the primary output, although in the absence of such a fault, the logic value on line 7 can be fully controllable through primary inputs B, C, and D. Therefore, this circuit is *not fully testable*. The main cause of this difficulty in this circuit is the fact that input B fans out to lines 5 and 6, and then after the OR3 gate, both line signals are combined in the AND3 gate. Such a fan-out is called *reconvergent fan-out*. Reconvergent fan-outs make the testing of the circuit much more difficult.

If a large number of input vectors are required to set a particular node value to 1 or 0 (fault excitation) and to propagate an error at the node to an output (fault effect propagation), then the testability is low. The circuits with poor controllability include those with feedbacks, decoders, and clock generators. The circuits with poor observability include sequential circuits with long feedback loops and circuits with reconvergent fan-outs, redundant nodes, and embedded memories such as RAM, ROM, and PLA.

15.4 Ad Hoc Testable Design Techniques

One way to increase the testability is to make nodes more accessible at some cost by physically inserting more access circuits to the original design. Listed below are some of the ad hoc testable design techniques.

Partition-and-Mux Technique

Since the sequence of many serial gates, functional blocks, or large circuits are difficult to test, such circuits can be partitioned and multiplexors (muxes) can be inserted such that some of the primary inputs can be fed to partitioned parts through multiplexers with accessible control signals. With this design technique, the number of accessible nodes can be increased and the number of test patterns can be reduced. A

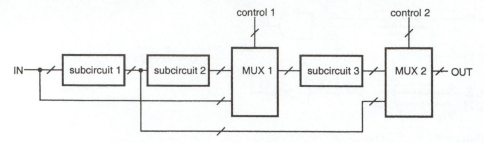

Figure 15.5 Partition-and-mux method for large circuits.

case in point would be the 32-bit counter. Dividing this counter into two 16-bit parts would reduce the testing time in principle by a factor of 2^{15}. However, circuit partitioning and addition of multiplexers may increase the chip area and circuit delay. This practice is not unique and is similar to the divide-and-conquer approach to large, complex problems. Figure 15.5 illustrates this method.

Initialize Sequential Circuit

When the sequential circuit is powered up, its initial state can be a random, unknown state. In this case, it is not possible to start the test sequence correctly. The state of a sequential circuit can be brought to a known state through *initialization.* In many designs, the initialization can be easily done by connecting asynchronous preset or clear-input signals from primary or controllable inputs to flip-flops or latches.

Disable Internal Oscillators and Clocks

To avoid synchronization problems during testing, internal oscillators and clocks should be disabled. For example, rather than connecting the circuit directly to the on-chip oscillator, the clock signal can be ORed with a disabling signal followed by an insertion of a testing signal as shown in Fig. 15.6.

Avoid Asynchronous Logic and Redundant Logic

The enhancement of testability requires serious trade-offs. The speed of an asynchronous logic circuit can be faster than that of the synchronous logic circuit counterpart. However, the design and test of an asynchronous logic circuit are more difficult than for a synchronous logic circuit, and its state transition times are difficult to predict. Also, the operation of an asynchronous logic circuit is sensitive to input test patterns, often causing race problems and hazards of having momentary signal values opposite to the expected values. Sometimes, designed-in logic redundancy is used to mask a static hazard condition for reliability. However, the redundant node cannot be observed since the primary output value cannot be made dependent on the value of the redundant node. Hence, certain faults on the redundant node cannot be tested or detected. Figure 15.7 shows that the bottom NAND2 gate is redundant and

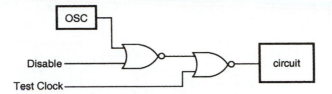

Figure 15.6 Avoid synchronization problems via disabling of the oscillator.

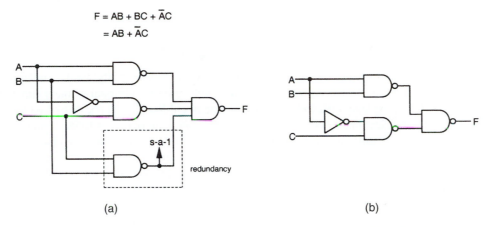

$$F = AB + BC + \overline{A}C$$
$$= AB + \overline{A}C$$

(a)   (b)

Figure 15.7 (a) A redundant logic gate example. (b) Equivalent gate with redundancy removed.

the stuck-at-1 fault on its output line cannot be detected. If a fault is undetectable, the associated line or gate can be removed without changing the logic function.

Although it is unessential to test redundant nodes when they are designed in as backup parts, either to enhance the circuit reliability or to increase the fabrication yield, the use of redundant circuits can make test generation much more complex and difficult. In fact, test generators, especially the random or deterministic test generators, would not be able to recognize such design intent. Some redundancy in circuits may be unintentional due to lack of design efficiency.

Avoid Delay-Dependent Logic

Chains of inverters can be used to design in delay times and use AND operation of their outputs along with inputs to generate pulses, as shown in Fig. 15.8.

Most automatic test pattern generation (ATPG) programs do not include logic delays to minimize the complexity of the program. As a result, such delay-dependent logic is viewed as redundant combinational logic, and the output of the reconvergent gate is always set to logic 0, which is not correct. Thus, the use of delay-dependent logic should be avoided in design for testability.

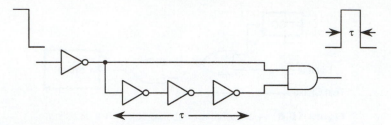

Figure 15.8 A pulse-generation circuit using a delay chain of three inverters.

15.5 Scan-Based Techniques

As discussed earlier, the controllability and observability can be enhanced by providing more accessible logic nodes with use of additional primary input lines and multiplexors. However, the use of additional I/O pins can be costly not only for chip fabrication but also for packaging. A popular alternative is to use scan registers with both shift and parallel load capabilities. The scan design technique is a structured approach to designing sequential circuits for testability. The storage cells in registers are used as observation points, control points, or both. By using the scan design techniques, the testing of a sequential circuit is reduced to the problem of testing a combinational circuit.

In general, a sequential circuit consists of a combinational circuit and some storage elements. In the scan-based design, the storage elements are connected to form a long serial shift register, the so-called *scan path,* by using multiplexors and a mode (test/normal) control signal, as shown in Fig. 15.9.

In the test mode, the scan-in signal is clocked into the scan path, and the output of the last stage latch is scanned out. In the normal mode, the scan-in path is disabled and the circuit functions as a sequential circuit. The testing sequence is as follows:

Step 1: Set the mode to *test* and let latches accept data from scan-in input.

Step 2: Verify the scan path by shifting in and out the test data.

Step 3: Scan in (shift in) the desired state vector into the shift register.

Step 4: Apply the test pattern to the primary input pins.

Step 5: Set the mode to *normal* and observe the primary outputs of the circuit after sufficient time for propagation.

Step 6: Assert the circuit clock for one machine cycle to capture the outputs of the combinational logic into the registers.

Step 7: Return to *test* mode; scan out the contents of the registers, and at the same time scan in the next pattern.

Step 8: Repeat steps 3–7 until all test patterns are applied.

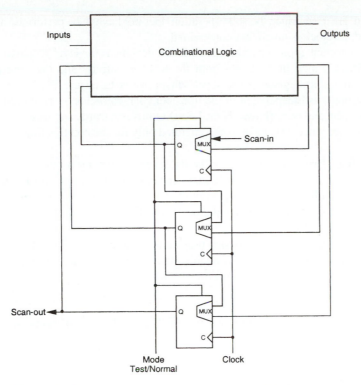

Figure 15.9 The general structure of scan-based design.

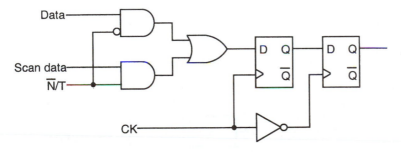

Figure 15.10 Scan-based design of an edge-triggered D flip-flop.

The storage cells in scan design can be implemented using edge-triggered D flip-flops, master-slave flip-flops, or level-sensitive latches controlled by complementary clock signals to ensure race-free operation. A detailed discussion of such latches and flip-flops is given in Chapter 8. Figure 15.10 shows a scan-based design of an edge-triggered D flip-flop. In large high-speed circuits, optimizing a single clock signal for skews, etc., both for normal operation and for shift operation, is difficult. To overcome this difficulty, two separate clocks, one for normal operation and one for shift operation, are used. Since the shift operation does not have to be performed at the target speed, its clock is much less constrained.

An important approach among scan-based designs is the level sensitive scan design (LSSD), which incorporates both the level sensitivity and the scan path approach using shift registers. The level sensitivity is to ensure that the sequential circuit response is independent of the transient characteristics of the circuit, such as the component and wire delays. Thus, LSSD removes hazards and races. Its ATPG is also simplified since tests have to be generated only for the combinational part of the circuit.

The boundary scan test method is also used for testing printed circuit boards (PCBs) and multichip modules (MCMs) carrying multiple chips. Shift registers are placed in each chip close to I/O pins in order to form a chain around the board for testing. With successful implementation of the boundary scan method, a simpler tester can be used for PCB testing.

On the negative side, scan design uses more complex latches, flip-flops, I/O pins, and interconnect wires and, thus, requires more chip area. The testing time per test pattern is also increased due to shift time in long registers.

15.6   Built-In Self Test (BIST) Techniques

In built-in self test (BIST) design, parts of the circuit are used to test the circuit itself. On-line BIST is used to perform the test under normal operation, whereas off-line BIST is used to perform the test off-line. The essential circuit modules required for BIST include:

- Pseudo random pattern generator (PRPG)
- Output response analyzer (ORA)

The roles of these two modules are illustrated in Fig. 15.11. The implementation of both PRPG and ORA can be done with Linear Feedback Shift Registers (LFSRs).

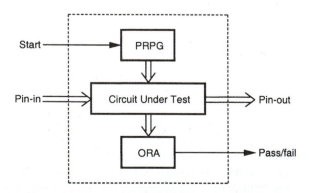

Figure 15.11 A procedure for BIST.

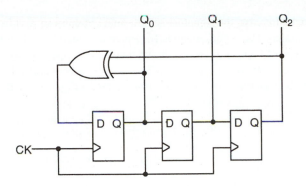

Figure 15.12 A pseudo-random sequence generator using LFSR.

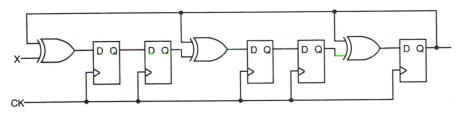

Figure 15.13 Polynomial division using LFSR for signature analysis.

Pseudo Random Pattern Generator

To test the circuit, test patterns first have to be generated either by using a pseudo random pattern generator, a weighted test generator, an adaptive test generator, or other means. A pseudo random test generator circuit can use an LFSR, as shown in Fig. 15.12.

Linear Feedback Shift Register as an ORA

To reduce the chip area penalty, data compression schemes are used to compare the compacted test responses instead of the entire raw test data. One of the popular data compression schemes is *signature analysis,* which is based on the concept of cyclic redundancy checking. It uses polynomial division, which divides the polynomial representation of the test output data by a characteristic polynomial and then finds the remainder as the signature. The signature is then compared with the expected signature to determine whether the device under test is faulty. It is known that compression can cause some loss of fault coverage. It is possible that the output of a faulty circuit can match the output of the fault-free circuit; thus, the fault can go undetected in the signature analysis. Such a phenomenon is called *aliasing.*

In its simplest form, the signature generator consists of a single-input linear feedback shift register (LFSR), as shown in Fig. 15.13, in which all the latches are edge-triggered. In this case, the signature is the content of this register after the last input bit has been sampled. The input sequence $\{a_n\}$ is represented by polynomial $G(x)$ and the output sequence by $Q(x)$. It can be shown that $G(x) = Q(x)P(x) + R(x)$, where $P(x)$ is the characteristic polynomial of LFSR and $R(x)$ is the remainder, the degree of which is lower than that of $P(x)$. For the simple case in Fig. 15.13, the characteristic polynomial is

$$P(x) = 1 + x^2 + x^4 + x^5$$

For the 8-bit input sequence $\{1\ 1\ 1\ 1\ 0\ 1\ 0\ 1\}$, the corresponding input polynomial is

$$G(x) = x^7 + x^6 + x^5 + x^4 + x^2 + 1$$

and the remainder term becomes $R(x) = x^4 + x^2$, which corresponds to the register contents of $\{0\ 0\ 1\ 0\ 1\}$.

Output Response Analyzer

The on-chip storage of a fault dictionary containing all test inputs with the corresponding outputs is prohibitively expensive in terms of the chip area. A simple alternative method is to compare the outputs of two identical circuits for the same input, with one of them regarded as reference. However, if both circuits have the same faults, their outputs can still match. Such faults cannot be detected with this technique, although the probability of two identical circuits having exactly the same faults would be very low.

In addition to the above circuits for built-in self test, self-checking design techniques can be used to detect faults autonomously during on-line operation. Usually a checker circuit is inserted such that the checker generates and sends out a signal when on-line faults occur. The distribution of checkers throughout a very large digital circuit or system can provide prompt detection of the fault location by tracing the checker which sent the fault signal. The use of self-checking circuits simplifies the development of software diagnostic programs. However, some additional hardware is required, and the checker itself needs to have self-checking capability. When self-checking capability of the checker itself is required, a single-output checker is not sufficient since that output may have a stuck-at fault, thus preventing the detection of actual faults in the circuit under test. A checker with a *pair* of outputs can be used instead to overcome this problem.

Built-In Logic Block Observer

The built-in logic block observer (BILBO) register is a form of ORA which can be used in each cluster of partitioned registers. A basic BILBO circuit is shown in Fig. 15.14, which allows four different modes controlled by C_0 and C_1 signals.

The BILBO operation allows monitoring of circuit operation through exclusive-ORing into LFSR at multiple points, which corresponds to the signature analyzer with multiple inputs.

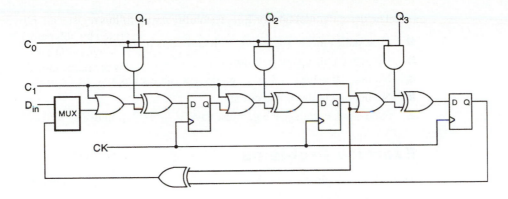

C_0	C_1	Mode
0	0	linear shift
1	0	signature analysis
1	1	data (complemented) latch
0	1	reset

Figure 15.14 3-bit built-in logic observer (BILBO) example.

15.7 Current Monitoring I$_{DDQ}$ Test

An often-used technique for testing fabrication defects is the I$_{DDQ}$ test. Under a bridging fault, the static currents drawn from the power supply in CMOS circuits can be noticeably high, well beyond the expected range of leakage currents. For example, if the drain node of the pMOS transistor in a CMOS inverter is shorted to the power supply rail due to a bridging fault, its I$_{DDQ}$ current can be very high even when the input is high. It can also detect other fabrication defects not easily detected by other test methods, including:

- Gate oxide short
- Channel punch-through
- p-n diode leakage
- Transmission-gate defect

The I$_{DDQ}$ test consists of applying the test vector and then monitoring the current drawn from the power supply rail in DC steady state. Although this test requires more testing time, the fault detection capability is greatly improved with small circuit overhead required to monitor the I$_{DDQ}$ in various parts of DUT.

While stuck-at tests require both fault sensitization and fault effect propagation, the I$_{DDQ}$ test requires only fault sensitization. However, its performance in open drain and open gate test is less effective. The I$_{DDQ}$ fault coverage is relatively easy to obtain and may potentially offer a full-chip coverage capability for large designs.

The design guidelines for I_{DDQ} testability are as follows:

- Low static current states; e.g., full CMOS is preferred
- No active pull-ups or pull-downs
- No internal drive conflicts; e.g., drivers share a bus
- No floating nodes in the circuit
- No degraded voltages; e.g., must have $V_{OH} = V_{DD}$ and $V_{OL} = 0$

Exercise Problems

15.1   Give a logic circuit example in which stuck-at-1 fault and stuck-at-0 fault are indistinguishable.

15.2   Show that the remainder of LFSR in Fig. 15.13 is indeed $R(x) = x^4 + x^2$.

15.3   Explain the merits or demerits of the bus structure in relation to testability. How would the bus structure impact the chip area overhead?

15.4   Determine whether the leakage current test for chips should be done prior to or after the functional test. What can you say about the test frequency of chips containing dynamic circuits designed to operate at very high frequency? Can it fail the functional test at much lower frequency? If so, explain why.

15.5   Show a few logic circuit examples whose logical fault coverage is dependent on the test vector sequence.

15.6   Find the set of all test vectors which detects the stuck-at-0 fault in line B in Fig. 15.2. Repeat for the stuck-at-1 fault in line C.

15.7   Show that if there are *undetectable* stuck-at faults in a combinational circuit then the circuit can be reduced according to the following rules (the rule set for OR gates is given below; prove it and find the rules for AND, NOR, NAND, and XOR gates).

Undetectable fault	Reduction rule for OR gates
Input x_i s-a-0	Remove input x_i
Input x_i s-a-1	Remove OR gate, connect output to 1
Output s-a-0	Remove OR gate, connect output to 0
Output s-a-1	Remove OR gate, connect output to 1

15.8   Apply the rules in Problem 15.7 to the circuit shown in Fig. 15.7.

REFERENCES

Abramovici, M., Breuer, M.A., and Friedman, A.D., *Digital Systems Testing and Testable Design,* New York, NY: Computer Science Press, 1990.

Alvarez, A.R., ed., *BiCMOS Technology and Applications,* second edition, Boston, MA: Kluwer Academic Publishers, 1994.

Annaratone, M., *Digital CMOS Circuit Design,* Norwell, MA: Kluwer, 1986.

Anner, G.E., *Planar Processing Primer,* New York, NY: Van Nostrand Rheinhold, 1990.

Athas, W.C., Swensson, L., Koller, J.G., and Chou, E., "Low-power digital systems based on adiabatic-switching principles," *IEEE Transactions on VLSI Systems,* vol. 2, pp. 398–407, December 1994.

Bakoglu, H.B., *Circuits, Interconnections and Packaging for VLSI,* Reading, MA: Addison-Wesley, 1990.

Barber, M.R., "Fundamental timing problems in testing MOS VLSI on modern ATE," *IEEE Design and Test,* pp. 90–97, August 1984.

Bellaouar, A. and Elmasry, M.I., *Low-Power Digital VLSI Design,* Norwell, MA: Kluwer Academic Publishers, 1995.

Bilardi, G., Pracchi, M., and Preparata, F.P., "A critique of network speed in VLSI models of computation," *IEEE Journal of Solid-State Circuits,* vol. DC-17, no. 4, pp. 696–702, August 1982.

Box, G.E.P. and Draper, N.R., *Empirical Model Building and Response Surfaces,* New York: John Wiley and Sons Inc., 1987.

Box, G.E.P., Hunter, W.G., and Hunter, J.S., *Statistics for Experimenters: An Introduction to Design, Data Analysis and Model Building,* New York: John Wiley and Sons Inc., 1978.

Breuer, M.A. and Friedman, A.D., *Reliable Design of Digital Systems,* Rockville, MD: Computer Science Press, 1976.

Brews, J.R., "A charge-sheet model of the MOSFET," *Solid-State Electronics,* vol. 21, pp. 345–355, 1978.

Brown, W.D. and Brewer, J.E., Nonvolatile Semiconductor Memory Technology: A Comprehensive Guide to Understanding and Using NVSM Devices, IEEE Press Series on Microelectronic Systems, 1997.

Chandrakasan, A.P. and Brodersen, R.W., *Low Power Digital CMOS Design,* Norwell, MA: Kluwer Academic Publishers, 1995.

Chang, C.Y. and Sze, S.M., *ULSI Technology,* New York, NY: McGraw-Hill, 1996.

Chen, H.Y. and Kang, S.M., "iCOACH: A circuit optimization aid for CMOS high-performance circuits," *Integration, the VLSI Journal,* 10 (1991), pp. 185–212.

Cheng, Y., Chan, M., Hui, K., Jeng, M., Liu, Z., Huang, J., Chen, K., Tu, R., Ko, P.K., Hu, C., *BSIM3v3 Manual,* Department of Electrical Engineering and Computer Science, University of California, Berkeley, 1996.

Cong, J. and He, L., "Optimal wiresizing for interconnects with multiple sources," *ACM Transaction on Design Automation of Electronic Systems,* vol. 1, no. 4, pp. 478–511, October 1996.

Cong, J., He, L., Koh, C.K., and Pan, Z., "Global interconnect sizing and spacing with consideration of coupling capacitance," *Proc. IEEE Int'l Conf. on Computer-Aided Design,* San Jose, California, pp. 628–633, November 1997.

De Los Santos, H.J. and Hoefflinger, B., "Optimization and scaling of CMOS bipolar drivers for VLSI interconnects," *IEEE Transactions on Electron Devices,* vol. ED-33, pp. 1722–1729, November 1986.

DeWilde, P., "New algebraic methods for modelling large-scale integrated circuits," *International Journal of Circuit Theory and Applications,* vol. 16, no. 4, pp. 473–503, October 1988.

Diaz, C.H., Kang, S.M., and Duvvury, C., *Modeling of Electrical Overstress in Integrated Circuits,* Norwell, MA: Kluwer Academic Publishers, 1994.

Diaz, C.H., Kang, S.M., and Leblebici, Y., "An accurate analytical delay model for BiCMOS driver circuits," *IEEE Transactions on Computer-Aided Design,* vol. 10, pp. 577–588, May 1991.

Digital/Analog Communications Handbook, Issue 9, *Mitel Semiconductor,* 1993.

Dillinger, T.E., *VLSI Engineering,* Englewood Cliffs, NJ: Prentice-Hall, Inc., 1988.

Dobberpuhl, D. et al., "A 200 MHz 64-b dual issue CMOS microprocessor," *IEEE J. Solid-State Circuits,* vol. 27, pp. 1555–1567, November 1992.

Embabi, S.H.K., Bellaouar, A., and Elmasry, M.I., *Digital BiCMOS Integrated Circuit Design,* Boston, MA: Kluwer Academic Publishers, 1993.

Enz, C., Krummenacher, F., and Vittoz, E., "An Analytical MOS transistor model valid in all regions of operation and dedicated to low voltage and low current applications," *Analog Int. Circ. and Signal Proc.*, vol. 8, pp. 83–114 (1995).

Ferris-Prabhu, A.V., "On the assumptions contained in semiconductor yield models," *IEEE Transactions on Computer-Aided Design,* vol. 11, pp. 955–965, August 1992.

Fey, C.F., "Custom LSI/VLSI chip design complexity," *IEEE Journal of Solid-State Circuits,* vol. SC-20, no. 2, April 1985.

Foty, D. *MOSFET Modeling with SPICE*, Prentice-Hall, 1997

Gabara, T.J. and Thompson, D.W., "High speed, low power CMOS transmitter-receiver system," *IEEE International Conference on Computer Design,* pp. 344–347, October 1988.

Glasser, L.A. and Dobberpuhl, D.W., *The Design and Analysis of VLSI Circuits,* Reading, MA: Addison-Wesley Publishing Co., 1985.

Goncalves, N.F. and De Man, H., "NORA: A racefree dynamic CMOS technique for pipelined logic structures," *IEEE Journal of Solid-State Circuits,* vol. SC-18, no. 3, pp. 261–266, June 1983.

Gray, P.R. and Meyer, R.G., *Analysis and Design of Analog Integrated Circuits,* fourth edition, New York, NY: John Wiley & Sons, Inc., 2001.

Greason, W.D., *Electrostatic Damage in Electronics: Devices and Systems,* Somerset, England: Research Studies Press, Ltd., 1987.

Grove, A.S., *Physics and Technology of Semiconductor Devices,* New York, NY: John Wiley & Sons, Inc., 1967.

Harris Semiconductor, *SC3000 1.5-Micron CMOS Standard Cells,* 1989.

Haznedar, H., *Digital Microelectronics,* Redwood City, CA: Benjamin/Cummings, 1991.

Hedenstierna, N. and Jeppson, K.O., "Comments on the optimum CMOS tapered buffer problem," *IEEE Journal of Solid-State Circuits,* vol. 29, no. 2, pp. 155–158, February 1994.

Hill, F.J. and Peterson, G.R., *Computer-Aided Logical Design with Emphasis on VLSI,* fourth edition, New York, NY: John Wiley & Sons, Inc., 1993.

Hollis, E.E., *Design of VLSI Gate Array ICs,* Englewood Cliffs, NJ: Prentice Hall, Inc., 1987.

Horowitz, M. and Dutton, R.W., "Resistance extraction from mask layout data," *IEEE Transactions on Computer-Aided Design,* vol. CAD-2, no. 3, pp. 145–150, July 1983.

Horst, E., Muller-Schloer, C., and Schwartzel, H., *Design of VLSI Circuits,* Heidelberg: Springer-Verlag, 1987.

Hu, T.C. and Kuh, E.S., *VLSI Circuit Layout: Theory and Design,* IEEE Press, 1985.

Hwang, I.S. and Fisher, A.L., "Ultrafast compact 32-bit CMOS adders in multiple-output domino logic," *IEEE Journal of Solid-State Circuits,* vol. 24, no. 2, pp. 358–369, June 1982.

Ikeda, T., Watanabe, A., Nishio, Y., Masuda, I., Tamba, N., Odaka, M., and Ogiue, K., "High-speed BiCMOS technology with a buried twin well structure," *IEEE Transactions on Electron Devices,* vol. ED-34, pp. 1304–1309, June 1987.

Itoh, K., *VLSI Memory Chip Design,* Springer Series in Advanced Microelectronics 5, 2001.

Jeng, M.C., Lee, P.M., Kuo, M.M., Ko, P.K., and Hu, C., *Theory, Algorithms, and User's Guide for BSIM and SCALP,* Electronic Research Laboratory Memorandum, UCB/ERL M87/35, Berkeley, CA: University of California, 1983.

Jha, N. and Kundu, S., *Testing and Reliable Design of CMOS Circuits,* Norwell, MA: Kluwer Academic Publishers, 1990.

Johnson, H.W. and Graham, M., *High-Speed Digital Design,* Prentice-Hall PTR, Englewood Cliffs, NJ, 1993.

Kang, S.M. "Accurate simulation of power dissipation in VLSI circuits," *IEEE Journal of Solid-State Circuits,* vol. SC-21, no. 10, pp. 889–891, October 1986.

Kang, S.M., Krambeck, R.H., Law, H.-F.S., and Lopez, A.D., "Gate matrix layout of random logic in a 32-bit CMOS CPU chip adaptable to evolving logic design," *IEEE Transactions on Computer-Aided Design,* vol. CAD-2, no. 1, pp. 18–29, January 1983.

Krambeck, R.H., Lee, C.M., and Law, H.-F.S., "High-speed compact circuits with CMOS," *IEEE Journal of Solid-State Circuits,* vol. SC-17, no. 3, pp. 614–619, June 1982.

Leblebici, Y. and Kang, S.M., *Hot-Carrier Reliability of MOS VLSI Circuits,* Norwell, MA: Kluwer Academic Publishers, 1993.

Lee, C.M. and Szeto, E.W., "Zipper CMOS," *IEEE Circuits and Devices Magazine,* pp. 10–16, May 1986.

Lee, K., Shur, M., Fjeldly, T.A., and Ytterdal, Y., *Semiconductor Device Modeling for VLSI,* Englewood Cliffs, NJ: Prentice-Hall, Inc., 1993.

Lee, S.-W. and Rennick, R.C., "A compact IGFET model-ASIM," *IEEE Transactions on Computer-Aided Design,* vol. 7, no. 9, pp. 952–975, September 1988.

Lin, H.C., Ho, J.C., Iyer, R.R., and Kwong, K., "Complementary MOS-bipolar transistor structure," *IEEE Transactions on Electron Devices,* vol. ED-16, pp. 945–951, November 1969.

Lopez, A.D. and Law, H.-F.S., "A dense gate matrix layout method for MOS VLSI," *IEEE Transactions on Electron Devices,* vol. ED-27, no. 8, pp. 1671–1675, August 1980.

Maly, W., *Atlas of IC Technologies,* Menlo Park, CA: Benjamin/Cummings, 1987.

Maly, W. and Director, S.W., Ed., *Statistical Approach to VLSI,* Amsterdam: North Holland, 1994.

Massobrio, G. and Antognetti, P., *Semiconductor Device Modeling with SPICE,* second edition, New York, NY: McGraw-Hill, 1993.

Matthys, R.J., *Crystal Oscillator Circuits,* John Wiley & Sons, New York, 1983.

McClurkey, E.J., *Logic Design Principles with Emphasis on Testable VLSI Circuits,* Englewood Cliffs, NJ: Prentice-Hall, 1986.

Mead, C. and Conway, L., *Introduction to VLSI Systems,* Reading, MA: Addison-Wesley Publishing Company, Inc., 1980.

Meyer, J.E., "MOS models and circuit simulation," *RCA Review,* 32, pp. 42–63, March 1971.

Miller, I. and Freund, E., *Probability and Statistics for Engineers,* second edition, Englewood Cliffs, NJ: Prentice Hall Inc., 1977.

Mokhari-Bolhassan, M.E. and Kang, S.M., "Analysis and correction of VLSI delay measurement errors due to transmission-line effects," *IEEE Trans. Circuits and Systems,* vol. 35, pp. 19–25, January 1988.

Momose, H., Shibata, H., Saitoh, S., Miyamoto, J., Kanzaki, K., and Kohyama, S., "1.0-mm n-well CMOS/bipolar technology," *IEEE Transactions on Electron Devices,* vol. ED-32, pp. 217–223, February 1985.

Muller, R.S. and Kamins T., *Device Electronics for Integrated Circuits,* second edition, New York, NY: Wiley, 1986.

Murphy, B.T., "Cost-size optima of monolithic integrated circuits," *Proceedings of IEEE,* vol. 52, pp. 1937–1945, December 1964.

Nagel, L.W., *SPICE2: A Computer Program to Simulate Semiconductor Circuits,* Memo ERL-M520, Berkeley, CA: University of California, 1975.

Najm, F., "A survey of power estimation techniques in VLSI circuits," *IEEE Transactions on VLSI Systems,* vol. 2, pp. 446–455, December 1994.

Osseiran, A., *Design for Testability,* Swiss Federal Institute of Technology (EPFL) Intensive Summer Course Note, 1993.

Patel, J.H., *ECE443 Class Notes,* University of Illinois at Urbana-Champaign, Spring 1994.

Plummer, J.D., Deal, M.D., and Griffin, P.B., *Silicon VLSI Technology,* Upper Saddle River, NJ: Prentice Hall, 2000.

Prince, B., *Semiconductor Memories: A Handbook of Design, Manufacture and Application,* second edition, New York, NY: John Wiley & Sons, Inc., 1996.

Rabaey, J.M. and Pedram, M., ed., *Low Power Design Methodologies,* Norwell, MA: Kluwer Academic Publishers, 1995.

Ross, P.J., *Taguchi Techniques for Quality Engineering,* New York: McGraw-Hill, 1988.

Rosseel, G.P. and Dutton, R.W., "Influence of device parameters on the switching speed of BiCMOS buffers," *IEEE Journal of Solid-State Circuits,* vol. 24, pp. 90–99, February 1989.

Ruehli, A.E. and Brennan, P.A., "Efficient capacitance calculations for three-dimensional multiconductor systems," *IEEE Transactions on Microwave Theory and Applications,* vol. MTT-21, no. 2, pp. 76–82, February 1973.

Sah, C.-T., *Fundamentals of Solid-State Electronics,* River Ridge, NJ: World Scientific Publishing Co., 1991.

Sakurai, T. and Kuroda, T., *Low Power CMOS Technology and Circuit Design for Multimedia Applications,* Lecture notes, Advanced Course on Architectural and Circuit Design for Portable

Electronic Systems, EPFL—Swiss Federal Institute of Technology, Lausanne, June 1997.

Sakurai, T. and Newton, A.R., "Alpha-power law MOSFET model and its application to CMOS inverter delay and other formulas," *IEEE Journal of Solid-State Circuits,* vol. 25, no. 2, pp. 584–594, April 1990.

Sakurai, T. and Newton, A.R., "Delay analysis of series-connected MOSFET circuits," *IEEE Journal of Solid-State Circuits,* vol. 26, no. 2, pp. 122–131, February 1991.

Sapatnekar, S.S. and Kang, S.M., *Design Automation for Timing-Driven Layout Sythesis,* Norwell, MA: Kluwer Academic Publishers, 1993.

Schichman, H. and Hodges, D.A., "Modeling and simulation of insulated-gate field-effect transistors," *IEEE Journal of Solid-State Circuits,* vol. SC-3, no. 5, pp. 285–289, September 1968.

Sechen, C. and Sangiovanni-Vincentelli, A., "The TimberWolf placement and routing package," *IEEE Journal of Solid-State Circuits,* vol. SC-20, no. 2, pp. 510–522, April 1985.

Sheu, B.J., Scharfetter, D.L., Ko, P.K., and Jeng, M.C., "BSIM, Berkeley short-channel IGFET model," *IEEE Journal of Solid-State Circuits,* vol. SC-22, pp. 558–566, 1987.

Shoji, M., "FET scaling in domino CMOS gates," *IEEE Journal of Solid-State Circuits,* vol. SC-20, no. 5, pp. 1067–1071, October 1985.

Shoji, M., *CMOS Digital Circuit Technology,* Englewood Cliffs, NJ: Prentice-Hall, 1988.

Shoji, M., *Theory of CMOS Digital Integrated Circuits and Circuit Failures,* Princeton, NJ: Princeton University Press, 1992.

Strojwas, A., ed., Selected Papers on Statistical Design of Integrated Circuits, IEEE Press, 1987.

Svensson, C. and Liu, D., "A power estimation tool and prospects of power savings in CMOS VLSI chips," *Proceedings of International Workshop on Low Power Design,* 1994.

Sze, S.M., *Physics of Semiconductor Devices,* second edition, New York, NY: Wiley, 1981.

Sze, S.M., *VLSI Technology,* New York, NY: McGraw-Hill, 1983.

Tsay, R.S., "An exact zero-skew clock routing algorithm," *IEEE Trans. Computer-Aided Design,* vol. 12, pp. 242–249, February 1993.

Tsividis, Y.P., *Operation and Modeling of the MOS Transistor,* New York, NY: McGraw-Hill, 1987.

Uehara, T. and Van Cleemput, W.M., "Optimal layout of CMOS functional arrays," *IEEE Transactions on Computers,* vol. C-30, no. 5, pp. 305–313, May 1981.

Uyemura, J.P., *Fundamentals of MOS Digital Integrated Circuits,* Reading, MA: Addison-Wesley, 1988.

Wadsack, R.L., "Fault modeling and logic simulation of CMOS and MOS integrated circuits," *Bell System Technical Journal,* vol. 57, no. 5, pp. 1449–1474, May-June 1978.

Weste, N.H.E. and Eshraghian, K., *Principles of CMOS VLSI Design—A Systems Perspective,* second edition, Reading, MA: Addison-Wesley Publishing Co., 1993.

Wolf, S. and Tauber, R.N., *Silicon Processing for the VLSI Era: Process Technology* (Volume 1), Lattice Press, 1986.

Wolf, S., *Silicon Processing for the VLSI Era: Process Integration* (Volume 2), Lattice Press, 1990.

Yang, P. and Chatterjee, P.K., "SPICE modeling for small geometry MOSFET circuits," *IEEE Transactions on Computer-Aided Design,* vol. CAD-1, no. 4, pp. 169–182, 1982.

Yoo, S.M. and Kang, S.M., "New high performance sub-1V circuit technique with reduced standby current and robust data handling," *IEEE International Symposium on Circuits and Systems,* May 28–31, 2000, Geneva, Switzerland.

Yuan, C.P. and Trick, T.N., "A simple formula for the estimation of capacitance of two-dimensional interconnects in VLSI circuits," *IEEE Electron Device Letters,* vol. EDL-3, no. 12, pp. 391–393, December 1982.

Yuan, J. and Svensson, C., "High-speed CMOS circuit technique," *IEEE Journal of Solid-State Circuits,* vol. 24, no.1, pp. 62–70, February 1989.

Zhou, D., Preparata, F.P., and Kang, S.M., "Interconnection delay in very high-speed VLSI," *IEEE Transactions on Circuits and Systems,* vol. 38, no. 7, pp. 779–790, July 1991.

Zimmermann, R. and Fichtner, W., "Low-power logic styles: CMOS versus pass-transistor logic," *IEEE Journal of Solid-State Circuits,* vol. 32, no. 7, pp. 1079–1090, July 1997.

Acceptability, 593, 594
Acceptable region, 602–604, 607
Access time, 405
    DRAM, 445
    ROM, 456, 460–464
    SRAM, 445
Accessibility, 627–630
Accumulation, 87–88
Active areas, 56, 59, 75(figure)
Adder. *See also* Full adder
    8-bit binary, 15–16, 305–307
    four-stage carry-lookahead, 389
Address, 408
Address decoder, 408, 410
Address multiplexing, 410–411, 423
Adiabatic logic, 512–516
Aging effects, 628
Aliasing, 597–598, 633
Aluminum, 64(figure), 87, 252
Amplifiers, 420, 429(figure), 432–433, 445–449
    current-mode sense, 450–451
AND function, 11, 294–295
AOI and OAI gates, 300, 336–337, 339
Area. *See* Active areas; Chip area; Width-to-
    length ratio
Aspect ratio, 236, 444
Astable circuits, 323
Automatic test pattern generation, 627,
    629, 632

Ball Grid Array (BGA), 42
Band gap, 85
Bandwidth improvement, 425
Base current
    and collector current, 526, 529, 553
    and emitter current, 530
    in forward active mode, 538
    and minority carriers, 538
    saturation β, 534
    in saturation mode, 540–541
Base region, 536–542
Berkeley Short-Channel IGFET Model (BSIM),
    158–159
Best-case analysis, 609
Bias, 87–90. *See also* Substrate bias effect

BiCMOS inverter circuit
    comparison with CMOS, 545–547
    delay times, 545–551
    (dis)advantages, 522–524, 543–544, 545
    logic gates, 551–553
    power dissipation, 543, 545–547
    voltage transfer characteristic, 544–551
Bidirectional buffer, 571
Bipolar circuits, 4–5
Bipolar junction transistors (BJTs)
    charge-control model, 536–541, 550
    current gain (β_F), 529
    cut-off mode, 531–532, 541, 544
    fabrication process, 523–524
    forward active mode, 524–525, 528–529, 538,
        541–542
    intrinsic delays, 541–542
    latch-up, 576–583
    operation, 524–526
    reverse active mode, 529–531, 538–539
    saturation mode, 532–536, 539–542
    SPICE version, 527
    structure, 524
Bird's beak, 58
Bistable circuits, 323, 324–330
Bit-line read circuitry, 553
Bit lines
    capacitance, 413, 414, 420, 436
    cells per, 410
    in DRAM, 410, 413, 415, 420
    folded, 434
    number of, 408
    in SRAM, 445–448
    voltage, 420, 434, 451
BJT inverter circuit
    delay times, 541–542
    dynamic behavior, 536–541
    static characteristics, 534–536
    VTC, 534–536
Blocks, 410
Body-effect coefficient, 95
Boltzmann constant, 85
Boolean functions, 10–11, 169, 311–314
Bootstrapping, 370–373
Boundary scan test, 632

Breakdown, oxide, 128
Bridging faults, 624, 635
Buffer circuit, 521–524
  bidirectional, 571
Buffers
  data input, 423
  data out, 408–409
  differential, 429(figure), 430(table)
  input address, 408, 410
  inverter-type, 428, 430(table)
  latch type, 428, 430(table)
  output, 410, 428–431
Built-in logic block observer, 634
Built-in self tests (BIST), 632–635
Buried contact, 194–195

Capacitance. *See also* Gate oxide capacitance;
    Interconnect capacitance
  bit line, 413, 414, 420, 436
  column, 441, 463
  depletion, 135, 159–160
  distributed, 129–134
  drain parasitic, 234
  in DRAM, 413–414
  equivalent, 131, 137
  equivalent large-signal, 136–137
  fan-out, 234
  in Flash memory, 464–465
  in FRAM, 474–475
  fringing field, 246–247, 250
  gate, 576
  junction, 134–140, 159–163, 369, 579
  large load, 545
  load (*see* Load capacitance)
  and NAND gate, 285–286
  overlap, 131–132, 163
  oxide, 131–134, 369
  parasitic (*see* Parasitic capacitance)
  and power consumption, 483, 485, 510–512
  row, 461, 463
  soft-node, 365–370
  source/drain, 67
  storage, 413, 420
  switched, 510–512
  well junction, 579
  zero-bias junction, 136, 138–140, 234
Capacitance ratio, 372
Capacitor Over Bit line (COB), 414
Carrier mobility. *See* Surface electron mobility
Carrier scattering, 120, 154
Carrier velocity, 120, 155–156
Carry lookahead adder, 389
Carry ripple adder, 15–16, 305–307
Carry_out signals, 10, 13, 304

CAS-before-RAS (CBR) refresh, 427, 428
Cells
  DRAM, 406, 410, 412–414, 436
  Flash memory, 467–471
  SRAM, 406, 438, 439–445
Central composite designs, 598
Channel(s), 32, 91
Channel current. *See* Drain current
Channel hot-electron (CHE) effect, 128–129
Channel length
  and drain depletion region, 127
  effective, 107, 150
  modulation, 107–109, 111, 113
  and n-channel MOSFET, 130–131
  short channel effect, 119–125
  in SPICE models, 150, 155–156
Channel voltage, 101, 103
Channel width, 125–126, 230, 232–238
  and propagation delay time, 521
Charge leakage, 365–370
Charge pump circuits, 471–472
Charge sharing, 378, 386–388, 391–393,
    395–397
Charge, stepwise, 517–519
Charge storage, 365–367, 375, 379, 436
Chemical vapor deposition, 61(figure), 62(figure)
Chemical-Mechanical Polishing (CMP), 58–59
Chip area, 117, 176–177, 182–186, 606. *See also*
    Width-to-length ratio
  DRAM, 414, 432
  in ROM, 453
Chip Carrier Packages (CCP), 43
Chip interface circuit, 408–409
Circuit noise, 173, 567–571
Clocked JK latch, 338–341
Clocked SR latch, 336–338
Clocks
  cross-talk prevention, 576
  and design for testing, 628
  distribution, 573–576
  and domino CMOS logic, 384–385
  and DRAM, 415–419, 424–425
  duty cycle, 576
  in Flash memory, 471–472
  generation, 571–573
  location, 558
  and parasitic capacitance, 576
  rise and fall times, 576
  skew, 381, 397, 572–573, 617–618
  transitions per cycle, 484–485
  in VLSI, 571–576
CMOS inverter, 197–214
  design criteria, 210–214
  layout, 67–72, 82(figure), 198–199, 213–214

power dissipation, 176, 193–195, 211–213, 260–267

    in SRAM, 440

    switching, 207–210, 220–230, 260–267

    voltage transfer characteristic, 198–207, 211–212

CMOS logic circuits (example), 8–18

CMOS NAND2, 290–293, 315, 391–393

CMOS NOR2, 287–290, 294, 315

CMOS n-well process, 59–65

CMOS transmission gates, 307–316, 379–381

Collector current

    amount of, 525–526

    and base current, 526, 529, 553

    control of, 526

    in cut-off mode, 532

    at edge of saturation, 536

    equation, 526

    and forward transit time, 549

    in forward-active mode, 529, 537, 541

    and minority carriers, 537

    and output voltage, 534

    in saturation mode, 533–534, 540

Column address decoder, 408, 458–460

Column capacitance, 441, 463

Combinational logic, 274–278

Common signal bus structure, 35–37

Complementary Pass-transistor logic (CPL), 314–316

Complex logic circuits, 294–316

Computer-aided design (CAD), 37, 43–45

Conductance, negative differential, 105

Conduction band, 85

Constant-field scaling, 116–117, 119

Constant-voltage scaling, 118–119

Contact windows, 63(figure)

Controllability, 626–627

Core, 408

Corners technique, 611

Cost per bit, 405

Cracking, 56

Critical path (timing), 305–307

Critical races, 508

Cross-coupled differential amplifier, 432–433

Crosstalk, 249, 576

Crystal oscillators, 572

Current gain

    base-to-collector, 576–577

    collector-emitter, 578, 580

    defined, 526

    in forward-active mode, 529

Current-mirror amplifier, 432, 433

Current-mirror amplifiers, 432, 433

Current-mode sensing, 450

D-algorithm, 626

Data

    representation of, 507

    storage of, 405–408

    throughput of, 425

Data input buffers, 423

Data out buffers, 408–409

DEC Alpha chips, 575

Decoders, 408, 410

    binary tree, 458–460

    pre- and main, 428–431

    row and column, 456–460

Delay constraints, 230–239

Delay times

    delay-dependent logic, 629

    and noise, 567–568

    and performance, 586

    and size, 390–391

    worst-case direction, 608–609

Depletion

    in BJTs, 539

    bulk charge, 158

    drain, 121–125

Depletion capacitance, 135, 159–160

Depletion region

    charge, 94, 95–96, 136, 539

    formation, 88

    and gate-to-source voltage, 93

    thickness, 89, 135

Depletion-load inverter

    design criteria, 192–195

    layout, 187–189

    load current, 223

    power dissipation, 193–195

    voltage transfer characteristic, 188–192

Depletion-type MOSFET, 91

    n-channel, 97

Design

    of CMOS inverters, 210–214

    of complex logic gates, 296

    of depletion-load inverters, 192–195

    full custom, 19, 38, 66–72

    hierarchical approach, 23–25

    high-speed VLSI, 576

    layout rules, 66

    and memory arrays, 405

    of NOR2 gates, 288–290

    and performance, 18–19

    process, 9–18

    and process variations, 586–588

    quality, 38–41, 239

    semi-custom, 19

    of transmission gates, 313–314

Design—*Cont.*
  verification stage, 13–15
  VLSI, 18–25, 28–38
Design cycle, 18–20
Design for manufacturability
  controllable *versus* noise parameters,
      588–590
  multi-criteria optimization, 615–618
  parametric yield, 593–594, 602–609
  process variations, 586–588, 615–618
  profitability, 594
  simulation, 594–601
  worst-case analysis, 609–614
Design for testability
  accessibility enhancement, 627–630
  built-in self testing, 632–635
  controllability and observability, 626–627
  current monitoring, 635–636
  defects and faults, 622–626
  delay-dependent logic, 629–630
  redundancy, 628–629
  scan-based techniques, 630–632
Design of experiments (DOE), 595–596
Design parameters
  clock gating, 509–510
  for combinational logic circuits, 275
  power, 481–486
Design reuse, 38
Design rule checkers (DRC), 13, 45
Design rules, 66–72
Designable parameters, 588–590, 599–600,
      607–609
  and worst-case analysis, 611
Device parameters, 586–588
di/dt reduction, 568
Dielectrics, 58, 414, 426
Differential amplifiers, 432
Differential tree search, 507
Diffusion current, 525–526
Diode leakage, 635–636
Distributed capacitances, 129–134
Divided bit-line NOR (DINOR), 467
D-latch, 343–349
  dynamic, 357–359
Domino CMOS logic
  charge-sharing, 386–389, 391–393
  clock distribution, 384–385
  NAND2, 391–393
  NORA, 393–396
  performance enhancement, 390–391
  precharge phase, 384
Doping
  in bipolar junction transistor, 524–525, 526, 531
  and channel region, 103, 128

  concentration, 84
  electrical behavior, 84
  and Fermi potential, 85
  and latch-up, 576, 581, 583
  and polysilicon resistivity, 54
  and process variations, 586, 587
  and scaling, 116
  of sidewalls, 138, 159–160
  and threshold voltage, 587
Double polysilicon, 438
Drain, 62(figure)
Drain current
  calculation, 102–104
  density, 118, 119
  and drain-to-source voltage, 104–105
  and gate-to-source voltage, 104–105, 112
  linear-mode, 116–117, 118
  in linear region, 158
  and load current, 171
  of nMOS transistor, 112
  and process variations, 587
  punch through, 127
  saturation-mode, 113, 117
  in SPICE models, 150–154, 158
  square root, 112
  in transmission gates, 309
Drain depletion, 121–125
Drain junction capacitance, 369
Drain-induced barrier lowering, 127
Drain-to-source voltage
  and BiCMOS inverters, 544
  and channel length, 107, 113
  and drain current, 104–105
  and drain depletion region, 121–122
  NAND2 gates, 282
  and nMOS inverters, 170
Driver transistors. *See* nMOS transistors
Drivers, pull-up and pull-down, 428
Dual Data Rate (DDR), 425
Dual In-line Packages (DIP), 42
Dual network, 11
Dynamic hazards, 508
Dynamic logic, 381–383
  ratioed, 377
  ratioless, 379
Dynamic pass transistor circuits, 374–379
Dynamic Random Access Memory (DRAM), 406
  access time, 445
  address multiplexing, 410–411, 423
  (a)synchronous, 423–426
  and capacitance, 414
  configuration, 410–411
  four-transistor, 412
  input and output buffers, 428–438

and leakage current, 411–412, 418,
    426–428, 436
low power, 432–433
one-transistor, 413–414, 419–423
page access mode, 423–424
three-transistor, 412–413, 414–419
two-transistor, 413–414
Dynamic recovery, 579

E-beam lithography, 52
Ebers-Moll model, 526–532
Edge of saturation, 536
Edge-triggered flip-flop, 346–349, 398–400
Effective channel length, 107, 150
Eight-bit binary adder, 15–17
Electrically Erasable PROM(EEPROM),
    406, 407
Electron affinity, 85
Electron concentration, 84
Electron drift, 119–120
Electron surface mobility, 103, 120, 154, 158
Electrostatic discharge (ESD)
    input circuits, 562–566
    and latch-up, 579
    models, 559
    output circuits, 567–571
    protection networks, 559–562
Elmore delay, 255, 390–391
Elmore time constant, 462
Emitter current
    components, 524–525
    in cut-off mode, 531–532
    equation, 526
    and minority charge, 538
    in reverse-active mode, 530–531
    in saturation mode, 533
Energy recycling, 512–516
Enhancement-load inverter, 186–187
Enhancement-type MOSFET, 91, 112, 134
Enz-Krummenacher-Vittoz (EKV) model, 159
Equilibrium Fermi level, 85–87, 89
Equivalent capacitance, 131, 137
Equivalent inverter approach, 268–269, 295–300
Equivalent large-signal capacitance, 136–137
Equivalent resistance, 309–311
Erasable PROM (EPROM), 406, 407
Etch-back, 58
Etched field-oxide isolation, 56
Etching, dry, 61
Euler-path, 297–299
Evaporation, 64
Exclusive OR (XOR) gate, 300
Experimental design, 595
Extended Data-Out access mode, 424

Fabrication
    lithography, 49–52
    LOCOS technique, 57–58
    metallization, 58–59, 64(figure), 452
    of nMOS transistor, 53–55
    n-well process, 59, 60–65(figures)
    simulation, 75–82(figures)
    transistor isolation, 56
Factorial design, 596–598
Fall time, 227, 488
FAN-out-oriented test generation (FAN),
    626–627
Fatigue, 475
Faults
    bridging, 624, 635
    catastrophic, 593
    delay, 626
    electrical, 623
    logical, 623
    parametric, 593
    pull-up delay, 623
    stuck-at, 623–626, 635–636
    stuck-on and stuck-open, 624
Feasible region, 602–603
Feature size, 3, 4, 449
Fermi potential, 85–87, 89, 95–96
Ferroelectric RAM (FRAM), 406, 407, 472–475
Field oxide
    defined, 53
    and DRAM leakage current, 426–427
    etching, 56
    growing, 57–58, 60(figure), 75(figure),
        77(figure)
    and narrow channel effects, 125
Field programmable gate arrays (FPGA),
    28–30, 625–626
Flash memory, 406, 451
    charge pump circuits, 471–472
    multilevel cell, 471
    NAND cell, 468–470
    NOR cell, 467–468, 470(table)
    programming mechanisms, 464–467
Flatband voltage, 87, 153–154, 587
Flip-flop
    and design for testability, 631–632
    DRAM amplifier, 420
    edge-triggered D type, 346–349, 398–400,
        631–632
    JK, 338–341
    master-slave, 341–345, 346–349
    SR, 330
    in SRAM, 445
Floating gate. *See* Flash memory
Forced β, 534

Forward active mode, 524–525, 528–529, 538, 541–542
Four-bit adder, 23–25
Four-stage carry-lookahead adder, 389
Fowler-Nordheim mechanisms, 464, 467, 468
Fringing field capacitance, 246–247, 250
Full adder. *See also* Adder
  and CPL, 315–316
  design example, 9–18
  layout optimization, 15
  n bit, 305–307
  one bit, 304–305
Full custom design, 19, 38, 66–72
Full latch amplifier, 432, 433
Full scaling, 116–117, 119
Full search, 507
Functional yield, 40
Fuse ROM, 406, 407
Fuses, 625–626

Gallium arsenide (GaA) circuits, 4–5
Gate, 83–84, 327–328, 564–565
  floating (*see* Flash memory)
Gate arrays, 30–33
  programmable, 625–626
Gate bias, 87–90
Gate oxide
  criticality, 60(figure)
  and process variation, 586
  short detection, 635–636
  and substrate interface, 94
  thickness of, 127–128, 587, 591
Gate oxide capacitance
  and depletion region charge, 94
  distributed, 131–134
  and oxide layer thickness, 587
  per unit area, 96
  and process variations, 587
  and scaling, 116, 117, 118
  in SPICE models, 159, 162
Gate voltage, 127, 154, 327
Gated clock signals, 509–510
Gate-to-drain capacitance, 132
Gate-to-source capacitance, 132
Gate-to-source voltage
  and channel approximation, 101
  and depletion region, 93
  and drain current, 104–105
  NAND2 gates, 202
  and nMOS inverters, 170
Gaussian distribution, 590, 600, 606, 612–613
Glass reflow, 58
Glitch reduction, 508–509
GND contacts, 71

Grading coefficient, 136
Gradual channel approximation (GCA), 101–104, 119
Guard rings, 582
Gummel-Poon model, 549

Hardware replication, 502–504
Heat dissipation, 483, 513
  ESD-induced, 560–562
High Capacitive-coupling Ratio cell (HICR), 467
Hold time, worst-case, 367, 370
Holding current, 578, 579–580
Holding (state), 330
Holding voltage, 578–579
Hole density, 84, 87
Hot-carrier effects, 128
Hot-electron injection, 464, 467
Hypercube sampling, 599–601
Hysteresis, 212, 472–473

I_{DDQ} test, 635–636
Impedance, 313, 543
  of clock distribution line, 576
  and DRAM, 428
  transmission line mismatches, 579
Implant-mask programming, 454, 460
Imprint, 475
Inductance, 242
Initialization, 628
Injection phenomena, 549
Input address buffers, 408, 410
Input buffers, 428–431
Input circuits, and ESD, 562–566
Input high voltage, 171–173, 180–181, 191, 205
Input low voltage
  calculation, 180, 190, 203–205
  and CMOS NOR2 gate, 287
  defined, 171–172
Insulation, 55, 85–86
Intercell routing, 32
Interconnect capacitance
  estimation, 245–252
  lumped, between inverters, 131, 218–219
  and performance, 586–587
  and propagation delay time, 521
  RC delay models, 252–260
  and submicron technology, 483
Interconnects
  and clock, 576
  composite layout, 65(figure)
  delays, 252–260, 626
  fabrication of, 58–59
  metallic, 58–59, 71, 252
  multilevel, 58–59

and noise, 173
    parasitics, 241–252
    resistance, 252, 256
Interface, gate-oxide and substrate, 94, 96
Inversion, 89–90, 294
    weak and strong, 156, 159
Inversion layer, 89–90, 101–103, 107, 132
Inverter gate, 564–565
Inverter threshold voltage, 169, 172, 207–210
Inverters
    chains of, 629
    chip area, 176–177
    generic, 33–34, 171(figure)
    ideal, 169–170
    power dissipation, 176–177
    transistor sizes, 576
    in TSPC dynamic circuit, 398
    voltage transfer characteristics, 171–176
Inverter-type buffer, 428
Ion implantation, 54, 96–97
Isolation techniques, 56

JK latch
    NAND-based, 338–339
    NOR-based, 339–341
    truth table, 340
Joint probability density function (jpdf),
        589, 612–613
Junction capacitance, 134–140, 159–163, 438
Junction potential, 135

Kirchhoff's Current Law
    and BJT inverters, 538
    and noise margins, 197, 205, 211
    and output low voltage, 179
    and threshold voltage, 205–206
    and VTC, 171
Knee current, 549

Lambda rules, 66
Latch
    D-type, 343–349, 357–359
    JK, 338–341
    SR, 330–338
Latch-up
    causes, 579
    description, 576
    prevention, 441, 581–583
Lateral diffusion, 147
Latin hypercube sampling, 599–601
Layout design rules, 66–72
Layout extraction, 45
$L(di/dt)$ drop, 567
Leaf cells, 22

Leakage current
    and depletion transistors, 187
    and DRAM, 411–412, 418, 426–428, 436
    and latch-up, 579, 580
    reverse diode, 490–491
    and soft-node capacitance, 365–366
    and SRAM, 449
    subthreshold current, 491–492
Least-square fitting, 601
Level sensitive scan design (LSSD), 632
Level shifting, TTL-to-CMOS, 562–566
Libraries, 32
Linear Feedback Shift Registers (LFSRs),
        632–634
Linear mode, 99
Linear region
    defined, 99
    drain current in, 158
    and enhancement-load inverter, 186
    of nMOS transistor, 227
    in SPICE, 149, 158
Linear regression, 601
Lithography, 50–53
Load capacitance
    and BiCMOS circuits, 522–523, 545
    of bit line, 436
    and BJT inverter, 542
    and clocks, 576
    components of, 231
    defined, 219
    NAND gate, 285–286
    and propagation delay time, 235–237
    and switching power, 261–263
    total, 234
Load current, 171, 188
Load device
    of depletion-load inverter, 187–188
    of enhancement-load inverter, 186–187
    and inverter characteristics, 170
    in pseudo-nMOS gates, 300–301
    in SRAM, 438
    voltage drop across, 171
Loads, off- and on-chip, 521
Local oxidation of silicon (LOCOS), 57–58, 126
Locality, 25–26
Logic block, 1–2
Logic chips, 3
Logic, dynamic, 377, 379, 381–383
Logic errors. *See* Charge sharing
Logic gates
    adiabatic, 514–516
    for AND and OR, 11
    BiCMOS applications, 551–553
    complex, 294–316

Logic gates—*Cont.*
  layout, 292–294
  mask layout, 66–67, 82 (figure)
  output transition probabilities, 504–507
  placement of, 22
  power dissipation, 483–484
  static *versus* dynamic, 357
  switching speed, 241
  two-input NAND, 290–292
  two-input NOR, 287–290
Logic high-voltage, 535, 541–542, 551
Logic low-voltage, 295, 302, 332, 542, 551
Logic "1" transfer, 360–362
Logic redundancy, 628–629
Logic "0" transfer, 363–365
Low power applications, 405
  DRAM, 432–433, 434
  SRAM, 185, 441, 445, 449

Majority carriers, 587
Manchester carry chain (circuit), 389
Manufacturability, 40. *See also* Design for
    manufacturability
Mask ROM. *See* Read-only memory
Masks, 62(figure)
  custom layout, 66–67
  defects, 623
  in fabrication process, 49–51
  of full-adder circuit, 304–305
    CPL, 315–316
  implant, 454, 460
  layout design, 66–72
  and power, 512
  and process variations, 586
  of standard-cells based chip, 37
  of TSPC dynamic circuit, 399
Mass Action Law, 84
Master-slave flip-flop, 341–343
  edge-triggered, 346–349
Mean forward transit time, 537–538
Memory arrays
  area efficiency, 405
  BiCMOS, 553–554
  ferroelectric RAM (FRAM), 472–476
  floating gate (Flash), 464–472
  mask read-only (ROM), 451–464
  organization, 408
  summary, 407(table)
Memory blocks, 410
Memory chips, 3
Memory read latency, 423
Metal, 58–59, 64(figure), 71, 252, 452–453. *See
    also* Aluminum
Metastability, 345

Micron rules, 66
Minority carriers, 525, 536–542
  and latch-up, 581–582
Model parameters, 148(table)
  SPICE, 165–167
Modularity, 21, 25
Module placement, 21
Monolithic Microwave Integrated Circuits
    (MMICs), 4–5
Monostable circuits, 323
Monte Carlo method, 599–600, 604–605, 606,
    607, 612
MOS field effect transistor (MOSFET)
  bias, 93, 109–111, 136
  capacitances, 129–140
  channel approximation, 101–102
  current-voltage characteristics, 100–114
  depletion-type, 91
    n-channel, 97
  enhancement type, 91, 134(figure)
  impedance, 543
  leakage current, 490–492
  models (*see* SPICE)
  n-channel, 91, 105, 111–114, 130(figure)
  operation principle, 91–92
  p-channel, 91
  pinch-off, 99
  saturation, 99, 100(figure), 107, 149
  scaling, 115–119
  short-channel, 119–125
  small geometry, 126–128, 156–157
  structure, 83–87, 90–93
MOSIS design rules, 66–67, 68, 80–81(figure)
Multi-Chip Modules (MCM), 43, 243
  testing, 632
Multiple-output domino CMOS circuit, 393
Multiple-threshold CMOS (MTCMOS), 498–499
Multiplexing, 410–411, 423
Multiplexors, 311–314, 627–628, 630

NAND decoder array, 457–458
NAND gates
  BiCMOS application, 551
  and CPL, 315
  design strategy, 284–285
  fault types, 623–624
  load capacitance, 285–286
  multiple input, 283–285
  and SR latch, 333–335, 337–338
  switching delays, 285–287
  two-input (NAND2), 202, 281–283, 285–287,
    291–292
    domino, 391–393
Narrow channel effects, 125–126

N-channel MOSFET
  cross section, 130(figure)
  current-voltage equations, 111
  defined, 91
  drain voltage characteristics, 105
  parameter measures, 111–114
Negative gate bias, 87–88
Newton-Raphson procedure, 612
nMOS transistors
  in BiCMOS inverters, 544, 550, 551,
    553–554
  and Boolean functions, 11
  channel width, 521
  in CMOS fabrication, 48
  in complex logic circuits, 295–296
  in composite layout, 65(figure)
  configuration of, 295–296
  in CPL, 314–315
  depletion-type, 275–287, 456
  in DRAM, 415
  and electrostatic discharge, 560
  fabrication, 53–55
  and Flash memory, 471
  and latch-up, 583
  linear region, 227
  and power, 486–490
  ratio to pMOS, 564
  in ROM layout, 452–454, 456, 459
  saturation, 112, 227
  in series-connected structures, 390–391
  size, 12
  in SRAM, 438, 445
  in standard-cells design, 35
  thick-oxide, 560
  in transmission gates, 308–309
N-net. *See* Pull-down network
Node transition factor, 485
Noise. *See also* Design for testability
  defined, 588
  effect of, 173–174
  $L(di/dt)$, 567–571
  statistical correlation, 589–590
Noise margins
  and depletion-load inverter, 187
  determining factors, 191
  for high and low signal, 172, 174–176
  of inverters, 191, 195–197, 210–211
  of resistive-load inverter, 181–182,
    185–186
  in SRAM, 439–440, 441
Noise parameters, 588–590
  worst-case, 609–613
Non-regenerative circuits, 323
NOR decoder array, 456–457

NOR gates
  BiCMOS application, 551
  and CPL, 315
  design strategy, 278
  faults, 623–624
  multiple input, 279–280
  in ROM, 462, 463
  and SR latch, 330–333
  two-input, 275–278, 280–281
NORA CMOS logic, 393–396
Normally on MOSFET, 97
N-well contact, 65(figure)
N-wells, 48, 59–65, 60(figure), 62(figure), 71, 582

Observability, 626–627
Ohmic contacts, 62(figure)
Ohm's law, 103
Open-loop gain, 451
OR function, 11, 294
  exclusive (XOR), 300
Orthogonal arrays, 598–599
Oscillator circuit, 572
Oscillators, testing, 628
Output buffers, 408–409, 428–431
Output high voltage
  and bistable circuits, 328
  calculations, 178, 189
  for low input, 171–172
  NAND gate, 282–283
  NOR gate, 276
Output low voltage
  and bistable circuits, 328–330
  calculation, 179–180, 190
  and complex logic gates, 295–296, 302
  defined, 172
  depletion nMOS load, 276–278
  and resistive-load inverters, 182
  and width/length ratio, 183–185
Output voltage
  of BJT inverter, 534–535
  and low input levels, 171–172
  rise and fall time, 221–222
Overdrive base charge, 540–542
Overlap capacitance, 131–132, 163
Oxidation, local (LOCOS), 57–58, 126
Oxide breakdown, 128
Oxide capacitance, 94, 96, 117, 118, 131–134
Oxide-interface charge, 96
Oxide-related capacitances, 94, 96, 117, 118,
  131–134

Packaging technology, 41–43
Packing density, 576
Page access mode, 423–424

Parallelism, 502–504
Parameter screening, 613–614
Parametric yield
  and design quality, 41, 593–594, 602–605
  estimation example, 605–607
  maximization, 607–609
  and worst-case analysis, 610–611
Parasitic capacitance
  and BiCMOS circuits, 545, 547
  and channel width, 521
  and chip dimensions, 117
  of clock receiver, 576
  in complex logic gates, 299, 314
  and delay constraints, 234, 238, 245–260
  and DRAM, 414, 431
  and dynamic logic gates, 357, 358, 379
  of interconnects, 245–252, 253, 483
  minimization of, 299
  and MOSFETs, 129–131
  NAND gate, 285
  and pass transistor, 369
  and performance, 586
  of polysilicon lines, 71
  and power consumption, 483, 485, 510–512
  and ROM, 463
  in SPICE models, 162
  and SRAM, 444
Parasitic extraction tool, 13
Parasitic resistance
  and chip dimensions, 117
  and interconnects, 252–260, 576
  and latch-up, 580
  and performance, 586
  of polysilicon lines, 71, 252
  and signal delay time, 252
Partition-and-mux technique, 627–628
Pass gates, 307–314, 511–512
Pass transistor circuits
  charge leakage, 365–367
  dynamic, 374–379
  logic "0" transfer, 363–365
  logic "1" transfer, 360–362
  and soft node, 359
  in SRAM, 438, 443
Passivation layer, 65(figure)
Pass-transistor logic, Complementary (CPL), 314–316
Path-Oriented DEcision Making (PODEM), 625–626
Patterning, 49–54, 61(figure), 64(figure)
P-channel MOSFET, 91
  current-voltage equations, 111
Performance
  and design, 18–19
  of DRAM, 410

measures of, 590–593
modeling, 594–601, 606
and parametric yield, 593–594
and parasitics, 586–587
and process variations, 586
SPICE modeling, 594
of SRAM, 409, 449
variability
  measurement, 594, 609
  minimization, 615–618
worst-case analysis, 609–613
Phase splitter, 570
Photoresist, 50–52
Physical defects, 623
Pierce crystal oscillator, 572
Pin Grid Array (PGA), 42–43
Pinch-off, 107, 132
  of MOSFET, 99
Pins, 410–411
Pin-through-hole packages (PTH), 42
Pipelining, 425, 499–502
Planarization, 58–59
Plasma etching, 61
pMOS transistors
  in BiCMOS inverter, 547–548, 550, 551
  channel width, 521
  in composite layout, 65(figure)
  and CPL, 314–315
  design, 11–12
  and Flash memory, 471
  and latch-up, 583
  in NOR ROM array, 462
  in NP-domino logic, 393–396
  placement, 71
  and power, 486–490
  ratio to nMOS, 564
  source diffusion regions, 582
  in standard-cells design, 38
  in transmission gates, 309
  in TSPC dynamic CMOS, 396
P-net. See Pull-up network
Poisson equation, 89, 116
Polycide, 252
Polysilicon
  column ordering, 297–299
  in complex logic gates, 297–299
  doped versus undoped, 53–54, 184
  double, 438
  functions, 61(figure)
  gate formation, 75(figure)
  line layout, 13
  long connections, 71
  and parasitics, 71
  patterning, 53–54
  resistivity, 54, 252

silicided, 252
  in SRAM, 438–439
  uses, 53–54
Positive gate bias, 88–90
Positive logic convention, 169, 274
Power. *See* Low power applications
Power density, 118–119
Power dissipation
  average, 261–262, 265
  of BiCMOS inverters, 543, 545–547
  of bipolar logic gates, 522, 545
  calculation, 492
  CMOS inverters, 176, 193–195, 211–213,
      260–267
  and constant-field (full) scaling, 117,
      118–119
  and DRAM, 418–419, 420
  dynamic, 260–266, 482–486, 493–494, 500
  leakage, 490–492
  and portability, 481–482
  reduction approaches, 496–517
  resistive-load inverter, 182–186
  short-circuit, 262–263, 486–490
  in SRAM, 440–441, 448–449
  static, 492
  switched capacitance reduction, 510–512
  and ULSI, 482
  worst-case direction, 609–610
Power meter, 263–265
Power-delay product, 267
Precharge, 381–382, 384–385, 388
Printed circuit boards, testing, 632
Probability density function, 589
Probability distribution function, 610
Process variations, 586–588
Product of sums, 300
Programmable gate arrays, 625–626
Programmable ROM (PROM), 406
Propagation delay time
  average current method, 220, 226–228, 230
  balanced, 228
  definitions, 220–222
  design verification, 13–15
  and interconnect capacitance, 521
  and load capacitance, 235–237
  NAND2 gate, 286–287
  and oscillation frequency, 241
  as performance measure, 590–591
  and RC network, 252–260
  step-input assumption, 229
  and tree structure, 508
  and voltage scaling, 494–495
Pseudo random pattern generator, 633
Pseudo-nMOS gates, 300–303
Pull-down current, 279

Pull-down driver, 428
Pull-down network, 296–297
Pull-down transistor, 562, 565
Pull-up, 439, 441, 550
Pull-up circuit, 547
Pull-up delay fault, 623–624
Pull-up driver, 428
Pull-up network, 296–297
Pulsed-sensing, 475
Punch-through, 127, 635–636
P-wells, 582

Quad Flat Packs (QFP), 43

Random access memory, 405–406
Random sampling, 599–601
RAS-only refresh (ROR), 427, 428
Ratioed logic, 377
Ratioless logic, 379
Read-only memory (ROM)
  access time, 456, 460–464
  dynamic, 452
  layout, 452–453
  Mask pattern, 407, 451–456, 460–464
  NAND-based, 454–456
  NOR-based, 452–454, 460–464
  row and column decoders, 456–460
  types of, 406–407 (*see also* Flash
      memory)
Read-write memory, 405–406
Reconvergent fanout, 627
Redundancy, 628–629
Reflection, 576
Refresh, 406, 411–412, 418, 426–428
  cycle constraint, 423
Regenerative circuits, 323–330
Regularity, 25
Reliability, 40–41
Replication, 28, 502–504
Resistance
  determining factors, 252
  equivalent, 309–311
  of interconnects, 252
  parasitic (*see* Parasitic resistance)
  in ROM, 459–460
  row, 461, 463
  and SRAM, 439–440
  substrate, 577
Resistive capacitive (RC) network, 242–243,
      252–260
Resistive-capacitive time constant, 560
Resistive-load inverter
  layout, 177
  load current, 223
  power dissipation, 182–186

Resistive-load inverter—*Cont.*
  in SRAM, 438
    voltage transfer characteristic, 177–182,
      184–185
Resistors, 184–185, 438–439
Response surface model (RSM), 595–596,
    597–598, 605, 606
  quadratic, 596, 601, 605
Restore, 420
Reverse active mode, 529–531, 538–539
Reverse conduction current, 366–367
Ring oscillator circuit, 239–241, 323
Ringing, transient, 623
Rise time, 488
Row address decoders, 408, 428–431, 456–458
Row capacitance, 461, 463
Row resistance, 461, 463

Saturated enhancement-load inverter, 186–187
Saturation
    BJT, 534, 536
    MOSFET, 99, 105, 149
      and channel length, 155
      current, 108, 117, 120, 154
      in SPICE model, 155
      voltage, 109, 155–156
Saturation β, 534
Scaled buffer chain, 521–522
Scaling
    of channel length, 127–128
    constant-field (full), 116–117, 119
    constant-voltage, 118–119
    and SRAM, 449
    supply voltage, 211–212, 493–495
Scan path, 630
Scan-based techniques, 630–632
Schmitt Trigger, 350–353, 565
Schottky source/drain contacts, 581
Sea-of-Gates (SOG) chips, 33, 34(figure)
Search algorithms, 507
Self refresh, 427, 428
Self-aligned process, 54
Self-checking, 634
Semiconductors. *See* Substrate
Semilatch amplifier, 433
Sense amplifiers, 432–433, 445–449, 475
Sensitivity analysis, 611
Sequential circuits, 323, 628
Setup time, 347–348
Shift register, dynamic
    CMOS, 379–381
    depletion-load nMOS, 375
    enhancement-load nMOS, 377–379
Short-channel effects, 119–125

Short-circuit current component, 486–488
Shorts. *See* Bridging faults
Sidewalls, 135, 138–140, 159–163
Signal distance, 576
Signal perturbation. *See* Noise
Signal routing, 293, 299
Signal timing, 508
Signature analysis, 633
Sign-magnitude representation, 507
Silicon compilation, 41
Silicon dioxide
    band-gap, 85
    electron affinity, 85
    in fabrication process, 49–52, 55, 58
    thickness, 84
Silicon, local oxidation of (LOCOS),
    57–58, 126
Silicon nitride, 57–58
Silicon-controlled rectifer (SCR), 576–581
Simplicial approximation, 607–608
Simulation, 75–82(figures), 591, 594–601
Simulation tools, 44–45. *See also* SPICE
Simulation waveforms, 421(figure), 435(figure)
Single bit access, 423
Sixteen-bit adder, 25
Size. *See* Feature size; Transistor size
Slewing, 579
Small-signal gate current, 327, 328
Soft node, 358, 359, 365–370, 395–397
Soft node capacitance, 365–370
Source, 62(figure)
    and depletion region charge, 94
    and doping, 54
    location of, 54
    in n-channel MOSFET, 91
Source potential, 109–110
Source/drain capacitance, 67
Source/drain contacts, 581, 583
Spacing, 576
Speed
    memory, 405, 409, 432, 439
    switching, 13, 241, 522–523
SPICE, 146–147, 148(table)
    BJT models, 527
    LEVEL 1, 149–153, 161–165
    LEVEL 2, 153–157, 163–164
    LEVEL 3, 157–158, 163
    LEVEL 4 (BSIM), 158–159
    mask layout, 67
    model parameters, 165–167
    and performance modeling, 594
    and power dissipation, 265
SR latch
    clocked, 336–338

NAND-based, 333–335, 337–338
NOR-based, 330–333
Standard-cells, 19, 33–38, 67
Static Random Access Memory (SRAM),
    406–407, 438–440
  access time, 445
  CMOS, 440–445
  current-mode sensing, 450–451
  high performance, 409
  low power, 185, 445, 449
  read, 443–444, 445–448
  write, 444–445, 448–449
Statistical design. *See* Design for
    manufacturability
Step-sensing, 473–475
Stepwise charge, 517–519
Storage capacitance, 413, 420
Stress-relief oxide, 57–58
Strobe signal, 568
Strong inversion, 156
Sub-micron technology, 230, 367, 483. *See also*
    Very large scale integration (VLSI)
Substrate
  defined, 83–84
  doping level, 576
  and DRAM, 436
  electron concentration, 84
  and gate oxide interface, 94
  inversion, 89–90
  latch-up prevention, 581–582, 583
  ohmic contacts, 62(figure)
  p-type, 84–85, 436
  and source voltage, 94
Substrate bias coefficient, 95, 98, 111–114
Substrate bias effect, 109–111
  and depletion-load nMOS, 188
  and depletion-type nMOS transistor, 280
  and NAND2 gate, 282
  and pass transistor, 360
  and threshold voltage, 98
  in transmission gates, 307–309
Substrate bias generator, 436–438
Substrate potential, 109–110
Substrate resistance, 577
Subthreshold conduction, 127, 156
Subthreshold current, 156, 366–367,
    491–492, 497
Sum-of-products, 300
Sum_out signals, 10, 304
Super buffer circuit, 267–269
Surface electron mobility, 102–103, 120,
    154, 158
Surface inversion, 89, 94
Surface, planarity of, 56–58

Surface potential, 87–89, 93
Switched capacitance, 510–512
Switching
  delay, in BiCMOS, 545–551
  power dissipation, 482–486
  reduction approaches, 507–510
  simultaneous, 302
  speed of, 13, 241, 522–523
  time, for SR latch, 332–333
Switching threshold
  of CMOS inverter, 206–207
  of NAND2 gate, 291–292
  of NOR2 gate, 288–290
Symmetry, 11
Synchronous complex logic, 378–379

Taguchi's method, 598–599
Taylor series, 157–158, 175
Temperature, 85, 428, 491. *See also* Heat
    dissipation
Testability, 38–40. *See also* Design for testability
Thermal problems, 41–42
Thermal voltage, 527
Thick-oxide, 560. *See also under* Gate oxide
Thin oxide, 586
Threshold voltage. *See also* Inverter threshold
    voltage
  adjustment, 96–97, 126, 181–182
  of BiCMOS inverter, 544
  of CMOS inverter, 205–207
  defined, 93, 169, 172
  and depletion region, 89
  and depletion-type inverters, 188
  and DRAM, 427, 428, 436, 438
  of enhancement n-channel MOSFET, 112–114
  excessive, 623
  in Flash memory, 464, 465–467, 468–470,
    471(figure)
  implant, 454–456, 460
  measurement of, 96
  narrow channel, 126
  and noise, 606
  and parametric yield, 606
  and pass transistor, 360–361
  physical components, 94–96
  and process variations, 587
  and propagation delay time, 591
  and resistive-load inverter, 177–178
  and ROM, 454–456, 460
  short channel effects, 120–125
  and small gate voltage, 105
  in small-geometry MOSFETs, 156–157
  in SPICE model, 154
  and SRAM, 443, 444–445, 449, 454–456

Threshold voltage—*Cont.*
  and substrate bias coefficient, 113–114
  and substrate bias effect, 98, 109–111
Time-critical operations, 25–28
Time-dependence, in BJTs, 539, 549–550
Time-domain behavior, 327–330
Timing
  critical path, 305–307
  and *di/dt* reduction, 568
  of DRAM, 420–423, 427–431
  failure causes, 626
  mismatched signal, 508
  SRAM core operations, 447(figure)
Toggle switch, 339–341
Tolerances, 609–613
Totem pole, 543
t_{RAC}, 423
Transconductance
  and bistable circuits, 327
  of n-channel MOSFET, 114
  and power dissipation, 488
  in pseudo-nMOS gate, 302
  and resistive-load inverter, 185
  and scaling, 116–118
  in SRAM, 451
Transient ringing, 623
Transistor size. *See also* Area; Scaling; Width-to-
    length ratio
  and clock, 576
  and delay, 234, 390–391
  and DRAM, 409–410, 428, 434, 438
  and full-adder layout, 15
  minimum, 12
  in series-connected structures, 390–391
  and switching speed, 13
Transistors. *See also* nMOS transistors;
    pMOS transistors
  in CMOS *versus* nMOS gates, 292
  in complex logic circuits, 292, 300–301
  for data out buffers, 408–409
  in memory- *versus* logic chips, 3
  number of, 3, 292, 300–301, 384
    in DRAM, 405, 412–413
  orientation of, 587
  parallel, 626
  size issues, 12, 13, 15, 234, 390–391
  uncommitted, 33
  variation in, 587–588
Transistor-Transistor Logic (TTL), 408, 562–566
Transit time constant, 328
Transition region, 175
Transmission gate, 307–314
Transmission lines, 242–244
Transmission-gate defect, 635–636

Tree structure, 458–460, 508
Trench capacitor, 414
Trends, 1–3
Triggering, 577–578
Tristability, 567–571
True single-phase clock (TSPC) dynamic CMOS,
    397–400
TTL-to-CMOS level shifting, 562–566
Tubs, 48
  twin, 48, 582
Tunneling current, 449, 464
Tunneling leakage, 426, 427
Twin-tub technology, 58, 582
Two-phase clocking, 374, 379–381

Ultra Large Scale Integration (ULSI), 2, 482
Ultraviolet light, 51, 406
Uncertain region, 175

Valence band, 85
Variable-threshold CMOS (VTCMOS),
    496–497
Variation range, 609
VDD contacts, 71
Velocity saturation, 120
Verification, 23, 44–45
Very large scale integration (VLSI)
  clock, 571–576
  CMOS, 4–5, 25
  design, 18–25, 28–38
    computer-aided, 21, 43–45
    high-speed, 576
  hot carrier effects, 128
  on-chip memory, 405
  power dissipation, 492–493
  submicron, 128, 243–244
  terminology, 2
Vias, 58
Voltage. *See also* Input high voltage; Input
    low voltage; Output voltage;
    Threshold voltage
  bias, 87–90, 93, 95, 102, 136
  in BJT, 533–534, 548
  bootstrapping, 370–373
  channel, 103
  critical points, 172
  drain-to-source (*see* Drain-to-source voltage)
  in DRAM, 415–416, 420–423, 433–438, 445
  drops in, 370–373
  in Flash memory, 464–467, 471–472
  in FRAM, 473–475
  gate-to-source (*see* Gate-to-source voltage)
  holding, 578–579
  on-chip, 433–434, 438

and pass transistor, 361–365
power supply reduction, 486
and ROM, 451–452
saturation, 105, 109, 155–156
    in BJT, 533–534, 548
scaling, 211–212, 493–495
sidewall equivalence factor, 138–139
soft-node, 363–365
in SRAM, 441, 443–448, 450–451
stepwise, 517–519
thermal, 527
Voltage generators, on-chip, 433–434
Voltage transfer characteristic (VTC)
    of BiCMOS inverter, 544
    in bistable circuit, 325
    of BJT inverter, 534–536
    CMOS inverter, 198–207, 209(figure)
    of combinational logic gate, 275
    critical voltages, 171–172
    depletion-load inverter, 188–192
    and depletion-load inverter, 187
    and logic levels, 173
    nMOS inverter, 169–170
    resistive-load inverter, 177–182, 184–185

Weak inversion, 156
Well junction capacitance, 579
Wells, 48
    n-, 59–65, 60(figure), 62(figure), 71, 582
    p-, 582
Width-to-length (W/L) ratio
    and CMOS SRAM, 441–443
    of complex logic circuits, 295–296
    for delay time, 231, 232–233
    and depletion-load inverter area, 194
    in digital circuits, 587–588
    and drain current, 587
    of gate, 176–177
    initial, 12

of NAND gate, 285
and output low voltage, 183–185
and parasitics, 13, 241
and power consumption, 183–184
and process variations, 587
and propagation delay time, 591
for ROM, 462
in series-connected structures, 390–391
and source/drain capacitance, 67
for SRAM, 441–443
and switching speed, 13
and transconductance, 67
Width-to-vertical separation ratio (w/h), 576
Wire
    length of, 241, 576
    width of, 576
Word lines
    cells per, 410
    in DRAM, 415
    driving, 521
    number of, 408
    in ROM array, 461
    selection of, 428–431
    and voltage, 420
Work function, 85–87, 94
Worst-case analysis, 609–613
    example, 613–614
    and parametric yield, 610–611
Worst-case direction, 609–610
Worst-case holding time, 367
Worst-case probability, 610

XNOR gate, 510
XOR gate, 300, 315

Y-chart, 21–23
Yield. *See* Parametric yield

Zipper CMOS logic, 396–397